# ARC FLASH HAZARD ANALYSIS AND MITIGATION

# ARC FLASH HAZARD ANALYSIS AND MITIGATION

Second Edition

J.C. Das

IEEE PRESS SERIES ON POWER ENGINEERING

Mohamed E. El-Hawary, *Series Editor*

IEEE PRESS

WILEY

Published by John Wiley & Sons, Inc., Hoboken, New Jersey.
Published simultaneously in Canada.



For general information on our other products and services or for technical support, please contact our Customer Care Department within the United States at (800) 762-2974, outside the United States at (317) 572-3993 or fax (317) 572-4002.

Wiley also publishes its books in a variety of electronic formats. Some content that appears in print may not be available in electronic formats. For more information about Wiley products, visit our web site at www.wiley.com.

*Library of Congress Cataloging-in-Publication data applied for*

ISBN 978-1-118-16381-8

Cover Design: Wiley
Cover Images: (Top image) © Sam Robinson/Getty Images, (Center image) © Chanin Wardkhian, © Ashwini H/Getty Images Design

Set in 10/12pt Times by SPi Global, Pondicherry, India

FSC
www.fsc.org
MIX
Paper from responsible sources
FSC® C013604

# CONTENTS

# FOREWORD

As is common with emerging technologies, the maturity of safety considerations for a technology may lag the momentum in applying the technology. This has been true with the industrial, commercial, and residential electrification of modern society that began in the late nineteenth century. While human contact with electricity was known to be hazardous, differentiating arc flash from electric shock did not receive significant attention until a century later. We now know that injuries from arc flash events in electric power systems are among the most traumatic and costly occupational injuries. The intense energy transfer occurring in a fraction of a second converts electrical energy into thermal, blast, acoustic, chemical, and electromagnetic components that individually have their own injury consequences. Collectively, these energy transfers to the human body produce complex physical, neurological, and emotional trauma that are very difficult and costly to treat and rehabilitate. The resulting tragedy not only impacts the injured person, it extends to family, friends, and coworkers. And that is just the human side. Arc flash events also damage vital infrastructure, disrupting operations and damaging critical equipment. Altogether, the consequences in human suffering, equipment damage, and disruption of essential electrical systems can be extraordinary. But they can be prevented.

Increasing awareness of arc flash hazards has inspired improvements in administrative control measures, including safe work practices and application of personal protective equipment. These are important components of a comprehensive solution in reducing the risk of injury, but they have limitations. Administrative control measures are susceptible to human error that occur in real time with little opportunity for recovery from gaps in knowledge, misinterpretation of conditions, or lapse in discipline. Personal protective equipment for arc flash events is currently limited to thermal and acoustic hazards and may only reduce severity of injury as opposed to completely protecting the individual from injury. Injuries from the blast forces and respiratory harm from toxic chemicals and hot gases have been difficult to address with personal protective equipment. There is a more comprehensive answer, one that includes engineering solutions that reduce the potential for an arc flash event, minimize total energy transfer, and reduce the frequency of exposure.

Increasing awareness of arc flash events and consequences has generated research and publication across engineering, science, health and safety, medical, and legal disciplines. J.C. Das has researched this body of knowledge and brought together innovative ideas and practical concepts with abundant references and real world case studies in arc flash analysis and mitigation. For the first time, design engineers, facility managers, safety professionals, and operating and maintenance personnel have a comprehensive reference for prevention methods. *Arc Flash Hazard Analysis and Mitigation* provides a comprehensive set of tools to aid in the design, evaluation, and redesign of electric power systems. There is no single "silver bullet" for arc flash mitigation, yet following the methodology and analyses discussed in the book, professional engineers and power system designers can design new industrial electrical systems and modify existing systems to limit arc flash hazard incident energy to no more than 8 cal/cm$^2$. Chapter 15 of the book presents innovative ideas for arc flash analyses in DC systems. The analysis tools enable comparison of mitigating technologies and choices in system design to optimize arc flash risk for the life of the facility. For the workers at risk, the engineering solutions serve to automatically reduce risk and function independently of their knowledge, skills, and vulnerability to human error.

We are on a journey in arc flash mitigation. Ongoing basic research will continue to explore the complexities of the arc flash phenomena. Equipment manufacturers will introduce more innovative products to eliminate or reduce exposure. Protection engineers will refine methods to sense and interrupt faults faster. Reliability engineers will help address the problem of hidden failures in circuit protection hardware, software, and schemes. Facility engineers will become more knowledgeable in demanding prevention through design. Workers will be better protected from arc flash hazards. *Arc Flash Hazard Analysis and Mitigation* provides a roadmap. For the next worker at risk of a permanently disabling, life-changing arc flash injury, we need to accelerate our journey.

H. Landis "Lanny" Floyd
*June 2012*

**H. Landis "Lanny" Floyd** is Principal Consultant, Electrical Safety and Technology with DuPont. He is a fellow of IEEE and recipient of many awards, including the 2002 IEEE Richard Harold Kaufman award for advancing the development and application of electrical safety technology and the 2004 IEEE Medal for Engineering Excellence for contributions in arc flash analysis and mitigation. He has written more than 70 papers and articles on workplace electrical safety. He is also the Editor of the *IEEE Industry Applications Magazine*. He is a nationally and internationally recognized safety expert.

# PREFACE TO SECOND EDITION

The authors and researchers all over the world expressed concerns on IEEE Guide 1584, 2002 methodology, test methods, and calculations of incident energy for arc flash hazard.

As a result, a joint venture by IEEE and NFPA was constituted. This addressed the concerns levied on 2000 edition of IEEE 584, and the 2018 edition is totally revised with respect to 2002 edition. This revision is being recognized all over the world. This second edition addresses the arc flash hazard calculations according to this revised edition. Major changes in concepts and methodology have occurred.

J.C. Das

# PREFACE TO FIRST EDITION

The arc flash hazard analysis has taken the industry by storm, as evidenced by a spate of technical papers in the current literature, especially in IEEE Industry Application Society Petroleum and Chemical Industry, Industrial and Commercial Power Systems, Pulp and Paper Industry Technical Conferences, and the IAS Safety Workshop. The concerns of worker safety in electrical environment are making new strides with respect to equipment innovations, electrical system designs, and arc flash analysis and its mitigation. This impetus has attracted the attention of the industry to bring forward new product innovations, and it has challenged the expertise of practicing and consulting engineers to innovate electrical power system designs and relay protections. The current technical papers and literature address one or the other aspect of this subject. There is no comprehensive published work on this important subject.

This book fulfills this gap. All the aspects of arc flash hazard calculations, which include short circuits, protective relaying, differential relays, arc flash detection relays, relay coordination, grounding systems, arc resistant equipment, current transformer performance, and the like, are included in easy-to-understand language with a number of case studies, practical applications, and references. Current technologies and arc flash mitigation strategies, such as coordination on instantaneous basis, current limiting devices, zone interlocking, and equipment innovations, are covered. Appendix B provides tabulated statements for quick look up of arc flash hazards in electrical power systems. Chapter 13 is devoted to secondary protection of substation transformers because of its importance in arc flash hazard reduction. The critique of IEEE 1584 Guide methodology by various authors and improvements in safety culture and work ethics are discussed. A new algorithm for the calculation of arc flash hazard accounting for the decaying nature of the short-circuit currents, first presented in IEEE Industry Application Transaction papers by the author, is included.

The IEEE 1584 Guide does not cover arc flash hazard calculations in DC systems. Chapter 15 provides detailed short-circuit calculations in DC systems and then their applications to arc flash hazard calculations in DC systems. Chapter 16 discusses application of Ethernet and IEC 61850 communication protocols in a large industrial system for control, diagnostics, and data accessibility.

The book is written for practicing engineers, consultants, electrical power systems managers, and operating personnel. Some sections require undergraduate-level or higher knowledge of electrical power systems. The book should attract a wide readership due to the ever-increasing importance of this subject in recent times.

J.C. Das

# ACKNOWLEDGEMENT

Special thanks go to the IEEE Standards Association for providing permission to use their content.

# ABOUT THE AUTHOR

J.C. Das is currently President of Power System Studies, Inc. Earlier, he headed the Power System Analysis Department at Amec Foster Wheller, Inc., Tucker, GA, for 30 years. He is specialist in conducting power system studies, including short-circuit, load flow, harmonics, stability, arc flash hazard, grounding, switching transients, and protective relaying. His interests include power system transients, EMTP simulations, harmonics, power quality, protection, and relaying.

He has authored or coauthored about 70 technical publications, nationally and internationally and has published 200 study reports for real-world power systems for his clients. He is author of the books:

*Power System Analysis*, Second Edition, CRC Press, 2011; *Transients in Electrical Systems,* McGraw-Hill, 2010; *Arc Flash Hazard Analysis and Mitigation* (second edition under publication), IEEE Press, Hoboken, NJ, 2012. *Power System Harmonics and Passive Filter Designs*, IEEE Press, Hoboken, NJ, 2015; *Understanding Symmetrical Components for Power System Modeling*, IEEE Press, Hoboken, NJ, 2017; *Power System Handbook in Four Volumes; Short-Circuit in AC and DC Systems*, ANSI/IEEE and IEC Standards; *Load Flow Optimization and Optimal Power Flow; Harmonic Generation Propagation and Control, Power System Protective Relaying*, CRC Press, Boca Raton, FL, 2018.

Mr. Das is a member of the IEEE Industry Applications and IEEE Power Engineering Societies. He is a member of TAPPI and CIGRE, a Fellow of Institution of Engineering Technology (UK), a Life Fellow of the Institution of Engineers (India), and a Member of the Federation of European Engineers (France). He is registered Professional Engineer in the States of Georgia and Oklahoma, a Chartered Engineer (C. Eng.) in the UK, and a European Engineer (Eur. Ing.) in the Europe. Mr. Das received IEEE, Pulp and Paper Industry Committee meritorious award in Engineering in 2005.

His highest education qualification is a PhD in Electrical Engineering.

# 1

# ARC FLASH HAZARDS AND THEIR ANALYSES

In the past, industrial electrical systems in the United States have been designed considering prevalent standards, that is, ANSI/IEEE, NEC, OSHA, UL, NESC, and the like, and arc flash hazard was not a direct consideration for the electrical system designs. This environment is changing fast, and the industry is heading toward innovations in the electrical systems designs, equipment, and protection to limit the arc flash hazard, as it is detrimental to the worker safety. This opens another chapter of the power system design, analysis, and calculations hitherto not required. There is a spate of technical literature and papers on arc flash hazard, its calculation and mitigation. References [1–8] describe arcing phenomena and arc flash calculations, sometimes commenting on the methodology of arc flash hazard calculations in IEEE Guide 1584 [9] (see Chapter 3).

These issues have become of great importance in the power system planning, designs and protective relay applications. "Safety by Design" is the new frontier (see Chapter 2).

Awareness of the various hazards caused by arc flash has increased significantly over the past decade. Arc flash is a dangerous condition associated with the unexpected release of tremendous amount of energy caused by an electric arc within electrical equipment [10]. This release is in the form of intense light, heat, sound, and blast of

*Arc Flash Hazard Analysis and Mitigation*, Second Edition. J.C. Das.

Published 2021 by John Wiley & Sons, Inc.

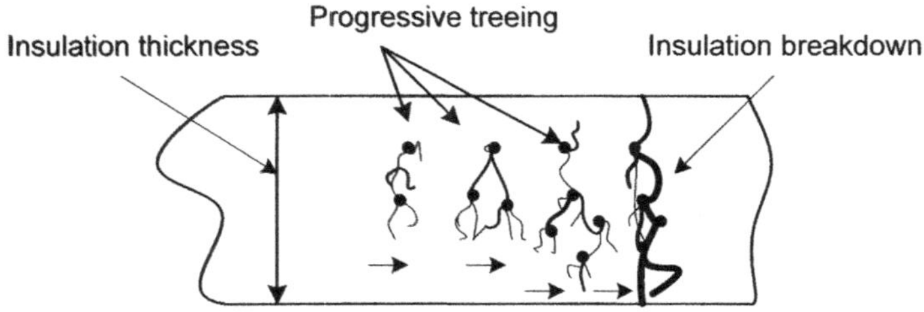


Figure 1.1. Treeing phenomena in nonself-restoring insulation, leading to ultimate breakdown of insulation.

arc products that may consist of vaporized components of enclosure material—copper, steel, or aluminum. Intense sound and pressure waves also emanate from the arc flash, which resembles a confined explosion. Arcing occurs when the insulation between the live conductors breaks down, due to aging, surface tracking, treeing phenomena, and due to human error when maintaining electrical equipment in the energized state. The insulation systems are not perfectly homogeneous and voids form due to thermal cycling. In nonself restoring insulations, treeing phenomena starts with a discharge in a cavity, which enlarges over a period of time, and the discharge patterns resemble tree branches, hence the name "treeing" (Figure 1.1). As the treeing progresses, discharge activity increases, and, ultimately the insulation resistance may be sufficiently weakened and breakdown occurs under electrical stress. Treeing phenomena is of particular importance in XLPE (cross-linked polyethylene) and nonself restoring insulations. Surface tracking occurs due to abrasion, irregularities, contamination, and moisture, which may lead to an arc formation between the line and ground. An example will be a contaminated insulator under humid conditions. Though online monitoring and partial discharge measurements are being applied as diagnostic tools, the randomness associated with a fault and insulation breakdown are well recognized, and a breakdown can occur at any time, jeopardizing the safety of a worker, who may be in close proximity of the energized equipment. Arc temperatures are of the order of 35,000°F, about four times the temperature on the surface of the sun. An arc flash can therefore cause serious fatal burns.

## 1.1 ELECTRICAL ARCS

Electrical arcing signifies the passage of current through what has previously been air. It is initiated by flashover or introduction of some conductive material. The current passage is through ionized air and the vapor of the arc terminal material, which has substantially higher resistance than the solid material. This creates a voltage drop in the arc depending upon the arc length and system voltage. The current path is resistive in nature, yielding unity power factor. Voltage drop in a large solid or stranded conductor is of the order of 0.016–0.033 V/cm, very much lower than the voltage drop in an arc, which can be of the order of the order of 5–10 V/cm of arc length for virtually all arcs in open air (Chapter 3). For low voltage circuits, the arc length consumes a substantial portion of the available voltage. For high voltages, the arc lengths can be considerably greater, before the system impedance tries to regulate or limit the fault current. The arc voltage drop and the source voltage drop are in quadrature. The length of arc in high voltage systems can be greater and readily bridge the gap from energized parts to ground.

Under some circumstances, it is possible to generate a higher energy arc from a low voltage system, as compared with a high voltage system.

In a bolted three-phase short circuit, the arcing resistance is zero, and there is no arcing, and no arc flash hazard. Sometimes, when short circuit occurs, it can be converted into a three-phase bolted short circuit by closing a making switch or circuit breaker, which solidly connects the three-phases. The fault current is then interrupted by appropriate relaying. This method, however, will subject the system to much greater short-circuit stresses and equipment damage, and, is, therefore, not recommended.

### 1.1.1 Arc as a Heat Source

The electrical arc is recognized as high-level heat source. The temperatures at the metal terminals are high, reliably reported to be 20,000 K (35,000°F). The special types of arcs can reach 50,000 K (about 90,000°F). The only higher temperature source known on earth is the laser, which can produce 100,000 K. The intermediate (plasma) part of the arc, that is, the portion away from the terminals, is reported as having a temperature of 13,000 K.

In a bolted three-phase fault, there is no arc, so little heat will be generated. If there is some resistance at the fault point, temperature could rise to the melting and boiling point of the metal, and an arc could be started. The longer the arc becomes, the more of the system voltage it consumes. Consequently, less voltage is available to overcome supply impedance and the total current decreases.

Human body can exist only in a narrow temperature range that is close to normal blood temperature, around 97.7°F. Studies show that at skin temperature as low as 44°C (110°F), the body temperature equilibrium starts breaking down in about 6 hours. Cell damage can occur beyond 6 hours. At 158°F, only a 1-second duration is required to cause total cell destruction.

### 1.1.2 Arcing Phenomena in a Cubicle

The arc formation in a *cubicle* may be described in four phases:

*Phase 1: Compression.* The volume of air is overheated due to release of energy, and the remaining volume of air inside the cubicle heats up due to convection and radiation.

*Phase 2: Expansion.* A piece of equipment may blow apart to create an opening through which superheated air begins to escape. The pressure reaches its maximum value and then decreases with the release of hot air and arc products.

*Phase 3: Emission.* The arcing continues and the superheated air is forced out with almost constant overpressure.

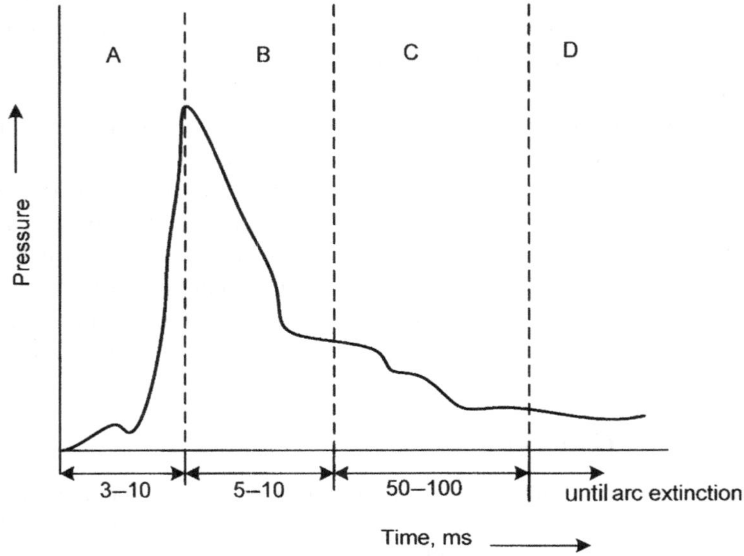


Figure 1.2. The various stages of pressure buildup and its release for an arc in a cubicle. A: Compression, pressure rises; B: Expansion, relief of pressure; C: Emission, gases exhausted; D: Thermal, pressure equalizes (not to scale).

*Phase 4: Thermal.* After the release of air, the temperature inside the switchgear nears that of an electrical arc. This lasts till the arc is quenched. All metals and insulating materials undergo erosion, may melt and expand many times, produce toxic fumes, and spray of molten metal.

Figure 1.2 shows these four phases.

## 1.2 ARC FLASH HAZARD AND PERSONAL SAFETY

The phenomenal progress made by the electrical and electronic industry since Thomas Edison propounded the principle of incandescent lighting in 1897 has sometimes been achieved at the cost of loss of human lives and disabilities. Although reference to electrical safety can be found as early as about 1888, it was only in 1982 that Ralph Lee [11] correlated arc flash and body burns with short-circuit currents. This article is considered by many as pioneering work on arcing phenomena in the open air. It quantified the potential burn hazards. Lee established the curable burn threshold for the human body as 1.2 cal/cm$^2$, which is currently used to define the arc flash boundary. Lee published a second article in 1987, "Pressure Developed from Arcs" [12].

Doughty et al., published two articles [13, 14], and Jones et al. published an article in 2000 [15]. The IEEE 1584 Guide can be considered a breakthrough for arc flash analyses. The previous methods in NFPA 70E were based upon theoretical concepts or drawn from limited testing. The new testing concentrated on arcing faults in a variety of electrical equipment enclosures, arcs in boxes, which is more typical of actual work locations. Yet some researchers are critical of the methodology of the IEEE 1584 Guide; for example, Stokes and Sweeting in "Electrical Arc Burn Hazards" [5], critique Lee's models and IEEE 1584 Guide equations and testing setup for arc flash burns. Yet the statistics collected on the prevention of arc flash hazard injuries shows that such injuries were prevented when the workers used the required personal protective equipment (PPE) calculated according to the IEEE Guide; see Chapter 3. Wilkins et al. published an article, "Effect of Insulating Barriers in Arc Flash Testing," in 2008 [16]. The authors used vertical conductors terminated in insulating barriers for their testing methodology. See Chapter 3 for further discussions and observations on these issues.

The OSHA definition of a recordable injury, TRIR, for 1 year of exposure, is as follows:

$$\text{TRIR} = \frac{\text{Total number of recordable injuries and accidents}}{200{,}000\ \text{hours}}. \tag{1.1}$$

Most insurance companies accept this parameter of definition because there is a cost associated with these incidents.

## 1.3 TIME MOTION STUDIES

Of necessity and for the continuity of processes, maintenance of electrical equipment in energized state has to be allowed for. If all maintenance work could be carried out in deenergized state, short circuits cannot occur and therefore there is no risk of arc flash hazard. For the continuous process plants, where the shutdown of a process can result in colossal amount of loss, downtime and restarting; it becomes necessary to maintain the equipment in the energized state. Prior to the institution of arc flash standards, this has been carried out for many years, jeopardizing worker safety, and there are documented cases of injuries including fatal burns.

The time/motion studies show that human reaction time to sense, judge, and run away from a hazardous situation varies from person-to-person. A typical time is of the order of 0.4 second. This means that 24 cycles is the shortest time in which a person can view a condition and *begin* to move or act. In all other conditions, it is not possible to see a hazardous situation and move away from it. As will be further demonstrated, this reaction time is too large for a worker to move away and shelter himself from an arc flash hazard situation.

## 1.4 ARC FLASH HAZARDS

Apart from thermal burns, an arcing phenomenon is associated with other hazards too, namely:

- electrical shock
- molten metal

- projectiles
- blast and pressure waves
- intense light
- intense sound
- fire
- effect of strong magnetic fields and plasma, of which not much is known
- toxic gases and vapors.

Thus, thermal burns due to arc flash *are only a part the picture* for overall worker safety. Figure F.1a,b in NFPA 70E [17], not reproduced here, provides hazard risk analysis procedure flowchart. It implies that each establishment must perform a number of tasks and establish training and safety procedures that should be implemented for workers' safety. The numbers of injuries from arc flash accidents are high (see Chapter 2). IEEE 1584 Guide documents many such cases.

This book is confined to the analysis of arc flash thermal damage and calculation of arc flash boundary, subsequently defined, according to IEEE 1584 Guide equations. The book concentrates on the various design, planning, and protection strategies by which the arc flash hazard can be reduced.

## 1.5 ARC BLAST

As opposed to arc flash, which is associated with thermal hazard and burns, arc blast is associated with extreme pressure and rapid pressure buildup. Consider a person positioned directly in front of an event and high pressure impinging upon his chest and close to the heart and the hazard associated with it.

The reports of the consequences of arc in air include descriptions of the rearward propulsion of personnel who were close to the arc. In many cases, the affected people do not remember being propelled away from the arc. The heat and molten metal droplet emanation from the arc can cause serious burns to the nearby personnel.

A substance requires a different amount of physical space when it changes state, say from solid to vaporized particles. When the liquid copper evaporates, it expands 67,000 times. This accounts for the expulsion of vaporized droplets of molten metal from an arc, which is propelled up to distance of 10 ft. It also generates plasma (ionized vapor) outward from the arc for distances proportional to the arc power. One cubic inch of copper vaporizes into 38.8 cubic feet of vapor.

The air in the arc stream expands in warming up from the ambient temperature to that of an arc, about 20,000 K. This heating is related to the generation of thunder by passage of lightning current through it. In documented instances a motor terminal box exploded as a result of force created by the pressure build-up, parts flying across the room [18]. Pressure measurement of 2160 lbs/ft$^2$ around the chest area and sound level of 165 dB at 2 ft have been made.

The pressure varies with the distance from the arc center and the short-circuit current. Figure 1.3 shows this relation based upon Lee's classical work [12].

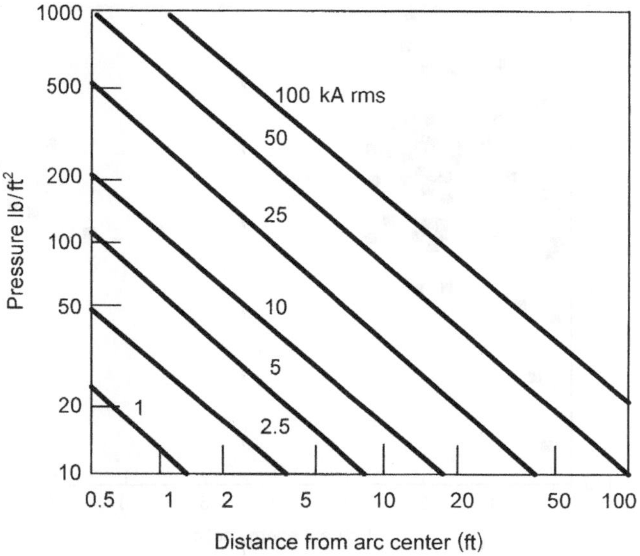


Figure 1.3. Pressure versus distance from the center of the arc, based on Lee's work. *Source*: Reference [12].

The hot air vapor from the arc starts to cool immediately; however, it combines with the oxygen of the air, thus becoming the oxide of the metal of the arc. These continue to cool and solidify, and become minute particles in the air, appearing as black smoke for copper and iron and gray smoke for aluminum. These are still hot and cling to any surface these touch, actually melting into many insulating surfaces that these may contact. The oxide particles are very difficult to remove because surface rubbing is not effective. Abrasive cleaning is necessary on plastic insulation. A new surface varnish should be applied, or surface current leakage could occur and cause failure within days.

Persons exposed to severe pressure from proximity of an arc are likely to suffer short-time loss of memory and may not remember the intense explosion of the arc itself. This phenomenon has been found true even for high-level electrical shocks.

The PPE is currently designed and tested to address the heat energy hazard. The arc-rated FR (fire resistant), including face hood shields window materials, have been observed to provide protection for the molten metal splatter hazard. There have been considerations of pressure-wave hazard [12, 19] and noise hazard [20]. This has resulted in NFPA 70E specifying hearing protection.

Noise has been monitored with microphones to understand its relationship with arc parameters. The noise results from initial explosive expansion of air and formation of a plasma region between conductors. The noise in single-phase arc events is assumed to behave similarly.

Figure 1.4 shows variations in noise level measurements, at a distance of 1.8 m from a variety of arc configurations—a scatter plot. These variations will narrow down

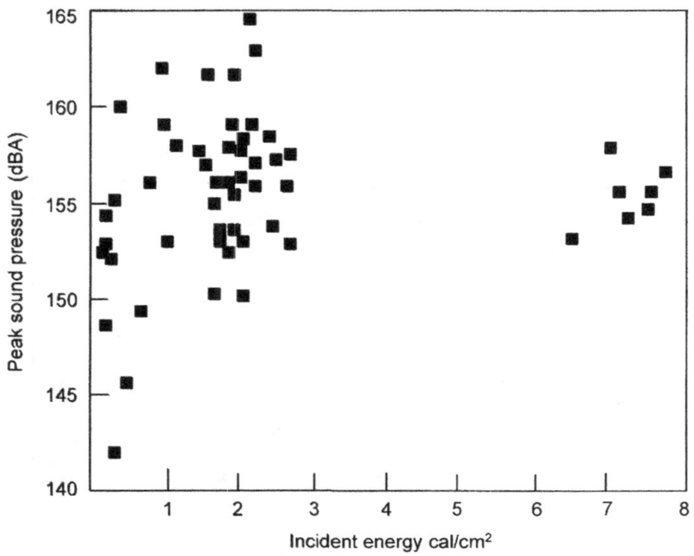


Figure 1.4. Peak sound pressure in dBA, at a distance of 1.8 m from a variety of arcing configurations. *Source*: Based on Reference [20].

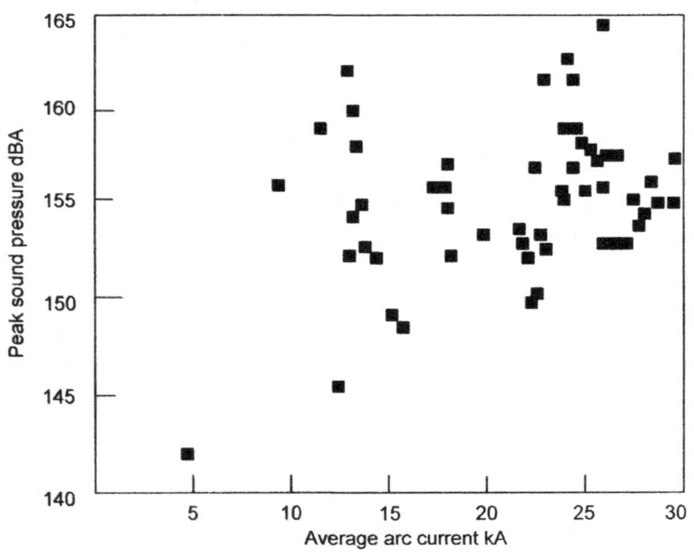


Figure 1.5. Average arcing current versus the peak sound pressure dBA. *Source*: Based on Reference [20].

if the test conditions were done in a fixed configuration. The arc ratings using PPE cannot be applied to hearing or pressure-wave protection. Figure 1.5 shows that for lower levels of arcing current, the noise levels can even be higher. This figure shows measurements at 0.61 m (2 ft) from a variety of three-phase arc configurations. NFPA 70E, table 130.7(C)(16), in 2009 was revised and recommends hearing protection (ear

TABLE 1.1. Ballistics $V_{50}$ Results for Arc-Rated Fabric Systems

| Multilayer Fabric Systems | Arc Rating, cal/cm$^2$ | Ballistic Layer | Fabric System Weight, g/m$^2$ | $V_{50}$, m/s | Fragment Diameter, mm |
|---|---|---|---|---|---|
| Cotton with flame retardant | 100 | No | 1424 | 186 | 5.6 |
| Aramid fabric system | 100 | No | 932 | 210 | 5.6 |
| Aramid fabric system | 80 | Yes | 780 | 280 | 5.6 |
| Aramid fabric system | 100 | No | 881 | 191 | 7.8 |
| Aramid fabric system | 100 | Yes | 922 | 240 | 7.8 |

Adapted from Reference [20].

canal inserts) even for category 0. In the 2002 edition, hearing protection was not specified for category 0 and 1 hazards. See also table 130.7(C)(15)a,b.

If current limiting fuses are used, which operate in about 1/2 cycle or less, the arcing time is reduced and so also the noise levels—this relation is not so well defined, and additional testing is recommended [20].

Figure 1.5 shows that noise sound pressure levels can exceed OSHA impulsive or impact noise level of 140 dB peak. Even at 1.8 m level the measured sound levels are well above small arms firing and without hearing protection, some individuals may suffer traumatic damage, including eardrum rupture [13]. A worker will be positioned closer than 1.8 m when working on energized equipment.

The shrapnel hazard has not been quantified or related to arc-flash parameters, but it is possible to measure shrapnel resistance of arc flash fabric systems and hood shield windows to standardized threats.

Arc flash hood windows and face shields must meet projectile impact requirements of ANSI Z87.1, which specifies that a 6.4-mm (0.25 in) steel ball projectile must not penetrate the shield window or face shield at a velocity of 91.4 m/s (300 ft/s). It does not consider irregular-shaped projectiles or velocities that may be from 150 to 180 m/s (500–600 ft/s) and accompany an arc fault event. Thus, testing of arc-flash PPE was conducted using fragments instead of bullets [20]. Table 1.1 provides the test results. $V_{50}$ signifies the velocity at which 50% of the projectiles penetrate the target specimen. This shows benefits of additional tightly woven para-armid ballistic fiber layer without weight increase.

## 1.6 ELECTRICAL SHOCK HAZARD

One of the most complete analyses of occupational electrical injuries in the United States are two papers by Jim Cawley [37, 38]. On an average, one person is electrocuted in work places *every day in United States.* There are a number of ways the exposure to shock hazard occurs. The resistance of the contact point, the insulation of the ground under the feet, flow of current path through the body, the body weight, the system voltage, and frequency are all important. A dangerous consequence is a heart condition, known as ventricular fibrillation, resulting in immediate arrest of blood circulation. Currents as small as a few milliamperes through the heart can cause disruption of electrical signals that the heart uses to perform its functions. Voltages as low as 50 V can cause fibrillation and can result in death.

The following synopsis of tolerable currents is from IEEE Standard 80, Guide for Safety in AC Substation Grounding [21]:

At 50 or 60 Hz, a current of 0.1 A can be lethal. The human body can tolerate slightly higher 25 HZ current and five times the DC current. At frequencies of 3000–10,000, even higher currents are tolerated. The most common physiological effect, stated in terms of increasing current, are: threshold of perception, muscular contraction, unconsciousness, fibrillation of heart, respiratory nerve blockage and burning [22], and IEC 604791 [23].

The perception level is 1 mA. Currents in the range of 9–25 A may be painful and may make it difficult or impossible to release energized objects. In the range 60–100 mA, ventricular fibrillation, stoppage of heart, or inhibition of respiration might occur, causing injury or death. As shown by Dalziel and others [24], the nonfibrillating current of magnitude $I_B$ at durations ranging from 0.03 to 3.0 seconds is related to energy absorbed by the body, given by:

$$S_B = (I_B)^2 \times t_s, \tag{1.2}$$

where $t_s$ is the time duration of the current in seconds, and $S_B$ is an empirical constant related to the energy through the body. Thus, reducing the arc flash incident energy through fast fault clearance times also reduces $S_B$.

Based upon the Dalziel and Lees' studies [25], it is assumed that 99.5% of all persons can safely withstand, without ventricular fibrillation, the passage of current $I_B$, given by:

$$I_B = \frac{k}{\sqrt{t_s}}, \quad k = \sqrt{S_B}. \tag{1.3}$$

Dalziel found that $S_B = 0.0135$ for a body weight of 110 lbs (50 kg). Then:

$$I_B = \frac{0.116}{\sqrt{t_s}}. \tag{1.4}$$

This gives 116 mA for 1 second and 367 mA for 0.1 second. For 70 kg weight, $S_B = 0.0246$ and $k = 0.157$. These values are adopted in IEEE Guide 80 [21]. Fibrillation current is assumed to be the function of body weight (Figure 1.6).

Other researchers have suggested different values of $I_B$. In 1936, Fwerris et al. [26] suggested 100 mA as fibrillation threshold; this value was derived by extensive experimentation at Columbia University. Some more recent experiments suggest

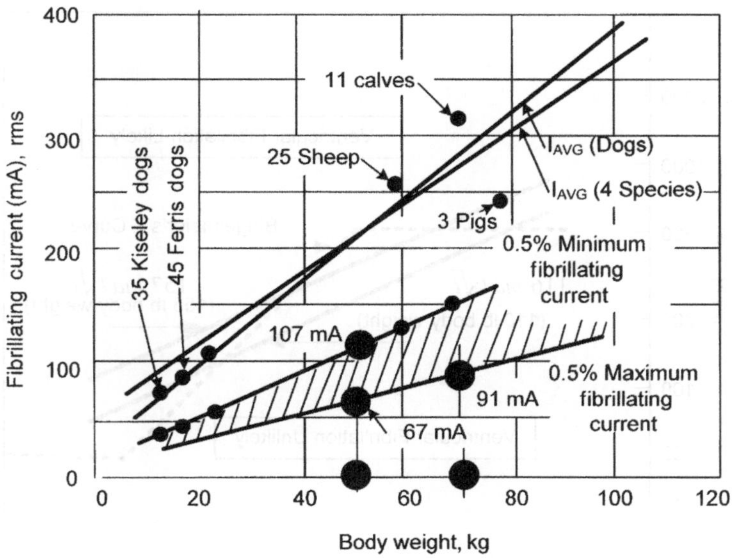


Figure 1.6. Fibrillating current (ma) rms, versus body weight. *Source*: Reference [21].

the existence of two thresholds: one for shock duration less than one heartbeat period, and the other for the current duration longer than one heartbeat period. For a /50 kg body weight, Biegelmeier [27] proposed threshold levels of 500 and 50 mA, respectively. Other studies were carried out by Lee and Kouwenhoven [28]. Figure 1.7 shows a comparison of Equation (1.4) and Z-shaped body current time developed by Biegelmeier.

### 1.6.1 Resistance of Human Body

For DC and AC 50 or 60 HZ currents, the human body can be approximated by a resistance. For the calculation of this resistance, the current path is considered from:

- one hand to both feet
- from one foot to another foot.

The internal resistance of the body is approximately 300 Ω, while the body resistance, including skin range from 500 to 3000 Ω. Based on Dalziel tests, using saltwater to wet hands and feet to determine let-go currents, hand-to-hand contact resistance is 2330 Ω, and hand-to-feet resistance equals 1130 Ω. Thus, the IEEE Guide for Safety in AC Substation Grounding considers that hand and foot contact resistances are zero, that glove and shoe resistances are zero, and a value of 1000 Ω is taken that represents the body from hand-to-feet and also from hand-to-hand resistance.

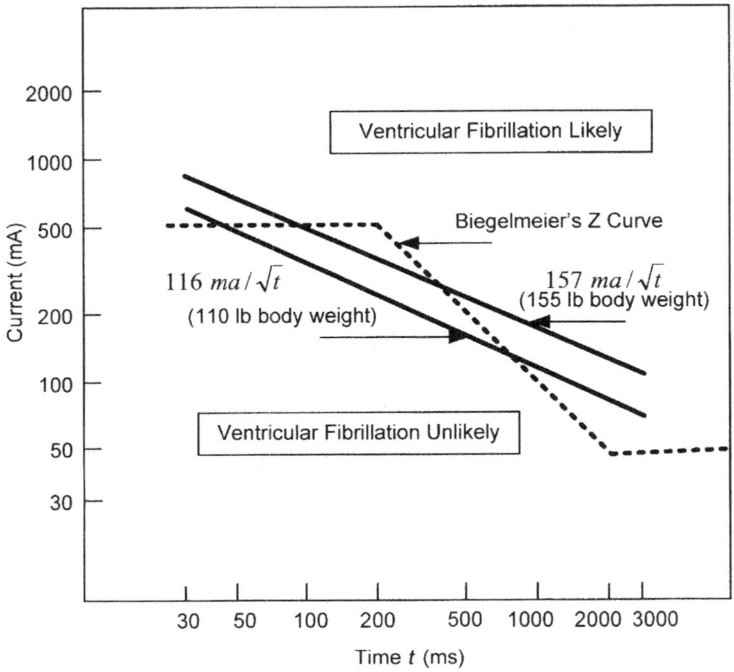


Figure 1.7. Ventricular fibrillation curves, current versus time. *Source*: Reference [21].

NFPA 70E states that energized parts operating at less than 50 volts are not required to be de-energized to satisfy an "electrical safe working condition." It further lays down that considerations should be given to the capacity of the source, any overcurrent protection between the source and the worker, and whether the work task related to the source operating at less than 50 volts increases exposure to electrical burns or to explosion from an electric arc.

Reference [29] contends that 50 V is inadequate and calculates the maximum and minimum body resistance for path from arm-to-arm and arm-to-leg of the order of 300–500 Ω. IEC standard 604791 [23] recommends shock voltages of less than 50 V in some situations. Some jurisdictions, for example, in France, the safe voltage limit is accepted as 35–50 V. However, NFPA 70E qualifies the 50 V limits by additional cautionary statements as indicated above.

Table 1.2 provides resistance values for 130 cm$^2$ areas of various materials. It is customary to overlay the natural soil with high resistivity materials to increase the step and touch potentials in utility substations [21]. For the grounding systems in industrial electrical distributions, generally, the concept of higher soil resistivity layers to increase step and touch potentials can be applied for the grounding installations around buildings, tanks, substations, fences, and motor and transformer pedestals.

Figure 1.8 from IEC standard [23] illustrates the time–current zones for AC currents of 15–100 Hz, and Table 1.3 provides the physiological effects. IEC considers that hand-to-hand body impedance for 125 V is between 850 and 2675 Ω, and grasping

TABLE 1.2. Resistance of 130-cm$^2$ Areas of Various Materials

| Material | Resistance in MΩ |
|---|---|
| Rubber gloves or soles | >20 0 |
| Dry concrete above grade | 1 0–5.0 |
| Dry concrete on grade | 0.2–1 0 |
| Leather sole, dry, including foot | 0.1–0.5 |
| Leather sole, damp, including foot | 0.05–0.2 |
| Wet concrete | 0.01–0.05 |

*Source*: Reference [23].

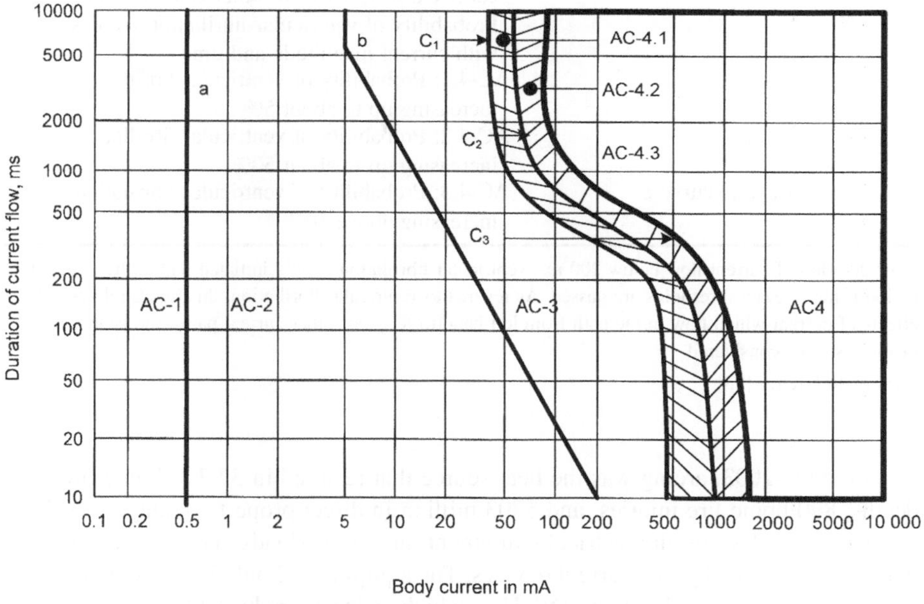


Figure 1.8. Shock hazard categories according to IEC. *Source*: Reference [23].

a conductor or faulty electric device rated 120 V can result in a current flow between 45 and 140 mA.

See also Section 2.4.

## 1.7 FIRE HAZARD

NFPA and National Fire Incident Reporting Systems (NFIRS) statistics of fire hazard can be viewed on websites. These statistics are based upon:

- heat source, that is, arcing
- contributing factors like electrical failure or malfunction
- equipment involved in electrical distribution, lighting, and power transfer.

TABLE 1.3. Time–Current Zones for 15–100 Hz AC Currents for Hand-to-Feet Pathway

| Zone | Boundaries | Physiological Effects |
|---|---|---|
| AC-1 | Up to 0.5 mA, curve a | Perception possible but usually no "startled" reaction |
| AC-2 | 0.5 mA up to curve b | Perception and involuntary muscular contractions likely but usually no harmful physiological effects |
| AC-3 | Curve b and above | Strong involuntary muscular contractions, difficulty in breathing. Reversible disturbances of heart function. Immobilization may occur. Effects increasing with current magnitude. Usually no organic damage to be expected. |
| AC-4[a] | Above curve $c_1$ | Pathophysiological effects may occur, such as cardiac arrest, burns, or other cellular damage. Probability of ventricular fibrillation increasing with current magnitude and time |
| | $c_1$–$c_2$ | AC-4.1: Probability of ventricular fibrillation increasing up to about 5% |
| | $c_2$–$c_3$ | AC-4.2: Probability of ventricular fibrillation increasing up to about 50% |
| | Beyond curve $c_3$ | AC-4.3: Probability of ventricular fibrillation increasing above 50% |

[a] For duration of current flow below 200 ms, ventricular fibrillation is only initiated within the vulnerable period if the relevant thresholds are passed. As regards to ventricular fibrillation, this figure relates to the effects of current which flow in the path from left hand to feet. For other current paths, the heart current factor has to be considered.

*Source*: Reference [23].

In 1999–2003, arcing was the heat source that resulted in 37,700 home fires, 240 deaths, 890 home fire injuries, and $703 million in direct property damage [30, 31].

Fires can develop in electrical equipment due to overloads and loose connections that are not cleared by overcurrent devices. The equipment should be listed by a nationally recognized test laboratory (NRTL), which helps to reduce the fire risk. Some precautionary and design measures are:

- Fire detection and suppression equipment should be permanently installed or readily accessible around the electrical equipment. Such equipment could possibly include smoke detectors, sprinkler systems, and portable fire extinguishers.
- The workplace should be designed so that escape routes are sufficiently wide, clear of obstructions, well marked and lighted. Normal and emergency lighting and exit signs are important.
- Special considerations should be applied to the electrical equipment located in hazardous areas, according to NEC.
- All conductors and wiring should be properly sized for protection against overheating (see Article 310 of NEC).
- Overcurrent protection should be provided to meet the requirements of NEC.

- Motors and generators should be properly protected so that these do not cause a fire hazard.
- The transformers should be protected and installed according to NEC, UL, and FM (factory mutual) guidelines. In general, all electrical equipment must be installed, operated, and maintained according to codes and standards (see Chapter 2).

The fire hazards are not further discussed in this book.

## 1.8 ARC FLASH HAZARD ANALYSIS

As early as December 1970, the Occupational Safety and Health Act required that each employer shall furnish to his employees, employment and place of employment that are free from recognized hazards that are causing or likely to cause death or serious physical harm to his employees. It was not till late 1991 that OSHA added words acknowledging arc flash as an electrical hazard. NFPA published the first edition of NFPA 70E in 1979.

Effective from January 1, 2009, the National Electric Safety Code (NESC) [32] requires that all power generating utilities perform arc flash assessments. The employer shall ensure that assessment is performed to determine potential exposure to an electric arc for employees who work on or near energized parts or equipment. If the assessment determines a potential employee exposure greater than 1.2 cal/cm$^2$ exists, the employer shall require employees to wear clothing or a clothing system that has an effective arc rating not less than the anticipated level of arc energy.

Currently, there are four major guides for arc flash calculations:

1. NFPA 70E, revised in 2018 [17]
2. IEEE 1584 Guide, 2018, which will undergo revisions [9]
3. IEEE 1584a, 2004, amendment 1 [33]
4. IEEE P1584b/D2 Draft 2, unapproved [34].

NFPA 70E 2012, in annex D, table D.1, provides limitations of various calculation methods. This is reproduced in Table 1.4. *The standard does not express any preference for which method should be used.* Reference [33] recognizes use of knowledge and experience of those who have performed studies as a guide in applying the standard. IEEE 1584 Guide also contains a theoretically derived model applicable for any voltage.

It is recognized that to construct an *accurate* mathematical model of the arcing phenomena is rather impractical. This is because of the spasmodic nature of the fault caused by arc elongation blowout effects, physical flexing of cables and bus bars under short circuits, possible arc reignition, turbulent flow of plasma, and high temperature gradients (the temperature at the core being of the order of 25,000 K, while at the arc boundary, of the order of 300–2000 K).

IEEE 1584 Guide equations are empirical equations based upon laboratory test results, though the standard includes some of Lee's equations also.

TABLE 1.4. Limitations of ARC Flash Hazard Calculation Methods

| Source | Limitations/Parameters |
|---|---|
| Ralph Lee [11] | Calculates arc flash boundary for arc in open air; conservative over 600 V and becomes more conservative as voltage rises |
| Doughty and Neal [14] | Calculated incident energy for three-phase arc on systems rated 600 V and below, applies to short-circuit currents between 16 and 50 kA |
| Ralph Lee [11] | Calculated incident energy for three-phase arc in open air on systems rated above 600 V, becomes more conservative as voltage rises |
| IEEE Standard 1584 [9] | Calculates incident energy and arc flash boundary for 208 V to 15 kV, three-phase 50–60 Hz; 700–106,000 A short-circuit currents and 13–152 mm conductor gaps.[a] |
| ANSI/IEEE C2, tables 410-1, 410-2 [32] | Calculates incident energy for open-air phase-to-ground arcs 1 kV to 500 kV for live line work. |
| Doan | Arc flash calculations for exposure to DC systems. Calculates incident energy for DC systems up to 1000 V DC. |

[a] Equations for higher voltages are included.
*Source*: NFPA 70E-2018.

If the equipment is maintained under deenergized condition, there is no arc flash hazard. NFPA 70E [17] states that energized electrical conductors and circuit parts that operate at less than 50 V to ground should not be required to be de-energized. Again, it is qualified that the capacity of the source and any overcurrent protection between the source and the worker should be determined and there should be no increased exposure to electrical burns or explosion due to electrical arcs. The IEEE 1584 Guide states that equipment below 240 V need not be analyzed for arc flash unless it involves at least one 125 kVA or larger low impedance transformer in its immediate power supply. The "low impedance" is not defined. Sometimes, the arc flash hazard can be high even in systems of 240 V. When incident energy exceeds 40 cal/cm$^2$, the equipment should only be maintained in the de-energized condition. There is no PPE (personal protective equipment) outfits specified for incident energy release >40 cal/cm$^2$; see Section 1.9 for definitions and discussions of PPE.

That an arc flash analysis shall not be required where all the following conditions exist has been deleted in NFPA 70E 2012:

- The circuit is rated 240 V or less.
- The circuit is supplied by one transformer.
- The transformer supplying the circuit is rated less than 125 kVA.

This qualification has now been removed in 1584, 2018 edition.

The user is referred to IEEE Guide 1584 for three-phase systems rated less than 240 V.

### 1.8.1 Ralph Lee's and NFPA Equations

Ralph Lee equations from Reference [11] are as follows:

Maximum power in a three-phase arc is:

$$P = \text{MVA}_{\text{bf}} \times 0.707^2\,\text{MW}, \tag{1.5}$$

where $\text{MVA}_{\text{bf}}$ is bolted fault mega-volt-ampere (MVA).

The distance in feet of a person from an arc source for a just curable burn, that is, skin temperature remains less than 80°C, is:

$$D_c = \left(2.65\,\text{MVA}_{\text{bf}}t\right)^{1/2}, \tag{1.6}$$

where $t$ is the time of exposure in seconds.

The equation for the incident energy produced by a three-phase arc in *open air* on systems rated above 600 V is given by:

$$E = \frac{793FVt_{\text{A}}}{D^2}\,\text{cal/cm}^2, \tag{1.7}$$

where:

$D$ = distance from the arc source in inches
$F$ = bolted fault short-circuit current, kA
$V$ = system phase-to-phase voltage, kV
$t_{\text{A}}$ = arc duration in seconds.

For the low voltage systems of 600 V or below and for an arc in the *open air*, the estimated incident energy is:

$$E_{\text{MA}} = 5271D_{\text{A}}^{-1.9593}t_{\text{A}}\left[0.0016F^2 - 0.0076F + 0.8938\right], \tag{1.8}$$

where $E_{\text{MA}}$ is the maximum open air incident energy in cal/cm$^2$, $F$ is short-circuit current in kA, range 16–50 kA, and $D_{\text{A}}$ is distance from arc electrodes, in inches (for distances 18 in and greater).

The estimated energy for an arc in a *cubic box of 20 in*, open on one side is given by:

$$E_{\text{MB}} = 1038.7D_{\text{B}}^{-1.4738}t_{\text{A}}\left[0.0016F^2 - 0.0076F + 0.8938\right], \tag{1.9}$$

where $E_{\text{MB}}$ is the incident energy and $D_{\text{B}}$ is the distance from arc electrodes, inches (for distances 18 in and greater).

### 1.8.2 IEEE 1584 Guide Equations

This is based on IEEE 2002 Guide. Included here for reference and completeness.

The IEEE equations are applicable for the electrical systems operating at 0.208 to 15 kV, three-phase, 50 or 60 Hz, available short-circuit current range 700–106,000 A, and conductor gap = 13–152 mm. For three-phase systems in open air substations,

TABLE 1.5. Classes of Equipment and Typical Bus Gaps

| Classes of Equipment | Enclosure Size (in) | Typical Bus Gaps (mm) |
|---|---|---|
| 15-kV switchgear | 45×30×30 | 152 |
| 15-kV MCC | 36×36×36 | 152 |
| 5-kV switchgear | 36×36×36 | 104 |
| 5-kV switchgear | 45×30×30 | 104 |
| 5-kV MCC | 26×26×26 | 104 |
| Low voltage switchgear | 20×20×20 | 32 |
| Shallow low voltage MCCs and panel boards | 14×12×≤8 | 25 |
| Deep voltage MCCs and panel boards | 14×12×>8 | 25 |
| Cable junction box | 14×12×≤8 or14×12×>8 | 13 |

*Source*: IEEE 1584-2018 Guide [9]. Also see Chapter 3.

open-air transmission systems, a theoretically derived model is available. For system voltage *below* 1 kV, the following equation is solved:

$$\log I_a = K + 0.662\log_{10} I_{bf} + 0.0966V + 0.000526G + 0.5588V(\log_{10} I_{bf}) - 0.00304G(\log_{10} I_{bf}), \quad (1.10)$$

where:

$I_a$ = arcing current in kA

$G$ = conductor gap in mm, typical conductor gaps are specified in [9] (see Table 1.5)

$K$ = -0.153 for open air arcs, -0.097 for arc in a box

$V$ = system voltage in kV

$I_{bf}$ = bolted three-phase fault current kA, rms symmetrical.

For systems of 1 kV and higher, the following equation is solved:

$$\log_{10} I_a = 0.00402 + 0.983\log_{10} I_{bf}. \quad (1.11)$$

This expression is valid for arcs both in open air and in a box. Use 0.85 $I_a$ to find a second arc duration. This second arc duration accounts for variations in the arcing current and the time for the overcurrent device to open. Calculate incident energy using both 0.85 $I_a$ and $I_a$ and use the higher value.

Equation (1.11) is a statistical fit to the test data and is derived using a least square method; see Appendix A for a brief explanation of least square method.

Incident energy at working distance, an empirically derived equation, is given by:

$$\log_{10} E_n = K_1 + K_2 + 1.081\log_{10} I_a + 0.0011G. \quad (1.12)$$

The equation is based upon data normalized for an arc time of 0.2 seconds, Where:

$E_n$ = Incident energy (J/cm$^2$) normalized for time and distance

$K_1$ = -0.792 for open air and -0.555 for arcs in a box

$K_2$ = 0 for ungrounded and high resistance grounded systems and -0.113 for grounded systems. Low resistance grounded, high resistance grounded, and ungrounded systems are all considered ungrounded for the purpose of calculation of incident energy.

$G$ = conductor gap in mm (Table 1.5).

Conversion from normalized values gives the equation:

$$E = 4.184 C_f E_n \left(\frac{t}{0.2}\right)\left(\frac{610^x}{D^x}\right), \tag{1.13}$$

where:

$E$ = incident energy in J/cm$^2$

$C_f$ = calculation factor = 1.0 for voltages above 1 kV and 1.5 for voltages at or below 1 kV

$t$ = arcing time in seconds

$D$ = distance from the arc to the person, working distance (Table 1.6)

$x$ = distance exponent as given in Reference [9] and reproduced in Table 1.7.

A theoretically derived equation can be applied for voltages above 15 kV or when the gap is outside the range in Table 1.5 (from Reference [9]).

$$E = 2.142 \times 10^6 V I_{bf}\left(\frac{t}{D^2}\right). \tag{1.14}$$

TABLE 1.6. Classes of Equipment and Typical Working Distances

| Classes of Equipment | Working Distance |
|---|---|
| 15-kV switchgear | 36 |
| 15-kV MCC | 36 |
| 5-kV switchgear | 36 |
| 5-kV switchgear | 36 |
| 5-kV MCC | 36 |
| Low voltage switchgear | 24 |
| Shallow low voltage MCCs and panel boards | 18 |
| Deep voltage MCCs and panel boards | 18 |
| Cable junction box | 18 |

*Source*: IEEE 1584-2018 Guide [9]. Also see Chapter 3.

TABLE 1.7. Factors for Equipment and Voltage Classes

| System Voltage, kV | Equipment Type | Typical Gap between Conductors | Distance × Factor |
|---|---|---|---|
| 0.208–1 | Open air | 10–40 | 2.000 |
| | Switchgear | 32 | 1.473 |
| | MCC and panels | 25 | 1.641 |
| | Cable | 13 | 2.000 |
| >1–5 | Open air | 102 | 2.000 |
| | Switchgear | 13–102 | 0.973 |
| | Cable | 13 | 2.000 |
| >5–15 | Open air | 13–153 | 2.000 |
| | Switchgear | 153 | 0.973 |
| | Cable | 13 | 2.000 |

*Source*: IEEE 1584 Guide [9].

For the arc flash protection boundary, defined further, the empirically derived equation is:

$$D_B = \left[4.184 C_f E_n \left(\frac{t}{0.2}\right)\left(\frac{610^x}{E_B}\right)\right]^{1/x}, \tag{1.15}$$

where $E_B$ is the incident energy in J/cm$^2$ at the distance of arc flash protection boundary.

For Lee's method:

$$D_B = \left[2.142 \times 10^6 V I_{bf} \left(\frac{t}{E_B}\right)\right]^{1/2}. \tag{1.16}$$

Due to complexity of IEEE equations, the arc flash analysis is conducted on digital computers. It is obvious that the incident energy release and the consequent hazard depend upon:

- The available three-phase rms symmetrical short-circuit currents in the system. The actual bolted three-phase symmetrical fault current should be available at the point where the arc flash hazard is to be calculated. In low voltage systems, the arc flash current will be 50–60% of the bolted three-phase current, due to arc voltage drop. In medium and high voltage systems, it will be only slightly lower than the bolted three-phase current. The short-circuit currents are accompanied by a DC component, whether it is the short circuit of a generator, a motor, or a utility source. However, for arc flash hazard calculations, the *DC component is ignored*. Also, any unsymmetrical fault currents, such as line-to-ground fault currents, need not be calculated. As evident from the cited equations, only three-phase symmetrical bolted fault current need be calculated.

- The time duration for which the event lasts. This is obviously the sum of protective-relay (or any other protection device) operating time plus the opening time of the switching device. For example, if the relay operating time is 20 cycles, and the interrupting time of the circuit breaker is 5 cycles, then the arc flash time or arcing time is 25 cycles.
- The type of equipment, that is, switchgear or MCC, or panel and the operating voltage
- The system grounding. This is deleted in 2018 edition, see Chapter 3

We can add to this list:

1. Electrical electrodes and potential arc lengths; spacing between phases, spacing between phases and ground, orientation-vertical or horizontal, insulated versus non-insulated buses.
2. Atmospheric conditions like ambient temperature, barometric pressure, and humidity.
3. Dissipation of energy in the form of heat, light, sound, and pressure waves.
4. Arc conditions like, randomness of arc, its interruption, arc plasma characteristics, size, and shape of enclosure.

For using the IEEE equations, the factors listed above need not be considered. As an example, there are many discussions about the gap distances specified in IEEE 1584 Guide and their effects on the incident energy release. While critique of IEEE equations and methodology does add to the technical aspects and paves the way for further revisions, this book limits the calculations according to current IEEE methodology. See also Chapter 3.

IEEE Standard 1584 Guide also provided equations for class L and RK1 fuses, not reproduced here.

## 1.9 PERSONAL PROTECTIVE EQUIPMENT

NFPA table 130.7(C)(16) describes the PPE characteristics for hazard risk category of 0, and 1 through 4. These are shown in Table 1.8.

The standard ASTM F1506 [35] calls for every flame-resistant garment to be labeled with an arc energy rating, ATPV (arc thermal performance exposure value). The rating of the garment is matched with the calculated incident energy release level. The test method of determining the ATPV specifies the incident energy on a multilayer system of materials that results in a 50% probability that sufficient heat transfer through the test specimen is predicted to cause onset of second-degree skin burn injury (see Reference [9]). Arc rating is reported as the minimum of ATPV or $E_{BT}$ (breakopen threshold). $E_{BT}$ is defined in ASTM F1959-06.

The maximum incident energy for which PPE is specified is 40 cal/cm$^2$ (167.36 J/cm$^2$). *It is not unusual to encounter energy levels much higher than 40 cal/cm$^2$ in*

TABLE 1.8. Protective Clothing Characteristics

| Hazard Category | Clothing Description | Range of Calculated Incident Energy | Arc Rating of PPE, cal/cm$^2$ |
|---|---|---|---|
| 0 | Nonmelting, flammable materials, that is, untreated cotton, wool, rayon, or silk, or blends of these materials, with a fabric weight of 4.5 oz/yd$^2$ | $0 \le E \le 1.2$ | N/A |
| 1 | Arc-rated clothing, minimum arc rating of 4 cal/cm$^2$ | $1.2 < E \le 4$ | 4 |
| 2 | Arc-rated clothing, minimum arc rating of 8 cal/cm$^2$ | $4 < E \le 8$ | 8 |
| 3 | Arc-rated clothing selected so that the system rating meets the required minimum arc rating of 25 cal/cm$^2$ | $8 < E \le 25$ | 25 |
| 4 | Arc-rated clothing selected so that the system rating meets the required minimum arc rating of 40 cal/cm$^2$ | $25 < E \le 40$ | 40 |

*Source*: NFPA 70E [17]. (All details in the NFPA table have not been reproduced.)

*actual electrical systems. Standards do not provide guidelines for higher incident energy levels.* Incident energy reduction techniques can be applied; otherwise, it is prudent not to maintain such equipment in energized state.

A category 4 PPE outfit looks like a space suit with face hood, eye shields, cover, and gloves. It restricts the mobility of a worker to perform delicate tasks, for example, maintenance work on terminals and wiring.

Thus, not only an accurate calculation of incident energy level, but its reduction in the planning and design stage and selection of appropriate protection and relaying of electrical systems are gaining importance [36].

A very significant change in NFPA70E 2012 from 2009 is the redefinition of PPE hazard risk categories. This will be based on developing a clothing system meeting a specific tested cal/cm$^2$ level and comprised of arc-rated materials. Suppose that an 8 cal/cm$^2$ protection is required; this can be achieved by:

- 8-cal/cm$^2$ arc-rated pants and shirts
- 4-cal/cm$^2$ treated pants and shirts and 4 cal/cm$^2$ arc-rated overall
- 8-cal/cm$^2$ arc-rated overall cotton shirts and pants.

It is the total level of arc-rated protection that matters. To achieve 8 cal/cm$^2$ of the example being discussed, any 8 cal/cm$^2$ system of one layer or more is acceptable. Natural fiber clothing like cotton, wool, and silk are still allowed but cannot add to the protection level because of risk of ignition. A note of caution is required in interpretation: a 4 cal/cm$^2$ arc-rated shirt and another 4 cal/cm$^2$ arc-rated shirt put together *do not add* to 8 cal/cm$^2$. *When layers are used, the total system must be tested to get the total arc rating, though most often the testing may prove that even greater protection may be provided than the arithmetic sum of the two layers.* Also, layering of clothing that is not arc-rated cannot be used to increase arc rating of the clothing. See appendix H of NFPA 70E.

In NFPA 70E 2012, the hazard risk category (HRC) 2* is eliminated. All HRC 2 tasks will require the use of a balaclava under an arc-rated face shield or the use of an arc-rated hood. Commercially, arc-rated outfits even up to 100 cal/cm$^2$ are available, still for category 4 applications.

## 1.10 HAZARD BOUNDARIES

The boundaries are defined in NFPA, and the following synopsis is relevant here (Figure 1.9).

The flash protection boundary is the distance at which the threshold of second-degree burns can occur and the incident energy release is 1.2 cal/cm$^2$ (5.0 J/cm$^2$). This is the boundary that is calculated by computer-based programs and IEEE Guide equations. Inside the boundary, the energy level will be higher. This boundary should not be crossed by any one, including a qualified person, without wearing the required PPE, (Table 1.8). The PPE outfits are designed to minimize the risk of sustaining energy greater than 1.2/cal cm$^2$. *That is, the threshold of second-degree burns can still occur even with appropriate PPE, and these burns are considered curable.*

Unqualified persons, that is, those not specifically trained to carry out the required tasks, are safe when they stay away from the energized part of a certain distance, which is the limited approach boundary. They should not cross the limited approach boundary and arc flash boundary unless escorted by a qualified person.

Crossing the restricted approach boundary means that special shock prevention techniques and equipment are required, and an unqualified person is not allowed to cross this boundary. Finally, the prohibited approach boundary establishes the space that can be crossed only, as if a live contact with exposed energized conductors or circuits was planned.

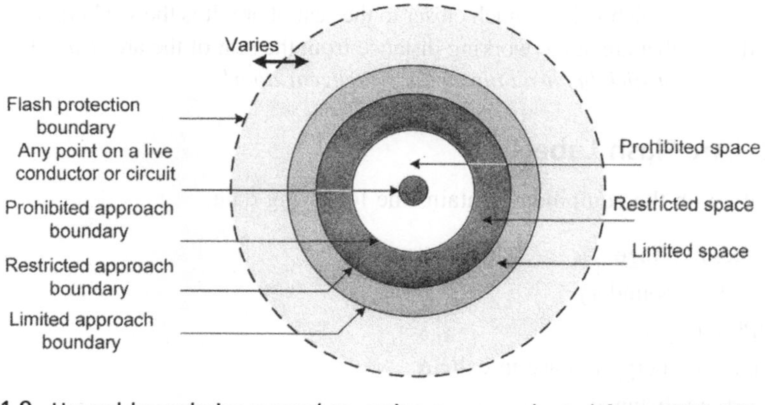


Figure 1.9. Hazard boundaries around an arcing source, adapted from NFPA 70E. *Source:* Reference [17].

TABLE 1.9. Limited Approach and Restricted Approach Boundaries

| System Voltage Range | Limited Approach Boundary | | |
|---|---|---|---|
| Phase-to-Phase | Exposed Movable Conductor (i.e., Overhead Line Conductors Supported on Poles) | Exposed Fixed Circuit Part | Restricted Approach Boundary, Included Inadvertent Movement Adder |
| Less than 50 V | Not specified | Not specified | Not specified |
| 50–300 V | 3.0 m | 1.0 m | Avoid contact |
| 301–750 V | 3.0 m | 1.0 m | 0.3 m |
| 751 V–15 kV | 3.0 m | 1.5 m | 0.7 m |
| 15.1–36 kV | 3.0 m | 1.8 m | 0.8 m |
| 36.1–46 kV | 3.0 m | 2.5 m | 0.8 m |
| 46.1–72.5 kV | 3.0 m | 2.5 m | 1.0 m |
| 72.6–121 kV | 3.3 m | 2.5 m | 1.02 m |
| 138–145 kV | 3.4 m | 3.0 m | 1.15 m |

*Source*: NFPA 70E [17]. (All details in the NFPA table have not been reproduced.)

The limited approach, restricted approach, and prohibited approach boundaries are all defined based upon the system voltage. *No calculations are required for establishing these boundaries.* Table 1.9 illustrates the limited approach and restricted approach boundaries based upon NFPA 70E.

### 1.10.1 Working Distance

Working distance is defined as the closest distance to a worker's body excluding hands and arms. IEEE 1584 Guide [9] specifies required working distances (Table 1.6). For the 15-kV switchgear, it is 36 in, while for a 480-V MCC, it is 18 in. A larger working distance reduces the incident energy and therefore the HRC. Recommendations of IEEE for the working distances are followed in the calculations in this book. The working distance does exclude hands and arms, which will be much closer to the seat of arc. It is the vital organs like eyes, chest, and heart that are at the working distance from the seat of the arc. *This assumes that a worker does not stick his head inside the switchgear door!*

### 1.10.2 Arc Flash Labels

The labeling on the equipment contains the following data:

- system voltage
- arc flash boundary
- PPE category
- incident energy release in cal/cm$^2$
- working distance
- restricted approach boundary
- prohibited approach boundary

TABLE 1.10. Arc Flash Boundary and Incident Energy Release for 30 kA of Bolted Fault Current (Arc Flash Current = 28.58 kA rms) in 13.8-kV Switchgear, 13.8-kV System Resistance Grounded, Working Distance = 36″, Gap = 153 mm

| Arc Duration in Seconds | Arc Flash Boundary in Inches | Incident Energy, cal/cm$^2$ |
|---|---|---|
| 0.058 | 74 | 3.0 |
| 0.5 | 851 | 26 |
| 1.0 | 1736 | 52 |
| 1.5 | 2633 | 78 |
| 2 | 3539 | 104 |

*Source*: Reference [17].

- equipment identification
- the protective device identification that clears the fault.

The labels can be generated on a variety of media, including plastic weatherproof laminates, and most commercial arc flash analysis program will allow custom designing the labels. A user can choose what goes on the label, including the description of PPE. Even the type fonts can be user selectable.

In NFPA 70E 2012, equipment labeling has been reworded to reflect Article 110.16 of the NEC. Each label must contain the incident energy at working distance or the arc rating of the required PPE, or the maximum HRC for that type of equipment. Also, arc flash boundary needs to be specified.

## 1.11 MAXIMUM DURATION OF AN ARC FLASH EVENT AND ARC FLASH BOUNDARY

A maximum duration of 2 seconds for the total fault clearance time of an arc flash event is considered, though, in some cases, the fault clearance time can be higher. In IEEE 1584 Guide Annexure B, it is stated that: "if the time is longer than 2 seconds, consider how long a person is likely to remain in the location of the arc-flash. It is likely that a person exposed to arc flash will move away quickly, if it is physically possible and 2 seconds is a reasonable maximum time for calculations. A person in a bucket truck or a person who has crawled into equipment may need more time to move away."

Tables 1.10 and 1.11 show the arc flash boundary calculations according to IEEE 1584 Guide equations for a bolted three-phase fault current of 30 kA, in 13.8-kV switchgear, 13.8-kV system resistance grounded, and also in a 480-V MCC, and 480-V system high resistance grounded.

For a 30-kA, 2-second fault in 13.8-kV switchgear, the incident energy boundary is 3539 in equal to *295* ft. For 1-second fault duration, it is *144.6* ft. The arc flash boundary at which a worker can be exposed to 1.2 cal/cm$^2$ of incident energy and sustain threshold of second degree burns seems to be very large.

The working space, in an electrical room switchgear installation in front of electrical equipment, is limited and may be as low as 5–6 ft, and must meet NEC requirements. Many companies, as a policy, keep the electrical rooms locked, and a worker must have the required PPE outfit before entering the electrical rooms.

TABLE 1.11. Arc Flash Boundary and Incident Energy Release for 30 kA of Bolted Fault Current (Arc Flash Current = 16.76 kA rms) in 480-V MCC, 480-V System High Resistance Grounded, Working Distance = 18″, Gap = 25 mm

| Arc Duration in Seconds | Arc Flash Boundary in Inches | Incident Energy, cal/cm$^2$ |
|---|---|---|
| 0.050 | 36 | 3.8 |
| 0.5 | 147 | 38 |
| 1.0 | 225 | 75 |
| 1.5 | 288 | 113 |
| 2 | 343 | 151 |

TABLE 1.12. Statistics of Arc Flash Incidents

| Accident Occurrence | Percentage |
|---|---|
| When the operator or worker is working with equipment doors open | 65 |
| When the operator happens to be in front of a closed door and the equipment is not arc resistant | 10 |
| When the operator is not present at all, and the equipment is not arc resistant | 25 |

### 1.11.1 Arc Flash Hazard with Equipment Doors Closed

There is some controversy in interpreting the intent of NFPA 70E, whether the arc flash hazard exists at all times, with the equipment door closed, or it exists only when the doors of an energized equipment are opened for maintenance? Let us first consider most common reasons for arc flash accidents:

- human error
- mechanical faults
- failed connections, loose connections, and terminals
- Adverse ambient conditions and pollution. This should consider pollution specific to plant operation, that is, corrosive gases and vapors may be present.
- rodents.

Table 1.12 shows statistical data of arc flash incidents, when these can occur. This shows that some arc flash hazard exists, even when someone is walking around the closed-door electrical energized equipment, though this probability is relatively small. *The IEEE Guide equations are based upon the hazard calculations with the door open.* With the door closed, the hazard level will be less. It is not so easy to calculate it. The manufacturers are reinforcing the latching mechanisms and strengthening the doors, yet the withstand capability of the doors in closed position under an arc flash event is a question mark. Only when the equipment is arc resistant is the incident energy level outside the equipment zero, so long as no doors and panels that are not intended to be opened are not opened (see Chapter 13).

To resolve this conflict, NFPA 70E 2012 adds:

It is the collective experience of the Technical Committee on Electrical Safety in the Workplaces that normal operation of the enclosed electrical equipment, operating at 600V or less, *that has been properly installed and maintained by qualified persons* is not likely to expose the employee to an electrical hazard.

It is also the opinion of the committee that there is little risk in performing normal operations of electrical equipment and devices, such as opening and closing circuit breakers, motor control centers (MCCs), or starters. When the committee states "interacting with equipment in a manner that could cause an arc flash hazard," it refers to operations, such as racking circuit breakers or installing and removing MCC buckets.

## 1.12 REASONS FOR INTERNAL ARCING FAULTS

The internal arcing fault can be caused by various factors, such as:

1. Insulation defects due to quality deterioration of components
2. Overvoltages of atmospheric origin or generated within the equipment due to switching transients
3. Incorrect operation due to not adhering to the procedures
4. Inadequate training of the persons involved with installation, maintenance, and operation
5. Breakage or tempering of safety interlocks
6. Highly polluted atmospheric conditions, high humidity, temperatures, corrosive gases, and salt-laden winds
7. Overheating of the contact area due to presence of corrosive agents, loose connections.
8. Faulty assemblies
9. Lack of proper protection and inadequate relaying. Though protective relaying by itself will not prevent an arcing fault, it will limit the arc fault energy and consequently equipment damage.
10. Insufficient maintenance and testing procedures, like insulation resistance records, infrared testing, and partial discharge measurements which can indicate a deteriorating insulation situation.
11. Incorrect operation of disconnects, switches, and grounding switches because of lack of interlocks
12. Problems with cable terminations, like inadequate design, faulty installation, and insulation failures
13. Ferroresonance voltages of instrument transformers
14. Inadequate protection for ground faults and improper selection of system grounding
15. Aging under electrical stress

16. Entrance of dust and rodents, corrosion at contact surfaces producing heat, loose contacts creating sparks. Snapping of the connections and wiring due to inadvertent force, human errors, and incorrect operation.
17. Closing on to an existing fault, without prior rectification and analyses of the faulty condition.
18. Sparking produced due to racking of circuit breakers, operation of fuses, and excessive current flows through loose contacts.
19. Human errors, that is, some parts or tools left or dropped inside the equipment during maintenance

This list may not be exhaustive and is indicative only. In the first place, if the electrical installations are inadequately designed and do not meet the requirements of national standards and safety codes, these will be more prone to higher incidents of arcing faults. Design and operational measures can be adopted for enhancing safety (see Chapter 2). The list also implies that improvements in arc flash hazard reduction can be achieved by avoiding the listed items and taking remedial measures. As an example, the incipient breakdowns can be predetermined by proper testing with infra red scans or partial discharge measurements (see Chapter 14).

## 1.13 ARC FLASH HAZARD CALCULATION STEPS

We can summarize the calculation procedures as follows:

1. Calculate bolted three-phase symmetrical short-circuit currents throughout the system where arc flash calculations are to be carried out. The data collection, single line diagrams, and switching conditions to be studied are all akin to the short-circuit calculations; see Chapters 5 and 6.

   It will be necessary to consider the various operating conditions, for example:

   - Normal operation
   - A tie circuit breaker is closed
   - Alternate operating situations that is transformers run in parallel, a generator is brought in or taken out of service
   - Dual feeds.

*It will be erroneous to conclude that if the arc flash calculations are based upon the maximum short-circuit currents in any operating conditions, the results will give maximum hazard levels, and other scenarios need not be considered. This is so because at reduced short-circuit currents, the protective relaying times may increase, giving rise to even greater arc flash hazard.*

2. The arcing currents can be calculated based upon the system voltage and the bolted three-phase currents and gap length. Calculate second arcing current at 85% $I_a$.

3. The working distance and gap lengths in all the examples and discussions in this book are according to IEEE standard. There impact on arc flash calculation results is, however, documented in Chapter 3.
4. Input the correct equipment type, switchgear, MCC or panels, or arc in the open air.
5. Conduct a rigorous relay coordination study throughout the distribution system. This is an important step, where alternate protective relay types, relay characteristics, and settings, protection strategies, current limiting devices, equipment selection, and alternate settings during maintenance modes become important. The protection and relaying is important, and for arc flash hazard reduction, it becomes an iterative calculation. With all other parameters well-defined, relaying can be manipulated to reduce HRC.
6. Calculate the incident energy, arc flash boundary, and the PPE level.

## 1.13.1 NFPA Table 130.7(C)(15)(a)

NFPA table 130.7(C)(9)(a) [17] relates the hazard risk category and PPE with respect to maintenance tasks to be carried out on energized equipment within arc flash boundary For example, if a worker is inserting or removing a starter "bucket" from MCC, the risk category is 3, but if he is just reading a panel meter while operating a meter switch, the risk category is zero. This table is applicable for specific three-phase short-circuit currents for specific durations as specified in the many foot-notes to the table. The IEEE equations are based upon an arc in an enclosure, with one side open, and it seems logical to relate the hazard level with respect to task activity; the following observations apply:

- All commercially available computer programs calculate the arc flash hazard for a bus fault. It is opined that for safety and conservatism, HRC for a bus fault should be considered. For example, the HRC for a fault downstream of a low voltage or medium voltage MCC (a fault on the load side of the motor starter) can be zero, while for a fault on the bus, cleared by an upstream device, it can be 3 or 4. MCC as a whole is labeled for the bus fault—a fault on the bus can shatter the isolation partition in the MCC on arc blast pressure, unless the equipment is arc resistant.
- It will be practically difficult to provide multiple arc flash hazard labels on the same equipment, and lay down procedures for various personal protective outfits based upon the maintenance task to be performed.
- A rigorous calculation procedure for task-based HRC is not available, though guidelines in this table can be applied.
- Multiple tasks are normally performed in one single instance.
- The recommendations of the table can be misapplied (see Section 3.13).
- *The table is judgmental; there are no mathematical calculations.*

The calculations, examples, and study cases in the book consider a bus fault, which gives maximum HRC on an equipment and ignore the nature of the task activity.

## 1.14 EXAMPLES OF CALCULATIONS

**Example 1.1**

This example is for calculations as per 2002 Guide. It is included here for reference. The results are compared with 2018 methodology in Chapter 3.

Calculate the incident energy using IEEE equations for a 30-kA three-phase bolted fault current in a 13.8-kV metal-clad switchgear, resistance grounded. The arcing time is 30 cycles for the arc fault current through the protective device.

Using IEEE Equation (1.11), calculate arcing current:

$$\log I_a = 0.00402 + 0.983 \log I_{bf}.$$

This gives $I_a$ = 28.578 kA. Note that the arc flash time is calculated based upon the arcing fault current through the protective device and not the bolted fault current. A further calculation for arc flash time is required at $0.85I_a$. Let us assume for this example that 30 cycles is applicable even for reduced $0.85I_a$. Then, calculate normalized incident energy from Equation (1.12):

$$\log E_n = (-0.555) + (-0.113) + 1.081 \log(28.578) + 0.0011(153).$$

Here, $K_1$ = -0.555 (arc in a box), $K_2$ = -0.113, resistance grounded system, and $G$ = 153 mm (Table 1.5). This gives log $E_n$ = 1.074. Calculate incident energy in J/cm$^2$ using Equation (1.13):

$$E = 4.184(1)(11.858)\left(\frac{0.5}{0.2}\right)\left(\frac{610^{0.973}}{910^{0.973}}\right).$$

Here, $t$ is the arcing time = 0.5 second, the distance $x$ is given in Table 1.7, and for 15 kV, it is 0.973, and $D$, the working distance from Table 1.6, is 910 mm.

This gives $E$ = 84.4 J/cm$^2$ = 20.08 cal/cm$^2$.

Calculate the arc flash boundary (see Equation 1.15):

$$D_B = \left[4.184(1)(11.858)\left(\frac{0.5}{0.2}\right)\left(\frac{610^{0.973}}{5}\right)\right]^{1/0.973}.$$

Here, $C_f$ = 1, $E_n$ as calculated before is 11.858 J/cm$^2$, $E_B$ = 5 J/cm$^2$ by definition. This gives $D_B$ = 16,541 mm = 54.267 ft. This is rather a large distance at which a worker can get threshold of second-degree burns.

**Example 1.2**

Repeat Example 1.1, but for arc in the open air. For high voltage applications, no distinction is made between open and box construction for calculation of arcing current. Therefore, arcing current = 28.578 kA, same as before.

In Equation (1.12), the factor $K_1$ is -0.792. This gives:

$$\log E_n = (-0.792) + (-0.113) + 1.081 \log(28.578) + 0.0011(153) = 0.837.$$

$E = 4.184(1)(6.871)(0.5/0.2)(610^{0.973}/910^{0.973})$ = 48.70 J/cm$^2$. This is equal to 11.60 cal/cm$^2$.

Also, arc flash boundary reduces to:

$$9437 \text{ mm} = 30.96 \text{ ft}.$$

This shows a considerable difference between arc in open air and a box.

**Example 1.3**

Repeat Example 1.2 using Lee's equations.

The incident energy is given by Equation (1.7). Substituting the values:

$$E = \frac{793 \times 30 \times 13.8 \times 0.5}{36^2} \text{ cal/cm}^2.$$

This gives 126.6 cal/cm$^2$, and from Equation (1.6), distance in feet for a curable burn is 30.82 ft. There is a vast difference in these calculations: 11.6 cal/cm$^2$ with IEEE method and 126.6 cal/cm$^2$ with Lee's method; his equations do not consider system grounding, and do not calculate arc flash currents.

**Example 1.4**

Calculate the incident energy and arc flash boundary based upon the IEEE simplified equations for a 480-V power circuit breaker (low voltage power circuit breaker [LVPCB]) of 800 A rating provided with electronic trip programmer, LS ( long time and short time). Also, calculate similarly for a 400-A molded case circuit breaker (MCCB) with thermal magnetic trip. The bolted fault current is 36.8 kA.

*800-A LVPCB*

The incident energy is given by the equation:

$$4.560 I_{bf} + 27.230 \text{ J/cm}^2.$$

Substituting, incident energy is = 195.04 J/cm$^2$ = 46.80 cal/cm$^2$

The arc flash boundary is given by the equation:

$$47.2 I_{bf} + 2660 \text{ mm}.$$

Substituting

Arc flash boundary = 173 in.

*400-A MCCB*

The incident energy equation is:

$$0.189 I_{bf} + 0.548 \text{ J/cm}^2.$$

Which gives 1.80 cal/cm$^2$

The arc flash boundary is:

$$9.16 I_{bf} + 194 \text{ mm}.$$

Substituting, this gives: 20.9 in.

**Example 1.5**
Repeat Example 1.4 with rigorous calculations.

Table 1.8 equations do not specify the gap length and also the system grounding. As the equations are for the maximum incident energy, it can be assumed that the system is ungrounded. Also for Table 1.8 equations, a working distance of 610 mm (2 ft) is specified in 1584 Guide. Thus, there are qualifications and compromises in using these equations.

To start the rigorous calculations, a time–current plot of settings on 800-A LVPCB and 400-A MCCB is shown in Figure 1.10. The maximum time on the trip band is used, shown by dots in this figure. This figure is based on circuit breakers of a certain manufacturer. This shows that the LVPCB *maximum* ST delay band is used, as the approximate equations are based upon this assumption. *The opening time of the low voltage circuit breakers is built in the time–current plots supplied by the manufacturer.* Thus, we consider the maximum operating times as shown by bold dots in Figure 1.10. These arcing times are 0.5 second for 800-A LVPCB and 0.017 second for the 400-A MCCB.

The results of the calculations are in Table 1.13. The calculations have been made for grounded and ungrounded systems, and with gaps and working distances as shown. A LVPCB will be provided in a lineup of LV switchgear, and a working distance of 610 mm and gap of 32 mm are applicable as per IEEE 1584 Guide. A 400-A MCCB may be installed in a MCC, where the working distance is reduced to 455 mm and the gap is 25 mm. The variations in the results are apparent and significant for an application.

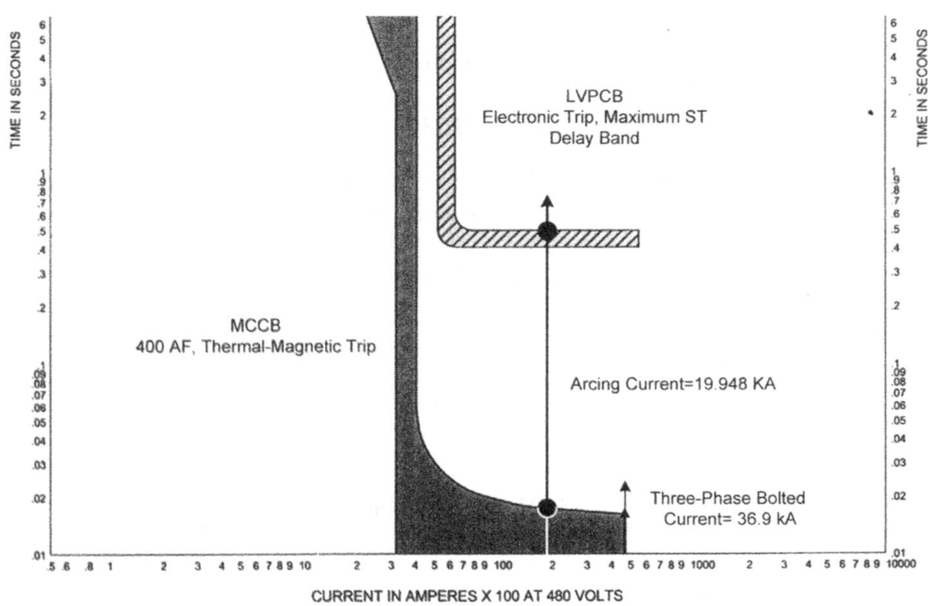


Figure 1.10. To illustrate the arcing time for low voltage circuit breakers, at the calculated arcing current.

TABLE 1.13. Arc Flash Calculations, Example 1.5

| Breaker | Bolted Fault (kA) | Arcing Fault (kA) | Trip Time (sec) | Opening Time (sec) | Arc Time (sec) | Ground | Gap (mm) | Working Distance (in) | Arc Flash Boundary (in) | Incident Energy Cal/cm$^2$ | PPE |
|---|---|---|---|---|---|---|---|---|---|---|---|
| 800-A LVPCB | 36.8 | 19.95 | 0.5 | 0 | 0.5 | No | 32 | 610 | 198 | 27 | 4 |
| | | | | | | Yes | 32 | 610 | 166 | 21 | 3 |
| 400-A MCCB (MCC) | 36.8 | 19.95 | 0.017 | 0 | 0.017 | No | 25 | 455 | 21 | 1.5 | 1 |
| | | | | | | Yes | 25 | 455 | 18 | 1.2 | 0 |

## 1.15 REDUCING ARC FLASH HAZARD

Electrical arc flash hazard cannot be reduced to HRC zero even in small distribution systems. Electrical power systems vary in complexity—an industrial power system may have a number of in-plant generators in cogeneration mode with utility tie and consume 200 MW of power with hundreds of unit substations and extensive power distribution at medium voltages and low voltage levels, containing motors of thousands of horsepower, a single motor exceeding ratings of 25,000 hp or more and multiple utility interconnections. Conversely, a system with a one or two 480 V transformers of 1000 KVA will also be classified as industrial systems. Thus, it is difficult to lay down all the general rules for reducing arc flash hazard in power systems.

This book contains many study cases and examples, and it is recommended that a reader may do a first reading of the book to grasp an overall picture, followed by a subsequent study. The following common parameters can be stated for arc flash hazard reduction:

1. Reduce the short-circuit currents by proper system designs and load distribution. The arcing current is a function of the bolted three-phase current, and its reduction in the first place will reduce the arc flash hazard (see Chapters 5 and 6).
2. Manipulate the protection and protective device coordination without sacrificing selectivity. Generally, it can be said that reducing the short-circuit currents in an existing system will be practically impossible, but the protective relaying can be manipulated to reduce the arcing time. See Chapters 7 through 14, which discuss differential protections, arc flash detection protection, zone selective interlocking, use of maintenance mode switches to alter the protective relay settings, application of modern multifunction microprocessor-based protective relays, and coordination on instantaneous basis with current limiting devices.
3. Consider high resistance grounding systems—though the IEEE equations give a *lower* incident energy for solidly grounded systems, as compared with resistance grounded or high resistance grounded systems; because it takes

a safe stance that single line-to-ground faults will quickly escalate to three-phase faults; yet approximately 70% of the faults start as single line-to-ground faults. The single line-to-ground faults in solidly grounded systems can be even higher than the three-phase faults. See Chapter 4.

The IEEE Guide equations reveal that the incident energy and the resulting arc flash hazard for a certain equipment type depends only upon these three parameters as stated above. Yet there are many aspects to be considered for a workers safety, discussed above and in the chapters to follow:

- Remote operation, avoiding human machine interface to avoid human errors during routine operation
- Remote racking of circuit breakers
- Proper system designs and maintenance. See 10 bulleted items describing design features and maintenance in Chapter 2.
- Proper worker training and documentation of safety procedures, and safety awareness
- Proper testing, online monitoring, and diagnostics of the systems
- Proper record keeping
- Applications of new equipments and technologies for arc flash reduction
- Proper rigorous arc flash hazard studies, labeling, and use of appropriate PPE
- Adherence to safety codes, standards, and recommended practices
- Remote operation, avoiding human machine interface to avoid human errors during routine operation
- Remote racking of circuit breakers

## 1.15.1 Reduction

The protection system, coordination and suitable selection of additional protective devices can reduce the arcing times. Examples are differential protection, light sensing supervised by current signature, zone interlocking, arc flash maintenance switches which bring alternative relay settings into operation.

In some cases the arc fault current can be low resulting in increased fault clearing time. For example, the fault on the secondary side of a transformer, which is used by the transformer primary protection device. Due to ratio of transformation this current will be reduced resulting in prolonged clearing time. A secondary protective relay can be applied which will transfer trip primary proactive device.

Well-planned procedures, work practices and training can avoid arc flash hazards. Remote racking, remote control panels for switching, supervisory controls can help worker for safe operation from a distance outside the arc flash boundary. Infra-red windows, partial discharge sensors can be very helpful diagnostic tools.

Arc-resistant equipment can be used.

TABLE 1.14. Measures to Protect Against Arc Flash

| Mitigation | Type | Advantages | Disadvantages | Recommendations |
|---|---|---|---|---|
| Compartmentalization (IEC type-4b or ANSI equivalent barriers and boots | Prevention | Reduces the likelihood of arc flash by compartmentalization and insulating energized parts | Equivalent to a guard on a saw. Hazard is still present. Maintenance and testing may require exposure to energized parts | It does not reduce the energy available, it does reduce the probability of a fault. Compartmentalization or change incident energy calculation. |
| Arc-resistant equipment | Containment (passive) | Protects the worker by directing energy away from the worker when required covers are on. Tested per IEEE Std. C37.20.7 | Maintenance and testing may require removal of covers. Equipment is exposed to damage and has a larger footprint. May require second migration means to stay within rated arcing duration especially at LV | This measure offers protection for specific tasks, but does not reduce available energy. This measure is recommended where active mitigation methods are not acceptable. |
| High-speed shorting switch | Active (reduction) | Fights fire with fire. Re-directs arc flash energy into the bus by creating an intentional bolted fault; may use optical, differential and current waveform recognition. | Unconventional method and requires additional equipment. | May qualify as arc resistant per IEEE Std. 37.20.7. Needs robust controls to minimize chances of nuisance operation. |
| ZSI (blocking schemes) | Active (reduction) | Communication-aided transfer trip scheme that reduces breaker trip time and amount of energy released | Requires communication wiring between breakers. Trip units or relays must be installed with ZSI capability | Recommended for LV applications. Also available with some MV relays. |

TABLE 1.15. Matrix of Sample Arc Flash Mitigation Measures

| Technology | Arc-Flash Mitigation | Incremental Commissioning Complexity | Protection for Intentional Exposure | Protection for Incidental Exposure | Foot-print | Capital Expense | Operational Expenditure | Equipment Survivability | Arc Flash Incident Energy Reduction | Typical Application |
|---|---|---|---|---|---|---|---|---|---|---|
| Prevention | Barriers | Low | Limited | Limited | No change | Low | Low | Moderate impact | None | LV switchgear switchboard |
| | ANSI compartmentalization | Low | Limited | Limited | No change | Low | Low | Moderate impact | None | LV switchgear switchboard |
| | IEC-4b compartmentalization | Low | Limited | Limited | No change | Low | Low | Moderate impact | None | LV switchgear switchboard |
| | HRG | High | Limited | Limited | Increased | Medium | Low | Moderate impact | SLG fault only | 480V to 5 kV |
| | Shielding barriers | Low | Limited | Limited | No change | Low | Low | No impact | None | LV arc-resistant switchgear |
| | Adm. controls | N/A | Limited | Limited | No change | Low | Low | No impact | None | All, training requirement |
| Containment | Arc-resistant equipment | Low | No | Yes | Increased | High | Low | No impact | None | LV/MV switchgear |
| | Solid dielectric bus | Medium | Yes | Yes | No change | Low | Low | Moderate | None | MV switchgear |
| Reduction | Differential protraction | High | Yes | Yes | No change | High | Medium | impact | Significant | LV/MV switchgear |
| | ZSI | Medium | Yes | Yes | No change | Low | Medium | Moderate | Significant | LV switchgear/ switchboard |
| | Transfer trip | High | Yes | Yes | Increased | High | Medium | impact | Significant | LV switchgear fed from transformers of >750 kVA |

TABLE 1.16. Measures for Mitigating Arc Flashes at Key Locations in Data Centers

| System Location | Issue | Recommendations |
|---|---|---|
| LV service entrance equipment | Protection on HV side of transformer does not respond quickly for LV faults | Transfer trip, differential protection along with ZSI |
| UPS input/output switchboards | Selectivity requirements lead to extended clearing times | ZSI or optical detection |
| PDU secondary (480 V to 208/120 V) | Low fault levels and P combined with transformer inrush currents extend arc-clearing times | Maintenance switch, compartmentalization of PDU with administrative controls |
| Generator paralleling equipment | Low fault levels combined with multiple sources extend arc-clearing times | Bus differential or optical detection, adaptive relay settings |
| MV distribution equipment | Multiple utility sources and/or generator sources result in high fault currents | High-speed shorting switch, bus differential, and optical relaying |

Figure 1.11. An outfit for PPE HRC4.

Figure 1.11 shows a PPE outfit for HRC=4. It looks like a space suit and hampers the mobility of a worker to render fine tasks. The arc flash energy must be reduced to no more than 8 cal/cm$^2$ with appropriate measures.

## 1.15.2 Arc Flash Labels

Computer-based software can generate a variety of arc flash labels. The format, size of the label, and even colors can be modified according to user's choice. NFPA specifies minimum data that should be included in an arc flash label. Generally, the labels are laminated to withstand weathering effects. Figure 1.12 shows a specimen arc flash label.

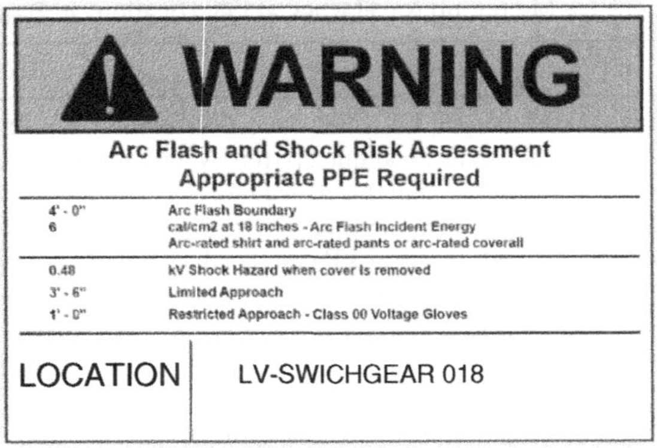


Figure 1.12. A specimen arc flash label.

## REVIEW QUESTIONS

1. The calculated symmetrical three-phase bolted fault in a 480-V switchgear assembly is 40 kA. Using IEEE equations, calculate the arc fault current. Assuming that the fault is cleared in 0.1 seconds, both for $I_a$ and 85% $I_a$, calculate the incident energy release and arc flash boundary. Use the gap length and working distance as per IEEE tables. The 480 V three-phase system is high resistance grounded.
2. Repeat with Lee's equations.
3. Repeat using Equation (1.9) of the text.
4. Consult NFPA 70 E and the text in this paper and specify the PPE outfits for category 3 and 4 hazard levels in detail.
5. What is the IEEE intent of calculating a second arcing fault current at 85% of $I_a$?
6. List five points in order of their importance for worker's safety.

## REFERENCES


1. W.A. Brown and R. Shapiro, "Incident energy reduction techniques," IEEE Industry Applications Magazine, vol. 15, no. 3, pp. 53–61, May/June 2009.
2. T. Gammon and J. Mathews, "Conventional and recommended arc power and energy calculations and arc damage assessment," IEEE Trans. Ind. Appl., vol. 39, no. 3, pp. 197–203, May/June 2003.
3. T. Gammon and J. Mathews, "Instantaneous arcing fault models developed for building system analysis," IEEE Trans. Ind. Appl., vol. 37, no. 1, pp. 197–203, Jan./Feb. 2001.
4. V.V. Terzija and H.J. Koglin, "On the modeling of long arc in still air and arc resistance calculations," IEEE Trans. Power Deliv., vol. 19, no. 3, pp. 1012–1017, July 2004.
5. A.D. Stokes and D.K. Sweeting, "Electrical arcing burn hazards," IEEE Trans. Ind. Appl., vol. 42, no. 1, pp. 134–142, Jan./Feb. 2006.

6. H.B. Land, III, "Determination of the case of arcing faults in low-voltage switchboards," IEEE Trans. Ind. Appl., vol. 44, no. 2, pp. 430–436, March/April 2008.
7. H.B. Land, III, "The behavior of arcing faults in low-voltage switchboards," IEEE Trans. Ind. Appl., vol. 44, no. 2, pp. 437–444, March/April 2008.
8. R. Wilkins, M. Allison, and M. Lang, "Effects of electrode orientation in arc flash testing," in Proc. IEEE IAS Annual Meeting, Hong Kong, pp. 459–465, 2005.
9. IEEE 1584, IEEE Guide for Performing Arc-Flash Hazard Calculations, 2002.
10. J.C. Das, "Arc flash hazard," in *McGraw-Hill Year Book of Science and Technology*, pp. 18–20, McGraw-Hill, New York, 2008.
11. R. Lee, "The other electrical hazard: Electrical arc blast burns," IEEE Trans. Ind. Appl., vol. 1A-18, no. 3, pp. 246–251, May/June 1982.
12. R. Lee, "Pressure developed by arcs," IEEE Trans. Ind. Appl., vol. IA-23, no. 4, pp. 760–764, July/Aug. 1987.
13. R.L. Doughty, T.E. Neal, T.A. Dear, and A.H. Bingham, "Testing Update on Protective Clothing and equipment for arc exposure," IEEE Industry Applications Magazine, vol. 5, no. 1, 37–49, Jan./Feb. 1999.
14. R.L. Doughty, T.E. Neal, and H.L. Floyd, II, "Predicting incident energy to better manage the electric arc hazard on 600-V power distribution systems," IEEE Trans. Ind. Appl., vol. 36, no. 1, pp. 257–269, Jan./Feb. 2000.
15. R.A. Jones, D.P. Liggett, M. Capelli-Schellpfeffer, T. Macalady, L.F. Saunders, R.E. Downey, L.B. McClung, A. Smith, S. Jamil, and V.J. Saporita, "Staged tests to increase awareness of arc-flash hazards in electrical equipment," IEEE Trans. Ind. Appl., vol. 36, no. 2, pp. 659–667, March/April 2000.
16. R. Wilkins, M. Lang, and M. Allison, "Effect of insulating barriers in arc flash testing," in conf. record, IEEE Pulp and Paper Industry Technical Conf. pp. 120–125, Seattle, 2008.
17. NFPA 70E, Electrical Safety in Workplace, 2012.
18. M.G. Drouet and F. Nadeau, "Pressure waves due to arcing faults in a substation," IEEE Trans. PAS, vol. PAS-98, no. 5, pp. 1632–1635, Sep. 1979.
19. M. Wactor, G.H. Miller, J. Bowen, and M. Capelli-Schellpfeffer, "Modeling of pressure wave associated with arc fault," IEEE PCIC conf. Rec., pp. 333–341, Sept. 2000.
20. T.E. Neal and R.E. Parry, "Sharpnel, pressure and noise," IEEE Industry Applications Magazine, vol. 11, no. 3, 49–54, June 2005.
21. IEEE Std. 80, Safety in AC Substation Grounding, 2000.
22. L.A. Geddes and L.E. Baker, "Response of passage of electrical current through the body," J. Assoc. Adv. Med. Instrum., vol. 2, pp. 13–18, Feb. 1971.
23. IEC 604791-1, Effect of Current on Human Beings and Livestock Part 1: General Aspects, 2005.
24. C.F. Dalziel, J.B. Lagen, and J.L. Thruston, "Electrical shock," AIEEE Trans. PAS, vol. 60, pp. 1073–1079, 1941.
25. C.F. Dalziel and R.W. Lee, "Lethal electric currents," IEEE Spectrum, pp. 44–50, Feb. 1969.
26. L.P. Fewrris, B.G. King, P.W. Spence, and H. Williams, "Effect of electrical shock on the Heart," AIEEE Trans. PAS, vol. 55, pp. 498–515 and 1263, May 1936.
27. U.G. Biegelmeier and W.R. Lee, "New considerations on the threshold of ventricular fibrillation for AC shocks at 50–60 Hz," Proc. IEEE, vol. 127, pp. 103–110, 1980.
28. W.B. Kouwenhoven, et al., "AC shocks of varying parameters affecting the heart," AIEEE Trans. PAS, vol. 78, part 1, pp. 163–169, 1959.

29. D. Roberts, "50-V shock hazard threshold," IEEE Trans. Ind. Appl., vol. 46, no. 1, pp. 102–107, Jan./Feb. 2010.
30. Society of Fire Protection Engineers (SPFE) Task Force, Engineering Guide to Predicting First and Second Degree Skin Burns, Boston, MA, 2000.
31. NFPA 921, Guide for Fire and Explosion Investigations, 2011.
32. ANSI/IEEE C2, National Electric Safety Code, 2007.
33. IEEE 1584a, IEEE Guide for Performing Arc-Flash Hazard Calculations—Amendment 1, 2004.
34. IEEE P1584b/D2, Draft 2, Guide for Performing Arc Flash Hazard Calculations—Amendment, July 2009 (unapproved).
35. ASTM F1506, Specifications for Textile Materials for Wearing Apparel for Use by Electrical Workers Exposed to Momentary Electrical Arc and Related Thermal Hazards, 1998.
36. J.C. Das, "Design aspects of industrial distribution systems to limit arc flash hazard: Part II-Analysis," IEEE Trans. Ind. Appl., vol. 41, no. 6, pp. 1467–1482, Nov./Dec. 2005.
37. J.C. Cawley and T. Homce, "Trends in electrical injury in the U.S., 1992–2002," IEEE Transactions on Industry Applications, vol. 44, no. 4, pp. 962–972, July/August 2008.
38. J.C. Cawley and B.C. Brenner, "Occupational electrical injuries in the U.S., 2003–2009," IEEE Paper No. ESW-2012-24, 2012.

# 2

# SAFETY AND PREVENTION THROUGH DESIGN: A NEW FRONTIER

The electrical installation should be designed and constructed to be safe by complying with standards, codes, and accepted engineering practices. A number of criteria apply to the planning and design of electrical systems, namely:

1. All switching devices must be selected to interrupt and withstand system short-circuit currents.
2. All interlocks must guarantee that human errors of operation are eliminated, for example, when switching between sources that may be out of phase, when paralleling two different sources of power, and when bringing another standby source in service.
3. The protective systems should isolate the faulty section without escalating the fault to nonfaulted areas.
4. All system loads must be served without excessive voltage dips and voltage instability under various switching conditions.
5. The steady state and transient stability of the systems must be ensured under the studied upset conditions.

*Arc Flash Hazard Analysis and Mitigation*, Second Edition. J.C. Das.

Published 2021 by John Wiley & Sons, Inc.

6. The starting of large motors should not create excessive voltage dips and instability.
7. The system and equipment grounding should receive adequate considerations and protections. Proper grounding of equipment and system neutrals is essential to protection from electrical shock and ground fault relaying.
8. Proper mitigation of harmonics from nonlinear loads should be provided.
9. In industrial environment all electrical equipment should be heavy duty to endure more frequent usage and for reliability. The listed equipment should be provided,
10. A reliability analysis of the distribution system in the design stage should be conducted.

This list of tasks for effective planning and development and for meeting the specific performance requirements of electrical power systems can be long—depending upon the electrical power system under consideration. In general, all electrical power systems should be:

- secure
- safe
- expandable
- maintainable
- reliable
- economical.

The integrity of the electrical equipment should be maintained with emphasis on type of enclosures, insulation coordination, operating mechanisms, grounding, and protective relaying. Yet the power system designs may fall short from personnel safety considerations. Even the functionality for which these systems are designed for adequate performance may be compromised. It is not unusual to see inadequately designed systems, lacking in some respect or the other. Competition and economical constraints can make even expert designers and planners cut corners, which may ultimately result in spending more funds for the short-term fixes and long-term upgrades.

## 2.1 ELECTRICAL STANDARDS AND CODES

There are a number of current standards and guidelines for planning and design of electrical power systems for commercial, industrial, and utility applications. The IEEE Industry Applications Society and Power Engineering Society have an immense database of publications exploring the new technology, drafting new standards, revising the existing standards, and providing guidelines for specifying, engineering, maintenance, equipment selection, and applications. The IEEE website [1] compiles more than two million documents that a user can access—this includes standards, technical papers in conferences, transactions, and journals. Reference [2] provides titles of IEEE "color books." The electrical requirements of protection, wiring, grounding, control, and communication systems, contained in NEC [3] (National Electrical Code) are important for electrical engineers. In most instances, the NEC is adopted by local ordnance as a part of building code. Then there are number of NFPA standards containing requirements on electrical equipment and systems [4], which include NFPA 70E 2012.

Legislation by U.S. federal government has the effect of giving certain ANSI standards the impact of law. Not all standards are ANSI approved. OSHA (Occupational Safety and Health Administration) requirements for electrical systems are contained in 29 CFR Part 1910 of the Federal Register [5]. The subparts of 1910 that are of particular interest relative to electrical safety are:

- *Subpart S.* Electrical—all sections.
- *Subpart R.* Special industries, section 1910.269—electrical power generation, transmission and distribution.
- *Subpart I.* Personal protective equipment, section 1910.137—electrical protective devices.
- *Subpart J.* General environmental controls, section 1910.147—hazard energy control (lockout/tagout).

The U.S. National Institute of Occupational Safety and Health (NIOSH) publishes Electrical Alerts to warn unsafe practices of hazardous electrical equipment [6]. The U.S. Department of Energy has advanced energy conservation standards. These include ASHRA/IES legislation embodying various energy conservation standards, such as ASHRAE/IES 90.1P. These standards impact architectural, mechanical, and electrical designs.

Underwriters Laboratories (UL) and other independent-testing laboratories may be approved by an appropriate jurisdictional authority like OSHA. A product may be UL labeled or listed. Generally, the designers insist for UL labels on the electrical components in a design and planning process and specifications for the electrical equipment. The UL publishes an Electrical Construction Materials Directory, an Electrical Appliance and Utilization Directory, and other standards. The Electrification Council (TEC) representative of investor-owned utilities publishes several informative handbooks [7].

The National Electrical Manufacturer's Associations (NEMA) [8] represents equipment manufacturers, and their publications standardize certain design features of electrical equipment and provide testing and operating standards. NEMA publishes a number of standards on electrical equipment, which contain important application and selection guidelines.

Then, there are many handbooks, which, over the course of years, have established reputations in the electrical field; some of these are cited in Reference [9].

Safety for operating personnel is achieved through proper design. National Electric Safety Code (NESC) [10] covers basic provisions for safeguarding from hazards arising out of conductors in electric supply substations, overhead and underground electric supply, and communication lines. It also covers work rules and safe clearances from live parts to ground and between phases. The Electrical Generating System Association [11] publishes performance standards for emergency, standby, and cogeneration equipment.

System protection is a fundamental requirement for all electrical systems. All switching devices must be applied safely within their interrupting ratings, faulted circuits must be isolated, and relay protection should be properly designed and coordinated. Physical protection of equipment from tempering, damage, and environment must be provided. The operating personnel must be trained for the specific jobs. Practically all states in the United States require that the system designs be performed under the seal of a licensed professional engineer registered in that particular state.

In spite of this spate of standards, codes and guidelines, the electrical power and distribution systems as designed and implemented currently may be inadequate with respect to personnel safety. For example, designing the systems to limit arc flash hazard is a new frontier. Even a well-designed electrical power system, meeting all the standards and code requirements, may fall short with respect to arc flash hazard mitigation. Also, mitigating or reducing the arc flash hazard is not the only criteria for worker's safety (see Chapter 1).

## 2.2 PREVENTION THROUGH DESIGN

Prevention through design (PtD) logo, as shown in Figure 2.1, is a new initiative sponsored by NIOSH. This initiative was launched in a July 2007 workshop held in Washington, DC to create a *national strategy* for PtD [12]. It may be defined as: "PtD involves addressing the occupational safety and health needs and redesign processes to prevent or minimize work-related hazards and risks associated with the construction, use, maintenance and disposal of facilities, materials and equipment."

This definition has a much wider base than merely the system designs of electrical installations. There is a tendency in United States to focus on administrative controls and PPE, as opposed to engineering solutions to protect workers from hazards in the

Figure 2.1. Logo for the U.S. NIOSH PtD initiative.

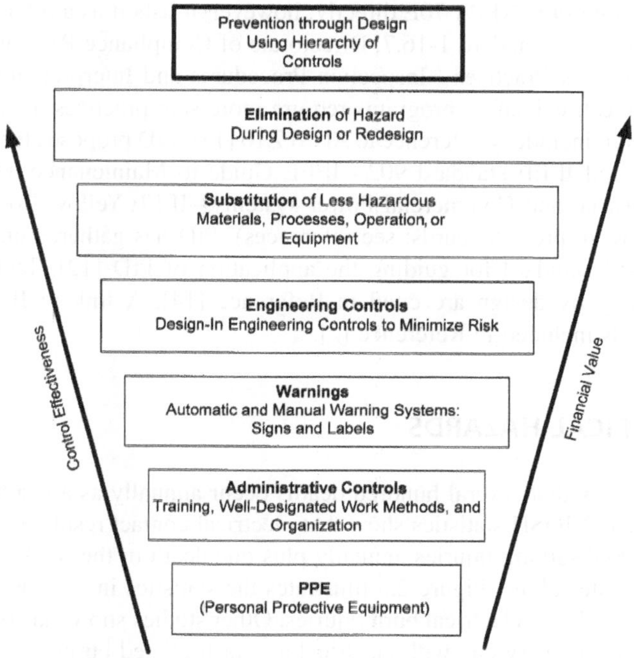


Figure 2.2. Key elements of safety by design.

workplace. Figure 2.2 illustrates the measures that are ranked down in descending order regarding relative effectiveness in helping to assure workers safety. The last three layers of Figure 2.2 are voluntary. Within realm of electrical safety, PtD initiative holds the promise of bringing an enhanced focus on understanding, acceptance and application of these hazard control measures.

In 2011 the American Society of Safety Engineers published ANSI/ASSE Standard Z3590.3, Prevention through Design: Guidelines for Addressing Occupational Hazards and Risks in Design and Redesign Processes. This standard provides guidance on applying prevention through design concepts and life cycle assessments relative to safety over the lifespan of a facility, process, or product.

## 2.3 LIMITATIONS OF EXISTING CODES, REGULATIONS, AND STANDARDS

NFPA 70E is the most comprehensive electrical safety standard available and addresses protection from electrical hazards. NEC and OSHA 29 CFR 1910.331-0.335 and 1910.269 are also important. Regulations are mainly focused on administrative controls and PPE, except for OSHA 1910.119, which covers both worker and public safety with regard to operation of highly hazardous manufacturing processes involving flammable, toxic, or explosive materials. It has requirements spanning all controls in Figure 2.2.

OSHA does not enforce NFPA 70E directly; however, it lists it as a reference standard in OSHA Instruction Standard 1-16.7, Directorate of Compliance Programs, Electrical Safety Related Work Practices—Inspection Procedures and Interpretation Guidelines. The effective electrical safety programs require more safe practices than contained in NFPA 70E, and it includes a reference to ANSI Z10 [13]. PtD proposes further revisions to NFPA 70E and IEEE standard 902—IEEE Guide to Maintenance, Operation and Safety of Industrial and Commercial Power systems—IEEE Yellow Book [2] (this is now replaced with three standards; see references). PtD has gathered enough support for a new ANSI standard for guiding the application of PtD [12]. Technical papers promoting safety by design are cited in Reference [14]. A link to IEEE standards announcement is included in Reference [2].

## 2.4 ELECTRICAL HAZARDS

The statistics show that several hundred deaths occur annually as a result of electrical shock, Table 2.1. NIOSH statistics show that electrical contact results in 4000 nondisabling and 3600 disabling injuries annually plus one death in the workplace everyday in the United States alone. Figure 2.3 illustrates the statistics in various sectors of the industry, which indicate electrical burn injuries. Other studies show that 10–15 employees are hospitalized every day with electrical arc flash-related burns.

It is surprising that 50% of those killed in electrical accidents are not in traditional electrical fields, but are from professions like contractors, outside service personnel, painters, laborers, and drivers.

As shown in Figure 2.3, the construction industry accounts for 45% of occupational electrical fatalities, but represents only 7% of the workforce [15]. Figure 2.4 illustrates the total fatalities in the construction industry, broken down with respect to type of electrical contact. Fatality rate with contact with overhead lines is twice that of contact

TABLE 2.1. Workplace Electrical Deaths, 1992–2000

| Industry | Fatalities | Percentage |
|---|---|---|
| Construction | 1292 | 45 |
| Agriculture, forestry, and fishing | 341 | 12 |
| Transportation, communications, and utilities | 307 | 11 |
| Manufacturing | 300 | 11 |
| Service industries | 269 | 10 |
| Retail trade | 81 | 3 |
| Public administration | 66 | 2 |
| Wholesale trade | 63 | 2 |
| Finance, insurance, and real estate | 24 | 1 |
| Nonclassifiable | 88 | 3 |
| Total | 2831 | |

*Source*: Based on Reference [15].

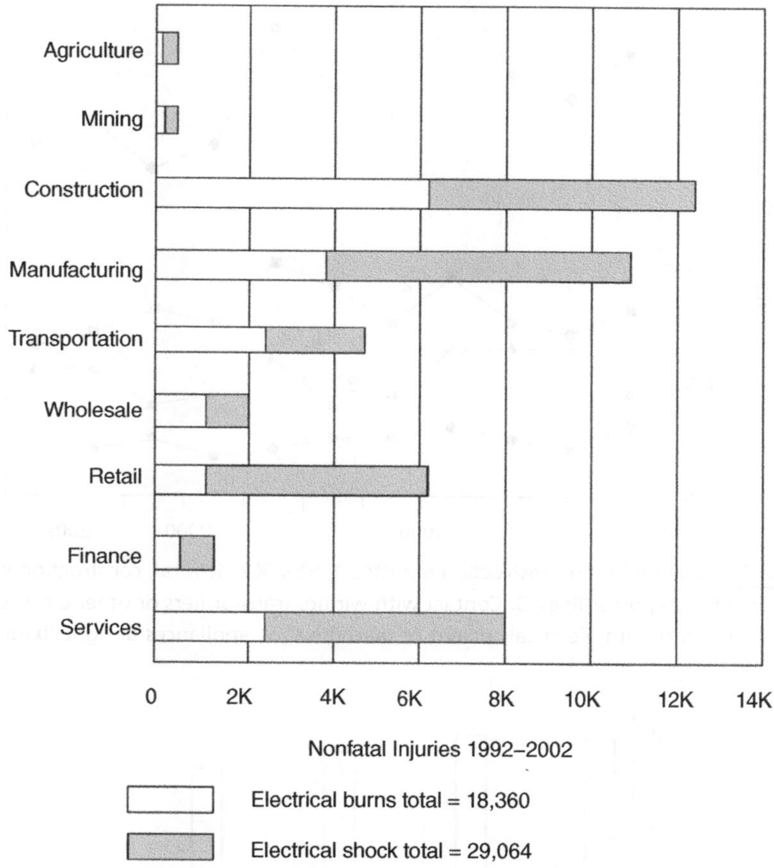


Figure 2.3. Statistics of electrical nonfatal injuries, industrywise, from 1992 to 2002.

with wiring of electrical components and contact with electrical current of machines, tools, appliances, or light fixtures. Construction industry also has the largest number of nonfatal electrical injuries—more than three times the overall industry rate.

Figure 2.5 shows the burn injury statistics—probability of survival, which decreases with the increasing age of the arc flash victim. The impact of the arc flash burns are:

- the treatment can cost upwards of $1,000,000 per case
- litigation fees
- production loss.

The safety control measures to protect the construction worker are not addressed in the design stage in the same manner as operating and maintenance workforce [16]. A construction worker is sometimes complacent to accept a more hazardous design, and this mentality must be changed.

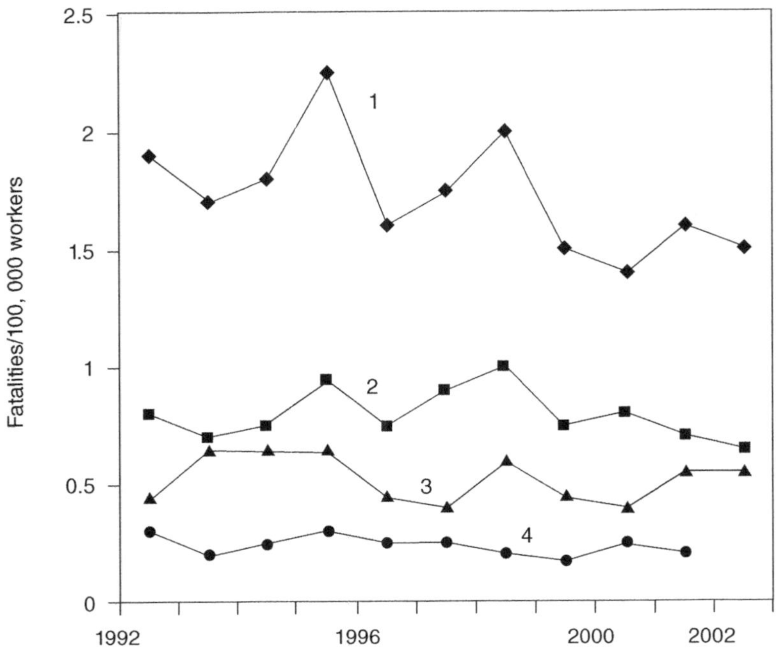


Figure 2.4. Fatalities in the construction industry, 1992–2002. 1: Total, construction industry; 2: Contact with OH power lines; 3: Contact with wiring, transformers or other electrical components; 4: Contact with electrical current of machine tool, appliances or light fixture.

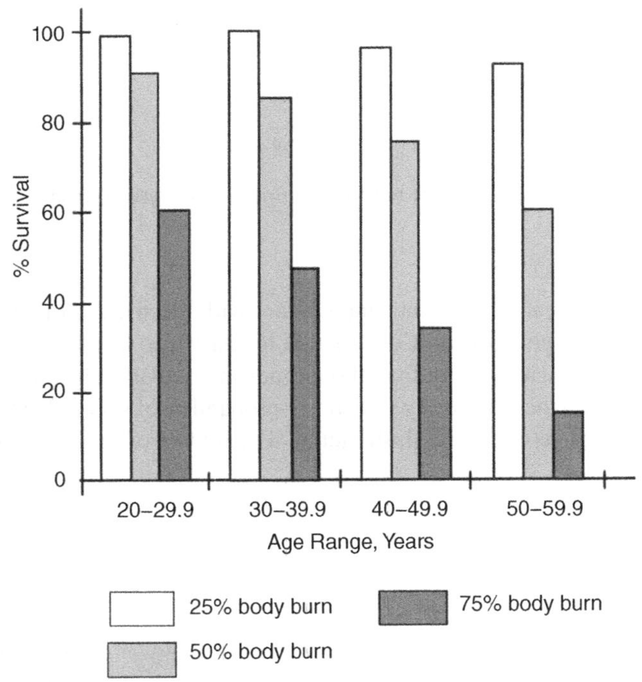


Figure 2.5. Percentage survival rate with respect to percentage body burns and age of the victim.

In 1980s, the European Commission performed a study that indicated 60% of fatal accidents in the construction have a contribution factor related to action or decision before the start of construction. This study led to the 1992 development of the European Union Directive on Temporary or Mobile Worksites [17]. The directive placed legal responsibilities on owners and on those associated with the design of the work. The directive required the member countries to put a requirement in place within 5 years. In 1994, the United Kingdom developed and implemented the Construction Design and Management (CDM) regulations [18]. These regulations require construction companies, project owners, and architects to address safety and health during the design phase of the project. Australia developed the Australian Occupational Health and Safety (OHS) Strategy (2000–2012) [19].

## 2.5 CHANGING THE SAFETY CULTURE

The safety culture is changing. This can be evidenced by many technical papers published in IEEE conferences, safety workshops, and other publications. Industry is coming around to demand better design practices and safe equipments. Some impediments to overcome are:

- A belief that it can be fixed later in the field.
- A notion that qualified and experienced workers do not make mistakes [6].
- Possibilities that consultants and experts are not fully aware of safety management principles and standards.
- Electrical designers may not be familiar with workplace hazards and safe work practices, especially so if they do not have a first-hand field experience.
- Consultants and design firms may be concerned that addressing worker safety in design stages will increase their liability [20].
- Facility owners may not understand the cost of retrofitting safe work practices and facility designs after it is complete and in place [21, 22].

## 2.6 RISK ANALYSIS FOR CRITICAL OPERATION POWER SYSTEMS

NEC 2008 added another article, Article 708—Critical Operation Power Systems (COPSs). The methodology of evaluating risk and performing reliability analyses applies to any critical facility. COPS are defined as the power systems for facilities or parts of facilities that require continuous operation for the reasons of public safety, emergency management, national security, and business continuity. The article also defines designated critical operation areas (DCOAs)—areas within a facility or site designated as requiring critical operation power. According to Article 708, COPS are classified by municipal, state, federal, or other codes by any governmental agency or authority having jurisdiction (AHJ).

NFPA 1600-2007, "Standard for Disaster/Emergency Management and Business Continuity Programs" is referred in number of places in Article 708. The sections in 708 include:

- circuit wiring and equipment
- power sources and connections
- overcurrent protection
- system performance and analysis.

A significant requirement is conducting and documenting a risk assessment:

1. Identify hazards, the likelihood of their occurrences, and vulnerability of the electrical system to these hazards. Naturally occurring hazards are geographical, metrological, and biological. Human caused events are intentional and accidental; equipment failures can be due to electrical or mechanical reasons.
2. Developing a mitigation strategy based upon the results of the analysis is the next step. Hazards listed in NEC Article 708.4 are not the only items that could cause electrical systems to fail.

Annex F of NEC provides some basic data on availability and reliability along the directions of implementation of functional performance tests (FPTs). Availability, reliability, and maintainability are the three interrelated aspects. Availability is calculated by:

$$A = \frac{\text{MTBF}}{\text{MTBF} + \text{MTTR}}, \tag{2.1}$$

where:

MTBF = mean time between failures
MTTR = mean time to repair.

The methods of improving reliability discussed in appendix F of NEC are below.

### 2.6.1 Existing Systems

To improve, selectively add redundant units, like generators, chillers, fuel supply systems, and the like to eliminate single-point failure. Optimize maintenance using a reliability-centered maintenance (RCM) approach, see Section 2.8.2 below.

### 2.6.2 New Facilities

(1) Develop and determine a reliability strategy; and (2) develop a reliability program. This is the application of reliability strategy to a specific system process or function.

Each step requires specific methods and tools. For example, various tools can be used to develop the requirements and analyze failures. A failure mode and effects criticality analysis (FMECA) and failure tree analyses (FTA) are two methods for evaluating possible failures (3) Identify reliability requirements.

The four steps discussed in development and implementation of functional performance tests (FPT) for COPS are:

1. Submit FPTs: System/component tests or FPT are developed from submitted drawings, system operating documents, and system operating and maintenance manuals. This includes equipment testing, that is, transformers, cables, UPS systems, generators, and the like. As the equipments/components/systems are installed, quality insurance procedures are administered to verify that the components are installed in accordance with manufacturer's recommendations, safety codes, and acceptable installation practices.
2. Review FPTs: The tests must be reviewed by the customer, electrical contactors, quality assurance personnel, and other key persons.
3. Make Changes to FTPs as required: The commissioning authority then makes the agreed corrections.
4. FPTs Approval: After the changes are made to the FPTs, these are submitted to the commissioning team. When it is acceptable, the customer, or the designated approval authority, approves the FPTs.

## 2.7 RELIABILITY ANALYSIS

Reliability is the probability of successful operation and is time dependent. For a system that has components with a relatively constant failure rates, reliability is an exponentially decaying function with time; the longer the time interval, the longer the reliability irrespective of system design. The reliability curve will be flatter for a well-designed system as compared with a poorly designed system.

System reliability assessment and evaluation methods are based upon probability theory that allows the reliability of a proposed system to be assessed quantitatively. This is finding a wide application. Alternative system designs, redundancy, impact on cost of changes, service reliability, protection and switching, and system maintenance policy can be quantitatively studied. Using reliability evaluation methods, system reliability indexes can be computed. The two basic system reliability indices are the load interruption frequency and expected duration of load interruption events. These can be used to compute other indexes, that is, total expected average interruption time per year, system availability or unavailability at the load supply point, expected energy demanded, but unsupplied per year. Reliability also addresses emergency and standby power systems. Preventive maintenance has a large impact not only on the availability of systems but also on arc flash hazards—a poorly maintained system is more prone to failures, including personal hazards. Some basic definitions of terms in reliability analysis are shown in Table 2.2. If the time $t$ over which a system must operate and the

TABLE 2.2. Definition Summary for the Reliability Analysis

| Parameter | Equation | Explanation of Symbols |
|---|---|---|
| $A_i$, inherent availability | MTBF/(MTBF + MTTR) | |
| $A_o$, operational availability | MTBM/(MTBM + MDT) | |
| λ, failure rate (f/h) | $T_f/T_p$ | $T_f$ = total number of failure during a given $T_p$ |
| MDT, mean down time | $(R_{dt} + R_{lt} + M_{dt})/T_{de}$ | $T_{de}$: total downtime events. $R_{dt}$, total downtime for unscheduled maintenance, $R_{lt}$, total logistic time for unscheduled maintenance and $M_{dt}$, maintenance down time |
| MTBF, mean time between failure (h) | $T_p/T_f$ | |
| MTBM, mean time between maintenance (h) | $T_p/T_{de}$ | |
| MTTM, mean time to maintain (h) | $M_{dt}/T_{ma}$ | $T_{ma}$, total number of schedule maintenance in $T_p$ |
| MTTR Mean time to repair(h) | $R_{dt}/T_f$ | |
| $R(t)$ reliability | $e^{-\lambda t}$ | |

*Source*: Based on IEEE 493 [2].

underlying distributions of failures of its constituent elements are known, then the system reliability can be calculated by taking the integral, essentially the area under the curve defined by the probability density function, PDF, from $t$ to infinity:

$$R(t) = \int_0^t f(t)dt, \tag{2.2}$$

where $f(t)$ is the PDF. See Appendix A.

### 2.7.1 Data for Reliability Evaluations

IEEE 493 [2] contains the reliability data collected from equipment reliability surveys and a data collection program over a period of 35 years or more. Data needed for a reliability analysis will depend upon the nature of the system being studied and the details of the study. Usually, data on individual system components and the times required to do various switching operations will be required. System component data required are as follows.

1. failure rates or forced outage rates associated with various modes of component failure
2. expected average time to repair or replace a component

3. scheduled maintenance outage rate of the component
4. expected duration of a scheduled outage event.

Switching time data needed includes expected times to open and close a circuit breaker, disconnect or throw-over switch, replace a fuse link, and performing emergency operations, such as installing jumpers.

The service reliability requirements of the loads and processes supplied is assessed to decide a proper definition of service interruption. It is not only the total collapse of voltage, but voltage sag or swell, that may also cause a shutdown. A failure modes and effects analysis is carried out. This means listing all component outage events or combinations of component outages that result in an interruption of service at the load point being studied. Component outages are categorized as

1. forced outages and failures
2. scheduled or maintenance outages
3. overload outages.

Component failure can be categorized by physical mode or type of failure.

### 2.7.2 Methods of Evaluation

Methods of evaluation include the following:

- cut-set method
- network reduction
- go algorithms
- state-space
- Monte Carlo simulations.

The details of these analyses will be beyond the scope of this book. In the cut-set method, the computation of the quantitative reliability indexes can proceed once the minimal cut sets of the system have been found. The first step is to compute the frequency, expected duration, and expected downtime of each minimal cut-set. Statistical methods and expressions for frequency and expected duration of the most commonly considered interruption events associated with first-, second-, and third-order cut sets are then applied to assess overall reliability. Digital computers are used for the reliability analysis as the statistical manipulations become complex, even for small systems. Some examples of calculations are included in IEEE 493 [2].

### 2.7.3 Reliability and Safety

A system designed for high MTBF is not necessarily the safest from the point of view of injuries and hazards to the worker. For example, an industrial distribution system served from a single source of power; one transformer connected to the utility source will not be very reliable. In such a system, any failure of a component connected in radial circuit can result in total failure of the served loads; for example, the failure of main transformer, circuit breaker, or interconnecting cable. This will call for more human intervention to attend to the failures. Yet this simple radial system can be designed safe from arc flash hazard and worker safety considerations. A system with redundant parallel running transformers, alternate sources of power, standby generation, and UPS systems, and autotransfer switches of power between sources will be more reliable with respect to reliability indices, but may not have the required worker safety features; for example, a proper interlocking of the sources may be missing or improperly designed. *It should not be construed that complex systems designed with higher reliability cannot be made equally or even safer than simpler systems.* For properly designed automated systems, lesser interruptions and human interventions will be needed. This will improve both—the reliability as well as safety.

The maintenance aspects will be common for reliability and safety. Poorly maintained systems are neither safe nor reliable. The aspects of reliability and safety should converge for the effective system design. This trend is gaining importance in the modern electrical power systems designs.

The following five aspects of safety program are important (IEEE 902, The Yellow Book):

1. To make personnel aware that there are rules, responsibilities, and procedures for safety in an electrical environment
2. To demonstrate the employer's intentions to fully comply with the federal law
3. To document general guidelines for providing workplace facilities that are free from unauthorized exposure to electrical hazards
4. To document general requirements and guidelines that direct the activities of personnel who could be deliberately or accidentally exposed to electrical hazards
5. To encourage and make it easier for each employee to be responsible for his or her own electrical safety and self-discipline.

## 2.8 MAINTENANCE AND OPERATION

Even the best designed systems may not perform reliably and safely if the quality and capability of its operation and maintenance is poor. As the systems become large in size and complexity, the problems of system management increase, including training of the qualified persons responsible for the management.

Generally, it can be said that when a problem occurs, primarily, there is tendency to blame it on the design of the electrical equipment, while in fact it is due to human error. The programs to reduce human intervention and provide automated systems of remote indications and controls are more cost-effective for the workers safety than system modifications or additional preventive maintenance. Thus, reducing the operating error in a given situation is an important factor in inherent system reliability and safety. The basic aspects to be looked into are:

- Development of maintenance strategy to ensure long-term reliability
- Development of record keeping and documentation files, which will include single-line diagrams, operating instructions, manufacturers data, inventory data, proper equipment labeling, tag down procedures, and the like. As an example, without proper and updated single line diagrams, wiring drawings and instruction manuals that should be readily available under a fault conditions, the operation of the industrial power system will be difficult and potentially unsafe.
- Development and implementation of testing and inspection methods
- Development of procedures for auditing maintenance and operation performance
- Development of procedures for personal safety
- Development of relaying and logic diagrams for load flow, impedance diagrams for short-circuit calculations, switching function diagrams to show alternate switching actions, say for a tie line
- Development and upkeep of all power system study reports, the adherence to the recommendations contained there in and implementation of modifications and upgrades recommended to overcome the shortcomings of the power systems, for example, replacement of underrated switching devices from short-circuit considerations
- Development of strategies to secure equipment from unwanted operations and a management strategy to clearly define the responsibilities of all mangers and operating personnel. It should be insured that all parties perform at a level that is consistent with their responsibilities.
- Development of strategies for the maintenance programs and seriously following it through over the course of time.

### 2.8.1 Maintenance Strategies

***Breakdown Maintenance.*** The maintenance efforts required to restore the equipment and systems to operating conditions after a failure or fault.

***Predictive Maintenance.*** Predictive maintenance is the practice to conduct diagnostic tests and inspections during normal equipment operations in order to detect incipient weaknesses or impeding failures.

***Preventive Maintenance.*** This is the practice of conducting routine inspections, tests, and servicing so that impeding troubles can be detected and reduced or eliminated. There is always a debate: how much maintenance is required and at what cost benefits? The maintenance may require a process shutdown and the cost of the loss of production has to be balanced with cost reductions obtained through improved safety and performance. It depends upon the type of facility to be maintained—a shutdown for maintenance in an electronic manufacturing plant will be more expensive that in a small industrial facility. The effort starts with a survey of each equipment and then adopting some guide lines, for example, NFPA 70B [4] and NETA MTS-1993 [23]. Some common tests performed are:

- Insulation resistance tests
- High potential testing, using AC and DC voltage
- Medium and high voltage cable testing
- Dielectric loss angle testing (DLA)
- Protective device testing and relay calibrations
- Low voltage and high voltage circuit breaker testing. This may include a time-travel analysis, consisting of applying a voltage to the trip coil or close coil and measuring the movement of the breaker mechanism.
- Saturation tests on current transformers
- Dielectric tests on insulating oil
- Grounding resistance measurements
- Contact resistance testing, that is, of circuit breaker contacts
- Infrared scanning; see Chapter 14.

### 2.8.2 Reliability-Centered Maintenance (RCM)

Reliability-centered maintenance is a systematic methodology that establishes initial preventive maintenance requirements or optimizes existing preventive maintenance requirements based on the consequences of equipment failure. RCM was first developed by the airline industry and has since been adopted by many other industries. The criticality and failure modes are analyzed and then the effective and preventive maintenance for each component is determined. All possible failure modes and preventive measures are evaluated for each component, thus establishing a program that is proven to result in improved facility reliability. It focuses on system components that are critical to maintain system functions.

Testing and repairs are integral part of any maintenance activity. The repairs may be warranted immediately, as in the case of a breakdown, or the identified equipment can be earmarked for future upgrades and repairs depending upon its existing operating condition.

## 2.9 SAFETY INTEGRITY LEVEL AND SAFETY INSTRUMENTED SYSTEM

IEC 61511 [24] has been developed as process sector implementation of the IEC 61508: "Functional Safety of Electrical/Electronic/Programmable Electronic Safety Related Systems" [25]. It has two basic concepts: the safety lifecycle and safety integrity levels (SIL). The safety lifecycle forms the central framework, and it is a good engineering procedure for safety instrumented systems (SIS) design in process industry (sensors, logic solvers, and final elements are a part of SIS). In the safety lifecycle process, risks are evaluated, and SIS performance requirements are established. Layers of protection are designed and analyzed. Then a SIS is optimally designed to meet the particular process risk. SIL levels indicate order of magnitude of risk reduction. Table 2.3 shows safety integrity levels. SIL 1 has the lowest level of risk reduction and SIL 4 has the highest. The standard suggests that applications that require the use of SIS of SIL4 are rare in the process industry; an exception can be nuclear plants. The standard mainly deals SIS and interface between SIS and other safety systems in requiring that a process hazard and risk assessment be carried out.

The difference between SIL related to *a single piece of equipment* and *a set of processes* must be clearly distinguished. In the United States, the onerous of specifying a particular equipment meeting the required safety levels (i.e., OSHA, UL, and ANSI/IEEE standard requirements) lies with the user. In Europe, the machine manufacturers must meet the specified SIL level. Relevant references are Standards: IEC 62061 also ISO 13849-1 [26, 27].

IEC 61511 part 3 details guidelines for selecting SIL in hazards and risk analysis. The information is intended to provide a broad range of global methods used to do such analysis. There are several informative annexes, of which annex A covers ALARP (as low as reasonably practical) principle and tolerable risk concepts (Figure 2.6).

This book mainly addresses arc flash hazards and their reduction in electrical distribution systems. This is an important aspect, but there are many related aspects of worker safety and system designs as briefly outlined in this chapter and also Chapter 1.

TABLE 2.3. SIL for Safety Functions Operating on Demand or in a Continuous Demand Mode

| Safety Integrity Level | Demand Mode of Operation (Average Probability of Failure to Perform Its Design Function on Demand PFD) | Continuous/High Demand Mode of Operation (Probability of a Dangerous Failure per Hour) |
|---|---|---|
| 4 | $\geq 10^{-5}$ to $<10^{-4}$ | $\geq 10^{-9}$ to $10^{-8}$ |
| 3 | $\geq 10^{-4}$ to $10^{-3}$ | $\geq 10^{-8}$ to $10^{-7}$ |
| 2 | $\geq 10^{-3}$ to $10^{-2}$ | $\geq 10^{-7}$ to $10^{-6}$ |
| 1 | $\geq 10^{-2}$ to $10^{-1}$ | $\geq 10^{-6}$ to $10^{-5}$ |

*Source*: IEC 61508-1, tables 2 and 3.

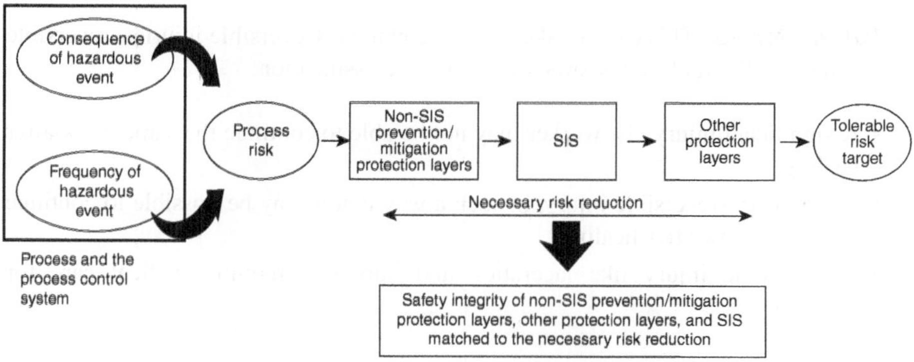


Figure 2.6. Risk and safety integrated concepts. *Source*: IEC 61511 [24].

## 2.10 ELECTRICAL SAFETY IN THE WORKPLACES

### 2.10.1 Risk Assessment

Risk assessment is an analytical process consisting of a number of discrete steps intended to endure that hazards are properly identified and analyzed with regard to their severity and likelihood of occurrence. Once these are identified, the risk can be estimated using the detailed flow chart in Figure 2.7. The following tasks are performed:

Identifying and analyzing electrical hazards
Identifying tasks to be performed
Documenting hazards associated with each task
Estimating the risk for each hazard/risk pair
Determining protective measures to reduce the level of risk.

### 2.10.2 Responsibility

Electrical system designers and planners have the responsibility for defining and achieving tolerable risk. The responsibilities of supplier and user differ over the life-cycle of electrical equipment, each entity should use risk assessment process.
Suppliers and users should involve qualified persons for meeting their responsibilities. The compliance with standards and regulations should be ensured.

### 2.10.3 Risk Parameters

An initial analysis of risk estimation should be carried out for each hazard. This is shown in Figure 2.8 and includes the following:

Likelihood of occurrence of injury or damage to health
Severity of injury or damage to health (Se)
Frequency and duration of the exposure of persons to health (Fr)
Likelihood of occurrence of a hazardous event (Pr)
Likelihood of avoiding or limiting the injury or damage to health (Av).

The parameters used in risk management can then be compiled into Table 2.4. The parameters should be based on worst case consideration for the electrical system.

***Parameter Se.*** This can be taken into account by irreversible injuries, reversible injuries, and death. Table 2.5 shows values of Se classification.

- A significant injury, the worker may not be able to continue the same work after healing.
- A major or irreversible injury, in such a way that it may be possible to continue the same work after healing.
- A reversible injury like laceration and burs that require medical care for recovery.

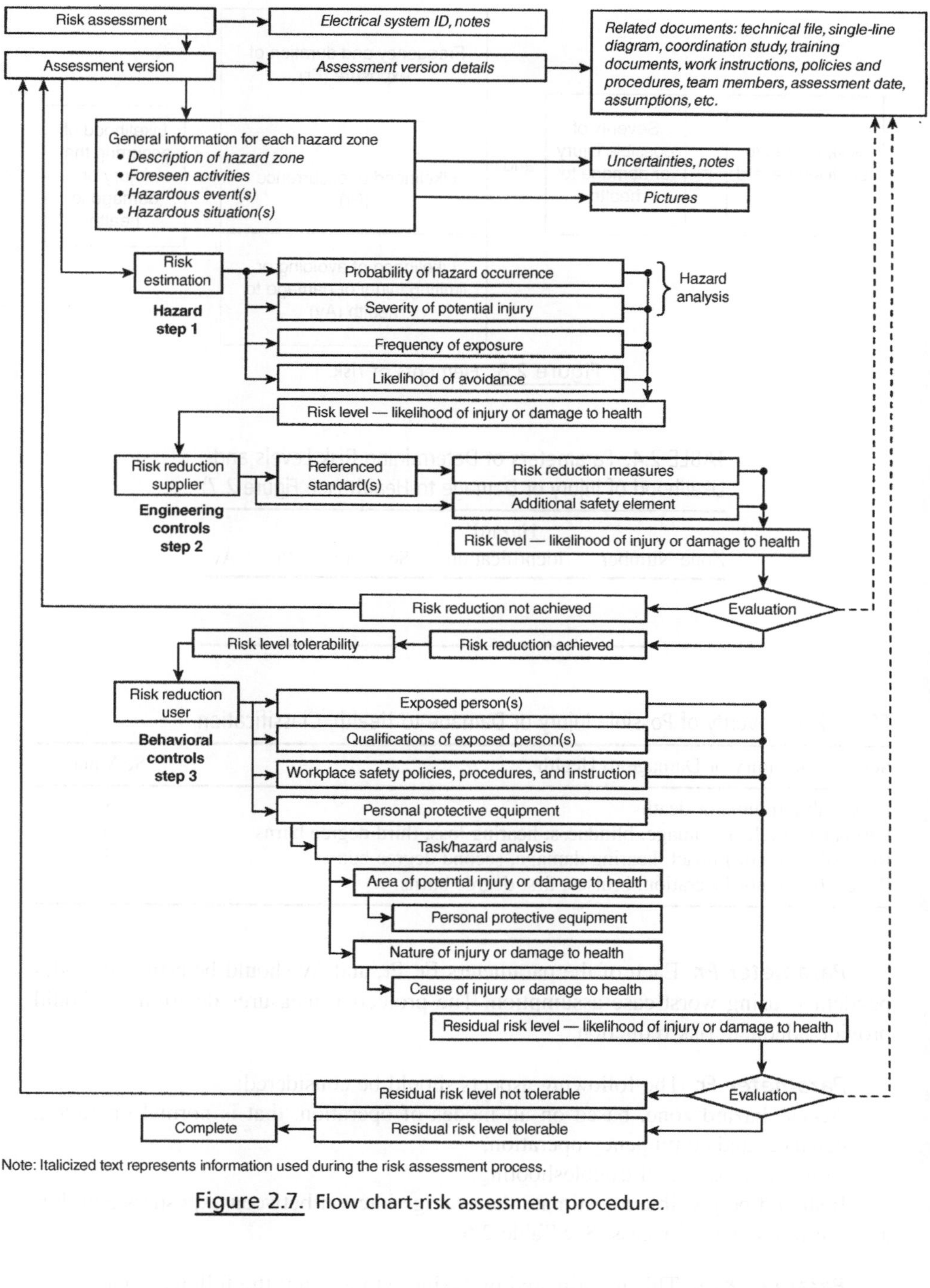


Note: Italicized text represents information used during the risk assessment process.

Figure 2.7. Flow chart-risk assessment procedure.

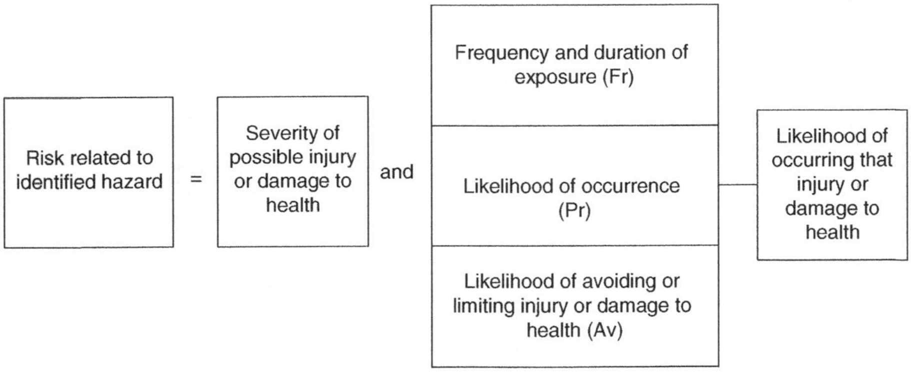


Figure 2.8. Elements of risk.

TABLE 2.4. Parameters of Determining Risk Levels and Likelihood of Injury or Damage to Health (see Figure 2.7)

| Zone Number | Hazard Identification | Se | Fr | Pr | Av |
|---|---|---|---|---|---|
| | | | | | |

TABLE 2.5. Severity of Possible Injury or Damage to Health, Classification

| Severity of Injury or Damage to Health | Se Value |
|---|---|
| Irreversible-trauma or death | 8 |
| Permanent-skeletal damage, blindness, hearing loss, third degree burns | 6 |
| Reversible-minor impact, hearing damage, second degree buns | 3 |
| Reversible-minor laceration, bruises, first degree burns | 1 |

***Parameter Pr.*** Each of the parameters Fr, Pr, and Av should be estimated independently, using worst-case assumption. The protective measures determined should provide adequate risk reduction.

***Parameter Fr.*** The following aspects should be considered:

Access hazard zone, based on all modes of operation, that is normal operation, maintenance, and contingency operation.

Consider repairs and troubleshooting.

It should be possible to estimate the average interval between exposures and thus average frequency of access. See Table 2.6.

***Parameter Pr.*** This is estimated by taking into account the following factors:

The predictability of performance of component parts of the system relevant to hazard in different modes, like normal operation, maintenance mode, contingency operation

TABLE 2.6. Frequency and Duration of Exposure

| Frequency of Exposure | Fr for Duration > 10 min |
|---|---|
| ≤1 per hour | 5 |
| >1 per hour to ≤1 per day | 5 |
| >1 per day to ≤1 every 2 weeks | 4 |
| >1 every 2 weeks to ≤1 per year | 3 |
| >1 per year | 2 |

TABLE 2.7. Likelihood of a Hazardous Event (Pr) Classification

| Frequency of Exposure | Pr for Duration > 10 min |
|---|---|
| Very high | 5 |
| Likely | 5 |
| Possible | 3 |
| Rare | 2 |
| Negligible | 1 |

TABLE 2.8. Likelihood of Avoiding or Limiting Injury or Damage to Health (Av) Classification

| Likelihood of Avoiding or Limiting Injury or Damage to Health | Av Value |
|---|---|
| Impossible | 5 |
| Rare | 3 |
| Possible | 1 |

The foreseeable characteristics of human behavior with regard to component parts of the system relevant to hazard. The factors are like stress and lack of awareness relevant to the hazard.

Table 2.7 shows the classification.

***Parameter Av.*** The parameter is estimated by taking into account aspects of electrical system design and its application that can help to avoid injury or damage to health.

Sudden or gradual appetence of a hazard, for example explosion caused by high fault values

Spatial possibility to withdraw from hazard

Nature of components and systems, for example use of inappropriate components

No indication of presence of a voltage or energized circuit

See Table 2.8.

## 2.11 RISK REDUCTION

The following parameters are of concern:

- Protective measures
- Engineering controls
- Awareness devices
- Procedures
- Training
- PPE

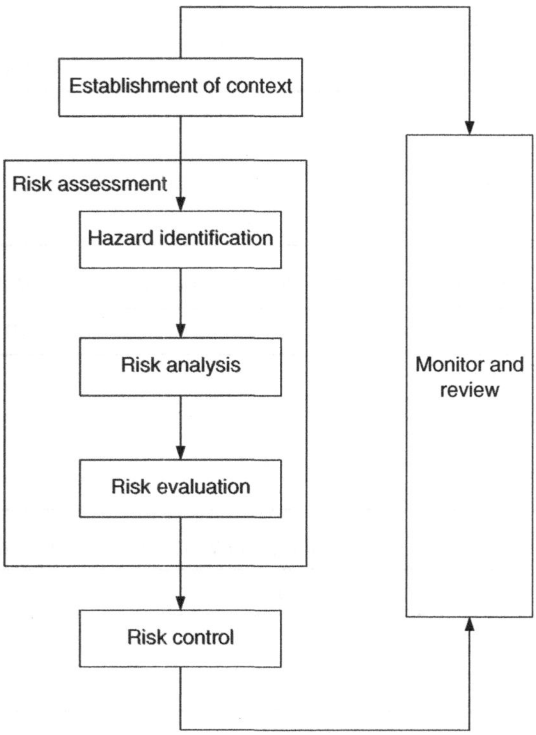


Figure 2.9. Risk management process. *Source*: Based on ISO 31000.

## 2.12 RISK EVALUATION

Each of protective measures can impact one or more elements that contribute to risk (Figure 2.9).

Some factors are:

- Design: Consider component failure, incorrect construction or manufacturing specification, inaccurate studies and calculations, inadequate procurement control, insufficient maintenance.
- Engineering controls: Inaccurate application and construction
- Unanticipated tasks
- Protective system failures
- Systems that increase awareness of potential hazards
- Organization and application of safe systems of work: Personal training. Identified hazard is not clearly communicated
- Training material not current
- Training not consistent with instructions
- Access restrictions
- Work permit system does not exist
- Insufficient monitoring, control or corrective actions
- Safe work procedures
- Inconsistent with current practices

TABLE 2.9. Sample Auditing Form

| Hazard | Risk Reduction Strategy | Confirmation |
|---|---|---|
| Human factors (mistakes) | Training and instructions include details regarding hazardous situations that could exist | Yes/no |
| Human factors (willful disregard) | Policies and procedures are in place to ensure that instructions are followed | Yes/no |
| Unqualified persons performing electrical work | Work permit procedures are in place to control personal activities | Yes/no |
| Inappropriate protection | Proper protective relaying and settings are ensured | Yes/no |
| Testing equipment faulty or inadequate | Testing equipment is properly calibrated | Yes/no |
| Meter malfunction | Proper testing and applications are ensured | Yes/no |
| Meter misapplications | Proper testing and applications are ensured | Yes/no |
| Qualified persons performing electrical work that exceeds their qualifications | Wok permit and assignment of responsibilities in place | Yes/no |

TABLE 2.10. The Hierarchy of Risk Control Methods

| Risk Control Methods | Examples |
|---|---|
| Arc flash and PPE | Appropriate studies and estimation of arc flash risks |
| Elimination | Conductors and equipment in an electrically unsafe condition |
| Substitution | Energy reduction strategies |
| Controls | Proper interlocks and safety measures to avoid human errors and faulty operation |
| Awareness | Signs and procedures to alert presence of hazards |
| Administrative controls | Procedures and written instructions |

- Does not consider all tasks and hazardous situations
- Insufficient monitoring and controls.

## 2.13 RISK REDUCTION VERIFICATION

Once the assessment has been done and protective measures are implemented prior to imitating electrical work.

Auditing: It is necessary to audit the risk reduction strategy that is applicable—for example, proper PPE and its use See Table 2.9.

## 2.14 RISK CONTROL

Risk management shares six risk management systems:

Leadership
Policy
Plan
DO: The plan is executed
Checking and monitoring
Review.

Table 2.10 shows hierarchy of risk control methods.
Table 2.11 is compiled from IEEE standard 493-2007.

TABLE 2.11. Failures per Year-IEEE 493-2007

| Equipment | IEEE 493, Table 10-2 | IEEE-493, Table 10-4 |
|---|---|---|
| Liquid transformers | 0.0062 | 0.00289 |
| Fixed circuit breakers | 0.00520 | |
| 0–600 V | 0.0042 | 0.0000338 |
| Above 600 V | 0.0276 | |
| 0–600 A | 0.0035 | |
| >600 A | 0.0096 | |
| Draw-out circuit breakers | 0.0030 | |
| 0–600 V | 0.0027 | 0.000509 |
| Above 600 V | 0.0036 | |
| 0–600 A | 0.0023 | |
| >600 A | 0.0030 | |
| Starters 0–600 V | 0.0139 | 0.00180 |
| Starters 600 V to 15 kV | 0.0153 | 0.00313 |
| Disconnects | 0.0061 | 0.00015 |
| Insulated bus | 0.001129 | 0.00303 |
| Bare bus 600 V | 0.00802 | 0.00349 |
| Bare bus >600 V | 0.001917 | 0.0013 |
| Cable in conduit above ground/1000 ft | 0.049180 | 0.000054 |
| Aerial cable/mile | 0.014370 | 0.01169 |
| 600 V cable above ground | 0.001430 | 0.000096 |
| 601 V to 15 kV cable above ground | 0.014300 | 0.000538 |
| Buried cable 0–600 V | 0.00380 | 0.00189 |
| Buried cable 600 V to 15 kV | 0.006170 | 0.0201 |
| Buried cable 15 kV | 0.00360 | |
| Cable terminations above ground 0–600 V | 0.000127 | |
| Cable terminations above ground 601 V to 15 kV | 0.000879 | |
| Aerial cable terminations | 0.00848 | |
| Underground | 0.000303 | |

## REVIEW QUESTIONS

**1.** Write five major considerations for starting PtD initiative.

**2.** Write a one-page strategy for reducing electrical hazards in the construction industry, citing five specific areas.

**3.** Write 10 points for design of safe electrical power systems.

## REFERENCES

1. IEEE-Xplore, http://ieeexplore.ieee.org/xplore/guesthome.jsp.
2. IEEE color books.
   - 141: *IEEE Recommended Practice for Electric Power Distribution for Industrial Plants* (IEEE Red Book), 1993

- 142: *IEEE Recommended Practice for Grounding of Industrial and Commercial Power Systems* (IEEE Green Book), 1991
- 241: *IEEE Recommended Practice for Electric Power Systems in Commercial Buildings*, (IEEE Gray Book) 1997
- 242: *IEEE Recommended Practice for Protection and Coordination of Industrial and Commercial Power Systems* (IEEE Buff Book), 2001
- 399: *IEEE Recommended Practice for Industrial and Commercial Power Systems Analysis* (IEEE Brown Book), 1997
- 446: *IEEE Recommended Practice for Emergency and Standby Power Systems for Industrial and Commercial Applications* (IEEE Orange Book), 2000
- 493: *IEEE Recommended Practice for the Design of Reliable Industrial and Commercial Power Systems* (IEEE Gold Book), 2007
- 602: *IEEE Recommended Practice for Electric Systems in Health Care Facilities* (IEEE White Book), 1996
- 739: *IEEE Recommended Practice for Energy Management in Commercial and Industrial Facilities* (IEEE Bronze Book), 1995
- 902: *IEEE Guide for Maintenance, Operation and Safety of Industrial and Commercial Power Systems* (IEEE Yellow Book), 1998. Now replaced with IEEE 3007.1, 2, and 3.
- 1015: *IEEE Recommended Practice for Applying Low-Voltage Circuit Breakers Used in Industrial and Commercial Power Systems* (IEEE Blue Book)
- 1100: *IEEE Recommended Practice for Powering and Grounding Electronic Equipment* (IEEE Emerald Book), 1999
- 551: *Recommended Practice for Calculating Short-Circuit Currents in Industrial and Commercial Power Systems* (IEEE Violet Book), 2006
- IEEE Standards announcement: http://standards.ieee.org/news/2012/3000collect.html

3. National Electrical Code (NEC) NFPA-70-2011.
4. NFPA Publications.
   - HFPE and Society of Fire Protection Engineers, SFPE Handbook of Fire Protection Engineering
   - 101H, Life Safety Code Handbook
   - 20. Centrifugal Fire Pumps
   - 70B. Electrical Equipment Maintenance
   - 70E. Electrical Safety Requirements for Employee Workplaces, 2012
   - 72. National Fire Alarm Code
   - 75. Protection of Electronic Computer/Data Processing Equipment
   - 77. Static Electricity
   - 78. Lightning Protection Code
   - 79. Electrical Standards for Industrial Machinery
   - 92A. Smoke Control Systems
   - 99. Health Care Facilities
   - 110. Emergency and Standby Power Systems
   - 130. Fixed Guide-Way Transit System.
5. Federal Register, *Superintendent of Documents*, U.S. Government Printing Office, Washington, DC.
6. NIOSH, *Publications Dissemination*, Cincinnati, OH.
7. TEC, 1111, Washington, DC.
8. NEMA, Washington, DC.

9. Popular handbooks:
   - D.G. Fink and H.W. Beaty, *Standard Handbook for Electrical Engineers*, 15th ed., McGraw-Hill, New York.
   - T. Croft, C.C. Carr, and J.H. Watt, *American Electricians Handbook*, 12th ed., McGraw-Hill, New York.
   - J.G. Webster (ed.), *Wiley Encyclopedia of Electrical and Electronics Engineering*, 21 Vols., John Wiley, New York, 1999.
   - *Illuminating Engineering Society (IES) Handbook*, Vols. 1 and 2. IES, New York.
   - *Electrical Transmission and Distribution Reference Book*, 4th ed., Westinghouse Electric Corporation, East Pittsburgh, PA, 1964.
   - *Applied Protective Relaying*, Westinghouse Electric Corporation, Coral Springs, FL, 1982.
   - R.S. Smeaton (ed.), *Motor Applications and Maintenance Handbook*, McGraw-Hill, New York, 1987.
   - D.L. Beeman, *Industrial Power Systems Handbook*, McGraw-Hill, New York, 1955.
   - Edison Electrical Institute, *Underground Systems Handbook*, 1957.
   - R.S. Smeaton (ed.), *Switchgear and Control Handbook*, McGraw-Hill, 1987.
   - J.M. McPartland, *Handbook of Practical Electrical Design*, McGraw-Hill, New York, 1984.
10. NESC, *C-2, National Electrical Safety Code, C-2*, 1993.
11. EGSA, Coral Springs, FL.
12. E. Manuele, "Prevention through design: Addressing occupational risks in the design and redesign processes," in special issue of By Design, Engineering Practice Specialty of American Society of Safety Engineers, pp. 1–13, Oct. 2007.
13. ANSI Z10, *Occupational Safety and Health Management Systems*, 2004.
14. Some publications on PtD
   - H.L. Floyd, "The NIOSH prevention through design initiative," in conf. Record, IAS Electrical Safety Workshop, pp. 77–79, March 2008.
   - D. Mohla, L.B. McClung, and N.R. Rafferty, "Safety by design," in Conf. Record IEEE IAS Petroleum and Chemical Industry Committee Tech. Conference, pp. 363–369, Sept. 1999.
   - B. McClung, D. Mohla, and L.B. McClung, "Electrical design refined for safety," in Proc. IEEE Industrial and Commercial Power Systems Tech. Conf., pp. 89–98, May 2005.
   - J. Brown, "Introduction to safety by design," presented at 2001 IEEE IAS Electrical Safety Workshop, Toronto, 2001.
   - H.L. Floyd, "Product safety opportunities in reducing occupational electrical injuries and Fatalities," IEEE Product Safety Engineering Society Symp., Santa Clara, 2004.
15. J.C. Cawley and G.T. Homce, "Trends in the electrical injuries in the U.S. 1992–2002," *IEEE Industry Applications Magazine*, vol. 44, no. 4, pp. 962–972, July/Aug. 2008.
16. J.A. Gambatese, J. Hinze, and M. Behm, *Investigations of the Viability of Designing for Safety*, The Center to Protect Workers' Rights, Silver Spring, MD, Tech Rep., May 2005.
17. Council Directive 92/57/EEC, *The Implementation of Minimum Safety and Health Requirements at Temporary or Mobile Construction Sites*, European Union Regulation, Official Journal L245, pp. 6–22, June 1992.
18. Construction Design and Management Regulations, *UK Health and Safety Executive, Statutory Instrument No. 320*, pp. 1–9, 2007.
19. Commonwealth of Australia, *National OHS Strategy 2002–2012*, Tech. Rep., pp. 1–9, 2002.

20. S. Jamil, A. Golding, H.L. Floyd, and M. Capelli-Schellpfeffer, "Human factors in electrical safety," in Proc. IEEE IAS Petroleum and Chemical Industry Tech. Conf. pp. 349–356, Sept. 2007.
21. W.C. Christensen, "Safety through design: Helping design engineers: 10 key questions," J. Am. Soc. Saf. Eng. Prof. Saf., pp. 32–39, pp. 1–28, March 2003.
22. ASSE: TR-Z790.001, *Prevention through Design Guidelines for Addressing Occupational Risks in Design and Redesign Processes*, American Society of Safety Engineers, Tech. Rep., pp. 1–28, Oct. 2009.
23. NETA, International Electric Testing Association. *MTS1993. Maintenance Testing Specifications*, 1993.
24. IEC 61511-SER Ed. 1.0, *Functional Safety-Safety Instrumentation Systems for Process Industry Sector—All Parts*, 2004.
25. IEC 61508, Ed. 2.0, *Functional Safety of Electrical/Electronic/Programmable Electronic Safety—Related Systems—All Parts*, 2010.
26. IEC 62061-1, Ed. 1.0, *Guidance for Application of ISO 13849-1 and IEC 62601 in the Design of Safety Related Control Systems for Machinery*, 2010.
27. IEC 62061, *Safety of Machinery—Functional Safety of Safety in Related Electrical Electronic and Programmable Electronic Control Systems*, 2005.

3

# CALCULATIONS ACCORDING TO IEEE GUIDE 1584, 2018

The authors and researchers all over the world expressed concerns on IEEE Guide 1584, 2002 methodology, test methods, and calculations of incident energy for arc flash hazard [1–10]. Also see Ref. [11], Chapter 3 of the first edition of this book.

As a result, a joint venture by IEEE and NFPA was constituted. This addressed the concerns levied on the 2000 edition of IEEE 584, and the 2018 edition. Ref. [12] is totally revised with respect to the 2002 edition.

This chapter details the fundamental aspect, the conceptual background, the calculation procedures and finally some examples of calculation.

## 3.1 MODEL FOR INCIDENT ENERGY CALCULATIONS

The model is a statistically derived model, with curve fitting program. It is applicable to the following systems:

- Voltages 208–15,000 V, three-phase line-to-line
- Frequency 50 or 60 Hz.
- Bolted fault currents:

*Arc Flash Hazard Analysis and Mitigation*, Second Edition. J.C. Das.

Published 2021 by John Wiley & Sons, Inc.

- 208–600 V: 500–106,000 A
- 601–13,000 V: 20065,000 A
- Gaps between conductors
  - 208–600 V: 6.35–76.2 mm (0.25–3 in)
  - 601–15,000 V: 19.05–254 mm (0.75–10 in)
- Working distances greater than or equal to 305 mm (12 in)
- Fault clearing time—no limit.

The enclosures tested at 600, 2700, and 14,300 V has certain dimensions as specified in 1584.

Electrode configuration, as described in Section 3.2.

AC single-phase and DC system are excluded, though some references are provided.

## 3.2 ELECTRODE CONFIGURATION

Criticism was levied on the 2002 edition of 1584. Only one test setup was rigged up, which considered vertical bus-bar configuration. Most electrical equipment has horizontal bus bars. Furthermore, barriers may be present. The 2018 edition considers the following electrode configurations, and each of which is tested:

**VCB**: Vertical conductors/electrodes inside a metal box/enclosure. This can be equated to the 2000 edition configuration.

**VCBB**: Vertical conductors/electrodes terminated in insulating barrier inside a metal box/enclosure.

**HCB**: Horizontal conductors/electrodes inside a metal box/enclosure.

**VOA**: Vertical conductors/electrodes in open air.

**HOA**: Horizontal conductors/electrodes in open air.

It is essential that a proper electrode configuration depending upon actual equipment involved is selected. See Figure 3.1.

## 3.3 IMPACT OF SYSTEM GROUNDING

The 2002 edition of 1584 divided the system grounding into two categories: ungrounded and grounded. The high-resistance and low-resistance grounded systems were considered ungrounded and solidly grounded systems as grounded. The arc flash hazard was calculated higher in ungrounded systems. See factor $k = -0.153$ or $-0.097$ in Equation (1), in 2002 Edition. Note that the test box was grounded in both cases. One explanation levied was that the turbulence with ungrounded system had a different pattern.

The 2018 edition of 1584 removes this distinction. The arc flash hazard in grounded or ungrounded systems is the same.

| | Electrode configuration in test |
|---|---|
| VCB | 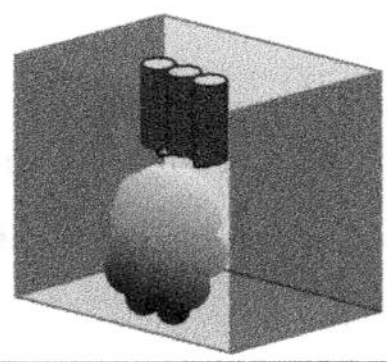 |
| VCBB | 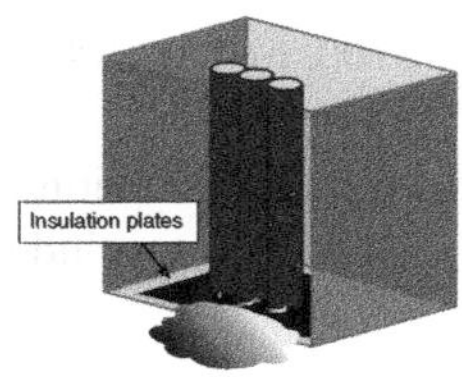  |
| HCB | 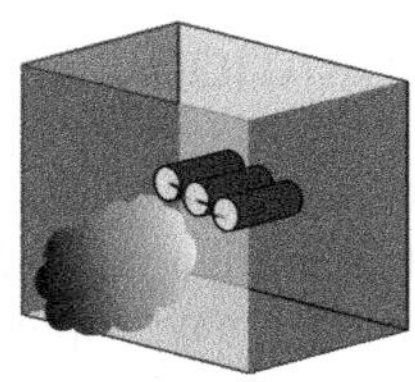 |
| VCB |  |
| VCBB |  |

Figure 3.1. Electrode configurations.

## 3.4 INTERMEDIATE AVERAGE ARCING CURRENT

The 2002 edition provided one equation for the calculation of arcing current (. 0. The variation in arcing current was accounted for by calculating arcing time at 80% of the calculated arcing current.

The 2018 edition provides elaborate arcing current calculations. Calculate the intermediate arcing current from the following Equation (3.1). This equation is used for the system voltage: $600\ \text{V} < V_{ac} \le 15,000\ \text{V}$

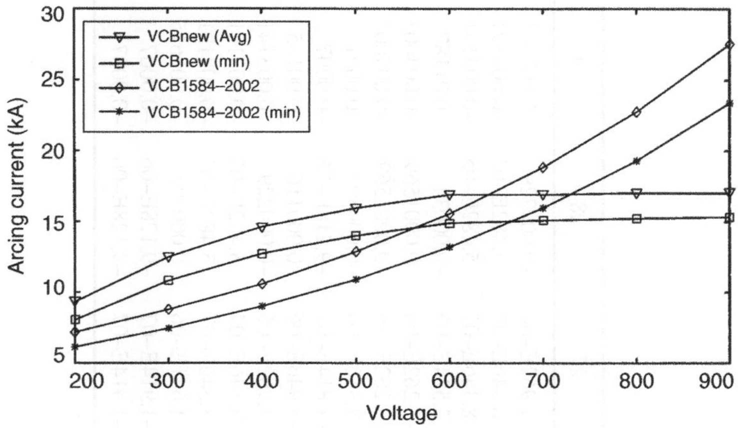


Figure 3.2. $I_{arc}$ versus $V_{oc}$ for 208–1000 V, Comparison of IEEE 2002 versus 2018.

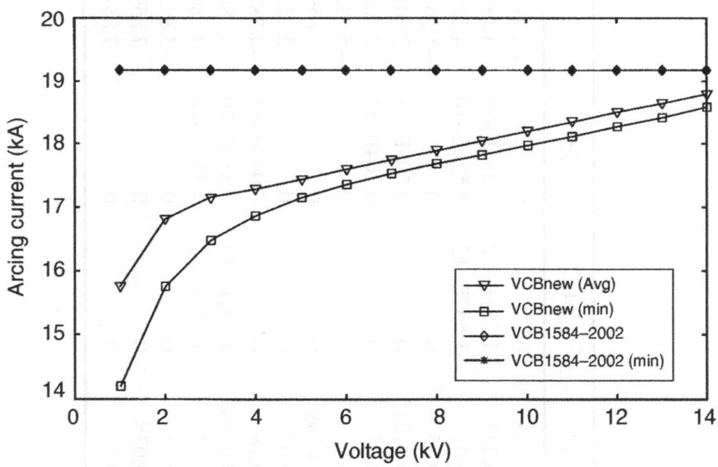


Figure 3.3. $I_{arc}$ versus $V_{oc}$ for 1–15 kV, Comparison of IEEE 2002 versus 2018.

$$I_{\text{arc}\_V_{\text{oc}}} = 10^{(k1 + k2\lg I_{\text{bf}} + k3\lg G)}\left(k4I_{\text{bf}}^6 + k5I_{\text{bf}}^5 + k6I_{\text{bf}}^4 + k7I_{\text{bf}}^3 + k8I_{\text{bf}}^3 + k9I_{\text{bf}} + k10\right) \quad (3.1)$$

where $I_{\text{bf}}$ is bolted current for three-phase faults (kA); $I_{\text{arc\_600}}$ is average rms arcing current at $V_{\text{oc}} = 600$ V (kA); $I_{\text{arc\_2700}}$ is average rms arcing current at $V_{\text{oc}} = 2700$ V (kA); $I_{\text{arc\_14,300}}$ is average rms arcing current at $V_{\text{oc}} = 14{,}300$ V (kA); and $G$ is gap between electrodes, mm.

See factors $k1$ to $k10$ in Table 3.1.

## 3.5 ARCING CURRENT VARIATION FACTOR

A second set of arc duration is calculated, using the reduced arcing current $I_{arc\_min}$ to examine the effect on protective device timing and therefore, incident energy. This

TABLE 3.1. Coefficients for Equation (3.1)

| E.C./Vac | | *k*1 | *k*2 | *k*3 | *k*4 | *k*5 | *k*6 | *k*7 | *k*8 | *k*9 | *k*10 |
|---|---|---|---|---|---|---|---|---|---|---|---|
| VCB | 600 V | −0.04287 | 1.035 | −0.083 | 0 | 0 | −4.783E−09 | 1.962E−06 | −0.000229 | 0.003141 | 1.092 |
| | 2700 V | 0.0065 | 1.001 | −0.224 | −1.557E−12 | 4.556E−10 | −4.186E−08 | 8.346E−07 | 5.482E−05 | −0.003191 | 0.9729 |
| | 14,300 V | 0.005795 | 1.115 | −0.011 | −1.557E−12 | 4.556E−10 | −4.186E−08 | 8.346E−07 | 5.482E−05 | −0.003191 | 0.9729 |
| VCBB | 600 V | −0.017432 | 0.98 | −0.05 | 0 | 0 | −5.567E−09 | 2.524E−06 | −0.00034 | 0.01187 | 1.013 |
| | 2700 V | 0.002823 | 0.995 | −0.01215 | 0 | −9.204E−11 | 2.901E−08 | −3.262E−06 | 0.0001569 | −0.004003 | 0.9825 |
| | 14,300 V | 0.014827 | 1.01 | −0.01 | 0 | −9.204E−11 | 2.901E−08 | −3.262E−06 | 0.0001569 | −0.004003 | 0.9825 |
| HCB | 600 V | 0.054922 | 0.988 | −0.1 | 0 | 0 | −5.382E−09 | 2.316E−06 | −0.000302 | 0.0091 | 0.9725 |
| | 2700 V | 0.001011 | 1.003 | −0.0249 | 0 | 0 | 4.859E−10 | −1.814E−07 | −9.218E−06 | −0.0007 | 0.9881 |
| | 14,300 V | 0.008693 | 0.999 | −0.02 | 0 | 0 | 2.233E−08 | −3.946E−06 | 0.000116 | −0.00145 | 0.9839 |
| VOA | 600 V | 0.043785 | 1.04 | −0.18 | 0 | −5.043E−11 | −4.783E−09 | 1.962E−06 | −0.000229 | 0.003141 | 1.092 |
| | 2700 V | −0.02395 | 1.006 | −0.0188 | −1.557E−12 | 4.556E−10 | −4.186E−08 | 8.346E−07 | 5.482E−05 | −0.003191 | 0.9729 |
| | 14,300 V | 0.005371 | 1.0102 | −0.029 | −1.557E−12 | 4.556E−10 | −4.186E−08 | 8.346E−07 | 5.482E−05 | −0.003191 | 0.9729 |
| HOA | 600 V | 0.111147 | 1.008 | −0.24 | 0 | 0 | −3.895E−09 | 1.641E−06 | −0.000197 | 0.002615 | 1.1 |
| | 2700 V | 0.000435 | 1.006 | −0.038 | 0 | 0 | 7.859E−10 | −1.914E−07 | −9.128E−06 | −0.0007 | 0.9981 |
| | 14,300 V | 0.000904 | 0.999 | −0.02 | 0 | 0 | 7.859E−10 | −1.914E−07 | −9.128E−06 | −0.0007 | 0.9981 |

applies to all systems from 208 to 15,000 V, but the impact will be most prominent at 208 and 600 V. Note that in the 2002 edition, a minimum arcing current of 85% was considered.

$$I_{\text{arc\_min}} = I_{\text{arc}}\left(1 - 0.5 \times \text{VarC}_{\text{f}}\right)$$
$$\text{VarC}_{\text{f}} = k1V_{\text{oc}}^{6} + k2V_{\text{oc}}^{5} + k3V_{\text{oc}}^{4} + k4V_{\text{oc}}^{3} + k5V_{\text{oc}}^{2} + k6V_{\text{oc}} + k7 \tag{3.2}$$

where $\text{Var}C_f$ = arcing current variation correction factor; $I_{\text{arc}}$ = final or intermediate arcing current. (for systems 208 V ≤ $V_{\text{oc}}$ ≤ 600 V, final current only, for systems 600 V < $V_{\text{oc}}$ ≤ 15, 000 V, intermediate average arcing current; $I_{\text{arc\_min}}$ = a second rms arcing current, based on the variation correction factor; $V_{\text{oc}}$ = open circuit voltage; and $k1$ to $k7$ = factors from Table 3.2.)

## 3.6 CALCULATION OF INTERMEDIATE INCIDENT ENERGY

$$E_{600} = \frac{12.552}{50} T \times 10^{\left[k1 + k2\lg G + \frac{k3 I_{\text{arc-600}}}{k4 I_{\text{bf}}^{7} + k5 I_{\text{bf}}^{6} + k6 I_{\text{bf}}^{5} + k7 I_{\text{bf}}^{4} + k8 I_{\text{bf}}^{3} + k9 I_{\text{bf}}^{2} + k10 I_{\text{bf}}} + k11\lg I_{\text{bf}} + k12\lg D + k13\lg I_{\text{arc\_600}} + \lg\frac{1}{\text{CF}}\right]} \tag{3.3}$$

$$E_{2700} = \frac{12.552}{50} T \times 10^{\left[k1 + k2\lg G + \frac{k3 I_{\text{arc-600}}}{k4 I_{\text{bf}}^{7} + k5 I_{\text{bf}}^{6} + k6 I_{\text{bf}}^{5} + k7 I_{\text{bf}}^{4} + k8 I_{\text{bf}}^{3} + k9 I_{\text{bf}}^{2} + k10 I_{\text{bf}}} + k11\lg I_{\text{bf}} + k12\lg D + k13\lg I_{\text{arc-2700}} + \lg\frac{1}{\text{CF}}\right]} \tag{3.4}$$

$$E_{14,300} = \frac{12.552}{50} T \times 10^{\left[k1 + k2\lg G + \frac{k3 I_{\text{arc-600}}}{k4 I_{\text{bf}}^{7} + k5 I_{\text{bf}}^{6} + k6 I_{\text{bf}}^{5} + k7 I_{\text{bf}}^{4} + k8 I_{\text{bf}}^{3} + k9 I_{\text{bf}}^{2} + k10 I_{\text{bf}}} + k11\lg I_{\text{bf}} + k12\lg D + k13\lg I_{\text{arc\_14,300}} + \lg\frac{1}{\text{CF}}\right]} \tag{3.5}$$

$$E_{\le 600} = \frac{12.552}{50} T \times 10^{\left[k1 + k2\lg G + \frac{k3 I_{\text{arc-600}}}{k4 I_{\text{bf}}^{7} + k5 I_{\text{bf}}^{6} + k6 I_{\text{bf}}^{5} + k7 I_{\text{bf}}^{4} + k8 I_{\text{bf}}^{3} + k9 I_{\text{bf}}^{2} + k10 I_{\text{bf}}} + k11\lg I_{\text{bf}} + k12\lg D + k13\lg I_{\text{arc}} + \lg\frac{1}{\text{CF}}\right]} \tag{3.6}$$

Take care to use appropriate $k1$ to $k13$ from Tables 3.3 and 3.4.

$D$ = working distance in mm

CF is the correction factor for enclosure size. CF = 1 for VOA and HOA configurations.

$G$ = gap between electrodes in mm.

$T$ = arc duration in ms.

$I_{\text{bf}}$ = bolted three-phase current in kA.

$I_{\text{arc}}$ = is defined in Equation (3.1).

TABLE 3.2. Coefficients for Equation (3.2)

| E.C. | *k*1 | *k*2 | *k*3 | *k*4 | *k*5 | *k*6 | *k*7 |
|---|---|---|---|---|---|---|---|
| VCB | 0 | −0.0000014269 | 0.000083137 | −0.0019382 | 0.022366 | −0.12645 | 0.30226 |
| VCBB | 1.138E−06 | −6.287E05 | 0.0012758 | −0.013778 | 0.080217 | −0.24066 | 0.33524 |
| HCB | 0 | −3.097E−06 | 0.00016405 | −0.0033609 | 0.033308 | −0.16182 | 0.34627 |
| VOA | 9.5606E−07 | −5.1543E−05 | 0.0011161 | −0.01242 | 0.075125 | −0.23584 | 0.33696 |
| HOA | 0 | −3.1555E−06 | 0.0001682 | −0.0034607 | 0.034124 | −0.15999 | 0.34629 |

TABLE 3.3. Coefficients for Equations (3.3), (3.6), (3.7), and (3.10)

| 600 V | *k*1 | *k*2 | *k*3 | *k*4 | *k*5 | *k*6 | *k*7 | *k*8 | *k*9 | *k*10 | *k*11 | *k*12 | *k*13 |
|---|---|---|---|---|---|---|---|---|---|---|---|---|---|
| VCB | 0.753364 | 0.566 | 1.752636 | 0 | 0 | −4.783E−09 | 0.000001962 | −0.000229 | 0.003141 | 1.092 | 0 | −1.598 | 0.957 |
| VCBB | 3.068459 | 0.26 | −0.098107 | 0 | 0 | −5.767E−09 | 0.000002524 | −0.00034 | 0.01187 | 1.013 | −0.06 | −1.809 | 1.19 |
| HCB | 4.073745 | 0.344 | −0.370259 | 0 | 0 | −5.382E−09 | 0.000002316 | −0.000302 | 0.0091 | 0.9725 | 0 | −2.03 | 1.036 |
| VOA | 0.679294 | 0.746 | 1.222636 | 0 | 0 | −4.783E−09 | 0.000001962 | −0000229 | 0.003141 | 1.092 | 0 | −1.598 | 0.997 |
| HOA | 3.470417 | 0.465 | −0.261863 | 0 | 0 | −3.895E−09 | 0.000001641 | −0.000197 | 0.002615 | 1.1 | 0 | −1.99 | 1.04 |

TABLE 3.4. Coefficients for Equations (3.4) and (3.8)

| 600 V | *k*1 | *k*2 | *k*3 | *k*4 | *k*5 | *k*6 | *k*7 | *k*8 | *k*9 | *k*10 | *k*11 | *k*12 | *k*13 |
|---|---|---|---|---|---|---|---|---|---|---|---|---|---|
| VCB | 2.40021 | 0.165 | 0.354202 | −1.557E−12 | 4.556E−10 | −4.186E−08 | 8.346E−07 | 5.482E−05 | −0.003191 | 0.9729 | 0 | −1.569 | 0.9778 |
| VCBB | 3.870592 | 0.185 | −0.736618 | 0 | −9.204E−11 | 2.901E−08 | −3.262E−06 | 0.0001569 | −0.004003 | 0.9825 | 0 | −1.742 | 1.09 |
| HCB | 3.486391 | 0.177 | −0.193101 | 0 | 0 | 4.859E−10 | −1.814E−07 | −9.128E−06 | −0.0007 | 0.9881 | 0.027 | −1.723 | 1.055 |
| VOA | 3.880724 | 0.105 | −1.906033 | −1.557E−12 | 4.556E−10 | −4.186E−08 | 8.346E−07 | 5.482E−05 | −0.003191 | 0.9729 | 0 | −1.515 | 1.115 |
| HOA | 3.616266 | 0.149 | −0.761561 | 0 | 0 | 7.859E−10 | −1.914E−07 | −9.128E−06 | −0.0007 | 0.9981 | 0 | −1.639 | 1.078 |

## 3.7 INTERMEDIATE ARC FLASH BOUNDARY (AFB)

Calculate using following equations:

$$\mathrm{AFB}_{600} = 10^{\frac{\left[k1 + k2\lg G + \frac{k3 I_{\mathrm{arc-600}}}{k4 I_{\mathrm{bf}}^{7} + k5 I_{\mathrm{bf}}^{6} + k6 I_{\mathrm{bf}}^{5} + k7 I_{\mathrm{bf}}^{4} + k8 I_{\mathrm{bf}}^{3} + k9 I_{\mathrm{bf}}^{2} + k10 I_{\mathrm{bf}}} + k11\lg I_{\mathrm{bf}} + k12\lg D + k13\lg I_{\mathrm{arc\_600}} + \lg\frac{1}{\mathrm{CF}} - \lg\frac{20}{T}\right]}{-k12}} \tag{3.7}$$

$$\mathrm{AFB}_{2700} = 10^{\frac{\left[k1 + k2\lg G + \frac{k3 I_{\mathrm{arc\_2700}}}{k4 I_{\mathrm{bf}}^{7} + k5 I_{\mathrm{bf}}^{6} + k6 I_{\mathrm{bf}}^{5} + k7 I_{\mathrm{bf}}^{4} + k8 I_{\mathrm{bf}}^{3} + k9 I_{\mathrm{bf}}^{2} + k10 I_{\mathrm{bf}}} + k11\lg I_{\mathrm{bf}} + k12\lg D + k13\lg I_{\mathrm{arc\_2700}} + \lg\frac{1}{\mathrm{CF}} - \lg\frac{20}{T}\right]}{-k12}} \tag{3.8}$$

$$\mathrm{AFB}_{1430} = 10^{\frac{\left[k1 + k2\lg G + \frac{k3 I_{\mathrm{arc-14,300}}}{k4 I_{\mathrm{bf}}^{7} + k5 I_{\mathrm{bf}}^{6} + k6 I_{\mathrm{bf}}^{5} + k7 I_{\mathrm{bf}}^{4} + k8 I_{\mathrm{bf}}^{3} + k9 I_{\mathrm{bf}}^{2} + k10 I_{\mathrm{bf}}} + k11\lg I_{\mathrm{bf}} + k12\lg D + k13\lg I_{\mathrm{arc-14,300}} + \lg\frac{1}{\mathrm{CF}} - \lg\frac{20}{T}\right]}{-k12}} \tag{3.9}$$

$$\mathrm{AFB}_{\le 600} = 10^{\frac{\left[k1 + k2\lg G + \frac{k3 I_{\mathrm{arc\_600}}}{k4 I_{\mathrm{bf}}^{7} + k5 I_{\mathrm{bf}}^{6} + k6 I_{\mathrm{bf}}^{5} + k7 I_{\mathrm{bf}}^{4} + k8 I_{\mathrm{bf}}^{3} + k9 I_{\mathrm{bf}}^{2} + k10 I_{\mathrm{bf}}} + k11\lg I_{\mathrm{bf}} + k12\lg D + k13\lg I_{\mathrm{arc\_6000}} + \lg\frac{1}{\mathrm{CF}} - \lg\frac{20}{T}\right]}{-k12}} \tag{3.10}$$

All the symbols are as defined above.

TABLE 3.5. Coefficients for Equations (3.5) and (3.9)

| 600 V | *k*1 | *k*2 | *k*3 | *k*4 | *k*5 | *k*6 | *k*7 | *k*8 | *k*9 | *k*10 | *k*11 | *k*12 | *k*13 |
|---|---|---|---|---|---|---|---|---|---|---|---|---|---|
| VCB | 3.825917 | 0.11 | −0.999749 | −1.557E−12 | 4.556E−10 | −4.186E−08 | 8.346E−07 | 5.482E−05 | −0.00319 | 0.9729 | 0 | −1.568 | 0.99 |
| VCBB | 3.644309 | 0.215 | −0.585522 | 0 | −9.204E−11 | 2.901E−08 | −3.262E−06 | 0.0001569 | −0.00403 | 0.9825 | 0 | −1.677 | 1.06 |
| HCB | 3.044516 | 0.125 | 0.245106 | 0 | −5.043E−11 | 2.33E−08 | −3.046E−06 | 0.000116 | −0.00145 | 0.9839 | 0 | −1.655 | 1.084 |
| VOA | 3.405454 | 0.12 | −0.93245 | −1.557E−12 | 4.556E−10 | −4.186E−08 | 8.346E−07 | 5.482E−05 | −0.00319 | 0.9729 | 0 | −1.534 | 0.979 |
| HOA | 2.04049 | 0.177 | 1.005092 | 0 | 0 | 7.859E−10 | −1.914E−07 | −9.128E−06 | −0.0007 | 0.9981 | −0.05 | −1.633 | 1.151 |

TABLE 3.6. Guidelines to Determine Equivalent Height and Width

| EC | Range | <508 mm | ≥508 and ≤660.4 mm | >660.4 and ≤1244.6 mm | >1244.6 mm |
|---|---|---|---|---|---|
| VCB | Width | =20 (if typical)<br>=0.03937 × width (if shallow) | =0.03937 × width | Obtained from Equation 3.11 and actual width | Obtained from Equation 3.11 and actual width = 1244.6 mm |
| | Height | =20 (if typical)<br>=0.03937 × width (if shallow) | =0.03937 × width | =0.03937 × height | =49 |
| VCBB | Width | =20 (if typical)<br>=0.03937 × width (if shallow) | =0.03937 × width | Obtained from Equation 3.11 and actual width | Obtained from Equation 3.11 and actual width = 1244.6 mm |
| | Height | =20 (if typical)<br>=0.03937 × width (if shallow) | =0.03937 × width | =0.03937 × height | Obtained from Equation 3.12 and actual width = 1244.6 mm |
| HCB | Width | =20 (if typical)<br>=0.03937 × width (if shallow) | =0.03937 × width | Obtained from Equation 3.11 and actual width | Obtained from Equation 3.11 and actual width = 1244.6 mm |
| | Height | =20 (if typical)<br>=0.03937 × width (if shallow) | =0.03937 × width | =0.03937 x height | Obtained from Equation 3.12 and actual width = 1244.6 mm |

## 3.8 ENCLOSURE SIZE CORRECTION FACTOR

For VCB, VCBB, and HCB, the equations are normalized or 508 mm × 508 mm × 508 mm enclosure, correction factors must be applied for larger and smaller enclosures.

### 3.8.1 Shallow and Typical Enclosures

An enclosure is shallow if system voltage is less than 600 V, both height and width are less than 508 mm (20 in), and its depth is less than 203.2 mm (8 in). Otherwise, it is considered typical.

## 3.9 DETERMINE EQUIVALENT HEIGHT AND WIDTH

$$\text{Width}_e = \left[660.4 + (\text{width} - 660.4) \times \left[\frac{V_{oc} + A}{B}\right]\right] \times 25.4^{-1} \tag{3.11}$$

$$\text{Height}_e = \left[660.4 + (\text{Height} - 660.4) \times \left[\frac{V_{oc} + A}{B}\right]\right] \times 25.4^{-1} \tag{3.12}$$

$A = 4$ for VCB and 10 for VCBB and HCB
$B = 20$ for VCB 24 for VCBB and 22 for HCB

The equivalent enclosure size (EES) is:

$$\text{EES} = \frac{\text{Height}_e + \text{Width}_e}{2} \tag{3.13}$$

## 3.10 DETERMINE ENCLOSURE SIZE CORRECTION FACTOR

For a typical enclosure, it is given by:

$$\text{CF} = b1 \times \text{EES}^2 + b2 \times \text{EES} + b3 \tag{3.14}$$

And or a shallow enclosure

$$\text{CF} = \frac{1}{b1 \times \text{EES}^2 + b2 \times \text{EES} + b3} \tag{3.15}$$

The factors $b1$ to $b3$ are given in Table 3.7.

TABLE 3.7. Coefficients for Equations (3.14) and (3.15)

| Bus Type | E.C. | *b*1 | *b*2 | *b*3 |
|---|---|---|---|---|
| Typical | VCB | −0.000302 | 0.03441 | 0.4325 |
| | VCBB | −0.0002976 | 0.032 | 0.479 |
| | HCB | −0.0001923 | 0.01935 | 0.6899 |
| Shallow | VCB | 0.002222 | −0.025556 | 0.6222 |
| | VCBB | −0.002778 | 0.1194 | −0.2778 |
| | HCB | −0.0005556 | 0.03722 | 0.4778 |

## 3.11 DETERMINATION OF $I_{ARC}$, $E$, AND AFB (600 V < $V_{OC}$ ≤ 15,000 V)

At first, the intermediate values are calculated.

### 3.11.1 Arcing Current

$$I_{arc\_1} = \frac{I_{arc\_2700} - I_{arc\_600}}{2.1}\left(V_{oc} - 2.7\right) + I_{arc\_2700} \tag{3.16}$$

$$I_{arc\_2} = \frac{I_{arc\_14,300} - I_{arc\_2700}}{11.6}\left(V_{oc} - 14.3\right) + I_{arc\_14,300} \tag{3.17}$$

$$I_{arc\_3} = \frac{I_{arc\_1}\left(2.7 - V_{oc}\right)}{2.1} + \frac{I_{arc\_2}\left(V_{oc} - 0.6\right)}{2.1} \tag{3.18}$$

where $I_{arc\_1}$= First $I_{arc}$ interpolation between 600 and 2700 V (kA); $I_{arc\_2}$= Second $I_{arc}$ interpolation when $V_{oc}$ is >2700 V (kA); and $I_{arc\_3}$= Third $I_{arc}$ interpolation when $V_{oc}$ is <2700 V (kA).

When 0.600 < $V_{oc}$ ≤ 2.7, the final value of acing current is:

$$I_{arc} = I_{arc\_3}$$

When 0.600 < $V_{oc}$ ≤ 2.7, the final value of acing current is:

$$I_{arc} = I_{arc\_2}$$

### 3.11.2 Incident Energy (*E*)

$$E_1 = \frac{E_{2700} - \mathbf{E_{600}}}{2.1}\left(V_{oc} - 2.7\right) + \mathbf{E_{2700}} \tag{3.19}$$

$$E_2 = \frac{\mathbf{E}_{14300} - E_{2700}}{11.6}\left(V_{oc} - 14.3\right) + E_{14,300} \tag{3.20}$$

$$E_3 = \frac{E_1\left(2.7 - V_{oc}\right)}{2.1} + \frac{E_2\left(V_{oc} - 0.6\right)}{2.1} \tag{3.21}$$

where $E_1$ is the first interpolation term between 600 and 2700 V (J/cm²); $E_2$ is the second interpolation term when $V_{oc}$ is >2700 V (J/cm²); and $E_3$ is the third interpolation term when $V_{oc}$ is <2700 V (J/cm²).

### 3.11.3 Arc Flash Boundary (AFB)

$$\text{AFB}_1 = \frac{\text{AFB}_{2700} - \text{AFB}_{600}}{2.1}\left(V_{oc} - 2.7\right) + \text{AFB}_{2700} \tag{3.22}$$

$$\text{AFB}_2 = \frac{\text{AFB}_{14,300} - \text{AFB}_{2700}}{11.6}\left(V_{oc} - 14.3\right) + \text{AFB}_{14,300} \tag{3.23}$$

$$\text{AFB}_3 = \frac{\text{AFB}_1\left(2.7 - V_{oc}\right)}{2.1} + \frac{\text{AFB}_2\left(V_{oc} - 0.6\right)}{2.1} \tag{3.24}$$

where $\text{AFB}_1$ is the first interpolation term between 600 and 2700 V (J/cm²); $\text{AFB}_2$ is the second interpolation term when $V_{oc}$ is > 2700 V (J/cm²); and $\text{AFB}_3$ is the third interpolation term when $V_{oc}$ is <2700 V (J/cm²).

$$\left(1\,\text{cal} = 4.184\,\text{J}\right)$$

When $0.600 < V_{oc} \leq 2.7$, the final value of incident energy and AFB are as follows:

$$E = E_1$$
$$\text{AFB} = \text{AFB}_1$$

When $V_{oc} > 2.7$, the final value of incident energy and AFB are as follows:

$$E = E_2$$
$$\text{AFB} = \text{AFB}_2$$

A second set of arc duration, $E$ AFB should be calculated for reduced arcing current.

## 3.12 DETERMINATION OF $I_{ARC}$, $E$, AND AFB ($V_{OC} \leq 600$ V)

### 3.12.1 Arcing Current

$$I_{arc} = \frac{1}{\sqrt{\left[\frac{0.6}{V_{oc}}\right]^2 \times \left[\frac{1}{I^2_{arc\_600}} - \left[\frac{0.6^2 - V^2_{oc}}{0.6^2 \times I^2_{bf}}\right]\right]}} \tag{3.25}$$

### 3.12.2 Incident Energy

The incident energy is given by:

$$E = E_{\leq 600}$$

where the incident energy for $V_{oc} \leq 600$ V is determined from Equation (3.6).

### 3.12.3 Arc Flash Boundary (AFB)

The AFB is given by:

$$\text{AFB} = \text{AFB}_{\leq 600}$$

where the AFB for $V_{oc} \leq 600$ V is determined from Equation (3.10).

## 3.13 A FLOW CHART FOR THE CALCULATIONS

A flow chart for the complete calculations is shown in Figure 3.4.

The three-phase bolted fault current is calculated with short-circuit study. The relay coordination and TCC plots are plotted before running arc flash calculations. The equipment type, gap length, working distance, dimensions, and electrode configuration is known. The enclosure correction factor is calculated. The intermediate values of arcing current at 600, 2700, and 14,300 V are calculated using equations provided. From the final, arcing current is known. The relay operating time is calculated based on TCC curves. The intermediate values of arc flash energy are calculated and then the final value. The variation in arc flash current is calculated and the relay operating time. The calculations of incident energy and arc flash are carried out next.

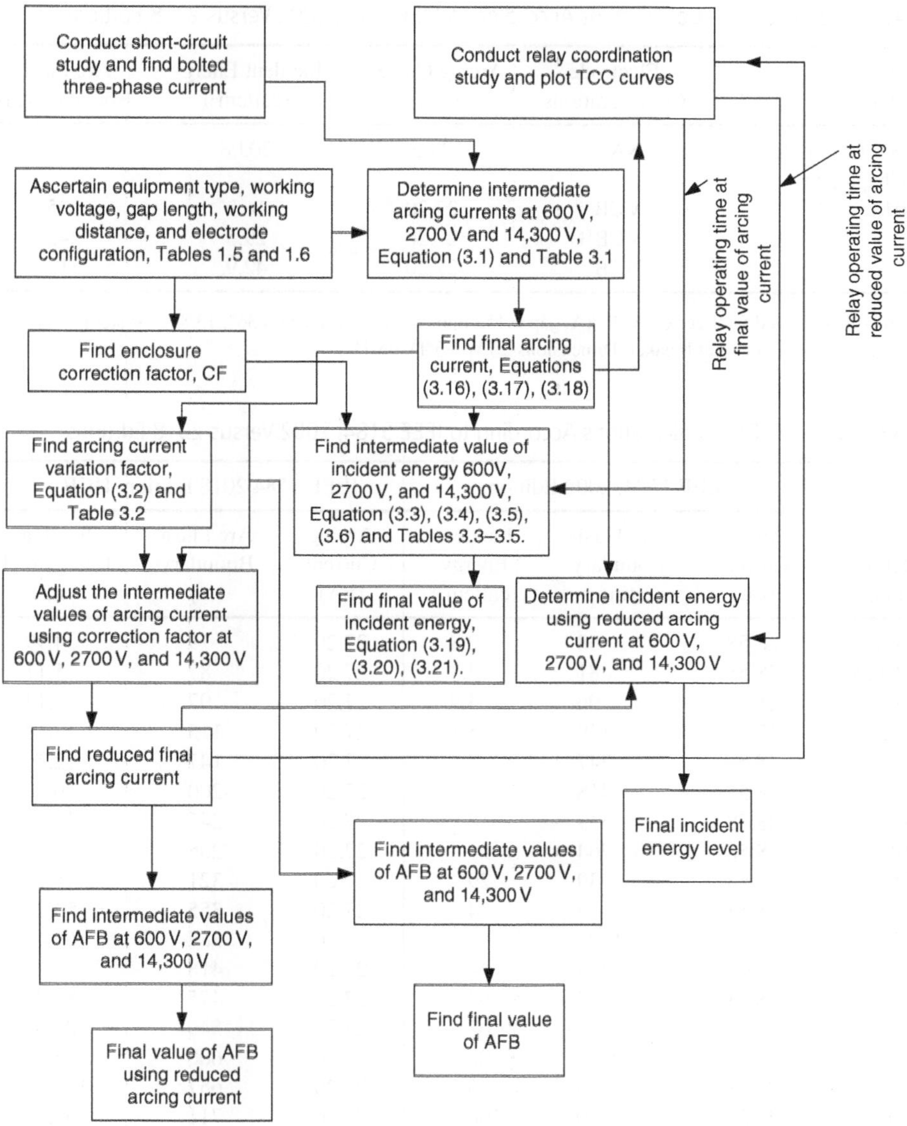


Figure 3.4. Flow chart for the calculation of arc flash hazard.

## 3.14 EXAMPLES OF CALCULATIONS

Two examples of calculations are shown in Tables 3.8 and 3.9. Table 3.8 shows the impact of electrode configuration. Note that arc flash boundary is much reduced according to the 2018 edition.

Table 3.9 shows the arc flash calculations 2002 versus 2018 guide for various fault clearing times from 0.0042 to 2 seconds. Again, note the vast difference in the arc flash boundary calculations.

TABLE 3.8. Arc Flash Calculations According to IEEE 1584, 2002 Versus 2018 Editions

| IEEE 1584 | Electrode Configurations | Arcing Current (kA) | Incident Energy (cal/cm$^2$) | Arc Flash Boundary (in) |
|---|---|---|---|---|
| 2002 (Example 1.1, Chapter 1) | NA | 28.6 | 20.08 | 651 |
| 2018 | VCB | 27.20 | 16.9 | 195 |
| | VCBB | 27.20 | 29.5 | 243 |
| | HCB | 27.20 | 42.9 | 312 |

Three-phase bolted fault current = 30 kA, gap = 152 mm, working distance = 36″, 13.8 kV system voltage, 13.8 kV metal clad circuit breaker, Dimensions 26′W, 95″D, 96″H.

TABLE 3.9. Arc Fault Calculations According to IEEE 5184, 2002 Versus 2018 Editions

| | IEEE 1584, 2002 Edition | | | IEEE 1584 2018 Edition, HCB | | |
|---|---|---|---|---|---|---|
| Relay Time (s) | Arcing Current (kA) | Arc Flash Boundary (in) | Incident Energy (cal/cm$^2$) | Arcing Current (kA) | Arc Flash Boundary (in) | Incident Energy (cal/cm$^2$) |
| 0.0042 | 28.58 | 68 | 2.8 | 27.20 | 79 | 4.99 |
| 0.0125 | 28.58 | 81 | 3.3 | 27.20 | 87 | 5.11 |
| 0.025 | 28.58 | 96 | 3.9 | 27.20 | 97 | 6.14 |
| 0.05 | 28.58 | 129 | 5.2 | 27.20 | 115 | 8.18 |
| 0.1 | 28.58 | 247 | 7.8 | 27.20 | 147 | 12.7 |
| 0.2 | 28.58 | 418 | 13 | 27.20 | 200 | 20.5 |
| 0.3 | 28.58 | 519 | 18 | 27.20 | 245 | 28.6 |
| 0.4 | 28.58 | 764 | 23 | 27.20 | 285 | 36.8 |
| 0.5 | 28.58 | 939 | 29 | 27.20 | 321 | 45 |
| 0.6 | 28.58 | 1115 | 34 | 27.20 | 355 | 53.2 |
| 0.7 | 28.58 | 1292 | 39 | 27.20 | 387 | 61.4 |
| 0.8 | 28.58 | 1469 | 44 | 27.20 | 418 | 69.6 |
| 1.0 | 28.58 | 1825 | 54 | 27.20 | 475 | 85.9 |
| 1.25 | 28.58 | 2273 | 67 | 27.20 | 540 | 106.4 |
| 1.5 | 28.58 | 2724 | 80 | 27.20 | 600 | 126.8 |
| 1.75 | 28.58 | 3176 | 93 | 27.20 | 657 | 147.3 |
| 2.0 | 28.58 | 3630 | 106 | 27.20 | 711 | 167.7 |

Bolted three-phase fault current 30 kA, system voltage 13.8 kV, gap length =152 mm, working distance =36", box dimensions 26"Wx 95" deep, 96" High, Circuit breaker opening time=3 cycles.

## REFERENCES

1. R. Wilkins, M. Allison, and M. Lang, "Effect of electrode orientation in arc flash testing," in Conf. Record, IEEE IAS Annual Meeting, vol. 1, pp. 459–465, 2005.
2. ASTM F2178. *Standard Test Method for Determining the Arc Rating and Standard Specification for Face Protective Devices*, 2006.

3. D. Sweeting, "Testing PPE for arc-hazard protection IEC 61482-1 test rig evaluation including proposed changes," IEEE Transactions on Industry Applications, vol. 47, no. 2, pp. 695–707, March/April 2011.
4. IEC 61482-1. *Live Working, Flame Resistant Material for Clothing for Thermal Protection of Workers—Thermal Hazards of an Electric Arc—Part I: Test Methods*, 2002.
5. D.R. Doan and T.E. Neal, "Field analysis of arc-flash incidents," IEEE Industry Applications Magazine, vol. 16, no. 3, pp. 39–45, June 2010.
6. V.P. Ignatko, "Electrical characteristics of AC open heavy-current arcs," in Proc. 3rd Int. Symposium on Switching Arc Phenomena, TU Lodz, Poland, pp. 98–102, 1977.
7. J. Paukert, "The arc voltage and the resistance of LV arc faults," in Proc. 7th Int. Symposium on Switching Arc Phenomena, TU Lodz, Poland, pp. 49–51, 1993.
8. R.A. Jones, et al., "Staged tests increase awareness of arc-flash hazard in electrical equipment," in Proc. 44th Annual IEEE Petroleum and Chemical Industry Technical Conference, Banff, AB, Canada, pp. 313–322, September 1997.
9. T. Gammon and J. Matthews, "Conventional and recommended arc power and energy calculations and arc damage assessment," IEEE Transactions on Industry Applications, vol. 39, no. 3, pp. 594–599, May/June 2003.
10. T. Gammon and J. Matthews, "Instantaneous arcing-fault models developed for building system analysis," IEEE Transactions on Industry Applications, vol. 37, pp. 197–203, January/February 2001.
11. J.C. Das, *Arc Flash Hazard Analysis and Mitigation*, IEEE Press, 2012.
12. IEEE 1584. *Guide for Performing Arc Flash Hazard Calculations*, 2002. Revised 2018.

# 4

# ARC FLASH HAZARD AND SYSTEM GROUNDING

IEEE Guide 1584, 2002 edition gave lower arc flash levels for high resistance or low resistance grounded systems, compared to solidly grounded systems. This qualification has been removed in 2018 edition, see Chapter 3. Much lower equipment damage and continuity of processes can be achieved with high resistance (HR) grounded systems. For low voltage distributions, generally, the solidly grounded systems are not applied, and ungrounded systems are extinct. This chapter on system grounding is included to highlight the current trends.

## 4.1 SYSTEM AND EQUIPMENT GROUNDING

The grounding systems can be studied under two classifications: (1) system grounding and (2) equipment grounding. System grounding refers to the electrical connection between the phase conductors and ground and dictates the manner in which the neutral points of wye-connected transformers and generators or artificially derived neutral systems through delta-wye or zigzag transformers are grounded. The equipment grounding refers to the grounding of the exposed metallic parts of the electrical equipment, which can become energized and create a potential to

*Arc Flash Hazard Analysis and Mitigation*, Second Edition. J.C. Das.

Published 2021 by John Wiley & Sons, Inc.

ground—say due to breakdown of insulation or fault—and can be potential safety hazard. In this chapter, we will discuss the system grounding with respect to arc flash considerations.

### 4.1.1 Solidly Grounded Systems

Figure 4.1 from IEEE standard 142 [1] illustrates various methods of system grounding. In a solidly grounded system, there is no intentional impedance between the system neutral and ground. A power system is solidly grounded when the generator, power transformer, or grounding transformer neutral is directly connected to the ground. A solidly grounded system is not a zero impedance circuit due to the sequence impedances of the grounded equipment, like a generator or transformer itself. These systems, in general, meet the requirements of an "effectively grounded" system in which ratio $X_0/X_1$ is positive and less than 3.0, and ratio $R_0/X_0$ is less than 1, where $X_1$, $X_0$, and $R_0$ are the positive-sequence reactance, zero-sequence reactance, and zero-sequence resistance, respectively. The coefficient of grounding (COG) is ratio of $E_{Lg}/E_{LL}$ in percentage, where $E_{Lg}$ is the highest rms voltage on an unfaulted phase, at a selected location, during a fault effecting one or more phases to ground, and $E_{LL}$ is the rms phase-to-phase power frequency voltage obtained at that location with the fault removed. These systems are, generally, characterized by COG of 80%. Approximately, a surge arrester with its rated voltage calculated on the basis of the system voltage multiplied by 0.8 can be applied.

The utility's systems at transmission, subtransmission, and distribution levels are solidly grounded. The main reason for this is that on occurrence of a ground fault, enough ground fault current should be available to selectively trip the faulty circuit.

The utility generators, connected in step up configuration to a generator transformer, are invariably HR grounded. If a generator neutral is left ungrounded, there is a possibility of generating high voltages through inductive–capacitive couplings (see section on ungrounded systems below). Ferroresonance can also occur due to presence of generator PT's.

The utility substations serving large chunks of power at high voltages for industrial plants through delta-wye transformers have low resistance grounded secondary wye windings. The most common voltages of distributions for the industrial plants are 13.8, 4.16, and 2.4 kV.

The low voltage systems in industrial power distribution systems used to be solidly grounded. However, this trend is changing, and HR grounding is being adopted.

The solidly grounded systems have an advantage of providing effective control of overvoltages, which become impressed on or are self-generated in the power system by insulation breakdowns and restriking faults. Yet these give the highest arc fault current and consequent damage and require immediate isolation of the faulty section. Single line-to-ground fault currents can be higher than the three-phase fault currents. These high magnitudes of fault currents have a twofold effect:

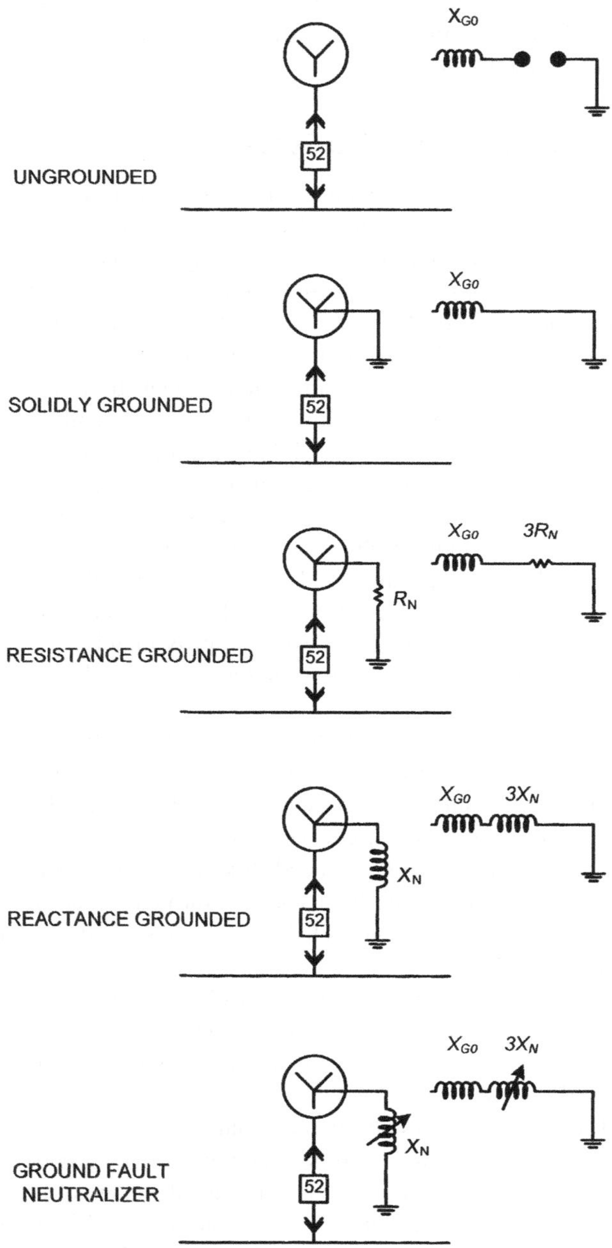


$X_{GO}$ =Zero sequence reactance of generator or transformer
$X_N$= Reactance of grounding reactor
$R_N$= Resistance of grounding resistor

Figure 4.1. Methods of system grounding. *Source*: Reference [1].

- higher burning or equipment damage; and
- interruption of the processes, as the faulty section must be selectively isolated without escalation of the fault to unfaulted sections.

The arc fault damage to the *equipment* for low voltage 480-V systems has been investigated using laboratory models [2]. Stanback reported that for single-phase 277-V arcing fault tests using spacing of 1–4 in from bus bars to ground and for currents from 3000 to 26,000 A, the burning damage can be approximated by the equation:

$$\text{Fault Damage} \propto (I)^{1.5}\,t, \tag{4.1}$$

where $I$ is the arc fault current, and $t$ is the duration in seconds.

$$V_{\mathrm{D}} = K_{\mathrm{s}}(I)^{1.5}t(\mathrm{in})^3, \tag{4.2}$$

where $K_s$ is the burning rate of material in $\mathrm{in}^3/\mathrm{As}^{1.5}$, $V_{\mathrm{D}}$ is acceptable damage to material in $\mathrm{in}^3$, $I$ is the arc fault current, $t$ is the duration of flow of fault current, and $K_{\mathrm{s}}$ depends upon type of material and is given by:

$$\begin{aligned} K_{\mathrm{s}} &= 0.72\times10^{-6} \text{ for copper,} \\ &= 1.52\times10^{-6} \text{ for aluminum,} \\ &= 0.66\times10^{-6} \text{ for steel.} \end{aligned} \tag{4.3}$$

NEMA [3] assumes a practical limit for the ground fault protective devices, so that:

$$(I)^{1.5}t \lhd 250I_{\mathrm{r}}, \tag{4.4}$$

where $I_r$ is the rated current of the conductor, bus, disconnect, or circuit breaker to be protected.

Combining these equations, we can write:

$$V_{\mathrm{D}} = 250K_{\mathrm{s}}I_{\mathrm{r}}. \tag{4.5}$$

As an example, consider a circuit of 4000 A. Then the NEMA practical limit is $1.0 \times 10^6(\mathrm{A})^{1.5}$ seconds and the permissible damage to copper, from Equation (4.5) is 0.72 $\mathrm{in}^3$. To limit the arc fault damage to this value, the maximum fault clearing time can be calculated. Consider that the arc fault current is 20 kA. Then, the maximum fault clearing time including the relay operating time and breaker interrupting time is 0.35 second. It is obvious that vaporizing 0.72 $\mathrm{in}^3$ of copper on a ground fault which is cleared according to established standards is still damaging to the operation of the equipment. A shutdown and repairs will be needed after the fault incidence.

Due to high arc fault damage and interruption of processes, the solidly grounded systems are not in much use in the industrial distribution systems. However, AC circuits of less than 50 V and circuits of 50–1000 V for supplying premises wiring systems

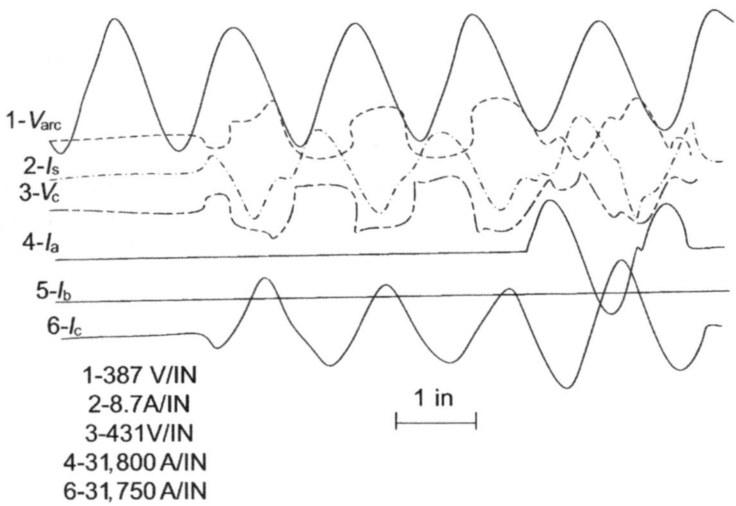


Figure 4.2. Arc fault in a 3/16″ gap, 480-V system.

and single-phase 120/240 Vcontrol circuits must be solidly grounded according to NEC [4].

Figure 4.2 shows a sustained arc fault current in a 3/16 in gap in a 480-V three-phase system [5]. Experimentally, an arc is established between phase *c* of the bus and ground, and a current of 1100-A flows. After three-cycle phase *a* is involved and the arc current for two-line to enclosure fault is 18,000 A, arc energy equals 7790-kW cycles.

Approximately 70% of the faults in the electrical systems are line-to-ground faults. Sometimes, these may be self-clearing and of transient nature (e.g., in OH line systems) or may evolve into three-phase faults over a period of time.

Thus, the probability of a worker being subject to arc flash due to ground faults is much higher. The IEEE 1584 Guide equations take a safe stance and state that if the hazard level is calculated for three-phase faults, it will be lower for the ground faults. In a single phase-to-ground fault, the arc tends to extinguish at natural current zero. In three-phase systems, the phases are displaced 120°, and at a current zero, even if the arc is extinguished in one phase, it will continue in the other two phases, arc columns repelling each other.

Consider 13.8–0.48 kV, delta-wye connected transformers of ratings from 500 kVA to 2500 kVA, percentage impedance 5.75%. For solidly grounded wye-windings, the line-to-ground fault current will vary from 10 kA to 50 kA approximately. For high-resistance grounding, the ground fault current can be limited from 2 to 5 A. The equipment damage as well as arc flash hazard will be practically nonexistent.

Thus, HR systems are recommended, though at medium voltages, these should be carefully evaluated. A discussion follows.

## 4.2 LOW RESISTANCE GROUNDING

An impedance grounded system has a resistance or reactance connected in the neutral circuit to ground (Figure 4.1). In a low resistance grounded system, the resistance in the neutral circuit is so chosen that the ground fault is limited to approximately full load current or even lower, typically 200–400 A. The arc fault damage is reduced, and these systems provide effective control of the overvoltages generated in the system by resonant capacitive-inductive couplings and restriking ground faults. Though the ground fault current is much reduced, it cannot be allowed to be sustained, and selective tripping must be provided to isolate the faulty section. For a ground fault current limited to 400 A, the pickup sensitivity of modern ground fault devices can be even lower than 5 A. Considering an available fault current of 400 A and the relay pickup of 5 A, approximately 98.75% of the transformer or generator windings from the line terminal to neutral are protected. This assumes a linear distribution of voltage across the winding. (Practically the pickup will be higher than the low set point of 5 A). The incidence of ground fault occurrence toward the neutral decreases as the square of the winding turns.

The low resistance grounded systems are adopted at medium voltages, 13.8, 4.16, and 2.4 kV for industrial distribution systems. Also industrial bus connected generators are commonly low resistance grounded. An industry practice has been, generally, to limit the fault current to 400 A, or even lower in some cases. Hybrid grounding systems are a recent trend in industrial bus-connected medium voltage generator grounding (see Section 4.13.2).

## 4.3 HIGH RESISTANCE GROUNDED SYSTEMS

High resistance grounded systems limit the ground fault current to a low value, so that an immediate disconnection on occurrence of a ground fault is not required. It is well documented that to control over voltages in the HR grounded systems, the grounding resistor should be so chosen that:

$$R_{\mathrm{n}} = \frac{V_{\mathrm{ln}}}{3I_{\mathrm{c}}}, \tag{4.6}$$

where $V_{ln}$ is the line to neutral voltage and $I_c$ is the stray capacitance current of each phase conductor. Figure 4.3 depicts transient voltage in percent of normal line-to-ground crest voltage versus the resistor kW/charging capacitive kVA [6]. The transients are a minimum when this ratio is unity. This leads to the requirement of accurately calculating the stray capacitance currents in the system. These calculations for a HR system are not documented here. Cables, motors, transformers, OH lines, surge arresters, and generators have distributed stray capacitances to ground—all contribute to the stray capacitance current. For the purpose of high resistance grounding (HRG), we can consider all these distributed stray capacitances lumped together. Surge capacitors connected line to ground must be considered in the calculations. References [6–8] provide

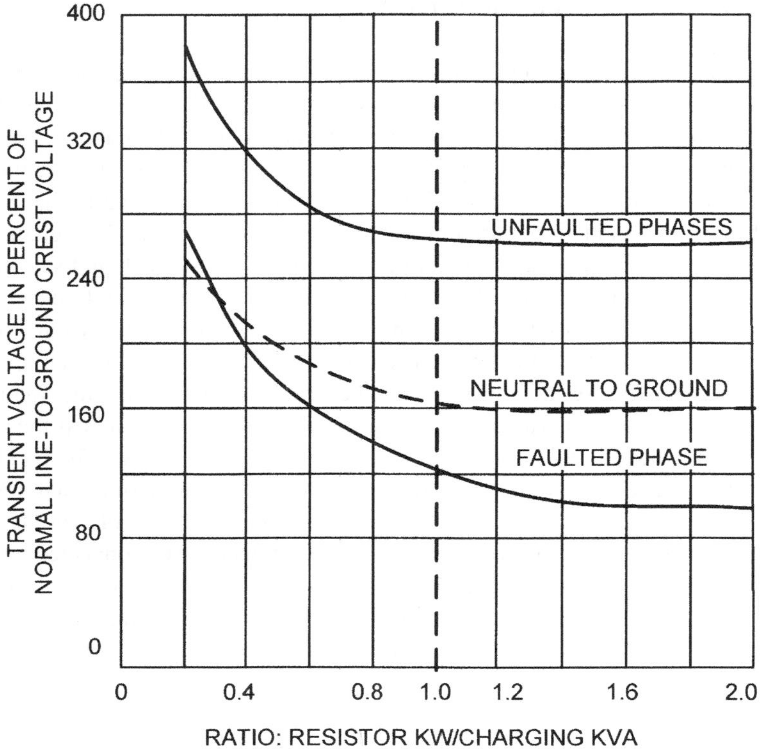


Figure 4.3. Overvoltage versus ratio of resistor kW/charging kVA. *Source*: Reference [6].

charging current (stray capacitance current) data of electrical system components. Once the system stray capacitance is determined, then the charging current per phase, $I_c$ is given by:

$$I_c = \frac{V_{ln}}{X_{co}}, \tag{4.7}$$

where $X_{co}$ is the capacitive reactance of each phase, stray capacitance considered lumped together.

This can be illustrated with an example. A HRG system for a wye-connected neutral of a 13.8 kV-0.48 transformer is shown in Figure 4.4a. This shows that the stray capacitance current per phase of all the distribution system connected to the secondary of the transformer is 0.21 A per phase. In a three-phase system, the three-phases are symmetrical, (though not perfectly) with respect to each other; we can assume that the charging currents of all three phases is equal. Figure 4.4b shows that under no fault condition, the vector sum of three capacitance currents is zero, as these are 90° displaced with respect to each voltage vector and therefore 120° displaced with respect to each other:

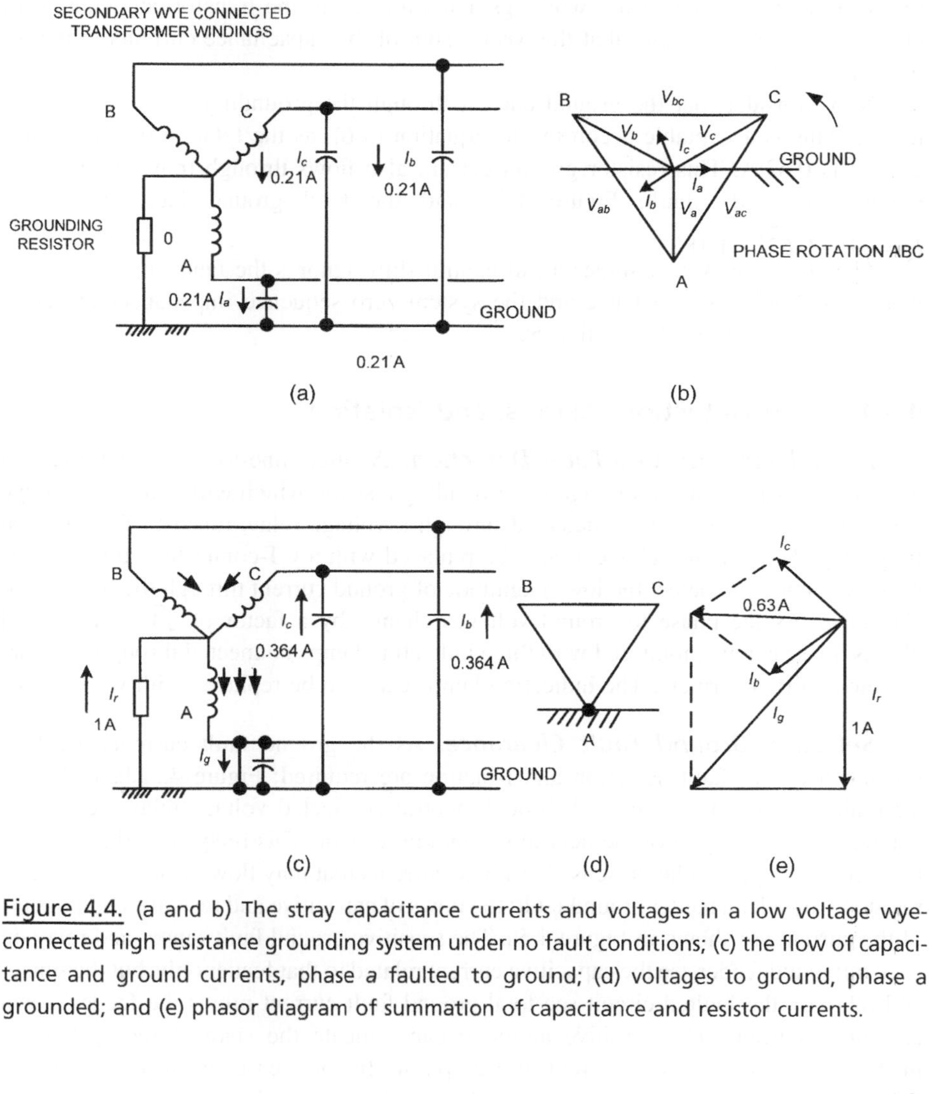


Figure 4.4. (a and b) The stray capacitance currents and voltages in a low voltage wye-connected high resistance grounding system under no fault conditions; (c) the flow of capacitance and ground currents, phase *a* faulted to ground; (d) voltages to ground, phase a grounded; and (e) phasor diagram of summation of capacitance and resistor currents.

$$\vec{I}_a + \vec{I}_b + \vec{I}_c = 0. \tag{4.8}$$

Thus, the grounded neutral does not carry any current, and the neutral of the system is held at the ground potential, *no capacitance current flows into the ground* or in the neutral connected grounding resistor. On occurrence of a ground fault, say in phase *a*, the situation is depicted in Figure 4.4c,d. The capacitance of faulted *a* phase is short-circuited to ground, and this phase does not contribute to any capacitance current. The faulted phase, assuming zero fault resistance is at the ground potential (Figure 4.4d), and the other two phases have line-to-line voltages with respect to ground. Therefore, the capacitance current of the unfaulted phases *b* and *c* increases proportional to the voltage, that is, $\sqrt{3} \times 0.21 = 0.365$ A. Moreover, this current in phase *b* and *c* reverses, flows through the transformer windings and sums up in the transformer winding of phase *a*. Figure 4.4e shows that this vector sum of the capacitance currents in phases *b* and *c* = 0.63 A.

Now consider that the ground current through the grounding resistor is limited to 1 A. This is acceptable according to Equation (4.6), as the total stray capacitance current is 0.63 A. This resistor ground current also flows through transformer phase winding *a* to the fault, Figure 4.4c, and the total ground fault current is: $I_g = \sqrt{1^2 + 0.63^2} = 1.182\ \text{A}$.

The above analysis assumes a full neutral shift, ignores the fault impedance, and that the ground grid resistance and the system zero sequence impedances are zero. Practically, the neutral shift will vary.

### 4.3.1 Fault Detection, Alarms, and Isolation

***Nonselective Ground Fault Detection.*** A single line-to-ground fault results in flow of resistive current through the grounding resistor, which will develop a voltage across it; this voltage can be detected through a voltage relay, tuned to fundamental frequency. The voltage relay can also be replaced with a CT-connected current relay, sensitive enough to detect the low-magnitude of ground current through the resistor for alarm/trip. As the phase to ground voltages change by a factor of $\sqrt{3}$ on unfaulted phases, these can be monitored with three indicating lamps connected through resistors or potential transformers. The indicating lamps can also be replaced with voltmeters.

***Selective Ground Fault Clearance.*** As the ground fault currents are low, special means of fault detection and isolation are required; Figure 4.5a,b. In Figure 4.5a, alarm/trip can be provided through neutral connected voltage relay, device 59, connected across a part of the neutral grounding resistor. This relay should preferably be a rectifier type of relay to sense harmonic currents that may flow through the ground resistor. Consider a feeder ground fault; the flows of capacitive and resistive components of the current are shown in Figure 4.5a. The sensitive ground fault sensor and relay on each unfaulted feeder see the capacitive current related to that feeder only, but the sensor and relay on the faulted circuit see total ground fault current minus the feeder's own capacitive current. The sensitive ammeter can indicate the state of the system by monitoring the capacitance current in each phase. In some cases of unequal loads on the feeders, the charging current on a feeder may be much reduced. Suppose that only two unequally loaded feeders are connected in Figure 4.5a, and a ground fault occurs on the heavily loaded feeder. The capacitance of the faulted phase is short circuited by the ground fault and the charging current flowing through the feeder may be only a fraction of the total charging current.

Ground fault sensors of sensitivity of the order of 10–20 mA are commercially available. These have to be installed carefully and tested for the application. The load cables have to be centrally installed through the window of the sensor occupying not more than 40–50% of the window diameter. For sensitive ground fault settings, effect of harmonic currents that may be returning through neutral ground must be accounted for.

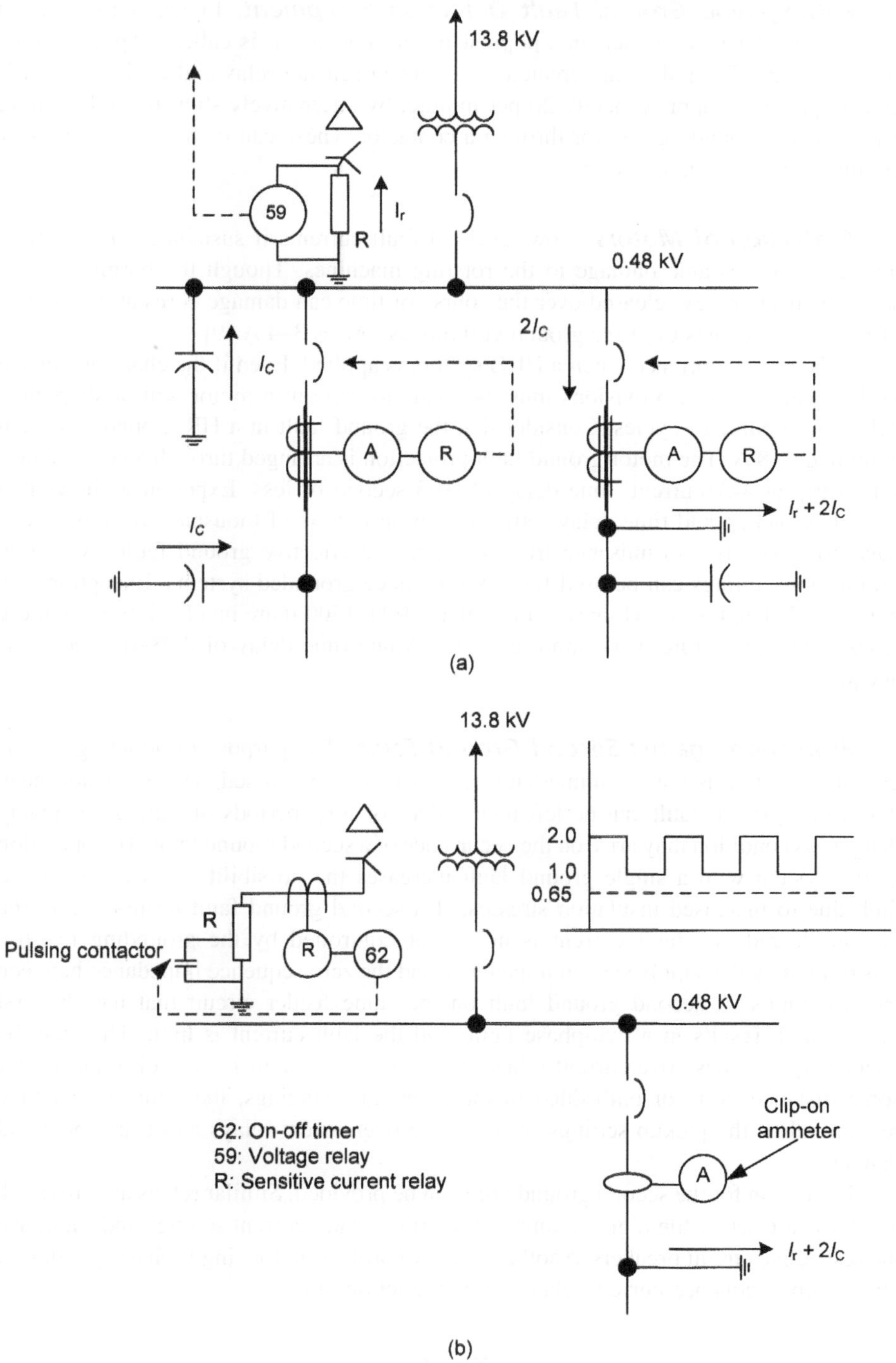


Figure 4.5. (a) Selective ground fault protection in a high resistance grounded system; (b) principal of pulsing type ground fault detection system in a high resistance grounded system.

***Pulsing-Type Ground Fault Detection Equipment.*** Figure 4.5b shows a ground fault localization scheme, popular in the industry. It is called a "pulsing-type HRG system." The pulses are created by a current sensing relay and cyclic timer, 62, at a frequency of approximately 20 per minute, by alternatively shorting and opening a part of the grounding resistor through a contactor. These can be traced to the faulty circuit with a clip on ammeter.

***Protection of Motors.*** Low levels of fault currents if sustained for long time may cause irreparable damage to the rotating machines. Though the burning rate is slow, the heat energy released over the course of time can damage cores and windings of rotating machines even for ground currents as low as 3–4 A [9].

This is of importance when a HRG system is applied. Even if the charging current in the system is low, provisions must be made to trip out a motor with a short time delay of a couple of cycles. Consider that the ground fault in a HR grounded system is limited to 8 A. The motor ground fault protection is arranged through a core balance CT to trip at 4-A current, time delay of 0.05 second or less. Experience shows that these low pickup and time delay settings even on motors of thousands of horsepower have not given rise to nuisance trips, and provide effective ground fault protection. Similar low settings can be used for low-resistance grounded systems. Exceptions are where cable lengths or OH lines of more than 1000–1500 ft are involved. Even in these cases, a pickup setting of no more than 7–8 A and time delay of 0.08–0.1 second is adequate.

***Protection against Second Ground Fault.*** The purpose of adopting an HR grounded system is that an immediate shutdown can be avoided. This does not mean that a first ground fault can be left unattended for long periods of time. A relatively dangerous condition may arise on the occurrence of a second ground fault. The operation of the system with a single ground fault increases the possibility of second ground fault due to increased insulation stresses. If a second ground fault occurs before the first is cleared, the fault current is no longer controlled by the grounding resistor. It is limited by the supply system impedance and the zero sequence impedance between the two faults. A second ground fault on the same feeder circuit that has the first ground fault results in a two-phase fault, and the fault current is high. This may be cleared by the phase overcurrent relays. However, if the fault is on a different feeder some distance away or embedded in the apparatus windings, the fault current may remain below the pickup settings of the phase overcurrent relays and cause potential damage.

Protection for the second ground fault can be provided. Similar relays are provided, and these are set a little higher than the first ground fault current settings and shunt trip the respective circuit breakers. Another effective protection for single-phasing is device 46, negative sequence current relays. In an induction motor:

$$\frac{Z_1}{Z_2} = \frac{I_s}{I_f}, \tag{4.9}$$

where $Z_1$ and $Z_2$ are the positive and negative sequence impedances, $I_s$ is the motor starting current or the locked rotor current, and $I_f$ is the full load current. For an induction motor with typical starting current of six times the full load current, the negative sequence impedance is 1/6th of the positive sequence impedance. A 5% negative sequence component in the voltage will give rise to 30% negative sequence currents, and practically even higher.

***Insulation Stresses and Cable Selection for HR Grounded Systems.*** As stated above, a ground fault gives rise to voltage stresses equal to $\sqrt{3}$ times the normal. The following review is of interest for cable systems.

Cables: ICEA publication S-61-401 and NEMA WCS5 [10] specify that:

- *100% Insulation Level*: Ground fault cleared as early as possible, but in any case within 1 minute.
- *133% Insulation Level*: This corresponds to that formerly designated for underground systems. Ground fault cleared within 1 hour.
- *173% Insulation Level*: Time required to deenergize is indefinite.

Thus, a 173% level of conductor insulation is required when operation with a single phase-to-ground exceeds 1 hour. But the insulation thicknesses specified in NEC for 600-V grade cables, table 310.13(A) of NEC for various insulation types, do not specify insulation levels.

Some manufacturers are of the opinion that the intrinsic strength of thin insulation section of insulation used for cable insulations is 500 V/mil. Consequently, thickness of insulations specified in NEC for 600-V grade cables is more than adequate to withstand any voltage encountered even in 600-V HR grounded systems. Thus, the mechanical considerations overweigh the electrical concerns. Other cable manufacturer's take a conservative approach and recommend 1000- or 2000-V grade cables. Considerations should also be given to the higher system operating voltages. NEMA standard WC5 specifies that operating voltages on cables should not exceed the rated voltage by more than 5% during continuous operation or 10% during emergencies lasting for not more than 15 minutes. This is of importance for three-phase 600-V systems. Further, the DC loads served through six-pulse rectifier systems, connected to 480-V or 600-V three-phase systems, will have a no-load voltage of 648 and 810 V, respectively. For 2.4-kV HR grounded systems, 5 kV grade cables, 100% insulation level normally applied, will be acceptable.

For all other electrical equipment like motors, controllers, switchgear, and transformers, NEMA and ANSI standards do not specify specific tests for grounded or ungrounded systems, and the dielectric and impulse test voltages should be acceptable for these systems. The higher voltages to ground may have minimal effect on the life of an equipment, which is difficult to predict.

Some observations pertaining to HR grounded systems are:

- The resistance limits the ground fault current, and, therefore, reduces burning and arcing effects in switchgear, transformers, cables, and rotating equipment.

- It reduces mechanical stresses in circuits and apparatus carrying fault current.
- It reduces arc blast or flash hazard to personnel who happen to be in close proximity of ground fault.
- It reduces line-to-line voltage dips due to ground fault, and three phase loads can be served, even with a single line-to-ground fault on the systems.
- Control of transient overvoltages is secured by proper selection of the resistor.

The limitation of the system is that the capacitance current should not exceed approximately 10 A to prevent immediate shutdowns. As the system voltage increases, so does the capacitance currents. This limits the applications to systems to rated voltages of 4.16 kV and below.

Though immediate shutdown is prevented, the fault situation should not be prolonged; the fault should be localized and removed.

The detection and alarming for the first line-to-ground fault can be easily implemented in HR grounded systems; however, it is difficult to provide selective ground fault clearance. Recently, many commercial HRG selective protection systems, especially for the low voltage systems, are available.

## 4.4 UNGROUNDED SYSTEMS

In an ungrounded system, there is no intentional connection to ground except through potential transformers or metering devices of high impedance. In reality, an ungrounded system is coupled to ground through distributed phase capacitances. It is difficult to assign $X_0/X_1$ and $R_0/X_0$ values for ungrounded systems. The ratio $X_0/X_1$ is negative and may vary from low to high values, and COG may approach 120%. These systems provide no effective control of transient and steady state voltages above ground. A possibility of resonance with high voltage generation, approaching five times or more of the system voltage, exists for values of $X_0/X_1$ between 0 and -40. For the first phase-to-ground fault, the continuity of operations can be sustained, though unfaulted phases have $\sqrt{3}$ times the normal line-to-ground voltage. All unremoved faults thus put greater than normal voltage on system insulation, and increased level of conductor and motor insulation may be required. The phenomena of intermittent grounds can occur, which can give rise to overvoltages equal to five to six times the normal rated voltage. This is not discussed in this book; see References [8, 11] for details. Under such high voltage stresses, the insulation can flashover, which can result in catastrophic failures. Even if this voltage escalation does not occur, systems above 8 kV can fail quickly after the first ground fault occurs, due to ionized gases created by the arcing at the ground fault. Systems with capacitive charging currents greater than 10 A can also fail due to prolonged thermal effects. Ungrounded medium voltage systems are extinct, though some old low voltage ungrounded systems are still in operation. Even in these systems, for the safety of the worker and equipment, it is important that the first ground fault is detected, alarmed, and cleared without delay to prevent a second ground fault on the system. The grounding practices in the industry are withdrawing from this method of grounding.

## 4.5 REACTANCE GROUNDING

In reactance grounding, a reactor is connected between the system neutral and ground; the magnitude of the ground fault current that will flow depends upon the size of the reactor. The ground fault current should be at least 25% and preferably 60% of the three-phase fault current to prevent serious transient overvoltages ($X_0$ < or equal to $10X_1$). This current is considerably higher than in a resistance grounded system, and the reactance grounding is not an alternative to resistance grounding. The system is generally used for the grounding of small generators, so that the generator ground fault current does not exceed three-phase fault current, and three-phase four-wire loads could be served. Reactance grounded systems are not common in the United States.

## 4.6 RESONANT GROUNDING

A reactor can be connected between the neutral of a system and ground and tuned to the system charging current so that the resulting current is resistive and of low magnitude. This current is in phase with the line-to-neutral voltage, so that the current and voltage zeros occur simultaneously. The system is used for voltages above 15 kV, consisting of overhead transmission or distribution lines. The system is rarely used for industrial and commercial establishments. A disadvantage of the system is that the resonant tuning can change due to switching conditions, that is, when a part of the system may be out of service or when the system expansion takes place. This grounding method is not common in the United States, though sometimes used in Europe and Russia.

## 4.7 CORNER OF DELTA-GROUNDED SYSTEMS

Low voltage systems that in the past were supplied from delta-connected secondary were inherently ungrounded. These were grounded by connecting one corner of the secondary delta windings to ground through an adequate connection, and there was no additional impedance inserted in this connection to obtain a grounded system (see Figure 4.6a). These are no longer in use. It is necessary to positively identify the grounded phase throughout the system to avoid connecting meters, instruments, and fuses in the grounded phase. A higher line-to-neutral voltage will occur on two phases than in a neutral grounded system. There is a possibility of exceeding the interrupting ratings of marginally applied circuit breakers, because for a ground fault, the interrupting duty on the affected circuit breaker pole may exceed the three-phase fault duty. This is discussed in Section 7.6.1. There is a high probability of sustaining arcing for 480-V or higher phase-to-phase, single-phase circuit extension without escalating to a three-phase fault.

The NEC Table 430-37 requires that for a three-phase system, three overloads are used, one in each phase for motor protection. The control circuit for motor starting has to be carefully designed, so that a ground fault in it does not automatically start or stop a motor.

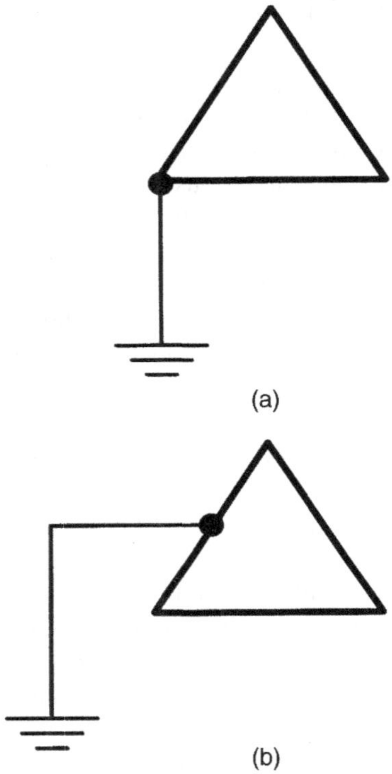


Figure 4.6. (a) Corner grounded delta grounded system; (b) mid-point grounded delta grounded system.

Some advantages of the system are low cost of establishing a grounded system, the effective control of transient overvoltages though a maximum of 1.73 times the normal phase-to-ground voltage can exist between two conductors and ground. A fault from phase-to-ground can be easily detected.

One phase of a delta system can be grounded at the midpoint, Figure 4.6b. The system is not recommended for voltages above 240 V. Serious arc flash hazard from a phase-to-ground fault can exist because of high fault levels [1]. The grounding of one phase of a delta system at midpoint of that phase for three-phase systems with phase-to-phase voltages over 240 V has little applications. The shock hazard of the high phase leg to ground is 1.73 times the voltage from the other two phases. These systems are no longer in use.

## 4.8 SURGE ARRESTERS

While selecting the ratings of surge arresters, system grounding should be carefully considered. A detailed discussion is provided in Reference [11]. A surge arrester should not flash over during continuous system voltages or temporary overvoltage. This requires the selection of MCOV (maximum continuous overvoltage) and TOV (temporary overvoltage) ratings. Proper applications require rigorous system studies, but it is obvious that for low-resistance or high-resistance grounded systems, the MCOV should be greater than line-to-line voltage with a safe margin, as on unfaulted phases, voltages rise equal to line-to-line voltage or more. A factor of interest is COG, previously defined, and it can be calculated with system sequence impedances.

## 4.9 ARTIFICIALLY DERIVED NEUTRALS

Many times, it is required to ground delta-connected transformer windings and other ungrounded separately derived systems. A neutral can be artificially derived in a delta-connected system with zigzag transformer or a wye-delta-connected grounding transformer. Figure 4.7b shows a zigzag transformer. Windings $a_1$ and $a_2$ are on the same limb and have the same number of turns, but are wound in opposite directions. The zero sequence currents in the two windings on the same limb have, therefore, canceling ampere turn effect:

$$i_{a1} = i_{b1} = i_{c1} = i_{a2} = i_{b2} = i_{c2}. \tag{4.10}$$

The impedance to the zero sequence currents is due to leakage flux of the windings. For the positive and negative sequence currents, neglecting magnetizing currents, the connection has infinite impedance. Figure 4.7a shows distribution of zero sequence currents on a ground fault ahead of a zigzag transformer. A wye-delta grounding transformer circuit can be similarly drawn (Figure 4.7c). This may consist of three-single phase transformers or a single three-phase transformer. The ground current in the faulted phase $a$ divides equally between the wye windings and returns to the faulted phase as shown, much similar to zigzag transformer. Again, during normal operation, with no ground fault, the impedance to three-phase currents is high, and only a small transformer magnetizing current flows.

The neutrals of large utility generators directly connected to the step-up transformers (GSUs) are not directly connected to ground through a high resistance but through a distribution transformer that is loaded on the secondary windings with a grounding resistor.

Figure 4.8 shows the grounding of neutrals, with available or artificially derived neutrals for low resistance or HR grounded systems. In Figure 4.8a, a neutral resistor is directly connected to ground; this method is used for the grounding of medium voltage or low voltage transformer neutrals. Figure 4.8b depicts grounding through a distribution transformer, with the secondary of the transformer loaded with a resistor of adequate size to limit the ground fault current to the desired level. Figure 4.8c illustrates grounding through a wye-delta transformer, artificially derived neutral, and, finally, Figure 4.8d shows grounding through a zigzag transformer.

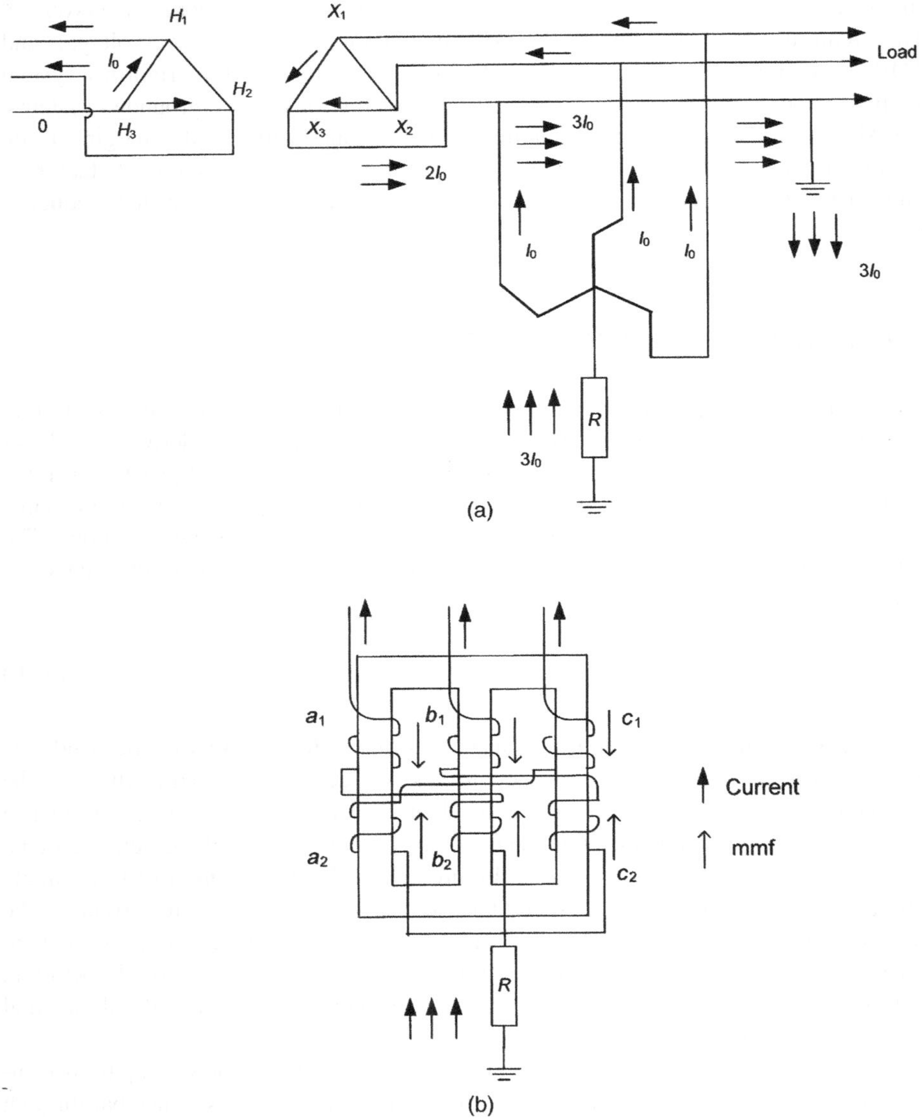


Figure 4.7. (a) Zigzag grounding transformer showing winding connections, currents, and flux relations. (b) Flow of ground currents for an external single phase-to-ground fault with zigzag grounding transformer artificially derived neutral in a delta connected secondary system. (c) Flow of ground currents for an external single phase-to-ground fault with delta-wye grounding transformer.

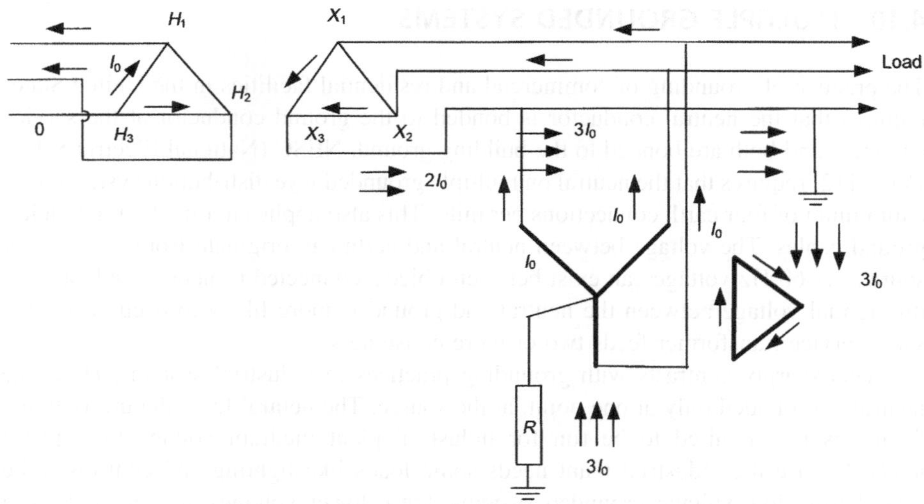


(c)

Figure 4.7. (*Continued*)

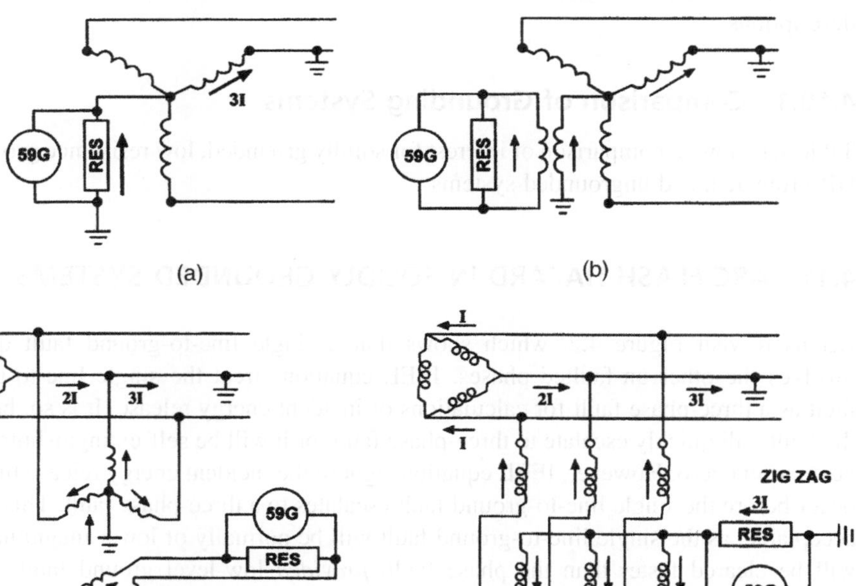


Figure 4.8. Methods of low resistance or high resistance grounding of neutrals. (a) Neutral connected directly to ground through a grounding resistor; (b) neutral connected through a distribution transformer, secondary loaded with appropriate rating of a grounding resistor; (c) artificially derived neutral with overvoltage protection through a wye–delta grounding transformer. (d) Artificially derived neutral with overvoltage protection through a zigzag grounding transformer.

## 4.10 MULTIPLE GROUNDED SYSTEMS

The practice of grounding of commercial and residential facilities in the United States requires that the neutral conductor is bonded to the ground conductor at the service entrance, and both are bonded to the building ground. NESC (National Electric Safety Code) [12] requires that the neutral on multiple grounded wye distribution systems have a minimum of four earth connections per mile. This also applies to direct buried underground cables. The voltage between neutral and earth can originate from a variety of sources. A 60-Hz voltage can exist between objects connected to neutral and earth. A differential voltage between the neutral and ground is more likely to occur when the same service transformer feeds two or more consumers.

This sharply contrasts with grounding practices in industrial systems. Here, the neutral is grounded only at one point, at the source. The neutral from the utility transformer is not required to be run for industrial plant medium voltage three-phase loads. In case the industrial plant needs some loads like lighting and controls to be served from low voltage grounded systems, these lower voltages are served from a separate transformer with artificially derived neutral. In case the service is at low voltage, a neutral may be run to supply phase-to-neutral loads, *but it is* not grounded anywhere in the plant except at the service transformer. NEC may be seen for further descriptions.

### 4.10.1 Comparison of Grounding Systems

Table 4.1 shows a comparison of interest for solidly grounded, low resistance grounded, HR grounded, and ungrounded systems.

## 4.11 ARC FLASH HAZARD IN SOLIDLY GROUNDED SYSTEMS

Let us re-visit Figure 4.2, which shows that a single line-to-ground fault quickly involves the other un-faulted phases. IEEE equations treat the single line-to-ground fault as a three-phase fault for calculations of incident energy release. It is so, because the fault will quickly escalate to three-phase fault, or it will be self-extinguishing at the next current zero. However, IEEE equations ignore the incident energy release that will occur before the single line-to-ground fault escalates to a three-phase fault. This seems acceptable, as the single line-to-ground fault will be normally of lower magnitude and will be cleared faster than the phase fault, *provided* low-level ground fault current protection is provided. Thus, the results of calculations will be conservative. Section 5.1 of the IEEE 1584 Guide says that the single phase faults, that is, the faults to ground or between phases are not tested. Reference [13] states that the escalation time of a single-phase fault to three-phase faults is rather small, of the order of one to two cycles. It also states that where this escalation of the line-to-ground fault is not possible or likely, the IEEE three-phase model will give conservative results.

Further testing results (see Chapter 3) confirm that single-phase faults are more difficult to sustain than three-phase faults. Dunki-Jacobs illustrated the arcing line-to

TABLE 4.1. Comparison of Grounding Systems

| System Property | Solidly Grounded Systems | Low-resistance Grounded | High Resistance Grounded Systems | Ungrounded Systems |
|---|---|---|---|---|
| Applicability | Has been popular for low voltage system in industrial applications. Some single-phase systems according to NEC must be solidly grounded. Utility EHV, HV, transmission, subtransmission, and primary and secondary distribution systems are solidly grounded, except at large industrial consumer substations. | Industrial systems at primary distribution voltages of 22, 13.8, 4.16, and 2.4 are low resistance grounded. Bus connected generators in industrial distributions are low resistance grounded; modern trend is hybrid grounding (see text). | Industrial low voltage systems are currently HR grounded. Utility unit connected generators are HR grounded through a distribution transformer. HR grounding can be applied with care to 2.4 and 4.16 kV systems also. The stray capacitance current should be no more than 10–15 A. | Not in current use, though some 4.16 kV, 2.4 kV and low voltage industrial distribution systems earlier designed are still in operation. These are gradually being replaced with HR (low voltages) and low resistance grounded (medium voltage) systems |
| Serve three-phase four wire loads | Yes | No | No | No |
| Immediate shutdown of facility on occurrence of first ground fault | Yes | Yes; though the fault current is reduced, it cannot be allowed to continue and immediate shutdown is required. | No, though a prolonged operation with a single ground fault has the potential of escalating into a second ground fault. | No, though a prolonged operation with a single ground fault has the potential of escalating into a second ground fault. |

(*Continued*)

TABLE 4.1. (*Continued*)

| System Property | Solidly Grounded Systems | Low-resistance Grounded | High Resistance Grounded Systems | Ungrounded Systems |
|---|---|---|---|---|
| Control of transient overvoltage | Yes | Yes | Yes | No; voltages of the order of 4–5 times the normal can occur due to arcing ground faults, imposing high insulation stresses and potential damage |
| Control of impressed or self-generated overvoltages | Yes | Yes | Controls series inductive–capacitive resonance voltages. Restrikes can still occur, though rare. | No |

| | | | | |
|---|---|---|---|---|
| Arc flash hazard to personnel assuming no escalation of ground fault | Severe; thousand of amperes can flow during a line-to-ground fault | Low, as the maximum ground fault current in low resistance grounded systems is generally 200–400 A, rarely exceeding 1000 A | Minimal; provided escalation of ground fault does not occur. Limited energy is released to arc. Little flash and spark are produced, or a slight "tack" welding action is produced. Double line to ground fault is a peril, and must be quickly removed. | Same as HR systems |
| Arcing fault damage to equipment, no escalation | Severe; the fault must be removed quickly | Much lower than solidly grounded systems, fault is removed quickly. | Minor, but prolonged fault clearance can escalate damage especially to motor windings and cores. | Minor, but transient overvoltages may cause fault escalation or multiple insulation failures. |
| Shock Hazard, unfaulted phases to ground | Line-to-neutral voltage | Approximately line-to-line voltage | Approximately line-to-line voltage | May escalate to four to five times the line-to-line voltage |
| Shock hazard, equipment frame to ground, during internal line-to-ground fault | Moderate; ground potential rise, step and touch voltages are a function of ground current | Much lower than solidly grounded systems | Minimum | Small |
| Detection of arcing faults | Readily detected | Readily detected | Special protective systems, as the currents are low. L-L faults may be high enough for phase relays to operate. | Ground detectors and special relaying required. Transient overvoltages may meanwhile cause additional insulation breakdown. L-L faults may be detected by phase overcurrent relays. |
| Selective ground fault protection | Moderately difficult and expensive to provide, see examples | Relatively easier as compared to solidly grounded systems | Very difficult | Very difficult, may not be possible at all. |

ground fault as a discontinuous sine wave. Three-phase arcing faults, as opposed to single phase faults, produce a constant source of plasma, and HRG is one means to limit the energy of the original single line-to-ground fault as stated above. It seems that safest type of 480-V grounded systems will be the ones that are selectively coordinated to isolate the low-level fault as soon as these occur. However, there are limitations in designing selectively coordinated HRG systems and solidly grounded systems as illustrated in the examples to follow.

Reference [14] contends that the arcs for ground fault can be sustained even at low values of ground fault current of the order of 800 A, which may not be any more than the load current, and the arcing will continue unless this low-level ground fault current is cleared by the protective relaying. This is based on the work of Land [15]: arcing ground faults on a low voltage system with 3.81-cm phase-to-phase spacing and 1.9 cm phase-to-ground spacing.

A significant finding of Land is that the arc started as a line-to-ground fault, does not escalate into a phase-to-phase arcing fault, but as a phase-to-ground-to-phase fault. The magnetic fields surrounding each bus constraint the arc to jump phase-to-phase, and this results in more slowly evolving phase-to-phase fault, and for a greater period of time, it remain as phase-to-ground fault. A phase-to-ground fault may be initiated with a conductor of larger cross section, that is, tools or hardware inadvertently left inside the enclosures.

Ground faults are more likely to occur than phase faults. IEEE 493-2007 [16] states that the ground faults are:

- 2.3 times more than phase faults in bus ducts
- 7.3 times more than phase faults in cables
- 7.8 times more than phase faults in cable joints.

The insulation of the bus system impacts the arc transfer to the other phases. With insulated buses, the arcing fault can be expected to remain stationary at the point of fault, and it can easily burn down 480/277-V systems. See also discussions in Chapter 3.

Thus, it becomes imperative that low-level ground faults are cleared quickly to prevent conductor and equipment damage. As the current reduces, the time to clear a fault on inverse characteristics relays will increase. The problem will be to precisely calculate the ground fault current as the return path involves a number of factors. Reference [14] lists these as:

- metallurgy of enclosure
- size and type of ground conductor
- corrosion on pressure contacts
- rectification from dissimilar materials
- mutual inductance
- skin effect

- phase-to-neutral voltage
- earth conductivity.

## 4.12 PROTECTION AND COORDINATION IN SOLIDLY GROUNDED SYSTEMS

NEC Article 230.95 requires that for solidly grounded wye electrical *services* 150–600 V, the service disconnect rated 1000 A or more, ground fault protection should be provided. The maximum permissible settings are 1200-A pickup and at 3000-A ground fault current, the maximum trip time is 1 second. For separately derived systems through a transformer in an industrial establishment, these provisions can be applied. These NEC recommendations are based on a number of incidents of ground fault damage where the ground fault protection was not provided.

In solidly grounded systems, as the ground fault currents can be even greater than three-phase fault currents, a selective ground fault protection cannot be achieved unless dedicated ground fault protection is included on all the circuits and subcircuits served from a low voltage transformer substation.

Delta connected windings of three-phase transformers act as a sink to the zero sequence currents, and these currents circulate in the delta winding, but do not appear in the line conductors feeding the transformer delta connected windings (see Chapter 11). The protection philosophy of solidly grounded systems is illustrated in the following examples.

**Example 4.1**

A simple low voltage system derived from a 2500-kVA, 13.8–0.48-kV delta-connected primary windings and the wye-connected secondary windings, solidly grounded, is shown in Figure 4.9. The secondary of the transformer is connected to a low voltage switchgear assembly, which in turn serves a number of low voltage motor control centers (only one shown). The main 4000-A low voltage power circuit breaker (LVPCB) is provided with electronic trip programmer, having LSIG functions (long-time pickup, long time delay, short-time pickup, short time delay, and instantaneous pickup settings for phase faults; also adjustable pickup and short-time delay settings for ground faults, see Chapter 7). Both phase and ground fault short-time functions are provided with $I^2t$ set-up that can be switched in or switched out. Similarly, the 800-A feeder LVPCBs to MCCs are provided with electronic trip programmers with LSIG functions. A three-phase three wire distribution is shown. The transformer wye connected winding neutral is grounded at one point only, close to the transformer. This means that any ground fault current must return through the single neutral connection. For solidly grounded neutrals, it is necessary to provide a neutral connected ground relay (see Chapter 11 for further discussions).

Ground fault settings and coordination using *minimum* pickup settings on the electronic trip programmers of the low voltage circuit breakers is shown in Figure 4.10. The low voltage circuit breaker trip programmers have a minimum pickup ground fault setting of 0.25 times the sensor rating, and in this case, these settings are 200 A for the

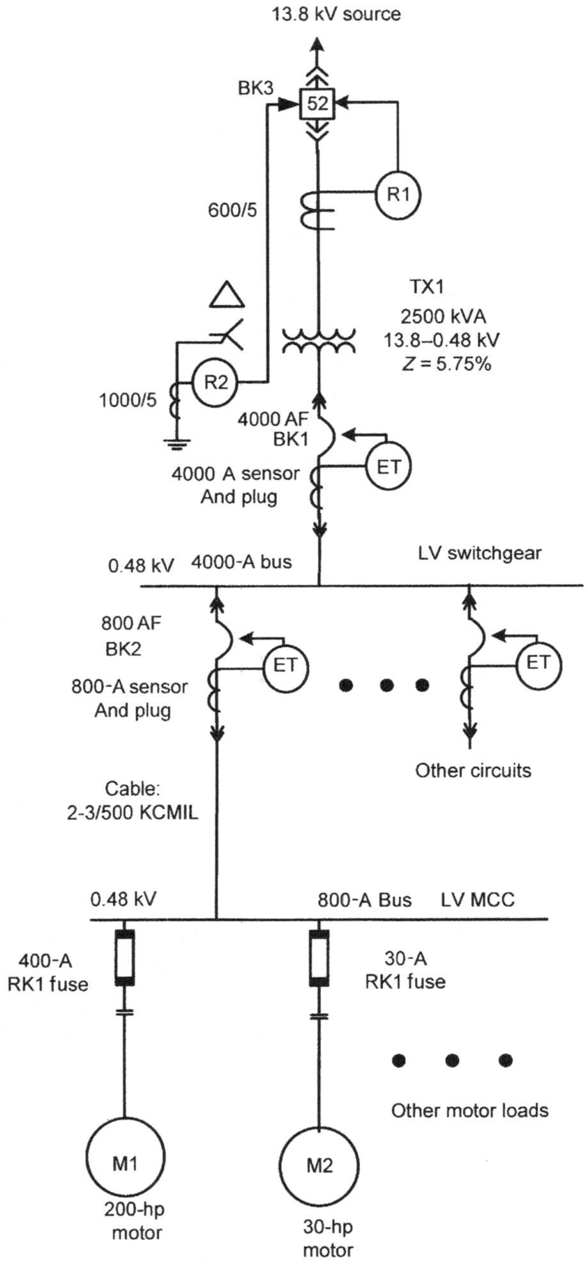


Figure 4.9. A solidly grounded distribution system for selective ground fault coordination.

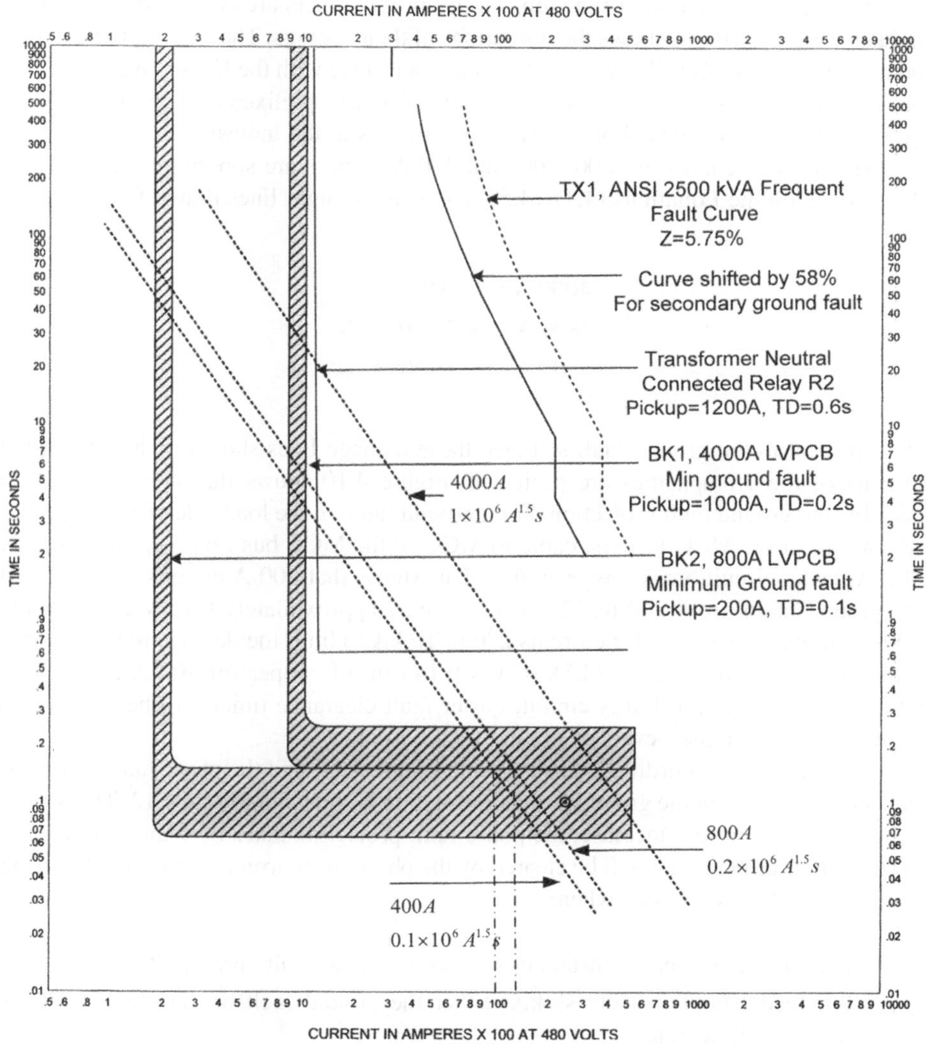


Figure 4.10. A three-step ground fault coordination of the distribution system in Figure 4.9, with diagonal tolerable damage lines.

800-A feeder breakers (BK2) and 1000 A for the main 4000-A breaker (BK1). Also, the neutral connected relay R2 is set up to coordinate with these settings, through a definite time characteristics, with a pickup setting of 1200 A and time delay setting of 0.6 second. The settings on R2 meet NEC requirements. Relay R2 must trip an upstream breaker as a ground fault maybe in the transformer windings, or the cable connection between the transformer and main secondary breaker at the low voltage switchgear. This fault cannot be cleared by tripping the secondary breaker.

Though the ground fault functions on trip programmers are commonly used in the industry, Figure 4.10 illustrates the limitations of their settings. The 4000-A main circuit breaker pickup could easily be set at 400 A to coordinate with the feeder circuit breaker, but it is not possible to do so as the minimum pickup is prefixed at 0.25 of the sensor rating, which is very typical of the trip programmers in the industry.

The damage curves of 4000, 800, and 400 A circuits are superimposed in Figure 4.10, based on the Equations (4.1)–(4.5). These are straight lines drawn for:

$$
\begin{aligned}
4000\ \text{A}:&\quad 1\times 10^{6}\,\text{A}^{1.5}\text{s}\\
800\ \text{A}:&\quad 0.2\times 10^{6}\,\text{A}^{1.5}\text{s}\\
400\ \text{A}:&\quad 0.1\times 10^{6}\,\text{A}^{1.5}\text{s}.
\end{aligned}
$$

With respect to the ground fault settings, these damage lines show to what extent the circuits of these ampacities are protected. Figure 4.10 shows that it is a three-step coordinated ground fault protection. For a ground fault on the load side of 800-A feeder (anywhere in the MCC feeders, cable to MCC or the MCC bus), pick up of 200 A on 800-A feeder circuit breaker is operative. This shows that 800-A circuits are protected for ground faults from 200 to 12 kA (time delay approximately 0.15 second), while 400-A circuit is protected for currents 200 to 10 kA to limit the damage to permissible limits according to Equation (4.5), that is 0.144 $\text{in}^3$ of copper for 800-A circuit, and 0.072 $\text{in}^3$ of copper for 400-A circuit. Faster fault clearance times will be required to protect these for higher levels of ground fault current.

The phase fault coordination is shown in Figure 4.11, and the phase fault coordination superimposed on the ground fault coordination is shown in Figure 4.12. This shows that as the fault current increase, the phase fault protection takes over the ground fault protection. A ground fault will be cleared by the phase overcurrent protection. Referring to Figure 4.12, this happens when:

- Feeder circuit breaker instantaneous takes over at fault current >9 kA.
- For feeder faults >4 and <9 kA, either the ground fault or the phase fault ST pickup will operate.
- For a bus fault on the low voltage switchgear, the main circuit breaker overcurrent operates at >22 kA.

### 4.12.1 Self-Extinguishing Ground Faults

Conservatively, faults below 38% of the bolted phase-to-ground fault will be self-extinguishing for a 480-V three-phase system, with 277 V to ground [2]. This means that ground fault protection is not necessary where bolted fault current at every point on a branch or feeder circuit is at least 263% of the trip setting protecting the circuit (Figure 4.12).

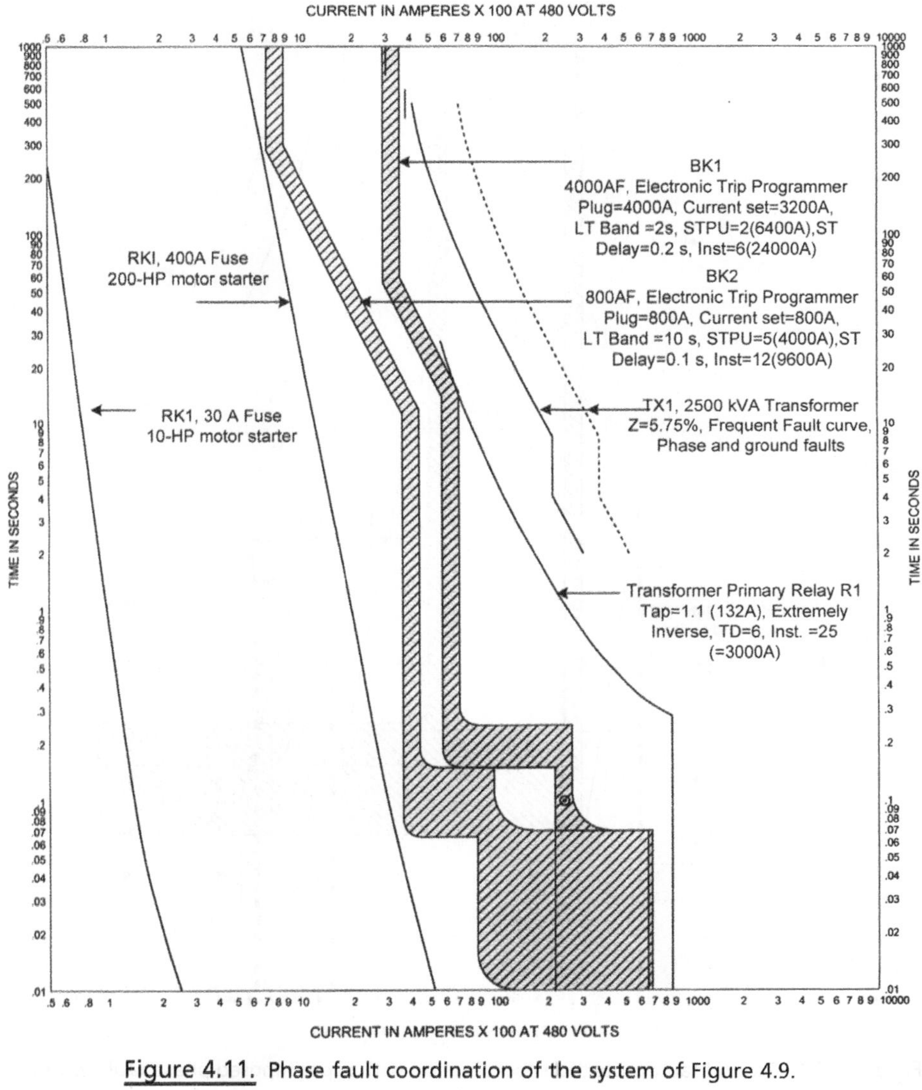


Figure 4.11. Phase fault coordination of the system of Figure 4.9.

This does not imply that no arc flash damage will occur, though it will be minimal. Figure 4.12 shows that a 200-hp motor starter protected with 400-A fuse has a self-clearing zone below 2.1 kA. See also Section 3.7.

As a safe stance, it is desirable that selective and sensitive ground fault protection is provided for the ground fault currents, from low level to the highest calculated level. If a fuse in one phase operates due to a single line-to-ground fault, this results in single phasing, and a fully loaded induction motor may stall. Both stator windings and rotor will be seriously overloaded due to negative sequence currents, which produce

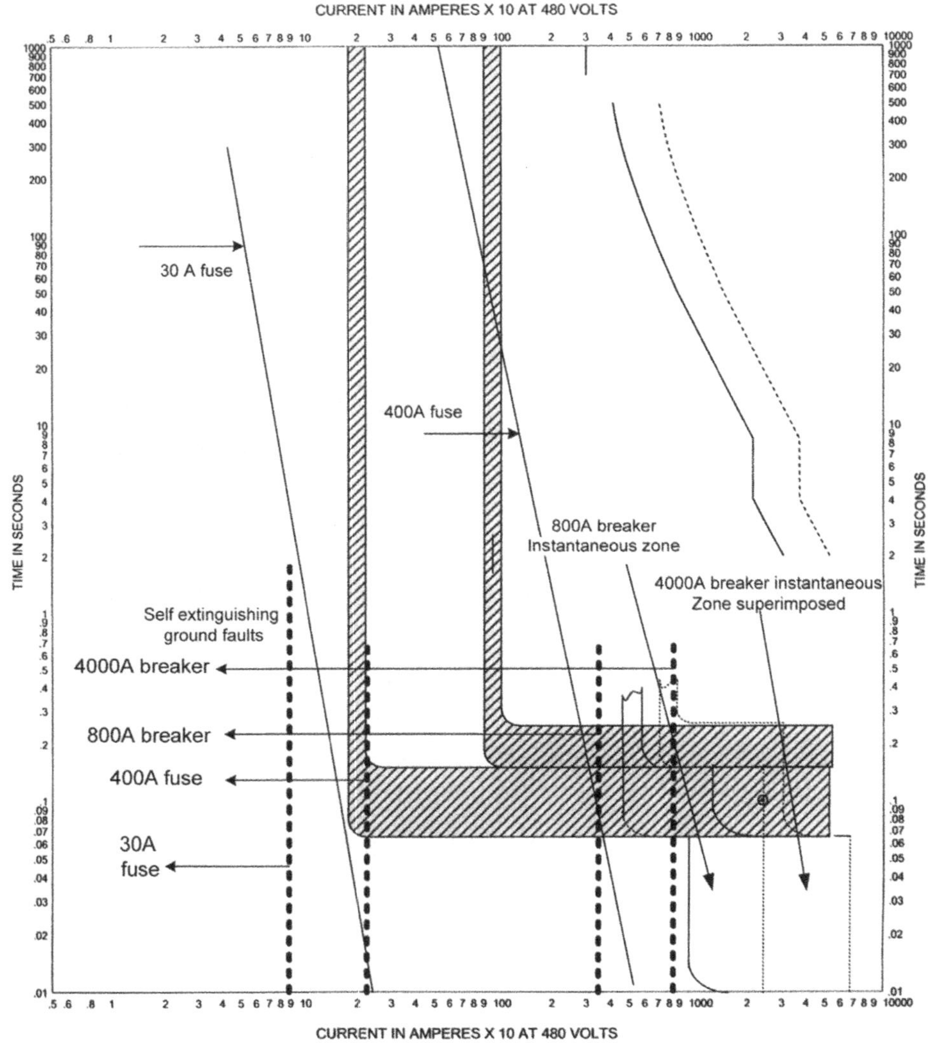


Figure 4.12. Ground fault protection coordination with superimposed phase fault coordination and self-clearing limits of ground faults, distribution system of Figure 4.9.

damaging overloads. The negative sequence impedance of an induction motor, with a locked rotor current of six times the full load current, is approximately 1/6 of the positive sequence impedance, and, therefore, even a 5% negative sequence component can produce 30% negative sequence current in the motor. Thermal relays generally used for low voltage motor protection will be insensitive to operate fast and the motor may be damaged. Also, an uncleared ground fault, even of low value, can be damaging to the motor windings, and core damage can occur.

Though the advantage of current limiting fuses is being extolled for arc flash reduction, this fact of single phasing should be borne in mind for grounded systems. The single phasing can be detected with negative sequence current relays (see Chapter 11).

### 4.12.2 Improving Coordination in Solidly Grounded Low Voltage Systems

Example 4.1 illustrates that:

1. A 200-A ground fault in the motor windings will not be cleared by its 400 A fuse in the 200-hp motor starter, and it can be very damaging to the motor insulation.
2. For a ground fault in a feeder or motor circuit in the MCC, the main feeder breaker BK2 at low voltage switchgear trips, resulting in complete shutdown of the MCC.

These limitations can be circumvented, as illustrated in Example 4.2.

**Example 4.2**

This example explores how the sensitivity of the ground fault settings in Example 4.1 can be improved. The ground fault protection is carried a step further and applied on every circuit from the source to the load. Each feeder or motor starter in the low voltage MCC is provided with a core-balance CT and a dedicated ground fault relay. Now, each feeder circuit breaker must be electrically tripped through a shunt trip coil. If we provide electrically operated circuit breakers with shunt trip coils and dedicated ground fault relays, as shown in Figure 4.13, the coordination can be arranged as shown in Figure 4.14. Each feeder on the MCC is provided with a ground fault pickup setting of 20 A and a short time delay. The opening times of low voltage circuit breakers is shown in Table 4.2, based upon recommendations in IEEE 1584 Guide, though practically, most manufacturers specify much shorter opening (tripping) times. The following ground fault settings are provided:

- Each feeder on low voltage MCC, controlled through a MCCB: 20 A, time delay 0.03 second.
- Each feeder from the 800-A LVPCB in the low voltage switchgear: 50 A, time delay 0.1 second.
- Main 4000-A LVPCB in the low voltage switchgear: 100 A, time delay 0.18 second.
- Transformer neutral connected ground relay: 150 A, time delay 0.26 second.

The neutral connected ground relays must trip a transformer primary protective device, in this case, breaker BK3 in Figure 4.9 or 4.13. This gives a well-coordinated and sensitive ground fault protection. Though this is an ideal situation for solidly grounded systems, many times, some lack of coordination is accepted, and a three-step

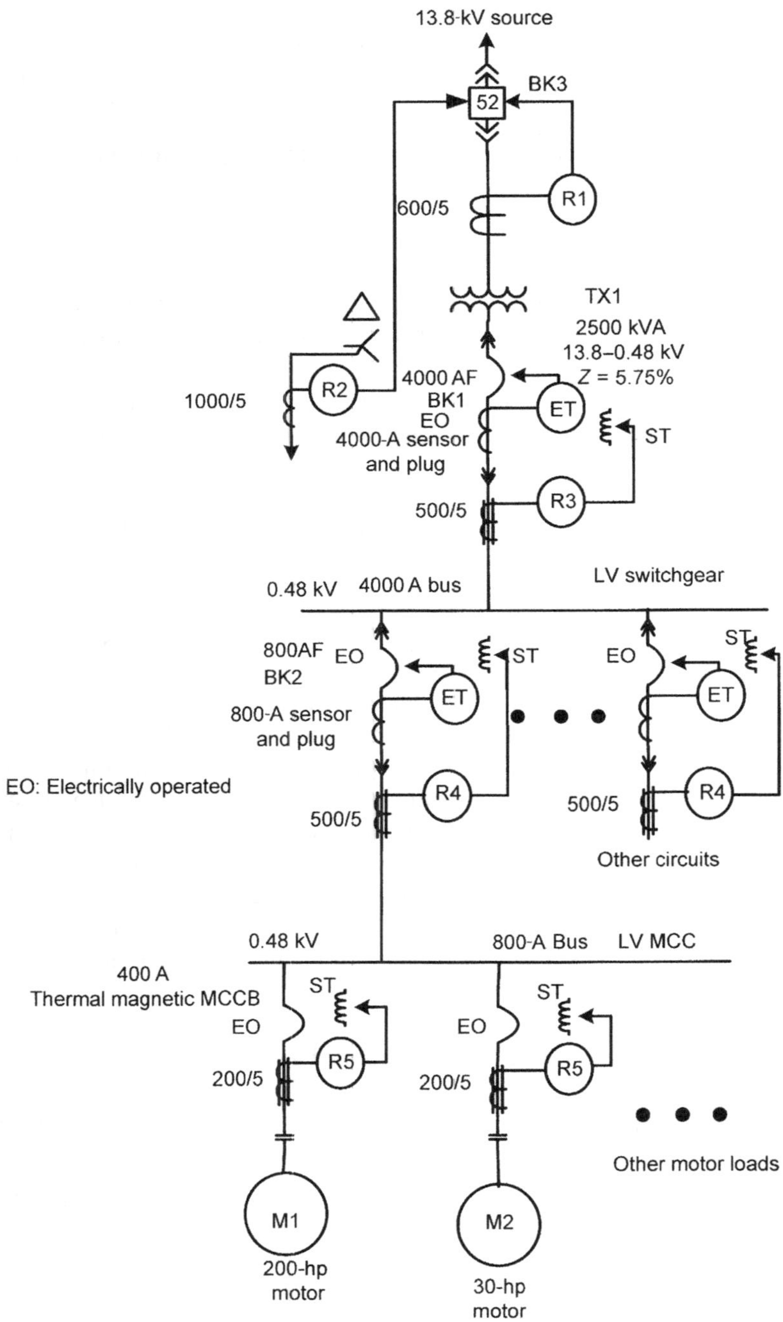


Figure 4.13. Distribution system in Figure 4.9, modified for improved selective ground fault protection.

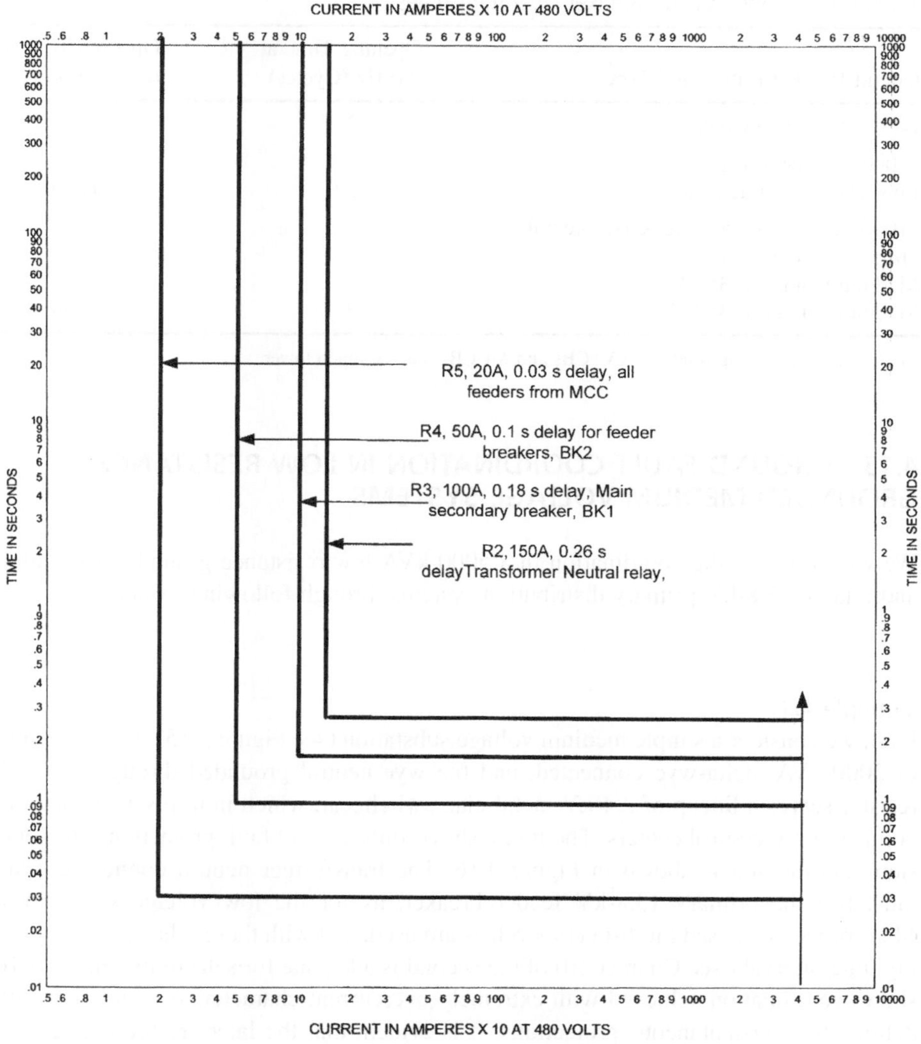


Figure 4.14. Ground fault coordination, with low ground fault pickup settings for the distribution system in Figure 4.13.

coordination, as shown in Figure 4.10, is provided, that is, the provision of dedicated ground fault protection on all circuits from MCC is omitted. This means that the ground faults may be cleared as phase faults, and if these are of low magnitude, these have to escalate or result in three-phase faults. This can cause increased equipment damage and arc flash hazard.

The saturation of zero-sequence CTs when applied in circuits with large short-circuit currents is a consideration, discussed in Chapter 12.

TABLE 4.2. Power Circuit Breaker Operating Times

| Circuit Breaker Rating and Type | Opening Time at 60 Hz (Cycles) | Opening Time at 60 Hz (Seconds) |
|---|---|---|
| Low voltage molded case | 1.5 | 0.025 |
| <1000 V, integral trip<br>Low voltage insulted case | 3.0 | 0.050 |
| <1000 V, power circuit breakers, integral trip or relay operated | | |
| Medium voltage 1–35 kV | 5 | 0.080 |
| Medium voltage 1–35 kV | 3 | 0.050 |

*Note*: Practical opening times for LVPCBs and MCCBs may be much lower.

## 4.13 GROUND FAULT COORDINATION IN LOW RESISTANCE GROUNDED MEDIUM VOLTAGE SYSTEMS

We will illustrate the coordination in a 5000-kVA low resistance grounded substation and a large 13.8-kV primary distribution systems through following examples.

**Example 4.3**

First, we consider a simple medium voltage substation (see Figure 4.15). A transformer of 5000-kVA, delta-wye connected, and the wye-neutral grounded through a 400-A resistor serves a lineup of 2.4-kV metal-clad switchgear, which in turn serves medium voltage motor control centers. The figure shows only ground fault protection. The four-step coordination is shown in Figure 4.16. The transformer neutral connected relay must trip the primary 13.8-kV feeder breaker, as for the low voltage substations. Microprocessor-based multifunction relays are used, and with these relays, a coordinating time interval ( see Chapter 10) of 0.3 second is adequate for selectivity. Figure 4.16 shows coordination achieved with extremely overcurrent characteristics and also with definite time instantaneous protections. It is evident that the later are preferable. The time delay increases in steps and the transformer neutral connected relay R2 has a time delay of 1 second at the maximum ground current of 400 A.

Observe that there are two neutral-connected relays: relay R2 trips the upstream breaker BK3, while relay R3 trips the downstream breaker BK1. These two relays are coordinated with respect to pickup settings and the time delay. If the ground fault lies on the MV switchgear bus and on the load side of the main secondary breaker BK1, relay R3 will operate faster to clear the fault and the primary breaker BK3 is not tripped. If the fault is in the transformer or upstream of breaker BK1; breaker BK3 is tripped with a further time delay.

The same operation could have been achieved by relay R6, connected to a core balance CT on the source side of the breaker BK1. In practical application, this has a problem that a large number of incoming cables from transformer to BK1 have to pass

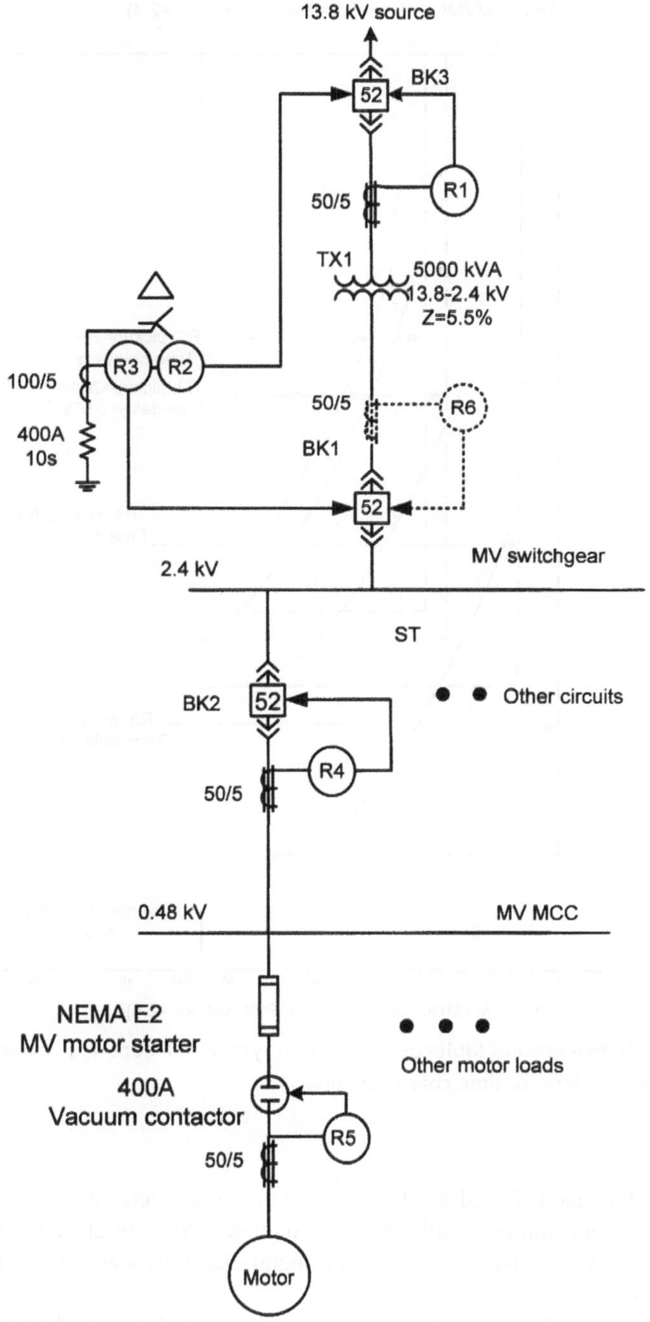


Figure 4.15. A medium voltage low resistance grounded distribution system for selective ground fault coordination.

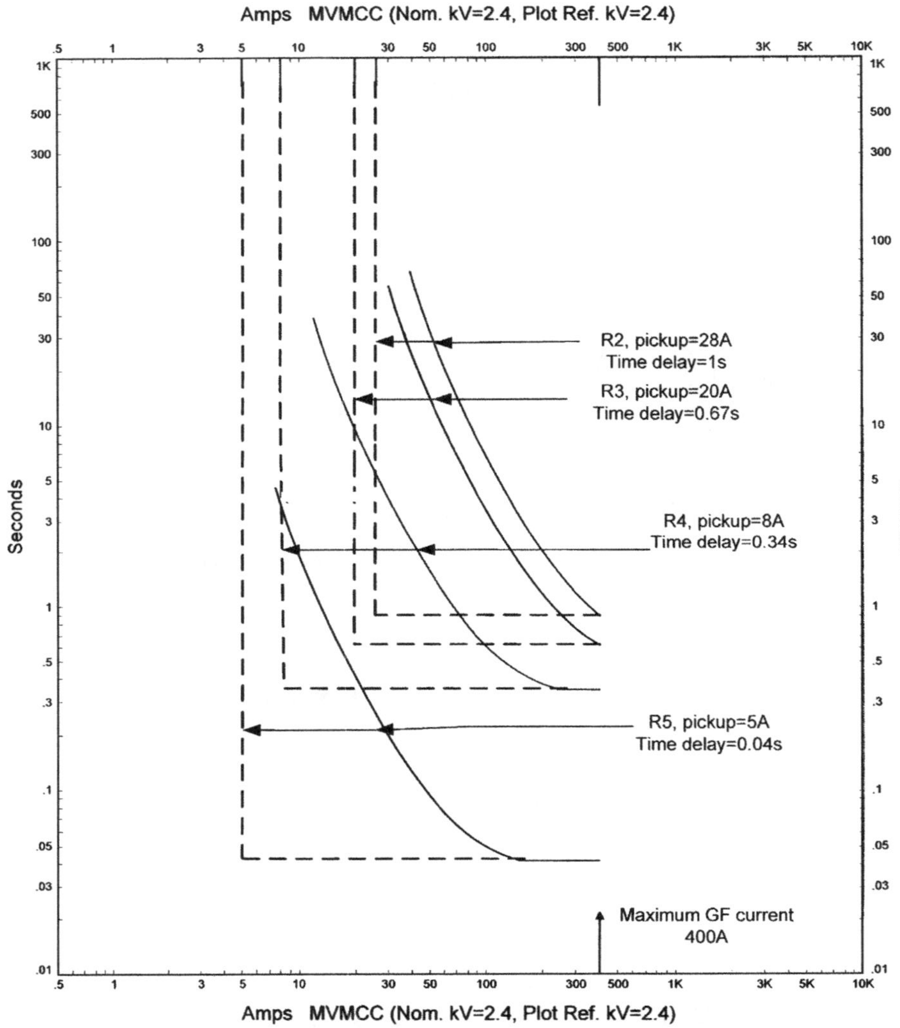


Figure 4.16. Selective ground fault coordination of system in Figure 4.15, extremely inverse characteristics versus definite time characteristics.

through a core balance CT, and a CT of large window diameter may not be commercially available to accommodate all the required cables. Also, in close-coupled substations connected to the transformer through a metal-clad bus, a zero sequence CT will be hard to install.

An arc flash calculation can be made; assuming that the ground faults are three-phase faults, the results will be conservative. This calculation reveals that for a 400 A current, using IEEE 1584 Guide equations, the incident energy release is only 1 cal/cm$^2$ (hazard risk category [HRC] = 0).

### 4.13.1 Remote Tripping

In an industrial distribution system, a primary distribution circuit breaker, say at 13.8 kV, serves a number of substations in a daisy chain. Sometimes as many as 10 unit substations over a considerable distance may be connected to one feeder breaker. The transformer primary protection is provided through current-limiting fuses (see Chapter 11). For the clearing of ground faults through transformer-neutral connected relays, a primary or source breaker is required to be tripped (see Figures 4.10, 4.13, and 4.15). This means that trip circuit wires and DC power from the primary breaker has to be extended to each unit substation over long distances. Though the DC control and trip cables can be continuously monitored and an open circuit or short-circuit condition can be alarmed, it is not desirable. Furthermore, lockout relays are used for remote tripping and blocking, and these take a considerable amount of inrush currents. Thus, the voltage drops on DC control system are of consideration. The following options exist:

- Relayed circuit breakers can be provided on the primary of each substation for local trips. These circuit breakers are also helpful for better primary protection of the transformers and when secondary arc flash protection is added (see Chapter 11).
- A capacitor trip device can be added for trip circuit control of the local transformer primary breakers; this can hold tripping charge for 48 hours or more on failure of normal AC power. Capacitor trip devices cannot be used for the powering of microprocessor relays due to high voltages these can generate.
- Redundant local tripping battery-charger systems can be provided.
- In every case, where the DC power from the main circuit breaker is extended, the circuit should be continuously monitored for open and short circuits.
- The possibility of adopting HRG systems can be explored, so that no ground fault tripping is required. Even then, a HR fault should be reliably alarmed for remedial actions. Practically, sometimes the alarms maybe ignored, causing extensive equipment damage.

### 4.13.2 Ground Fault Protection of Industrial Bus-Connected Generators

The ground fault protection of industrial bus-connected generators requires special considerations and ground fault protective relaying. This is illustrated through the following example.

**Example 4.4**

Figure 4.17 shows the system configuration of a large industrial distribution system with three plant generators and two utility interconnections and ground fault protection of the 13.8-kV distributions. Each generator is hybrid grounded and has a neutral breaker. The blocks in dotted lines show phase differential protection zones, which indicate that the entire system is covered in overlapping differential zones for phase

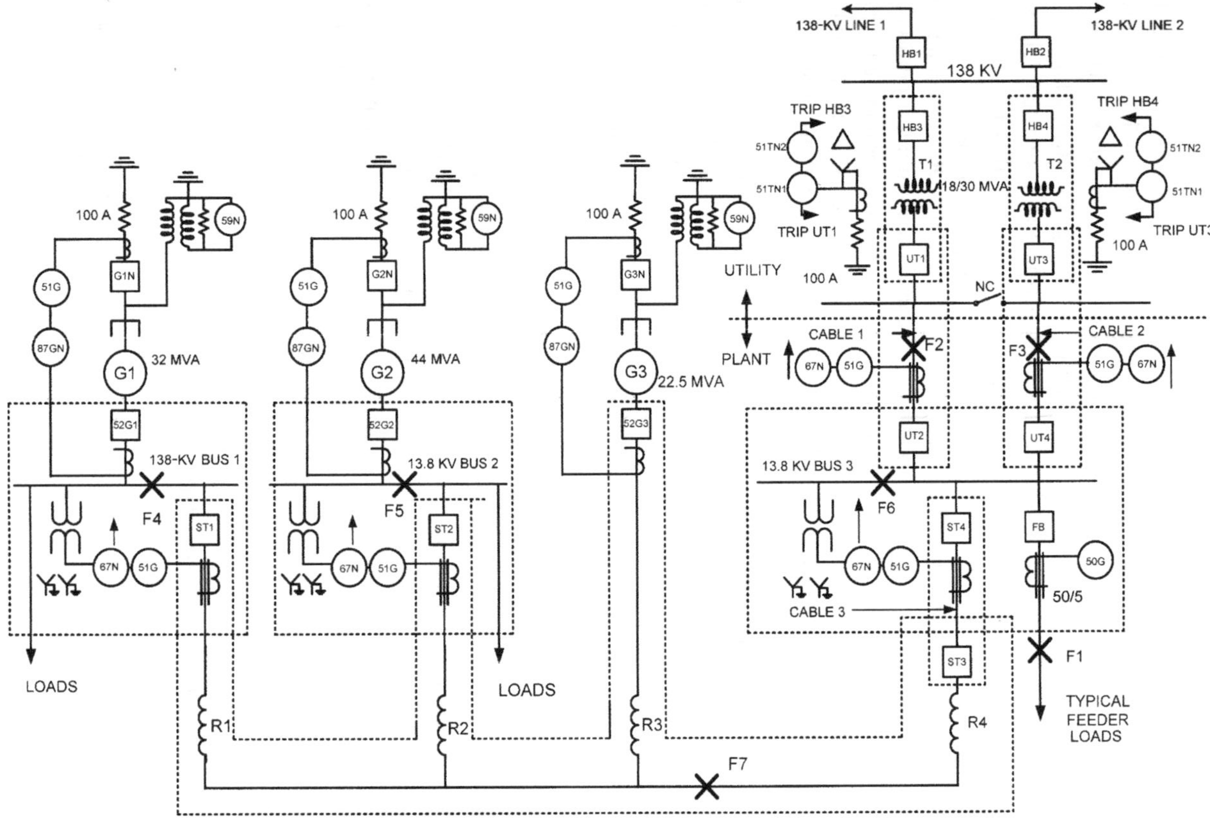


Figure 4.17. Ground fault coordination in a large interconnected 13.8-kV system, with three generators and two utility tie interconnections.

faults (except feeder circuits from the 13.8-kV buses). The phase differential protections on generators are not shown. See Chapter 8 for unit protection systems. *However, phase differential protection systems may not provide sensitive ground fault protection*; the minimum pickup current of a phase differential protection zone can be calculated, using manufacturer's data, CT accuracies, and number of protected circuits [17]. It may be too high for the low-level ground fault currents. In Figure 4.17, each generator is grounded through a 100-A resistor, and also two utility transformers T1 and T2 are grounded through a 100-A resistor, limiting the ground fault current anywhere in the 13.8-kV distributions to 500 A, which reduces the ground fault damage.

The ground fault damage to stator windings of industrial generators has been examined in Reference [18]. This reference considers no generator neutral breaker and assumes that source side and generator neutral currents are both 400 A. Considering that the generator breaker is opened in six cycles by instantaneous ground differential relaying in response to a generator ground fault, the 400-A current from the neutral source is not interrupted, because this reference did not consider a generator neutral breaker, which is mostly true in many generator installations. Greater fault energy (approximately four times) is released into the fault from the ground fault current of 400 A from the generator neutral circuit than the system's 400-A current from the source. This is because generator current decays slowly in about 0.8–1.0 second, depending upon the generator single line-to-ground fault time constant. The paper then suggests HRG of the generator, ground current limited to 10 A. Industrial bus connected generators cannot, however, be HR grounded because of two reasons: (1) the stray capacitance currents in the systems can be high. Reference [17] calculates a stray capacitance current of 80 A in a large distribution system; (2) the generator ground fault protection must coordinate with the ground fault protection on the feeder and tie circuits.

An IEEE Working Group was constituted and their findings are in Reference [19]; hybrid grounding systems for industrial bus-connected generators is recommended. As shown in Figure 4.17, this consists of conventional resistance grounded system connected through a neutral breaker, which could be simultaneously tripped with the generator breaker; for example, breakers 52G1 and 51N for generator 1. This low resistance grounding branch provides coordination with the system ground fault relays through backup ground relay 51G. Also, a HRG branch through a distribution transformer with secondary loading resistance is provided. This remains permanently connected to the generator neutral. On opening of the neutral breaker in response to the operation of generator differential ground fault relay, device 87GN, or standby ground fault relay 51G, the grounding system reverts to HRG, and throttles the ground fault current flow because high resistance is introduced in the neutral circuit.

It is imperative that a bus-connected industrial generator, even of a small rating, is provided with ground fault differential protection, device 87GN, because 51G device has to coordinate with other system ground relays and there will be a time delay associated with its coordination, resulting in some ground fault damage to the generator windings. The device 87GN will operate instantaneously, trip out generator and neutral breakers, and HRG branch is active to limit the ground fault damage. See also Chapter 8.

TABLE 4.3. Relay Operation for Selective Ground Fault Tripping, Example 4.4

| Fault at→<br>Relay↓ | F1 | F2 | F3 | F4 | F5 | F6 | F7 |
|---|---|---|---|---|---|---|---|
| Feeder 50G | X | NO | NO | NO | NO | NO | NO |
| Gen1, 51G | O | O | O | X | O | O | O |
| Gen2. 51G | O | O | O | O | X | O | O |
| Gen 3, 51G | O | O | O | O | O | O | X |
| ST1, 51G | O | O | O | O | O | O | X |
| ST1, 67N | NO | NO | NO | X | NO | NO | NO |
| ST2, 51G | O | O | O | O | O | O | X |
| ST2, 67N | NO | NO | NO | NO | X | NO | NO |
| ST4, 51G | O | O | O | O | O | O | X |
| ST4, 67N | O | O | O | NO | NO | X | NO |
| UT2, 51G | O | O | O | O | O | O | O |
| UT2, 67N | NO | X | NO | NO | NO | NO | NO |
| UT4, 51G | O | O | X | O | O | O | O |
| UT4, 67N | NO | NO | O | NO | NO | NO | NO |
| T1, 51TN1 | O | X | O | O | O | O | O |
| T2, 51TN2 | O | O | X | O | O | O | O |

The fault locations as marked in Figure 4.17.
X, relays that operate to clear the fault; O, relays that are operative, and coordinated; NO, relays that are not operative for the direction of fault current flow.

TABLE 4.4. Area of Shutdown for Fault Locations Shown in Figure 4.15, Example 4.4

| Fault Location | Area Isolated |
|---|---|
| F1 | Only feeder breaker tripped |
| F2 | Breakers UT2 and UT1 tripped |
| F3 | Breakers UT4 and UT3 tripped |
| F4 | Breakers 52G1 and ST1 tripped |
| F5 | Breakers 52G2 and ST2 tripped |
| F6 | Breakers ST4, UT1, UT2, UT3, and UT4 tripped |
| F7 | Breakers ST1, ST2, ST4, and 52G3 tripped |

Table 4.3 shows the tripping matrix for selective ground fault clearance, and Table 4.4 shows the area of shutdown for the respective fault locations in Figure 4.17. It can be seen that without the provision of directional ground fault relays, selective tripping cannot be obtained.

Figure 4.18 shows that coordination is achieved throughout the distribution system, including the utility ground fault relays, which are the electromechanical type, though these could be replaced with modern relays and provide similar characteristics as other ground fault relays. The low pickups of the protective devices are of interest:

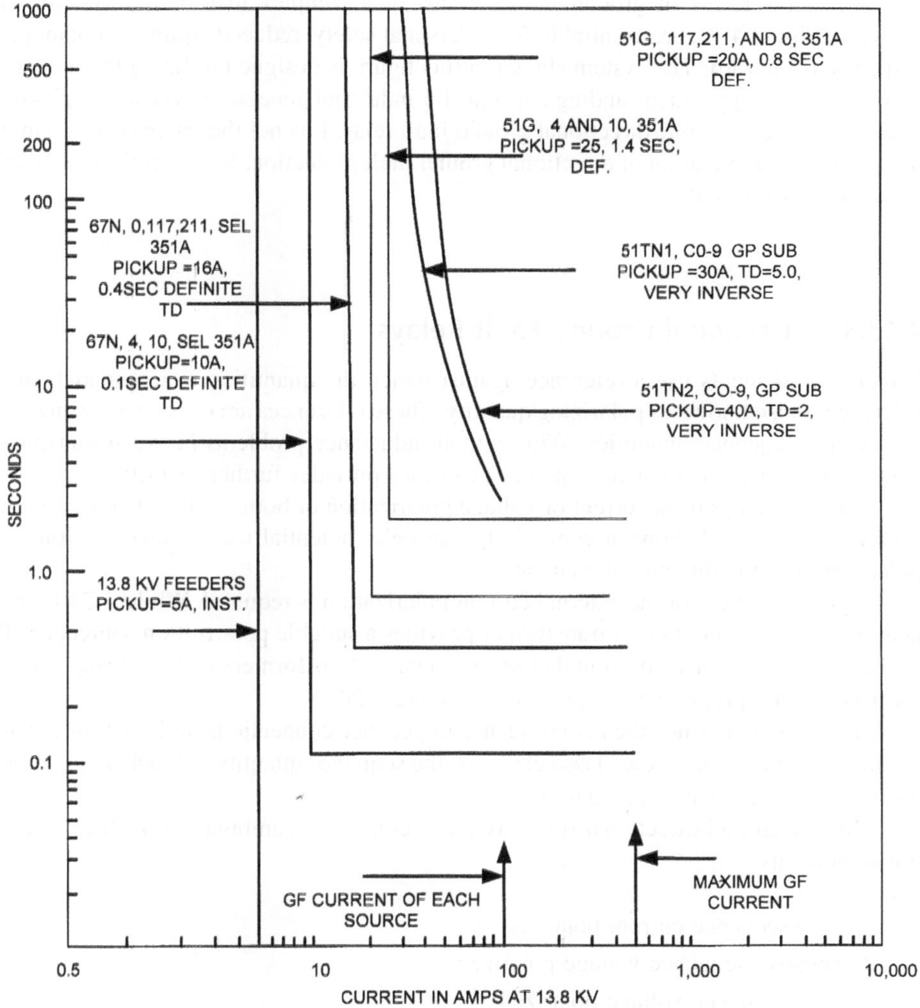


Figure 4.18. Selective ground fault coordination for the 13.8-kV distribution system of Figure 4.17.

- 50G, all 13.8-kV feeder breakers: 5 A, instantaneous (all transformers in the downstream distribution system have delta connected 13.8 kV windings, permitting these settings)
- 67N, breakers UT2, UT4: 10 A
- 67N, breakers ST1, ST2, ST4: 16 A
- 51G, Generator 3, breakers ST1, ST2, and ST4: 20 A
- 51G, Generators 1 and 2, UT2, UT4: 25 A.

These low levels of ground fault pickups in a complex distribution system, as shown in Figure 4.18, are admirable from personal safety, reduced equipment damage, and arc flash hazard. The system shown in this figure is designed utilizing the modern relaying technology and grounding concepts for industrial generators. These low pickup settings require a reliable directional ground fault relay. It is not the intention to go into the theory and application of directional ground fault protection, but the following brief description is relevant.

### 4.13.3 Directional Ground Fault Relays

Directional elements use a reference against which the quantities are compared; this reference is also known as polarizing quantity. These references are either zero sequence or negative sequence quantities. With mutual inductance problems between transmission lines, the use of negative sequence quantities provides further security.

Classical relays used current or voltage polarization or both, derived from external sources. Figure 4.19 shows a connected open delta potential transformer for voltage polarization of the directional element.

A proper source for the external current polarization is required. While a CT in the neutral circuit of a delta–wye transformer provides a suitable polarization source, a CT in the wye-wye connected grounded or ungrounded transformers and in zigzag transformers is not a proper current polarization source [20].

This is obvious when the zero sequence impedance connections of these transformers are examined. With these classical relays, the sequence quantity for each application must be selected for the application.

Modern ground directional relays (GDRs) consist of combination of three directional elements:

1. zero sequence current polarized
2. negative sequence voltage polarized
3. zero sequence voltage polarized.

The selection logic in the relay makes it possible to select one or two or the best choice for a particular application.

The zero sequence voltage polarization does not require external PTs in modern GDRs. This voltage can be internally generated from wye-wye PT inputs, (but not from open delta connected PT inputs).

External zero sequence current source is not necessarily required. The relay algorithm calculates the sequence components from the line current inputs. Table 4.5 shows the calculations of sequence currents for the GDR, based upon the real-world parameters of the distributions [17].

The phase fault protections, which are not completely shown in Figure 4.17, reduce the arc flash hazard anywhere in the 13.8-kV distribution to no more than HRC 2, even though the three-phase bolted fault currents are 35 kA on some of the buses.

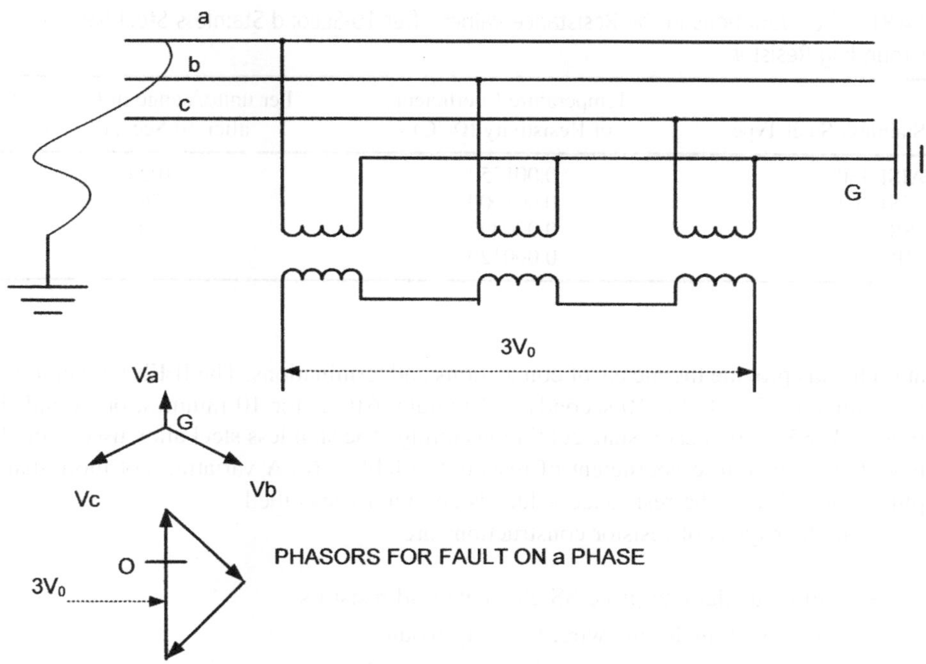


Figure 4.19. Derivation of a zero sequence polarizing voltage for directional ground fault overcurrent relay biasing, through wye-open delta connected PTs.

TABLE 4.5. Sequence Currents Seen by the Directional Ground Relays for Single Line-to-Ground Fault in Figure 14.17 (in the Direction of the Operation of the Relay)

| 67N Relay Location | $I_1$ (Amperes) | $I_2$ (Amperes) | $I_0$ (Amperes) |
|---|---|---|---|
| UT2 | $119 < -2.48^0$ | $119 < -2.48^0$ | $133 < -1.48^0$ |
| UT4 | $119 < -2.48^0$ | $119 < -2.48^0$ | $133 < -1.48^0$ |
| ST1 | $93.7 < -1.81^0$ | $93.7 < -1.81^0$ | $133 < -1.81^0$ |
| ST2 | $73.2 < -1.52^0$ | $73.2 < -1.52^0$ | $133 < -1.65^0$ |
| ST4 | $71.9 < -2.48^0$ | $71.9 < -2.48^0$ | $100 < -1.48^0$ |

*Note*: $I_1 + I_2 + I_0$ do not give 499 A as the sequence. Currents are partial currents.

## 4.14 MONITORING OF GROUNDING RESISTORS

Neutral grounding resistors should meet the requirement of IEEE 32 [21]. These are rated line to neutral voltage, though for HR grounding of generators, it is not unusual to see full line-to-line rated systems, grounded through distribution transformers. Generally, edge-wound design is used for neutral grounding systems. These resistors are constructed of flat stainless steel alloy, wound on its edge to form a helical resistive element. The element is then mounted on solid porcelain core, and welded terminals

TABLE 4.6. Variations in the Resistance Values after 10-Second Stainless Steel Neutral Grounding Resistor

| Stainless Steel Type | Temperature Coefficient of Resistivity (Ω/°C) | Per unit Change in Resistance after 10 Seconds (%) |
|---|---|---|
| AISI 430 | 0.001350 | 102.6 |
| AISI 304 | 0.001000 | 76 |
| 18SR | 0.000397 | 30.2 |
| 1JR | 0.000120 | 9 |

at both ends provide the means of connections and terminations. The IEEE temperature rise limits are 760°C for 10 seconds or I minute, 610°C for 10 minutes, or extended time and 385°C for steady-state continuous rating. The stainless steel alloy used should have low temperature coefficient of resistivity (Table 4.6). A variation not more than plus minus 10% in the resistance values is commonly specified.

The other types of resistor constructions are:

- continuous flat one piece SS elements grid resistors
- wire wound, nichrome wire, helically wound
- smooth wound, nickel chromium wire wound
- cast iron grid.

Whatever resistor type is used for grounding, it should be monitored for continuity and short-circuit continuously, under no-fault and under fault conditions. An open circuit of a grounding resistor in low-resistance or high-resistance grounding system will result in the system to become ungrounded, with all its perils. This is not acceptable. A short-circuit in the grounding resistor, though not so common, will result in solidly grounding the system. Without proper monitoring, these faults in the resistor may remain undetected for a long time, until next schedule routine maintenance.

Reference [22] describes how a combination of measurements of voltage across the resistor, current through the resistor and ohms measurement of the resistor to ground, can achieve an effective monitoring system, both under fault conditions and under normal operation. Many commercial products are available.

## 4.15 SELECTION OF GROUNDING SYSTEMS

The selection of grounding systems is summarized:

1. Ground low voltage distributions through HR grounding systems. The stray capacitance currents in these systems is low, typically no more that 0.1–2 A per 1000 kVA of system capacity at 480- or 600- V, three-phase, three-wire distributions.

2. Select low resistance grounded systems for primary and secondary medium voltage distribution voltages, that is, at 22, 13.8, 4.16, or 2.4 kV, which are the most common voltages in industrial power distribution systems.
3. Adopt hybrid grounding system for bus-connected plant generators. Provide a neutral breaker and ground fault differential relaying even for smaller machines.
4. Retain solid grounded for the systems as required by NEC. Derive solid grounded systems from three-phase three-wire low resistance of HR grounded systems through isolation transformers.
5. Do not adopt ungrounded systems for any applications.
6. Provide sensitive and selective ground fault protective systems. For solidly grounded systems, the selectivity and sensitivity of the ground fault protection is important to limit the equipment and arc flash damage.
7. Provide monitoring of neutral grounding resistors.
8. Provide adequate ratings of the grounding resistors. For low-resistance grounded systems, where the faults are selectively cleared, it is customary to rate the grounding devices on a short-time basis of 10 seconds. The 10-second rating may be only a small percentage of the continuous rating [21]. However, for HRG systems, where the faults are not immediately cleared, all components must be rated on a continuous basis.

This chapter provides an overview of system grounding, and is not exhaustive for design, engineering, and providing ground protections in practical applications. Nevertheless, it does point to the various factors and basic concepts that must be considered in such designs, with a special emphasis on reducing the arc flash damage. The grounding of systems derived through solid state converters or rectifiers is not covered, and Reference [23] provides a description of these systems. Also, equipment grounding and safety of substation grid grounding systems, step, touch, and transfer potentials is not discussed; see References [11, 24] for further reading.

In this chapter, we have used Stanbeck's equations for the calculations, which are conservative. The ground fault settings and coordination presented in this chapter for various grounding systems will limit the equipment damage and arc flash hazards, irrespective of the tested and theoretical differences in the arc flash currents and self-extinguishing nature of ground faults.

## REVIEW QUESTIONS

1. Design a four-step coordinated solidly grounded system for three-phase, three-wire 480-V system, served from a 2500-kVA, 13.8–0.48 kV, delta-wye connected transformer, impedance 5.75%, $X/R = 7$. The 13.8 kV system has a three-phase short-circuit level of 600 MVA, $X/R = 15$ and ground fault current is 400 A. (1) Construct a single line diagram through transformer to low voltage switchgear to MCCs and apply suitable composite loads not exceeding 90% of the transformer rating. Show

at least 3 MCCs, and largest motor size of 200 hp (2) Calculate all impedances, positive negative and zero sequence and conduct short-circuit calculations; establish three-phase and single-line to ground faults throughout the system see Chapter 5 (3) Show all protective devices, consult vendors catalogues and web sites, (4) construct TCC plots for phase-fault and ground-fault coordination. (5) Mark the self-extinguishing fault lines and also energy lines for main devices. (6) Calculate the arc fault damage based upon the coordination for line-to-ground faults.

**2.** Same as problem 1, but the transformer size is 3000-kVA, 13.8 - 2.4-kV, primary windings in delta, secondary in wye connections, wye windings grounded through 200-A, 10-second resistor. Write the specifications of the resistor.

**3.** Write a one page note on the self-extinguishing ground faults. Highlight the controversy as discussed in Chapter 3 and this chapter.

## REFERENCES


1. IEEE Standard 142. *IEEE Recommended Practice for Grounding of Industrial and Commercial Power Systems*, 1991.
2. H.I. Stanback, "Predicting damage from 277 volt single phase to ground arcing faults," IEEE Transactions on Industry Applications, vol. IA-13, no. 4, pp. 307–314, July/August 1977.
3. NEMA PB1-2, *Application Guide for Ground Fault Protective Devices for Equipment*, 1977.
4. NEC. *National Electrical Code*, NFPA 70, 2011.
5. R.H. Kaufman and J.C. Page, "Arcing fault protection for low voltage power distribution systems—Nature of the problem," IEEE Transactions on Industry Applications, vol. 79, pp. 160–167, June 1960.
6. Westinghouse, *Transmission and Distribution Handbook*, Westinghouse, East Pittsburgh, PA, 1964.
7. D.S. Baker, "Charging current data for guess-free design of high resistance grounded systems," IEEE Transactions on Industry Applications, vol. 15, no. 2, pp. 136–140, March/April 1979.
8. General Electric, *GE Industrial Power Systems Data Book*, General Electric, Schenectady, NY, 1964.
9. J.R. Dunki-Jacobs, "The reality of high resistance grounding," IEEE Transactions on Industry Applications, vol. 13, pp. 469–475, Sept./Oct. 1977.
10. ICEA Publication S-61-401, NEMA WC5. "Thermoplastic Insulated Wire and Cable for the Transmission and Distribution of Electrical Energy" (ICEA S-61.402, Third Edition), 1973.
11. J.C. Das, *Transients in Electrical Systems-Recognition, Analysis and Mitigation*, McGraw-Hill, New York, 2010.
12. National Electrical Safety Code, C-2, 2007.
13. J.R. Dunki-Jacobs, "The escalating arcing ground fault phenomenon," IEEE Transactions on Industry Applications, vol. 22, no. 6, pp. 1156–1161, Nov. 1986.

14. D.G. Lucks, "Calculating incident energy released with varying ground fault magnitudes on solidly grounded systems," IEEE Transactions on Industry Applications, vol. 46, no. 2, pp. 761–769, March/April 2010.
15. H.B. Land, III, "The behavior of arcing faults in low-voltage switchboards," IEEE Transactions on Industry Applications, vol. 44, no. 2, pp. 437–444, March/April 2008.
16. IEEE Std. 493 (Gold Book), *Recommended Practice for Design of Reliable Industrial and Commercial Power System*, 2007.
17. J.C. Das, "Ground fault protection of bus-connected generators in an interconnected 13.8 kV system," IEEE Transactions on Industry Applications, vol. 43, no. 2, pp. 453–461, March/April 2007.
18. L.J. Powell, "The impact of system grounding practices on generator fault damage," IEEE Transactions on Industry Applications, vol. 34, no. 5, pp. 923–927, Sept/Oct. 1998.
19. IEEE/IAS Working Group Report, "Grounding and ground fault protection of multiple generator installations on medium voltage industrial and commercial power systems," Parts I, II, III and IV, IEEE Transactions on Industry Applications, vol. 40, no. 1, pp. 11–28, Jan/Feb. 2004.
20. Westinghouse, *Applied Protective Relaying*, Westinghouse Electric Corporation, Coral Springs, FL, 1982.
21. ANSI/IEEE Std. 32. *IEEE Standard Requirements, Terminology, and Test Procedure for Neutral Grounding Devices*, 1972.
22. D. Selkirk, M. Savastianik, and K. Crawford, "The dangers of grounding resistor failure," IEEE Industry Application Magazine, vol. 16, no. 5, 53–59, Oct. 2010.
23. J.C. Das, "Grounding of AC and DC low-voltage and medium-voltage drive systems," IEEE Transactions on Industry Applications, vol. 34, no. 1, pp. 295–216, Jan./Feb. 1998.
24. ANSI/IEEE Std. 80, *IEEE Guide for Safety in Substation Grounding*, 2000.

# 5

# SHORT-CIRCUIT CALCULATIONS ACCORDING TO ANSI/IEEE STANDARDS FOR ARC FLASH ANALYSIS

The calculation of short-circuit currents is the first step in the process of arc flash calculations. We need to calculate only the bolted symmetrical three-phase short-circuits currents. A short-circuit calculation is called symmetrical when all three phases are involved, that is, a three-phase bolted faults. Unsymmetrical short-circuit calculations signify a line-to-ground fault or a double line-to-ground fault or phase-to-phase fault, and these are not required to be calculated for applications of equations in the IEEE 1584 Guide. Yet a synopsis of unsymmetrical calculations is provided to support the contents of Chapter 4. The IEEE 1584 Guide equations relate three-phase bolted fault currents with arcing currents. Calculations of short-circuit currents are also required for protective relaying and coordination. All commercially available programs on digital computers calculate arc flash hazard as a subroutine to short-circuit calculations and protective relay coordination.

The short-circuit current calculations are the basic requirements for any industrial distribution system, small or large. All switching devices must be selected to successfully interrupt the short-circuit currents at the point of their application. Also, static components like conductors and cables must be rated to withstand short-circuit currents for a specific duration, until the faults are cleared. To correlate the calculated

*Arc Flash Hazard Analysis and Mitigation*, Second Edition. J.C. Das.

Published 2021 by John Wiley & Sons, Inc.

short-circuit currents for selecting appropriate ratings on switching devices, a knowledge of the short-circuit rating structure of the switching devices is required, so that the calculated results are correlated with appropriate ratings of the switching devices. *This aspect is not related to arc flash calculations*, though for system integrity and safety, it is an important basic parameter.

## 5.1 TYPES OF CALCULATIONS

In a multivoltage system, four types of short-circuit calculations may be required. These are:

1. First-cycle (momentary) duties for fuses and low voltage circuit breakers (LVPCBs)
2. First-cycle (momentary) duties for medium or high voltage (HV) circuit breakers
3. Contact parting (interrupting) duties for HV circuit breakers (circuit breakers rated above 1 kV)
4. Short-circuit currents for time-delayed relaying devices.

The first-cycle, momentary duty or close and latch duty are all synonymous. However, the term "close and latch" is applicable to HV circuit breakers only, while the term "first cycle" can be applied to LVPCBs and fuses. Irrespective of the type of fault current calculation, the power system is reduced to single Thevénin equivalent impedance behind the source voltage.

### 5.1.1 Assumptions: Short-Circuit Calculations

The source voltage or prefault voltage is the system-rated voltage, though a higher or lower voltage can be used in the calculations. The worst short-circuit conditions occur at maximum loads, because the rotating loads contribute to the short-circuit currents. It is unlikely that the operating voltage will be above the rated voltage at maximum loading. Under light load conditions, the operating voltage may be higher, but the load contributions to the short-circuit currents will also be reduced. The effect of higher voltage at a reduced load is offset by the reduced short-circuit contributions from the loads. Therefore, the short-circuit calculations are normally carried out at the rated voltage. Practically, the driving voltage will not remain constant; it will be reduced, and varies with the machine loading and time elapsed subsequent to short-circuit. The fault current source is assumed sinusoidal; all harmonics and saturation are neglected. All circuits are linear; the nonlinearity associated with rotating machines, transformer, and reactor modeling is neglected. As the elements are linear, the theorem of superimposition is applicable.

Loads prior to short-circuit are neglected; short-circuit occurs at zero crossing of the voltage wave. At the instant of fault, the DC current value is equal in magnitude to the AC fault current value, but opposite in sign.

### 5.1.2 Short-Circuit Currents for Arc Flash Calculations

For calculations of the short-circuit currents, the IEEE 1584 Guide states that only symmetrical short-circuit currents need be considered. The short-circuit currents are asymmetrical by nature (see Chapter 6). There is always an exponentially decaying DC component associated with a three-phase symmetrical fault, which makes the short-circuit current wave asymmetrical about the zero axis. Both the AC and DC components decay. However, for arc flash calculations, we need not consider the DC component.

Fault arc resistance varies nonlinearly with the current, and due to its erratic nature, it is not a constant resistance during any one cycle. An expression for arc resistance per centimeter of the arc length is: $50P^{1/16}\ I^{3/4}$, where $P$ is the pressure in atmospheres and $I$ is the current in kA. The voltage across the arc is more constant. A discussion of the arc flash calculations vis-à-vis IEEE Guide 1584 is in Chapter 3.

## 5.2 RATING STRUCTURE OF HV CIRCUIT BREAKERS

In an attempt to harmonize ANSI/IEEE standards for HV circuit breakers with IEC, there has been considerable changes in the rating structures of the HV circuit breakers as reflected in many standards [1–8], and also Reference [9]. These changes are in the nomenclature, K-factor, peak (close and latch) ratings, TRV (transient recovery voltage) specifications, and definite purpose circuit breakers for capacitor switching, and these revisions to ANSI/IEEE standards impact the short-circuit calculations for application and selection. An interested reader may pursue the referenced material.

ANSI/IEEE standard [5] tabulates the short-circuit ratings of earlier $K > 1$ and $K = 1$ circuit breakers. For example, consider a 15-kV circuit breaker, rated short-circuit current = 37 kA rms symmetrical, maximum short-circuit current = 49 kA, K-factor = 1.3, and close and latch = 130 kA peak. The rated short-circuit current of 37 kA multiplied by 1.3 gives 49 kA. K-factor is voltage range factor, which signifies that the short-circuit interrupting ratings at a certain voltage of application can be proportioned. For example, if the circuit breaker is applied at 13.8 kV, the interrupting rating will be:

$$(37\times15)/13.8 = 40.21\ \text{kA}.$$

This proportionality holds down to 15/1.3 = 11.53 kV. Now if this circuit breaker is applied at 10 kV, the maximum short-circuit interrupting rating of 49 kA is applicable.

For $K = 1$ rated circuit breaker, only one short-circuit interrupting rating is specified. A 40-kA, $K = 1$ rated circuit breaker can be applied at 13.8, 10, or 2.4 kV, and its interrupting rating remains at 40 kA. This circuit breaker has a close and latch rating of 104 kA peak.

Earlier, $K > 1$ rated circuit breakers' close and latch rating was based on $X/R = 25$. If we consider asymmetry in the short-circuit current waveform, it translated to a multiplying factor of approximately 2.7. That is, the close and latch rating = 2.7 × 37 × 1.3 = 130 kA.

For $K = 1$ rated circuit breakers, the rating structure is changed more in line with IEC, and the X/R ratio is changed from 25 to 17. This gives a multiplying factor of 2.6 at 60 Hz in line with IEC standards. Thus, the close and latch for this 40 kA circuit breaker will be: 40 × 2.6 = 104 kA peak.

The multiplying factors must be adjusted for *actual* fault point *X/R*. The peak current need not be calculated for arc flash. However, this description is included because of its importance in the calculations of short-circuits currents and the theoretical concepts governing these calculations. The maximum peak current *does not occur* at 1/2 cycle in the phase that has maximum initial DC component, unless short-circuit occurs on a *purely inductive circuits*. The maximum peak occurs before 1/2 cycle and before the symmetrical current peak. For earlier K > 1 rated circuit breakers, 1/2 cycle peak has been used in the calculations given by:

$$\text{Half cycle}_{\text{peak}} = I_{\text{acpeak}}\left(1 + e^{-\pi/(X/R)}\right). \tag{5.1}$$

IEEE Standard-Violet Book [1] provides the following equations that give better approximation to peak:

$$I_{\text{peak}} = I_{\text{AC peak}} + I_{\text{DC}} = \sqrt{2}I_{\text{AC,rms}}\left(1 + e^{-\frac{2\pi\tau}{(X/R)}}\right) = I_{\text{AC peak}}\left(1 + e^{-\frac{2\pi\tau}{(X/R)}}\right), \tag{5.2}$$

where

$$\tau = 0.49 - 0.1e^{-\frac{(X/R)}{3}}. \tag{5.3}$$

For the rms first-cycle rms current, the equation

$$I_{\text{rms}} = \sqrt{I_{\text{ac,rms}}^2 + I_{\text{dc}}^2} \tag{5.4}$$

is accurate only if DC is constant, but it is not. Many times, the DC component is calculated at 1/2 cycle, though this does not correspond to real peak value. An accurate IEEE [1] equation is:

$$I_{\text{rms,total}} = I_{\text{AC,rms}}\sqrt{\left(1 + 2e^{\frac{-4\pi\tau}{(X/R)}}\right)}, \tag{5.5}$$

For older K > 1 rated circuit breakers, it can be calculated from:

$$I_{\text{asym}} = I_{\text{sym}}\sqrt{1 + 2e^{-2\pi/(X/R)}}. \tag{5.6}$$

This equation essentially calculated asymmetrical current at 1/2 cycle. An asymmetrical multiplying factor of 1.6 has been used that is applicable for *X/R* = 25.

Figure 5.1 depicts short-circuit current decaying transients. At the contact parting time of the circuit breaker, there will be some DC component that has to be interrupted. This makes the current to be interrupted at the contact parting time asymmetrical in

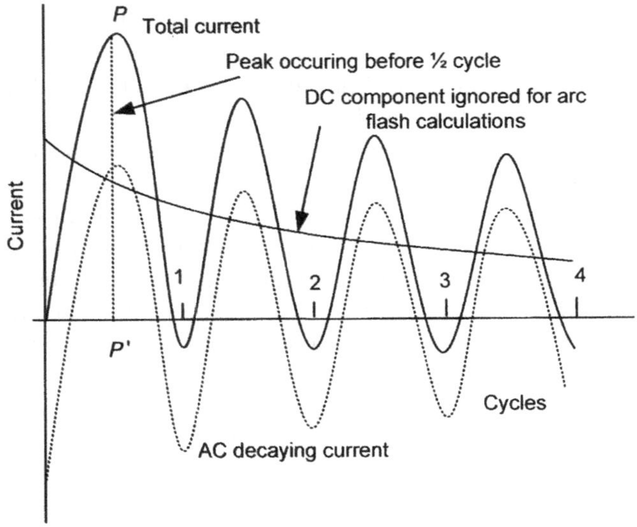


Figure 5.1. Decaying AC and DC components of the short-circuit current and the total current asymmetrical wave shape.

nature. This asymmetry is built in the rating structure of ANSI/IEEE circuit breakers rated on a symmetrical basis. The asymmetry is specified in terms of the DC component of the current that the circuit breaker must interrupt at its contact parting time. For example, a five-cycle symmetrical rated circuit breaker has a contact parting time of three cycles, and it must be capable of breaking 32.91% DC component [7]. It may be possible to reduce interrupting duties in some cases by increasing contact parting time [10]. The generator circuit breakers are required to interrupt much higher DC components, as the asymmetry is higher close to a generator (see IEEE Standard Reference [4]).

In the short-circuit calculations according to standards, a step-by-step account of the decay cannot be considered. Depending on the type of calculation, the ANSI/IEEE standards consider that the dynamic (rotating equipment) reactances are multiplied by factors given in Table 5.1 [2]. The static equipment impedances are assumed to remain time invariant, that is, harmonics and saturation are neglected. Maintaining a constant emf and artificially increasing the equivalent impedance to model a machine during short-circuit has the same effect as the decay of the flux trapped in the rotor circuit. In Table 5.1, manufacturer's data for the transient and subtransient and locked rotor reactances should be used in the calculations for accuracy. This modeling of short-circuit current decay from rotating machines, gives conservatism to the empirical calculations of short-circuit currents for switching devices, but is not accurate for arc flash calculations (see Chapter 6).

TABLE 5.1. Impedance Multiplier Factors for Rotating Equipment for Short-Circuit Calculations

| Type of Rotating Machine | Positive Sequence Reactance for Calculating | |
|---|---|---|
| | Interrupting Duty (per Unit) | Closing and Latching Duty (per Unit) |
| All turbogenerators, all hydrogenerators with amortisseur windings, and all condensers | $1.0X''_d$ | $1.0X''_d$ |
| Hydrogenerators without amortisseur windings | $0.75X'_d$ | $0.75X'_d$ |
| All synchronous motors | $1.5X''_d$ | $1.0X''_d$ |
| Induction motors | | |
| Above 1000 hp at 1800 rpm or less | $1.5X''_d$ | $1.0X''_d$ |
| Above 250 hp at 3600 rpm | $1.5X''_d$ | $1.0X''_d$ |
| From 50 to 1000 hp at 1800 rpm or less | $3.0X''_d$ | $1.2X''_d$ |
| From 50 to 250 hp at 3600 rpm | $3.0X''_d$ | $1.2X''_d$ |
| Neglect all three-phase induction motors below 50 hp and all single-phase motors. | | |
| Multivoltage level calculations including low-voltage systems | | |
| Induction motors of above 50 hp | $3.0X''_d$ | $1.2X''_d$ |
| Induction motors below 50 hp | $\infty$ | $1.67X''_d$ |

$X''_d$ of synchronous rotating machines is the rated voltage (saturated) direct-axis subtransient reactance. $X''_d$ of synchronous rotating machines is rated voltage (saturated) direct-axis transient reactance. $X''_d$ of induction motors equals 1.00 divided by per unit locked rotor current at rated voltage.

*Source*: IEEE Reference [2].

## 5.3 LOW-VOLTAGE MOTORS

For calculations of short-circuit duties for comparison with medium or HV circuit breakers closing and latching capabilities (or momentary ratings according to pre-1964 basis, now no longer in use), motors smaller than 50 hp can be ignored. However, these are required to be considered for low-voltage circuit breaker applications. To simplify multivoltage distribution system, short-circuit calculations and obviate the necessity of running two first-cycle calculations, one for the low-voltage circuit breakers and the other for HV circuit breakers, a single first-cycle network can replace the two networks according to the recommendation of IEEE standard [11]. This network is constructed by:

1. Including all motors <50 hp using a multiplying factor of 1.67 for subtransient reactance or an estimate of first-cycle impedance of 0.28 per unit based on the motor rating

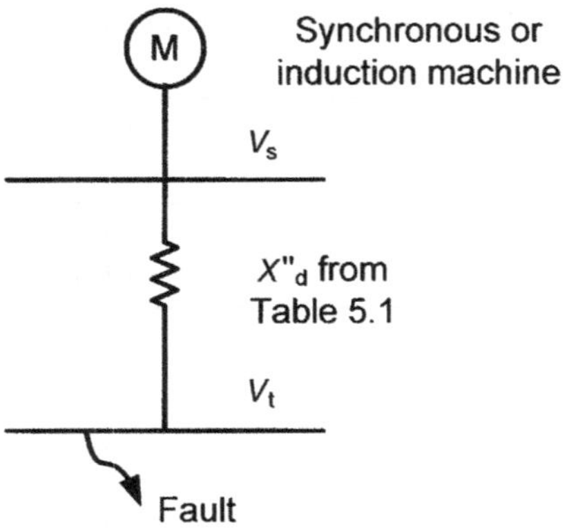


Figure 5.2. Equivalent machine model for short-circuit calculations.

2. Including all motors ≥50 hp and using a multiplying factor of 1.2 for the subtransient reactance or an estimate of first-cycle impedance of 0.20 per unit based on the motor rating.

## 5.4 ROTATING MACHINE MODEL

The rotating machine model for the short-circuit calculations is shown in Figure 5.2. The machine reactances are modeled with suitable multiplying factors from Table 5.1. The multiplying factors are applicable to resistances as well as reactances, so that the *X/R* ratio remains constant. The voltage behind the equivalent transient reactance at no load will be equal to the terminal voltage, as no prefault currents need be considered in the calculations. A justification of neglecting the prefault currents is the concept of constant flux linkages. The total current before and after the transition (pre- and postfault) should not change. The DC component is equal in magnitude to the AC component, but of the opposite polarity. Thus, the AC and DC components of the current summate to zero to maintain a constant flux linkage, that is, no load conditions. If preloading is assumed, this balance is no longer valid.

## 5.5 CALCULATION METHODS

Again, the procedures of calculations in IEEE standards are oriented for calculations of short-circuit duties on the switching devices. The calculation procedures are:

### 5.5.1 Simplified Method $X/R \leq 17$

The results of the *E/X* calculation can be directly compared with the circuit breaker symmetrical interrupting capability, provided that the circuit *X/R* ratio for three-phase faults and $(2X_1 + X_0)/(2R_1 + R_0)$ for single line-to-line ground fault is 17 or less. (For single phase-to-ground faults $X_1$ is assumed equal to $X_2$.) This is based on the rating structure of the circuit breakers. When the circuit *X/R* ratio is 17 or less, the asymmetrical short-circuit duty never exceeds the symmetrical short-circuit duty by a proportion greater than that by which the circuit breaker asymmetrical rating exceeds the symmetrical capability. It may only be slightly higher at four-cycle contact parting time.

### 5.5.2 Simplified Method *X/R* > 17

A further simplification of the calculations is possible when the *X/R* ratio exceeds 17. For *X/R* ratios higher than 17, the DC component of the short-circuit current may increase the short-circuit duty beyond the compensation provided in the rating structure of the circuit breakers. A circuit breaker can be immediately applied without calculation of system resistance, *X/R* ratio, or remote/local considerations, if the *E/X* calculation does not exceed 80% of the circuit breaker symmetrical interrupting capability. This means that resistance component of the system need not be determined.

### 5.5.3 *E/Z* Method for AC and DC Decrement Adjustments

Where a closer calculation is required, and the current exceeds 80% of the circuit breaker symmetrical rating, AC and DC decrement adjustments should be considered. This method is also recommended when a single line-to-ground fault supplied predominantly by generators at generator voltage exceeds 70% of the circuit breaker interrupting capability for single line-to-ground faults. For calculations using this method, the fault point *X/R* ratio is necessary. Two separate networks are constructed: (1) a resistance network, with complete disregard of the reactance, and (2) a reactance network with complete disregard of the resistance. The fault point *X/R* ratio is calculated by reducing these networks to an equivalent resistance and an equivalent reactance at the fault point. This gives more accurate results than any other reasonably simple procedure, including the phasor representation at the system frequency.

Once the *E/X* calculation is made and the *X/R* ratio is known, the interrupting duty on the HV circuit breakers can be calculated by multiplying the calculated short-circuit currents with a postappropriate multiplying factor given by curves in IEEE standard [2]. This multiplying factor is based on:

1. contact parting time of the circuit breaker
2. calculated *X/R* ratio
3. effects of DC decay (remote sources) and effects of AC and DC decay (local sources).

If the short-circuit current is fed from generators through (1) two and more transformations, or (2) a per-unit reactance external to the generator that is equal to or exceed 1.5 times the generator per unit subtransient reactance on a common MVA base, that is, it supplies less than 40% of its terminal short-circuit current, it is considered a remote source. In this case, the effect of AC decay need not be considered, and the curves of multiplying factors include only DC decay, in ANSI/IEEE standard [2]. The multiplying factor for the remote curves is calculable, and is given by

$$I_{\text{asym}} / I_{\text{sym}} = \sqrt{1 + 2e^{-4\pi C/(X/R)}}, \tag{5.7}$$

where $C$ is the contact parting time in cycles at 60 Hz. As an example, the remote multiplying factor for five-cycle circuit breaker, which has a contact parting time of three cycles, and for a fault point *X/R* of 40, from Equation (5.7), is 1.21, which can also be read from the curves in ANSI/IEEE standard [2]. These curves are not reproduced here.

When the short-circuit current is predominantly fed through no more than one transformation or a per unit reactance external to the generator, which is less than 1.5 times the generator per unit reactance on the same MVA base, that is, it supplies more than 40% of its maximum terminal fault current, it is termed a local source. The effect of AC and DC decrements should be considered. The multiplying factors are applied from curves in ANSI/IEEE standard [2]; these curves are not reproduced here.

The asymmetrical multiplying factors for the *local* curves are not a known equation. A number of sources may contribute to a fault through varying impedances. Each of these contributions has a different AC and DC decay rate. The impedance through which a fault is fed determines whether it is considered a local or remote source. The AC decay in electrically remote sources is slower, as compared with the sources closer to the fault. The time constant associated with the AC decay is a function of rotor resistance, and added external reactance prolongs it. For a system with several short-circuit sources, which may include generators that are classified local or remote, depending on the interconnecting impedances, neither the remote nor the local multiplying factors can be exclusively applied. It is logical to make use of both local and remote multiplying factors in a *weighting process*. This weighting consists of applying a remote multiplying factor to that part of the *E/Z* symmetrical short-circuits current that is contributed by remote sources. Similarly, the local multiplying factor is applied to the local component of the fault current contribution. The fraction of interrupting current that is contributed by remote sources is identified as the NACD (no AC decrement) ratio:

$$\text{NACD ratio} = \frac{\sum \text{NACD source currents}}{E / X \text{ for the interrupting network}}. \tag{5.8}$$

This computation requires additional calculations of remote and total current contributed at the fault point from various sources and is facilitated by digital computers. Figure 5.3 shows interpolated multiplying factors for various NACD ratios [12].

For the short-circuit current contribution from motors, *irrespective of their type and rating and location in the system,* the AC decay is built into the premultiplying impedance factors in Table 5.1. Thus, it is assumed that the motors, howsoever remote

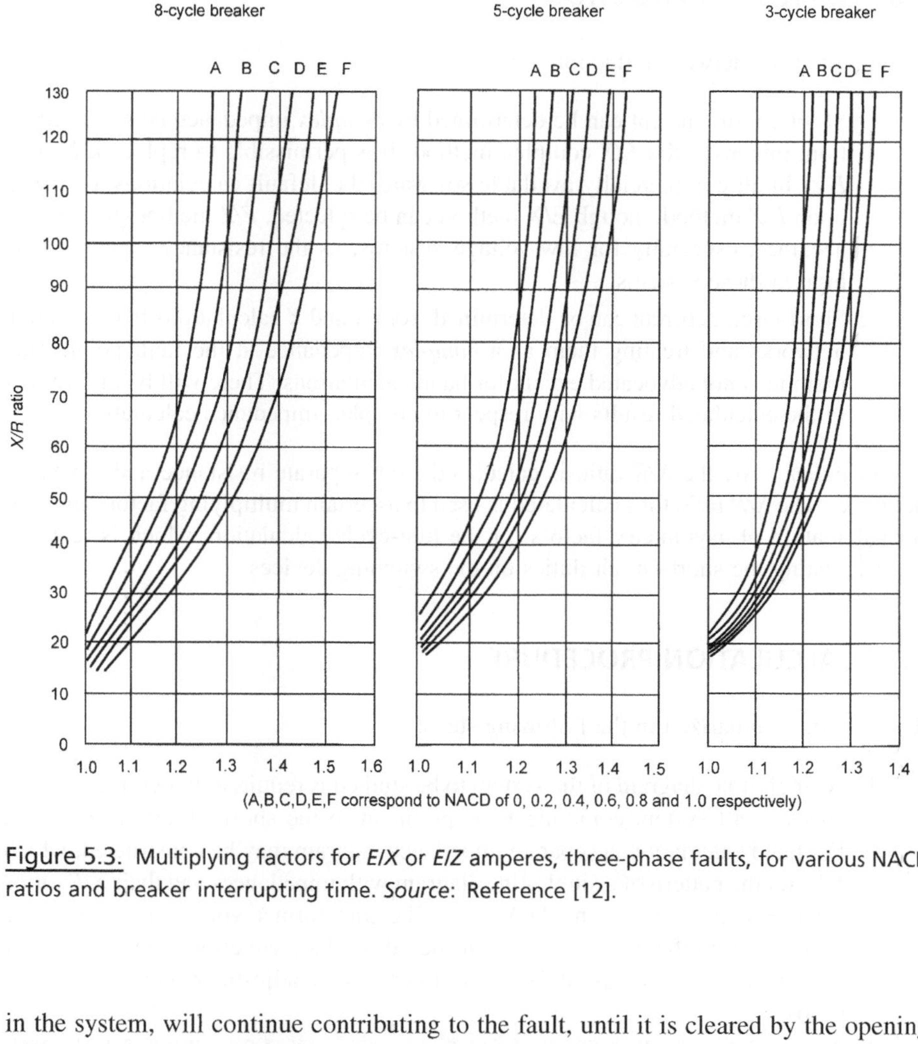


Figure 5.3. Multiplying factors for *E/X* or *E/Z* amperes, three-phase faults, for various NACD ratios and breaker interrupting time. *Source*: Reference [12].

in the system, will continue contributing to the fault, until it is cleared by the opening of the circuit breaker. The magnetic contactors controlling the low-voltage motors may drop in the first cycle of the fault voltage dip at 30–70% of their rated voltage, disconnecting the motors from service. This is conservative from the short-circuit calculation point of view, but may give increased arc flash hazard calculations (see Chapter 6). Practically, the motors contribute to the short-circuit currents only for a couple of cycles.

While this provides a gist of the short-circuit calculations in the standards, it is not exhaustive enough, and quoted references provide further reading. For arc flash calculations, *we are not calculating the short-circuit ratings on switching devices, therefore no post multiplying factors—local or remote or weighted are required to be calculated.* Yet two calculations will be required because in Table 5.1, the multiplying factors for the first cycle and interrupting duty calculations are different.

## 5.6 NETWORK REDUCTION

Two methods of network reduction are:

1. Short-circuit current can be determined by *complex* impedance network reduction; this gives the *E/X* complex method. It is permissible to replace E/X with *E/Z*. In all commercially available software, the default calculations are based upon *E/Z* method, though E/X method can be selected. *E/Z* method gives better accuracy, especially for low-voltage systems, as the resistance component is high in these systems.
2. Short-circuit current can be determined from *R* and *X* calculations from separate networks and treating them as a *complex* impedance at the fault point. This method is not advocated except for hand calculations. There will be differences in the calculated results with respect to complex impedance calculations.

*In either case,* the *X/R* ratio is calculated from separate resistance and reactance networks. The *X/R* ratio thus calculated is used to ascertain multiplying factors and also for calculation of asymmetry factors for the first-cycle calculation, which is required for calculating the short-circuit duties on the switching devices.

## 5.7 CALCULATION PROCEDURE

This can be summarized in the following steps:

1. A single-line diagram of the system to be studied is required. It identifies impedances of all system components as pertinent to the short-circuit calculations. For hand calculation, a separate impedance diagram may be constructed, which follows the pattern of a single-line diagram with impedances and their *X/R* ratios calculated on a common MVA base. The transformer voltage ratios may be different from the base voltages considered for data reduction. The transformer impedance can be adjusted for transformer voltage adjustment taps and voltage ratios.
2. Appropriate impedance multiplying factors are applied from Table 5.1, depending on the type of calculation. For HV circuit breakers, at least two networks are required to be constructed, one for the first-cycle (close and latch) calculations and the other for the interrupting duty calculations.
3. A fault-point impedance positive sequence network (for three-phase faults) is then constructed, depending on the location of the fault in the system. Both resistances and reactances can be shown in this network, or two separate networks, one for resistance and the other for reactance, can be constructed. Table 5.2 shows the resistance values to be used. Typical curves for estimating *X/R* ratios are provided in ANSI/IEEE standard [2]. All commercially available computer programs have built-in data for the *X/R* ratios for generators, motors, and transformers; though manufacturer's data can be used where available.

TABLE 5.2. Resistance of System Components for Short-Circuit Calculations

| System Component | Approximate Resistance |
|---|---|
| Turbine generators and condensers | Effective resistance |
| Salient pole generators and motors | Effective resistance |
| Induction motors | 1.2 times the DC armature resistance |
| Power transformers | AC load loss resistance (not including no-load losses or auxiliary losses) |
| Reactors | AC resistance |
| Lines and cables | AC resistance |

The effective resistance = $X_{2V}/(2\pi f T_{a3})$, where $X_{2V}$ is the rated-voltage negative-sequence reactance and $T_{a3}$ is the rated voltage generator armature time constant in seconds.
*Source*: IEEE Reference [2].

4. For *E/Z* complex calculation, the fault-point positive sequence network is reduced to single impedance using complex phasors. Alternatively, the resistance and reactance values obtained by reducing separate resistance and reactance networks to a single-point network to calculate the fault-point *X/R* ratio can also be used for *E/Z* calculation. This considerably simplifies hand calculations, compared with complex impedance reduction, but is not accurate.
5. If there are many sources in the network, NACD is required to be calculated, and this sets a limit to the complexity of networks that can be solved by hand calculations. The currents from NACD sources have to be traced throughout the system to the faulty node to apply proper weighting factors, and this may not be an easy calculation in interconnected networks. The calculation of the first-cycle duty does not require considerations of remote or local.
6. The adjusted short-circuit currents, thus calculated, can be used to compare with the short-circuit ratings of the existing switching equipment or selection of new equipment.

### 5.7.1 Analytical Calculation Procedure

***Hand Calculations.*** In simple systems, with limited buses not more than six to eight, the currents from the various sources can be vectorially summed at the fault point. No preimpedance multiplying factors and postmultiplying duty factors are required. The DC component of the short-circuit current is calculated. The time constants associated with AC and DC decay are required to calculate the currents at the contact parting time of the circuit breaker. This may not be always easy. Once each of the components is calculated, the theorem of superimposition applies, and the total currents can be calculated for circuit-breaker duties.

***Dynamic Simulation.*** Alternative methods of calculations of short-circuit currents that give accurate results are recognized. The short-circuit calculations can be conducted on EMTP or other digital computer programs that could emulate the transient

TABLE 5.3. Comparison of Short-Circuit Calculations, ANSI/IEC/EMTP, and Dynamic Simulation Methods

| Bus kV | ANSI | IEC | EMTP/Dynamic Simulation | Remarks | Reference |
|---|---|---|---|---|---|
| 13.8 | 23.39 | | 17.81* | Breaking/interrupting | [14] |
| | 24.87 | | 18.52* | current at 50 ms | |
| | 25.49 | | 20.16* | AC + DC | |
| 4.16 | 46.58 | | 45.70* | | |
| | 35.37 | | 36.44* | | |
| 13.8 | 15.45 | 16.31 | | *E/Z* symmetrical | [15] |
| | 15.36 | 15.21 | | Current | |
| 2.4 | 21.47 | 19.09 | | | |
| 220 | 20.99 | 22.57 | 20.30** | *E/Z* symmetrical | [16] |
| 21 | 74.35 | 74.43 | 60.23** | Current | |
| 10 | 25.98 | 23.68 | 17.75** | | |

All currents are in kA rms symmetrical.
* The dynamic simulation results.
** EMTP simulation results.

behavior of machines during short-circuit conditions. In this sense, these simulations become more like transient stability analyses carried out in the time domain. The behavior of machines as influenced by their varying electrical and mechanical characteristics can be modeled, and the calculation accuracy is a function of the machine models, static and dynamic elements, as well as the assumptions. Synchronous machines can be modeled in much greater details; the IEEE standard [13] details out the synchronous machine models for transient analysis studies. This may result in a number of differential equations so that harmonics and DC offset effects can be accounted for. The inclusion of voltage regulators and excitation systems is possible. Dynamic simulation puts heavy demand on the computing resources as well as on accurate data entries, yet this forms an important verification tool for the empirical calculations.

Variations in the results of the short-circuit calculations using different calculation techniques have been studied by many authors [14–16]. Table 5.3 shows these variations for the same system configurations. Generally, ANSI/IEEE calculations are more conservative when compared with dynamic simulation or EMTP. This table shows that in some cases, ANSI calculations are 30% higher than the corresponding results obtained with dynamic simulation. Due to large variations documented in this table, a question arises when evaluating the circuit breakers that are applied above their short-circuit ratings according to ANSI/IEEE empirical calculations—whether the application above a certain percentage of the calculated numbers can be ignored. This will not be a prudent approach, unless rigorous alternate techniques are explored and are acceptable to the owners of an establishment. It must be realized that short-circuit currents can be very damaging to the equipment, including fire and explosion hazards and personal safety. For the applications in new system designs, the calculated short-circuit duties should be preferably no more than 85% of the ratings; this allows some future expansions and growth, which almost always occurs. It is far more economical to provide for some future expansion of the electrical systems in the design stage; as subsequent replacements and retrofits can be expensive.

## 5.8 CAPACITOR AND STATIC CONVERTER CONTRIBUTIONS TO SHORT-CIRCUIT CURRENTS

A number of dynamic simulations of capacitors are presented in Reference [1]. During fault, the capacitor discharge takes place in a very short time of 1/30 to 1/8 cycles. The electromagnetically induced forces of discharge current are instantaneously proportional to the square of the current. The capacitor discharge currents will have no effect on circuit breaker contact parting or clearing operations. Some stresses may be imposed for circuit breaker closing and latching duty. The standard also considers rectifier, inverter, cycloconverter, and chopper circuits, and concludes that DC systems can contribute to the short-circuit currents on the AC side under the following conditions:

1. When the DC system has a fault current source, such as motors, batteries, or photovoltaic cells
2. When the converter operates as an inverter
3. Contributions occur to the first cycle short-circuit only, and magnitude is approximately three times the AC three-phase apparent power input of the converter transformer at rated machine DC load and an *X/R* ratio of 10. Interrupting short-circuit currents are not impacted.

## 5.9 TYPICAL COMPUTER-BASED CALCULATION RESULTS

Typical results of computer-based calculations are shown in Figure 5.4a–c.

### 5.9.1 First-Cycle or Momentary Duty Calculations

Figure 5.4a shows the results of calculations. The bus number is G1, on the leftmost corner. *E/Z* calculation gives 28.397 kA, and the angle is −87.51 DEG. This is the symmetrical rms current. Within parenthesis is shown 678.76 MVA. The bus voltage is 13,800 V, and the equivalent impedance is $0.0122 + j0.283\ \Omega$. Based upon this impedance value, the short-circuit current of 28.397 kA at −87.51 DEG can be calculated on 13.8 kV as a verification. The $X/R = 37.23$. If we calculate *X/R* from complex impedance, it is 23.19. This is different because we calculate *X/R* based upon separate *X* and *R* network reduction and not from the complex impedance.

Then the SYM*1.6 current is shown, which means that 28.397 kA is multiplied by 1.6 = 45.436 kA to calculate the rms asymmetrical rating. This factor of 1.6 is based on the rating structure of the earlier medium voltage circuit breakers $K > 1$ at $X/R = 25$. Higher factors will be obtained with higher *X/R*. *When selecting a circuit breaker, it is*

```
G1        E/Z:      28.397 KA AT  -87.51 DEG ( 678.76 MVA)  X/R:     37.23
          SYM*1.6:   45.436 KA        MOMENTARY BASED ON X/R:   46.570 KA
          SYM*2.7:   76.673 KA            CREST BASED ON X/R:   77.070 KA
          VOLTAGE:  13800.   EQUIV. IMPEDANCE=  0.0122 + J  0.2803  OHMS
          CONTRIBUTIONS:  GEN1                7.872 KA      ANG:   -88.36
          RG1             SYN                11.561 KA      ANG:   -88.83

          C1011           101A                0.322 KA      ANG:   -83.70
          C1021           102A                0.655 KA      ANG:   -84.58
          C1031           103A                1.432 KA      ANG:   -86.02
          C104A1          104AA               0.406 KA      ANG:   -84.51
          C104B1          104BA               1.560 KA      ANG:   -85.28
          C1051           105A                0.230 KA      ANG:   -83.54
          C1111           111A                1.093 KA      ANG:   -84.93
          C1121           112A                0.116 KA      ANG:   -83.56
          C1131           113A                0.238 KA      ANG:   -83.63
          C1151           115A                0.458 KA      ANG:   -83.51
          C116A1          116AA               1.839 KA      ANG:   -85.60
          C116B1          116BA               0.628 KA      ANG:   -84.96
```

(a)

```
G1        E/Z:      23.786 KA AT  -88.03 DEG ( 568.55 MVA)  X/R:     41.59
          VOLTAGE:  13800.   EQUIV. IMPEDANCE=  0.0115 + J  0.3348  OHMS
          CONTRIBUTIONS:  GEN1                7.872 KA      ANG:   -88.36
          RG1             SYN                11.099 KA      ANG:   -88.86
          C1011           101A                0.144 KA      ANG:   -83.76
          C1021           102A                0.295 KA      ANG:   -84.94
          C1031           103A                0.880 KA      ANG:   -86.61
          C104A1          104AA               0.180 KA      ANG:   -84.77
          C104B1          104BA               1.050 KA      ANG:   -85.66
          C1051           105A                0.106 KA      ANG:   -83.99
          C1111           111A                0.514 KA      ANG:   -85.25
          C1121           112A                0.053 KA      ANG:   -84.00
          C1131           113A                0.107 KA      ANG:   -84.04
          C1151           115A                0.211 KA      ANG:   -83.98
          C116A1          116AA               1.004 KA      ANG:   -86.16
          C116B1          116BA               0.278 KA      ANG:   -85.04

          GENERATOR NAME -- AT BUS --    KA     VOLTS PU  LOCAL/REMOTE
       GEN1                            7.872     0.00          L

       GEN2                            2.854     0.77          L
       GEN-3                           2.324     0.67          L
       utility                         3.155     1.00          R
          TOTAL REMOTE:      3.155 KA  NACD RATIO:  0.1326

                        SYM2      SYM3      SYM5      SYM8
          MULT. FACT:   1.109     1.133     1.114     1.120
          DUTY (KA) :  26.380    26.958    26.505    26.645

                        TOT2      TOT3      TOT5      TOT8
          MULT. FACT:   1.524     1.362     1.211     1.116
          DUTY (KA) :  36.251    32.400    28.816    26.541
```

(b)

Figure 5.4. (a) Typical computer-based calculation results, momentary (first cycle) short-circuit duty calculations, (b) interrupting duty calculations, and (c) low voltage circuit breakers duty calculations.

```
27364A      FAULT:   20.632 KA AT  -78.54 DEG (  17.15 MVA)  X/R:      5.17
            VOLTAGE:     480.   EQUIV. IMPEDANCE=  0.0027 + J  0.0132  OHMS
            LOW VOLTAGE POWER CIRCUIT BREAKER    20.632 KA
            MOLDED CASE CIRCUIT BREAKER > 20KA   20.875 KA
            CONTRIBUTIONS:  E29626AA             20.632 KA      ANG:  -78.54

29A366N     FAULT:   20.055 KA AT  -78.61 DEG (  16.67 MVA)  X/R:      5.21
            VOLTAGE:     480.   EQUIV. IMPEDANCE=  0.0027 + J  0.0135  OHMS
            LOW VOLTAGE POWER CIRCUIT BREAKER    20.055 KA
            MOLDED CASE CIRCUIT BREAKER > 20KA   20.327 KA
            CONTRIBUTIONS:  29AE365NA            20.055 KA      ANG:  -78.61

29AE365NA   FAULT:   21.458 KA AT  -81.12 DEG (  17.84 MVA)  X/R:      6.85
            VOLTAGE:     480.   EQUIV. IMPEDANCE=  0.0020 + J  0.0128  OHMS
            LOW VOLTAGE POWER CIRCUIT BREAKER    21.609 KA
            MOLDED CASE CIRCUIT BREAKER > 20KA   22.943 KA
            CONTRIBUTIONS:  MTRI-0180             1.731 KA      ANG:  -75.86
                            MTRI-0179             5.061 KA      ANG:  -84.38
            29AE365N        115A                 14.681 KA      ANG:  -80.62
```

(c)

Figure 5.4. *(Continued)*

*important that the current is adjusted for the actual fault point* X/R *ratio.* This is shown next in the computer calculations, "Momentary Based on X/R" = 46.570 kA. This means that at the calculated *X/R* = 37.23, the multiplying factor = 1.64.

Though the manufacturers have been specifying momentary ratings (close and latch capabilities) of indoor, oil-less circuit breakers in terms of rms asymmetrical, the ANSI/IEEE standards do not publish these ratings. The ratings are published only in peak current [5]. This peak factor is 2.7 based upon *X/R* = 25, and is again adjusted for actual *X/R*, which gives 77.07 kA peak. For older HV circuit breakers, calculate asymmetrical rms peak based upon 1/2 cycle peak, using Equation (5.1). Substituting the numerical values in this equation, the peak = 77.07 kA is checked, as shown in the computer outputs.

Next, the contributions from all the circuits connected to bus G1 are shown. If we vectorially sum up these contributions:

$$\vec{I}_1 + \vec{I}_2 + \vec{I}_3 + \ldots. + \vec{I}_n = \vec{I}_T, \tag{5.9}$$

this results in the total current of 28.387 kA at -87.51 DEG, as shown in the computer output. For arc flash, we will use 28.397 kA bolted three-phase symmetrical fault current. For coordination of the protective devices, which operate instantaneously, the asymmetrical rms current is used, in this case 46.57 kA.

### 5.9.2 Interrupting Duty Calculations

Figure 5.4b shows the results of calculations. The interrupting duty calculation shows symmetrical rms current based upon *E/Z* = 23.786 kA. This is lower than 28. 39.7 kA for the first cycle or momentary calculations in Figure 5.4a. Compare the multiplying factors for first cycle and interrupting duty calculations in Table 5.1. It can be seen that the interrupting duty contributions from the various rotating motor loads are lower, compared with the first cycle calculations. After the listing of contributions from the individual buses, the local and remote sources are identified. The letter "L" signifies local, and "R" the remote. There are two more generators in the system, generator G2 and G3, and the current contributed by these generators through system impedances meets the definition of local sources. Utility contribution is considered as remote source. The motor contributions are neither termed local nor remote, and then NACD is calculated based on Equation (5.8). This is 0.1326. Next, the weighted multiplying factors for various rating-structures of the circuit breakers are tabulated. SYM5 signifies a circuit breaker of interrupting time five cycles and rated on symmetrical current basis. SYM3 similarly signifies a circuit breaker rated on symmetrical current basis and of three-cycle interrupting time. TOT signifies "total current basis." This is prior to circa 1964, and these circuit breakers are no longer manufactured, but some may be still in service.

For arc flash, we will use 23.786 kA bolted three-phase symmetrical current irrespective of the circuit breaker type and its rating structure. For a five-cycle circuit breaker rated on symmetrical current basis, the multiplying factor is 1.114, and a circuit breaker should be selected so that its interrupting rating is >1.114 × 23.786 = 26.505 kA.

### 5.9.3 Low Voltage Circuit Breaker Duty Calculations

Figure 5.4c shows the duty calculations on three low voltage buses of 480 V, and provides three different tabulations for:

- LVPCB (low voltage power circuit breaker)
- molded case circuit breaker (MCCB) <20 kA
- MCCBs >20 KA.

Different multiplying factors apply to these types of circuit breakers because of different test power factors. The duties are corrected for the actual *X/R* ratio (power factor). See Chapter 7.

For arc flash, we will use 20.632-kA bolted three-phase current for bus 27364 A, and no multiplying factors need to be considered.

## 5.10 EXAMPLES OF CALCULATIONS

The short-circuit calculations in practical systems are demonstrated through examples.

**Example 5.1**

Figure 5.5 shows a low voltage distribution system. It is required to calculate short-circuit currents and low voltage circuit breaker duties on buses 2 and 3.

The raw impedance data of the system, as shown in Table 5.4, is converted to a common 100-MVA base in per unit. A familiarity with basic circuit theory and per unit system is assumed, and the details of these calculations are not shown. Motors M1, M2, and M3 have a rated voltage of 460 V, while the bus voltage is 480 V. Appropriate modifications to the per unit impedance values should be made.

As we are calculating the short-circuit currents in a low voltage system, the first cycle impedance multiplying factors should be applied to the calculated motor impedances. This is shown in Table 5.5.

For a fault at bus 3, a positive sequence network can be constructed as shown in Figure 5.6. Again, a familiarity with construction of sequence networks is required [9].

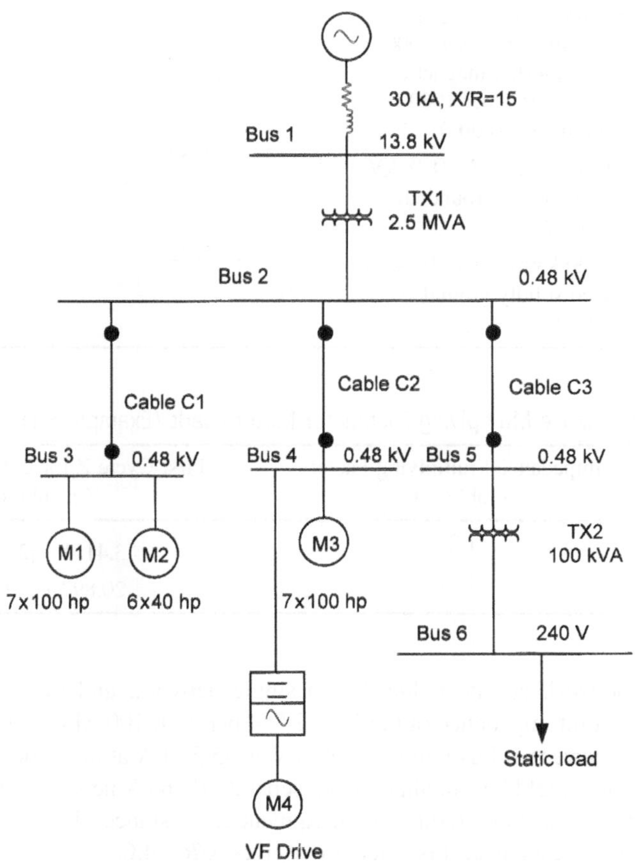


Figure 5.5. A low voltage distribution system for calculations of short-circuit currents (Example 5.1).

TABLE 5.4. Impedance Data for Example 5.1

| Description of Equipment | Per Unit Resistance on a 100-MVA Base | Per Unit Reactance on a 100-MVA Base |
|---|---|---|
| Utility's 13.8-kV source, three-phase fault level = 717.07 MVA, *X/R* ratio = 15 | 0.00928 | 0.13915 |
| 2.5-MVA transformer TX1, 13.8 - 0.48 kV, delta-wye connected solidly grounded, Z = 5.75%, X/R = 10.67 | 0.21462 | 2.2900 |
| Transformer TX2, 100 kVA, 0.48 - 0.24 kV, delta-wye connected solidly grounded, *Z* = 2.60%, *X/R* = 1.92 | 12.0103 | 23.0598 |
| 600 V cable C1, 2-3/C, 500 KCMIL in nonmagnetic conduit, 100 ft. *R* = 0.0276 Ω per 1000 ft and *X* = 0.03110 ohms per 1000 ft | 0.5990 | 0.67490 |
| 600-V cable C2, 2-3/C, 500 KCMIL in magnetic conduit, 100 ft. *R* = 0.0294 Ω per 1000 ft and *X* = 0.03490 ohms per 1000 ft | 0.6380 | 0.7574 |
| 600-V cable C3, 1-3/C, 4/0 in magnetic conduit, 50 ft. *R* = 0.0640 Ω per 1000 ft and X = 0.03810 ohms per 1000 ft | 1.3889 | 0.8268 |
| 2.5-MVA transformer TX1, 13.8 - 0.48 kV, delta-wye connected solidly grounded, Z = 5.75%, *X/R* = 10.67 | 2.8459 | 22.170 |
| Transformer TX2, 100 kVA, 0.48 - 0.24 kV, delta–wye connected solidly grounded, Z = 2.60%, *X/R* = 1.92 | 12.5132 | 61.690 |

TABLE 5.5. Impedance Multiplying Factors for Motor Loads (Example 5.1)

| Motor | Impedance Multiplying Factor (Table 5.1) | First-Cycle Z for L-V Short-Circuit Calculations |
|---|---|---|
| M1, M3 | 1.2 | 3.4151 + j26.604 |
| M2 | 1.67 | 20.897 + j103.022 |

The complex network can be reduced to a single network, and the results give an equivalent fault point impedance of $0.674 + j2.554$ per unit, 100-MVA base. Therefore, the short-circuit current at bus voltage of 480 V is 45.53 kA at an angle of −75.22°.

The *X/R* ratio should be calculated from separate *R* and *X* networks. In Figure 5.6, first ignore the reactance and reduce to an equivalent resistance, then ignore the resistance and calculate reactance; this calculation gives *X/R* = 4.0.

An equivalent circuit for the short circuit at bus 2 can be similarly drawn and reduced to an equivalent complex impedance. This is shown in Figure 5.7 and the

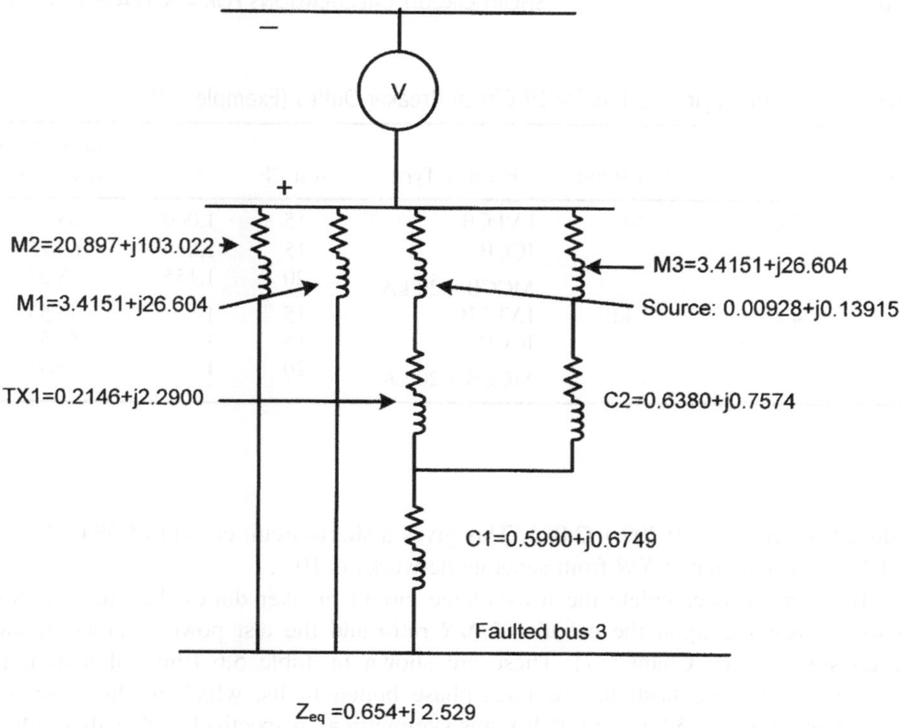


**Figure 5.6.** Equivalent network for a fault at bus 3 (Figure 5.5 and Example 5.1).

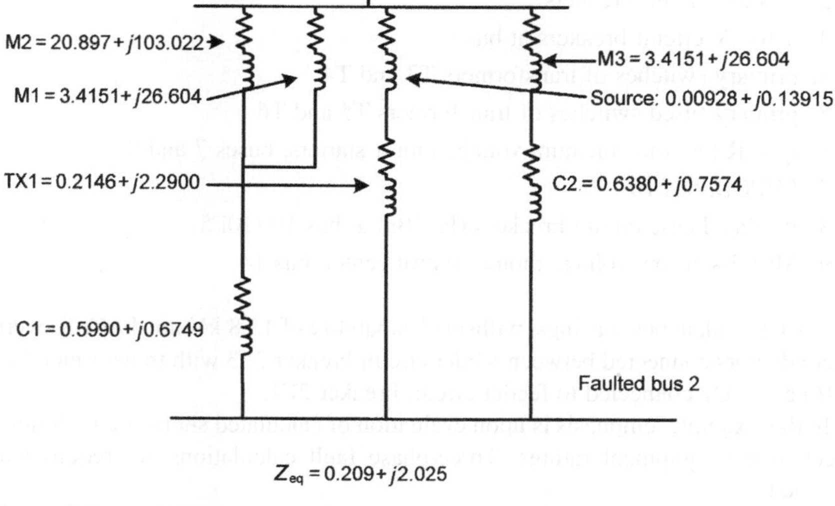


**Figure 5.7.** Equivalent network for a fault at bus 2 (see Figure 5.5 and Example 5.1).

TABLE 5.6. Multiplying Factors for LV Circuit Breaker Duties (Example 5.1)

| Bus | kA rms sym | X/R Ratio | Breaker Type | Test PF | MF | Adj sym kA Breaker Duty |
|---|---|---|---|---|---|---|
| 2 | 59.08 | 10.1 | LVPCB | 15 | 1.069 | 63.15 |
| | | | ICCB | 15 | 1.069 | 63.15 |
| | | | MCCB > 20 kA | 20 | 1.155 | 68.23 |
| 3 | 45.53 | 4.0 | LVPCB | 15 | 1 | 45.53 |
| | | | ICCB | 15 | 1 | 45.53 |
| | | | MCCB > 20 kA | 20 | 1 | 45.53 |

reduced impedance = 0.209 + *j*2.025. This gives a short-circuit current of 59.08 kA at -84.11°. The calculated *X/R* from separate networks is 10.1.

If we need to calculate the low-voltage circuit breaker duties then multiplying factors depending upon the calculated *X/R* ratio and the test power factors should be considered (see Chapter 7). These are shown in Table 5.6 (this calculation is not required for arc flash hazard three-phase bolted faults, which in this case for buses 2 and 3 are 45.53 and 59.08 kA rms symmetrical, respectively). See also Reference [6].

**Example 5.2**
Figure 5.8 illustrates a multivoltage level distribution system. Short-circuit duties are required to verify the adequacy of ratings of the switching devices shown in this single-line diagram. These devices are:

1. 13.8-kV circuit breakers at buses 1, 2, and 3
2. 138-kV circuit breakers
3. 4.16-kV circuit breakers at bus 6
4. primary switches of transformers T3 and T4
5. primary fused switches of transformers T5 and T6
6. type R fuses for medium voltage motor starters, buses 7 and 8
7. LVPCB, bus 10
8. insulated case circuit breakers (ICCBs) at bus 10 (10F5)
9. MCCBs at low voltage motor control center, bus 14.

Also, calculate bus bracings, withstand capability of 13.8 kV no. 4 ACSR overhead line conductors connected between feeder circuit breaker 2F3 with transformer T7, and no. 4/0 cable C1 connected to feeder circuit breaker 2F4.

In this example, emphasis is upon evaluation of calculated short-circuit duties with respect to the equipment ratings. Three-phase fault calculations are required to be performed.

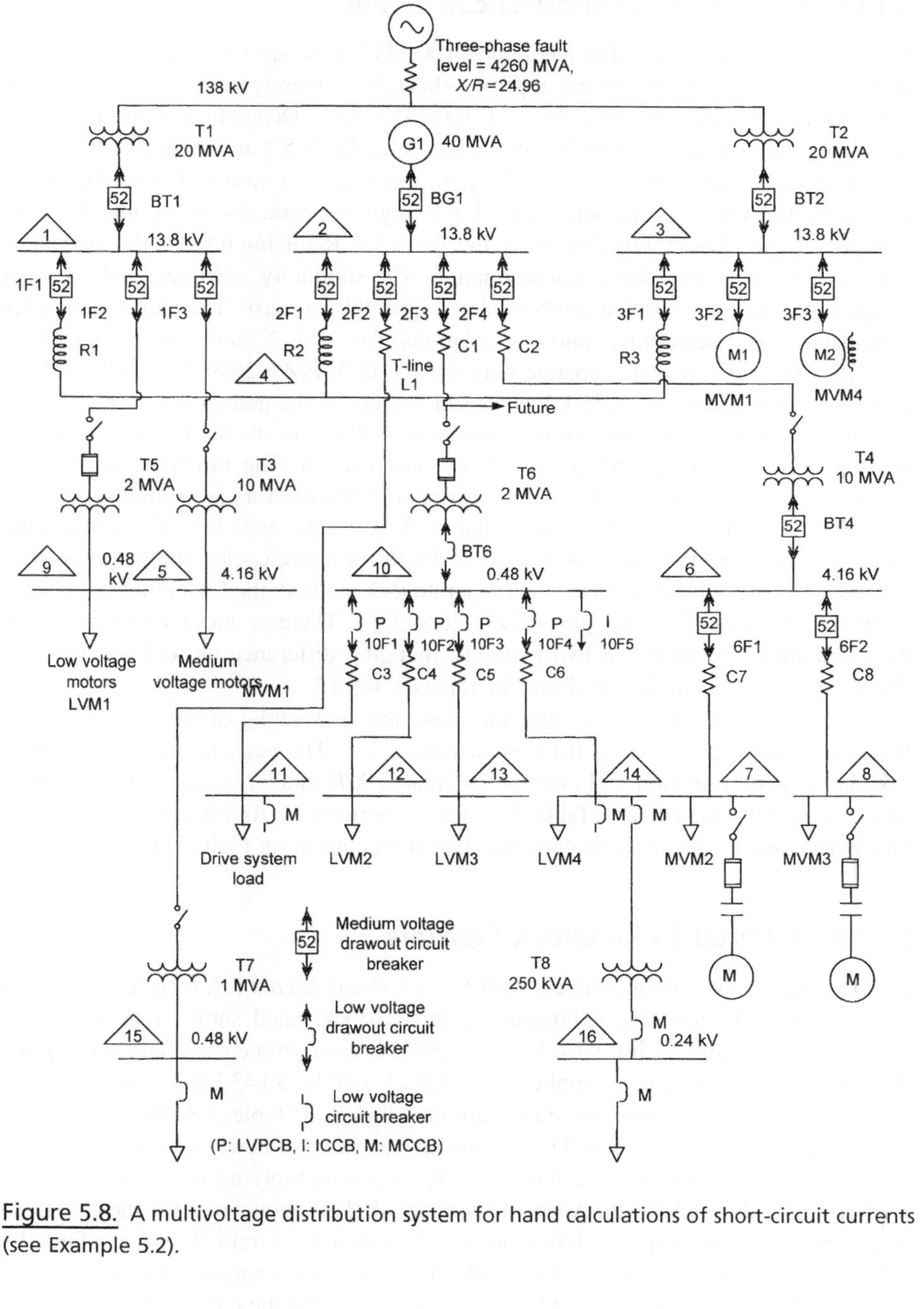


Figure 5.8. A multivoltage distribution system for hand calculations of short-circuit currents (see Example 5.2).

### 5.10.1 Calculation of Short-Circuit Duties

The impedance data reduced to a common 100-MVA base are shown in Table 5.7. No impedance multiplying factors are applied—this table is merely a conversion of the raw equipment impedance data to a common 100-MVA base. Depending upon the type of calculation, the impedance multiplying factors from Table 5.1 are applied to the calculated impedances in Table 5.7. The fault point networks for various faulted buses can be constructed, one at a time, and reduced to a single network. As an example, the fault network for the 13.8-kV bus 2 is shown in Figure 5.9. Reducing it to single impedance requires wye-delta impedance transformation. The simplicity, accuracy, and speed of computer methods of solution can be realized from this exercise. The reduced complex impedance for interrupting duty calculations for bus 2 fault is: $Z = 0.003553 + j0.155536$, and $X/R$ from separate networks is 47.9; $E/Z = 26.892 < -88.7°$. All the generator contribution of 14.55 kA is a local source, as the generator is directly connected to the bus. The remote (utility) source contributes to the fault at bus 2 through transformers T1 and T2 and synchronizing bus reactors. The utility's contributions through transformers T1 and T2 and synchronizing bus reactors are summed up. This gives 9.01 kA. The remote/total ratio, that is, NACD ratio is 0.335. The multiplying factor is 1.163, and the interrupting duty is 31.28 for a five-cycle symmetrical circuit breaker. If the calculation is based on a separate $R$–$X$ method, the fault point impedance is: $0.003245 + j0.15521$. This gives $E/Z = 26.895$ kA. There is not much difference in the calculations by using the two methods, though a difference up to 5% can occur. The results of calculations are shown in Tables 5.8–5.15.

Table 5.8 shows the interrupting and close and latch rating of the 13.8 kV circuit breakers according to revised IEEE standards, $K = 1$. The peak first cycle current is calculated using Equation (5.1) for the calculated $X/R$ ratio. *The interrupting duty is based upon bus fault current.* Table 5.9 shows contributions to this bus from all other buses. The vector sum of these currents gives total current on faulted bus 2.

### 5.10.2 *K*-Rated 15 kV Circuit Breakers

If old ratings of the circuit breakers with $K = 1.3$ is considered [5], then for a standard five-cycle symmetrical circuit breaker rating of 15 kV, rated short-circuit current is 28 kA, and close and latch is 97 kA peak or 58 kA rms asymmetrical. The interrupting duty current at the voltage of application, 13.8 kV will be 30.43 kA rms symmetrical. For the calculated interrupting duty currents shown in Table 5.8, the multiplying factors, the $X/R$ ratios, and NACD ratio remains the same as newly rated circuit breakers. The first cycle duty at the calculated X/R gives a multiplying factor of 1.659, and the first cycle duty of 46.48 kA rms asymmetrical. This shows that for these *K*-rated circuit breakers, interrupting duties on bus 2 exceed the circuit breaker ratings by 2.79%, though the close and latch duty of 46.48 kA rms asymmetrical is much below the circuit breaker ratings. It would be hasty to suggest that the entire bus 2 switchgear be replaced.

As the duties are calculated for a bus fault, it is necessary to calculate the duties on individual circuit breakers on this bus, by dropping the short-circuit contributions

TABLE 5.7. Impedance Data (100 MVA Base) Distribution System (Example 5.2)

| Symbol | Equipment Description | Per Unit Resistance | Per Unit Reactance |
|---|---|---|---|
| U1 | Utility source, 138 kV, 4260 MVA, $X/R$ = 25 | 0.00094 | 0.02347 |
| G1 | Synchronous generator, 13.8 kV, 40 MVA, 0.85 power factor, saturated subtransient reactance = 11.5%, saturated transient = 15%, $X/R$ = 56.7 | 0.00507 | 0.28750 |
| R1, R2, R3 | Reactors, 13.8 kV, 2 kA, 0.25 ohms, 866 KVA, $X/R$ = 88.7 | 0.00148 | 0.13127 |
| T1, T2 | 20/33.4 MVA, OA/FA, 138 - 13.8 kV, delta-wye transformers, $Z$ = 8.0% on 20 MVA OA rating, $X/R$ = 21.9, wye winding low resistance grounded through 400 A, 10 seconds resistor | 0.01827 | 0.39958 |
| T3, T4 | 10/14 MVA, OA/FA, 13.8 - 4.16 kV, delta-wye transformer, Z = 5.5%, $X/R$ = 15.9, wye-winding low-resistance grounded through 200 A 10 seconds resistor | 0.03452 | 0.54892 |
| T5, T6 | 2/2.58 MVA, OA/FA, 13.8 - 0.48 kV, delta-wye transformer, $Z$ = 5.75%, $X/R$ = 6.3, wye-winding high-resistance grounded | 0.44754 | 2.83995 |
| T7 | 1/1.29 MVA, AA/FA, 13.8 - 0.48 kV delta-wye transformer, $Z$ = 5.75%, $X/R$ = 5.3, wye-winding high-resistance grounded | 1.06494 | 5.65052 |
| T8 | 250 KVA, AA, 0.48 - 0.24 kV delta-wye transformer, $Z$ = 4%, $X/R$ = 2.7, solidly grounded | 5.49916 | 15.02529 |
| C1 | 1-3/C #4/0 15-kV grade shielded, MV-90, IAC (interlocked armor), XLPE cable laid in aluminum tray, 200 ft | 0.00645 | 0.00377 |

*(Continued)*

TABLE 5.7 (*Continued*)

| Symbol | Equipment Description | Per Unit Resistance | Per Unit Reactance |
|---|---|---|---|
| C2 | 1-3/C 500 KCMIL, 15-kV grade shielded, MV-90, IAC, XLPE cable laid in aluminum tray, 400 ft | 0.00586 | 0.00666 |
| C7, C8 | 2-3/C, 350 KCMIL, 5-kV grade, shielded, MV-90, XLPE, IAC cable, laid in aluminum tray, 100 ft | 0.01116 | 0.00790 |
| C3, C4, C5, C6 | 3-3/C, 500 KCMIL, 0.6-kV grade, THNN, 90°C cables, laid in tray, 60 ft | 0.23888 | 0.23246 |
| M1 | 1 × 12,000-hp, squirrel cage induction motor, 2 pole (10,800 kVA), locked rotor reactance = 16.7%, $X/R$ = 46 | 0.0336 | 1.54630 |
| M2 | 1 × 10,000-hp synchronous motor, 8-pole, 0.8 power factor (10,000 kVA), $X$ = 20%, $X/R$ = 34.4 | 0.05822 | 2.0000 |
| MVM1 | 2 × 1500-hp squirrel cage induction, 2-pole, 5 × 300-hp induction motors, 4-pole, and 1 × 3500-hp, 12 pole, 0.8 power factor synchronous motor | 0.21667<br>0.77958<br>0.20324 | 6.1851 (2 × 1500-hp)<br>11.719(5 × 300-hp)<br>5.7142 (3500-hp) |
| MVM2, MVM3 | 1 × 1800-hp, 12-pole, 0.8-power factor synchronous and 3 × 1100-hp, 6-pole induction motors | 0.45904<br>0.21675 | 11.111 (1800 hp)<br>5.6229 (3 × 1100 hp) |
| LVM1 | Grouped 1350-hp induction motors ≥50- and 640-hp induction motors <50 hp | 1.4829<br>5.4362 | 12.370 (>50 hp)<br>26.093 (<50 hp) |
| LVM2, | Grouped 300-hp induction motors ≥50 hp and | 6.0664 | 50.606 (>50 hp) |
| LVM3, LVM4 | 172 hp motors, <50 hp | 20.227 | 97.093 (<50 hp) |
| L1 | 5200 ft long, GMD = 4 ft, ACSR conductor A-AA class (SWAN), #4 AWG | 1.41718 | 0.41718 |

2-3/C: Abbreviated for two three-conductor cables per phase. MV-90, XLPE, THNN: Cable insulation types. The impedance of a cable is also a function of its construction and method of installation.

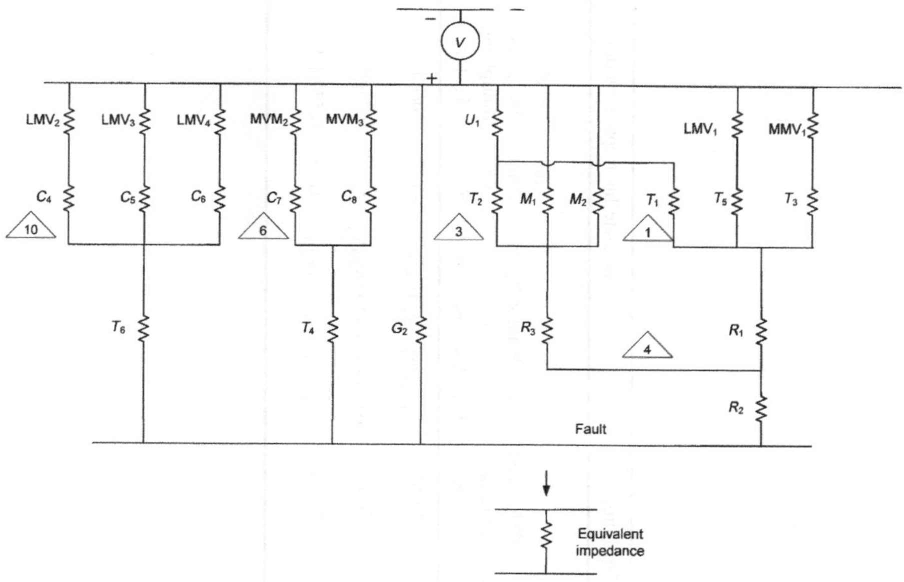


Figure 5.9. Equivalent network for a fault at 13.8 kV bus 2 (see Example 5.2).

from the loads connected to each of these circuit breakers. Neglecting the load contribution means that the fault point *X/R* ratio and duty multiplying factors will change. The short-circuit current component of the feeder circuit breaker can be vectorially subtracted from the bus fault current to calculate the duty of a circuit breaker connected to that bus. Table 5.9 shows contributions from each source and Table 5.10 shows the interrupting duties on the feeder circuit breakers connected to bus 2. It is observed that the feeder circuit breaker 2F2 to the transmission line and the feeder breaker 2F3 to the 2-MVA transformer are overdutied. As the feeder circuit breaker 2F2 has no rotating loads, its interrupting duty is the same as that for a bus fault. It is generally possible to retrofit these two circuit breakers with breakers of higher interrupting rating, depending on the manufacturer and age of the equipment. The newly rated K = 1 circuit breakers can usually be fitted in the same circuit breaker cubicles with minor modifications. Increasing bus tie reactor impedance by providing another reactor in series with the existing reactor could be another solution, especially if a system expansion is planned. Some remedial measures to the short-circuit problems are:

- retrofitting overdutied circuit breakers with new circuit breakers
- replacing with new K = 1 circuit breakers or entirely new switchgear of higher ratings
- adding current-limiting reactors
- redistribution of loads and reorganization of distribution system
- short-circuit current limiters [17]

TABLE 5.8. Calculated Duties on 13.8 kV Breakers (Example 5.2)

| | | | | Calculate First-Cycle Duties (Close and Latch Capabilities) | | | | Calculate Interrupting Duties | | | |
|---|---|---|---|---|---|---|---|---|---|---|---|
| Bus No. | Breaker Type | Breaker Close and Latch Capability (kA, Peak) | Breaker Interrupting Rating (kA sym rms) | Fault Point *Z* Based on Complex Impedance Reduction | Fault Point *X/R* Ratio Based upon Separate Networks | Multiplying Factor Based on *X/R* | Calculated Duty (kA Peak) | Fault Point *Z* Based on Complex Impedance Reduction | Fault Point *X/R* Ratio | NACD/ Weighted Multiplying Factor | Calculated Duty (kA rms sym) |
| 1 | 15 kV, indoor | 82 | 31.5 | 0.005767 + *j*0.175930 | 41.61 | 2.727 | 64.81 | 0.005831 + *j*0.18426 | 41.73 | 0.6/1.169 | 26.53 |
| 2 | oil-less, 5-cycle | Ref. [2], k = 1 | | 0.003578 + *j*0.149318 | 47.84 | 2.740 | 76.75 | 0.003553 + *j*0.155536 | 47.92 | 0.335/1.163 | 31.28 |
| 3 | sym. K = 1 | | | 0.004903 + *j*0.159347 | 41.05 | 2.726 | 71.54 | 0.005291 + *j*0.171520 | 41.16 | 0.571/1.162 | 28.33 |

TABLE 5.9. Flow of Component Interrupting Duty Short-Circuit Currents from Connected Buses to 13.8-kV Bus 2 (Example 5.2)

| From Bus | Current Amplitude and Angle | *X/R* |
|---|---|---|
| Gen. 2 | 24.550 < -88.988° | 56.61 |
| Bus 4 | 11.021 < -88.491° | 37.96 |
| Bus 6 | 1.244 < -87.338° | 21.51 |
| Bus 10 | 0.078 < -82.965° | 8.10 |
| Bus 15 | 0 < 0° | – |
| Vector sum of currents | 26.9 < -88.805 | 47.92 |

TABLE 5.10. Calculated Duties on Feeder Circuit Breakers on 13.8-kV Bus 2 (Example 5.2)

| Breaker ID | Breaker Service | Breaker Rating Interrupting (kA sym rms) | Breaker Calculated Interrupting Duty (kA sym rms) |
|---|---|---|---|
| BG2 | Generator breaker | 30.43 | 16.94 |
| 2F1 | Synchronous bus | | 18.68 |
| 2F2 | 13.8-kV distribution line | | 31.28 (overexposure = 2.79%) |
| 2F3 | 2-MVA transformer | | 31.18 (overexposure = 2.46%) |
| 2F4 | 10-MVA transformer | | 30.03 |

Circuit Breakers rated on 1979 Basis $K = 1.3$ [2].

- duplex reactors
- series connected devices for low-voltage systems (see Chapter 7).

A detailed discussion of these topics is not covered.

### 5.10.3 4.16-kV Circuit Breakers and Motor Starters

Table 5.11 tabulates 4.16-kV metal-clad circuit breaker ratings and calculated duties for K factor-rated circuit breakers, and Table 5.12 shows a similar comparison of R-type fuses in medium voltage motor starters, NEMA type E2. These devices are applied much below their short-circuit ratings. For the fuses in the medium voltage starters, it is not necessary to construct the interrupting duty network.

### 5.10.4 Transformer Primary Switches and Fused Switches

Short-circuit ratings of transformer primary switches (without fuses) are specified in terms of asymmetrical kA rms and 10-cycle fault closing. The former rating indicates

TABLE 5.11. Calculated Duties on 4.16-kV Breakers (Example 5.2)

| | | | | Calculate First-cycle Duties (Close and Latch Capabilities) | | | | Calculate Interrupting Duties | | | |
|---|---|---|---|---|---|---|---|---|---|---|---|
| Bus No. | Breaker Type | Breaker Close and Latch Capability (kA asym rms) | Breaker Interrupting Rating at Voltage of Application (kA sym rms) | Fault Point *Z* Based on Complex Impedance Reduction | Fault Point *X/R* Ratio | Multiplying Factor Based on *X/R* | Calculated Duty (kA rms asym.) | Fault Point *Z* Based on Complex Impedance Reduction | Fault Point *X/R* Ratio | NACD/ Weighted Multiplying Factor | Calculated Duty (kA rms sym.) |
| 6 | 4.16 kV, indoor, oil-less, 5 cycle sym. | 58 (=97 kA crest) K = 1.24, Ref. [2] | 30.43 | 0.029041 + *j*0.517160 | 18.41 | 1.649 | 41.70 | 0.032647 + *j*0.572088 | 18.03 | 0.475/1.00 | 24.22 |

TABLE 5.12. Short-Circuit Duties on 4.16-kV MCC (Example 5.2)

| | | | Calculated First-cycle Duties (sym and asym) | | | |
|---|---|---|---|---|---|---|
| Bus No. | Motor Starter Fuse Type | Fuse Interrupting Rating (kA rms sym/asym) | Fault Point *Z* Based on Complex Impedance Reduction | Fault Point *X/R* Ratio | Multiplying Factor Based on *X/R* | Calculated Duty (kA rms sym/ asym) |
| 7,8 | Current limiting type R | 50/80 | 0.037043 + *j*0.522930 | 15.09 | 1.523 | 26.47/40.31 |

TABLE 5.13. Calculated Duties 13.8-kV Transformer Primary Switches and Fuses (Example 5.2)

| Transformer | Transformer Primary Switch/ Fused Switch | Short-Circuit Ratings | Fault-point Impedance (100-MVA base) | Fault Point *X/R* | Symmetrical Current (kA rms) | Asymmetrical Current (kA rms) |
|---|---|---|---|---|---|---|
| 10 MVA | Switch only | Switch 61 kA rms asym. Fault closing 10 cycles = 61 kA rms asym. | 0.008709 + *j*0.153013 | 18.48 | 26.926 | 41.92 |
| 2 MVA | Fused switch, with current-limiting type class E fuse | Fuse: interrupting = 50 kA rms, sym. = 80 kA rms, asym. | 0.009926 + *j*0.153013 | 16.04 | 27.285 | 41.84 |

TABLE 5.14. Short-Circuit Duties on Low Voltage Circuit Breakers (Example 5.2)

| Breaker Identification | Breaker Interrupting Rating (kA sym) | Fault Point Z per Unit (100-MVA Base) | Fault Point *X/R* | Multiplying Factor | *E/Z* | Calculated Duty (kA sym) |
|---|---|---|---|---|---|---|
| Bus 10, LVPCB | 50 | 0.37625 + *j*2.491190 | 6.65 | 1.002 | 47.73 < -81.36° | 47.82 |
| Bus 10, ICCB | 65 | 0.37625 + *j*2.491190 | 6.65 | 1.063 | 47.73 < -81.36° | 50.73 |
| Bus 14, MCCB1 | 65 | 0.588317 + *j*2.697419 | 4.68 | 1 | 43.57 < -77.70° | 43.57 |
| Bus 11, MCCB | 35 | 0.617505 + *j*2.723650 | 4.44 | 1 | 43.07 < -77.23° | 43.07 |
| Bus 15, MCCB | 35 | 2.480748 + *j*6.217019 | 2.51 | 1 | 17.97 < -68.25° | 17.97 |
| Bus 16, 240 V MCCB | 10 | 6.087477 + *j*17.72271 | 2.92 | 1 | 12.84 < 71.04° | 12.84 |

TABLE 5.15. Calculated Duties on 138-kV Breakers (Example 5.2)

| | | | | Calculate First-Cycle Duties (Close and Latch ca Pabilities) | | | | Calculate Interrupting Duties | | | |
|---|---|---|---|---|---|---|---|---|---|---|---|
| Bus No. | Breaker Type | Breaker Close and Latch Capability (kA, Peak) | Breaker Interrupting Rating at Voltage of Application (kA sym rms) | Fault Point *Z* Based on Complex Impedance Reduction and Current kA | Fault Point *X/R* Ratio Based upon Separate Networks | Multiplying Factor Based on *X/R* | Calculated Duty (kA, Peak) | Fault Point *Z* Based on Complex Impedance Reduction and Current (kA) | Fault Point *X/R* Ratio Based upon Separate Networks | NACD/ Weighted Multiplying Factor | Calculated Duty (kA rms sym) |
| UTL | 138 kV-3 cycles, outdoor | 82 | 31.5 | 0.000888 + *j*0.022346<br>$I$ = 18.707 kA < -87.72$^0$ | 25.34 | 2.667 | 49.89 | 0.000890 + *j*0.022448<br>$I$ = 18.623 kA < -87.73$^0$ | 25.40 | 0.96/1.092 | 20.34 |

the maximum asymmetrical withstand current capability, and the latter signifies that the switch can be closed on to a fault for 10 cycles, with maximum fault limited to specified asymmetrical fault closing rating. The upstream protective devices must isolate the fault within 10 cycles. The short-circuit ratings on power fuses are discussed in Chapter 7. First-cycle calculations are required. Table 5.13 shows the comparison, and that the equipment is applied within its short-circuit ratings.

### 5.10.5 Low Voltage Circuit Breakers

The switching devices in low-voltage distribution should be categorized into LVPCBs, ICCBs, and MCCBs, as discussed in Chapter 7. These have different test power factors, depending on their type and ratings and interrupting duty multiplying factors are different. First-cycle calculation is required for ascertaining the duties. Table 5.14 depicts these calculations. It is observed that the short-circuit duties on MCCBs at buses 11 and 16 exceed the ratings. A reactor in the incoming service to these buses can be provided. Alternatively, the underrated MCCBs can be replaced, or series rated devices (Chapter 7) can be considered, as there are no downstream rotating motor loads.

Finally, Table 5.15 shows the calculations for the 138-kV circuit breaker. Note the slight increase in the short-circuit currents in the utility source due to rotating motors and generator in the distribution system.

## 5.11 THIRTY-CYCLE SHORT-CIRCUIT CURRENTS

Thirty-cycle short-circuit currents are required for overcurrent devices co-coordinated on a time–current basis. The 30-cycle short-circuit current is calculated on the following assumptions:

- The contributions from the utility sources remain unchanged.
- The DC component of the short-circuit current decays to zero.
- The contribution from the synchronous and induction motors decays to zero.
- The generator subtransient reactance is replaced with transient reactance or a value higher than the subtransient reactance.

The 30-cycle current of the synchronous generators and motors will vary depending upon the excitation systems. If the power for excitation system is taken from the same bus to which the motor is connected, then for a fault close to the motor, the voltage reduces practically to zero and the field current will quickly decay depending upon machine time constants. If the excitation power is taken from an independent source not affected by the short circuit, the excitation will be maintained. This can be an important factor for machine stability. For brushless excitation systems of synchronous motors, the excitation power required is small, which can be provided by a battery source or uninterruptible power system enhancing the motor stability.

TABLE 5.16. 30-Cycle Currents, Distribution System (Example 5.4)

| Bus Identification | 30-Cycle Currents in kA sym. |
|---|---|
| 13.8-kV buses | |
| 1 | 19.91 < -88.12$^0$ |
| 2 | 21.11 < -88.68$^0$ |
| 3 | 19.91 < -88.12$^0$ |
| 4 | 22.73 < -88.30$^0$ |
| 4.16-kV buses | |
| 5 | 18.26 < -86.88$^0$ |
| 6 | 18.38 < -86.59$^0$ |
| 7 | 18.17 < -85.79$^0$ |
| 8 | 18.17 < -85.79$^0$ |
| Low voltage buses | |
| 9 | 39.01 < -81.53$^0$ |
| 10 | 39.10 < -81.43$^0$ |
| 11 | 35.93 < -77.98$^0$ |
| 12 | 35.93 < -77.98$^0$ |
| 15 | 17.85 < -68.39$^0$ |
| 16 | 12.45 < -71.29$^0$ |

**Example 5.3**
Calculate 30-cycle currents in the distribution system of Figure 5.8 (Example 5.2).

The generator subtransient reactance is changed to transient reactance, and all motor contributions are dropped. Table 5.16 shows the results, which can be compared with interrupting duty currents. The decay varies from 8% to 28%. The buses that serve the motor loads show the highest decay. These currents are used in the relay coordination for overcurrent devices for durations exceeding approximately 0.5 second.

## 5.12 UNSYMMETRICAL SHORT-CIRCUIT CURRENTS

Unsymmetrical fault types involving one or two phases and ground are:

- a single line-to-ground fault
- a double line-to-ground fault
- a line-to-line fault.

Unsymmetrical faults are more common. The most common type is a line-to-ground fault. Approximately 70% of the faults in power systems are single line-to-ground faults. While applying symmetrical component method to fault analysis, we will ignore the load currents. This makes the positive sequence voltages of all the generators in the system identical and equal to the prefault voltage.

The impedance encountered by the symmetrical components depends on the type of power system equipment, that is, a generator, a transformer, or a transmission line. As discussed in Chapter 4, grounding plays a major role. While in a solidly grounded system, the line-to-ground fault current can vary from low values to slightly more than the three-phase short-circuit currents; in resistance grounded systems, the current is approximately given by line to neutral voltage divided by the grounding resistor. This is so because the impedance of grounding resistor predominates and the sequence impedances of the system components are relatively low (see Chapter 4). The sequence impedances are required for component modeling and analysis for unsymmetrical faults. Zero sequence impedance of overhead lines depends on the presence of ground wires, tower footing resistance, and grounding. It may vary between two and six times the positive sequence impedance. The line capacitance of overhead lines is ignored in short-circuit calculations. While estimating sequence impedances of power system components is one problem, constructing the zero, positive, and negative sequence impedance networks is the first step for unsymmetrical fault current calculations. The zero sequence impedance of transformer varies with their winding connections.

In the analysis to follow, $Z_1$, $Z_2$, and $Z_0$ are the positive, negative, and zero sequence impedances as seen from the fault point; $V_a$, $V_b$, and $V_c$ are the phase to ground voltages at the fault point, prior to fault, that is, *if the fault does not exist*; and $V_1$, $V_2$, and $V_0$ are corresponding sequence component voltages. Similarly, $I_a$, $I_b$, and $I_c$ are the line currents, and $I_1$, $I_2$, and $I_0$ their sequence components. A fault impedance of $Z_f$ is assumed in every case. For a bolted fault, $Z_f = 0$.

### 5.12.1 Single Line-to-Ground Fault

Figure 5.10a shows that phase $a$ of a three-phase system goes to ground through an impedance $Z_f$. The flow of ground fault current depends on the method of system grounding. A solidly grounded system with zero ground resistance is assumed. There will be some impedance to the flow of fault current in the form of an impedance of the return ground conductor or the grounding grid resistance. A ground resistance can be added in series with the fault impedance $Z_f$. The ground fault current must have a return path through the grounded neutrals of generators or transformers. If there is no return path for the ground current, $Z_0 = \infty$, and the ground fault current is zero. This is an obvious conclusion.

Phase $a$ is faulted in Figure 5.10b. As the load current is neglected, currents in phases $b$ and $c$ are zero, and the voltage at the fault point, $V_a = I_a Z_f$. The sequence components of the currents are given by

$$\begin{vmatrix} I_0 \\ I_1 \\ I_2 \end{vmatrix} = \frac{1}{3}\begin{vmatrix} 1 & 1 & 1 \\ 1 & a & a^2 \\ 1 & a^2 & a \end{vmatrix}\begin{vmatrix} I_a \\ 0 \\ 0 \end{vmatrix} = \frac{1}{3}\begin{vmatrix} I_a \\ I_a \\ I_a \end{vmatrix}. \tag{5.10}$$

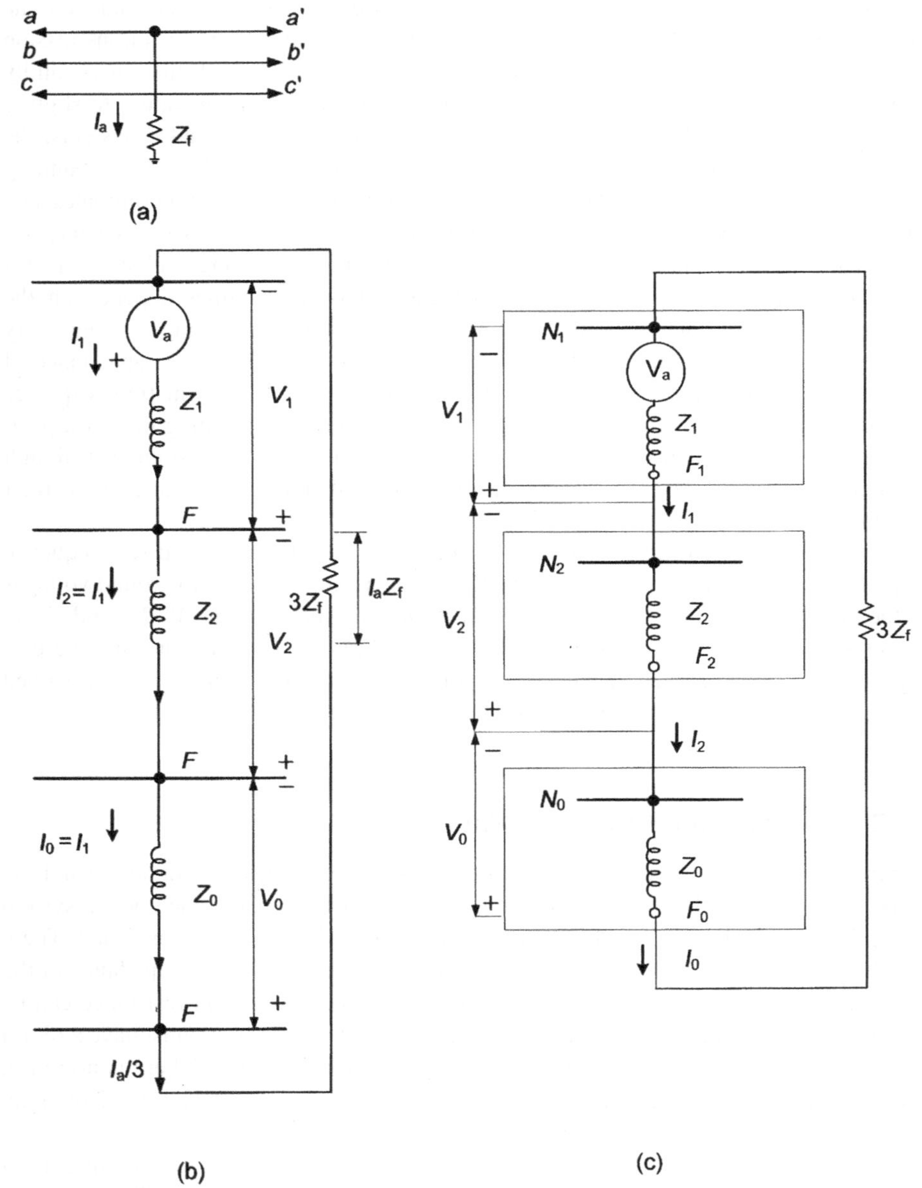


Figure 5.10. (a) Line-to-ground fault in a three-phase system, (b) line-to-ground fault equivalent circuit, and (c) sequence network interconnections.

Also,

$$I_0 = I_1 = I_2 = \frac{1}{3} I_a \tag{5.11}$$

$$3I_0 Z_f = V_0 + V_1 + V_2 = -I_0 Z_0 + (V_a - I_1 Z_1) - I_2 Z_2, \tag{5.12}$$

which gives

$$I_0 = \frac{V_a}{Z_0 + Z_1 + Z_2 + 3Z_f}. \tag{5.13}$$

The fault current $I_a$ is

$$I_a = 3I_0 = \frac{3V_a}{(Z_1 + Z_2 + Z_0) + 3Z_f}. \tag{5.14}$$

This shows that the equivalent fault circuit using sequence impedances can be constructed as shown in Figure 5.10b. In terms of sequence impedances' network blocks, the connections are shown in Figure 5.10c.

This result could also have been arrived at from Figure 5.10b:

$$(V_a - I_1 Z_1) + (-I_2 Z_2) + (-I_0 Z_0) - 3Z_f I_0 = 0, \tag{5.15}$$

which gives the same Equations (5.10) and (5.11). The voltage of phase $b$ to ground under fault conditions is

$$\begin{aligned} V_b &= a^2 V_1 + aV_2 + V_0 \\ &= V_a \frac{3a^2 Z_f + Z_2(a^2 - a) + Z_0(a^2 - 1)}{(Z_1 + Z_2 + Z_0) + 3Z_f}. \end{aligned} \tag{5.16}$$

Similarly, the voltage of phase $c$ can be calculated.

An expression for the ground fault current for use in grounding grid designs and system grounding is as follows:

$$I_a = \frac{3V_a}{(R_0 + R_1 + R_2 + 3R_f + 3R_G) + j(X_0 + X_1 + X_2)}, \tag{5.17}$$

where $R_f$ is the fault resistance; $R_G$ is the resistance of the grounding grid; $R_0$, $R_1$, and $R_2$ are the sequence resistances; $X_0$, $X_1$, and $X_2$ are sequence reactances.

### 5.12.2 Double Line-to-Ground Fault

A double line-to-ground fault is shown in Figure 5.11a. Phases $b$ and $c$ go to ground through a fault impedance $Z_f$. The current in the ungrounded phase is zero, that is, $I_a = 0$. Therefore, $I_1 + I_2 + I_0 = 0$.

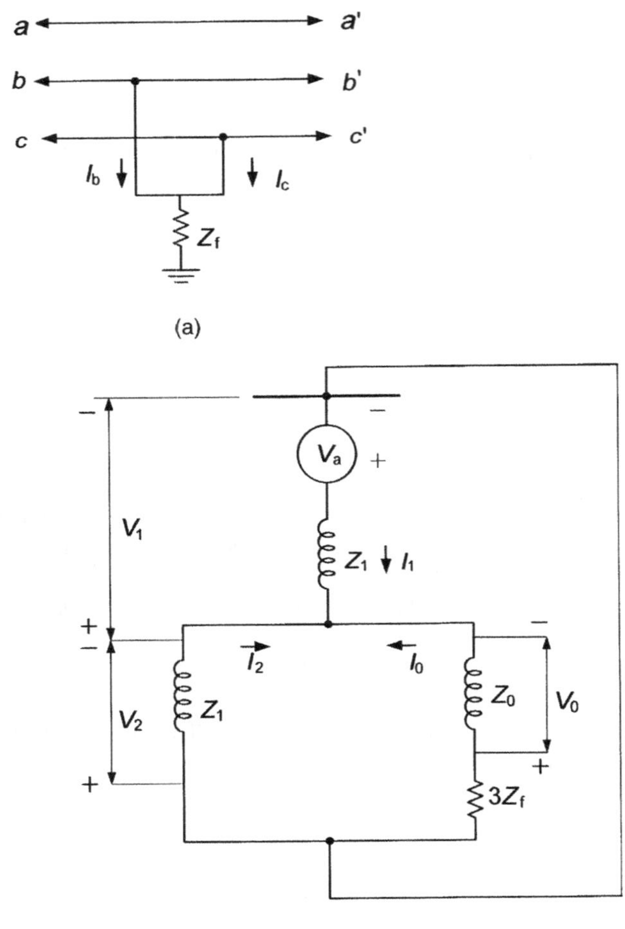


Figure 5.11. (a) Double line-to-ground fault in a three-phase system, (b) double line-to-ground fault equivalent circuit, and (c) sequence network interconnections.

$$V_b = V_c = (I_b + I_c)Z_f. \tag{5.18}$$

Thus,

$$\begin{vmatrix} V_0 \\ V_1 \\ V_2 \end{vmatrix} = \frac{1}{3}\begin{vmatrix} 1 & 1 & 1 \\ 1 & a & a^2 \\ 1 & a^2 & a \end{vmatrix}\begin{vmatrix} V_a \\ V_b \\ V_b \end{vmatrix} = \frac{1}{3}\begin{vmatrix} V_a + 2V_b \\ V_a + (a + a^2)V_b \\ V_a + (a + a^2)V_b \end{vmatrix}, \tag{5.19}$$

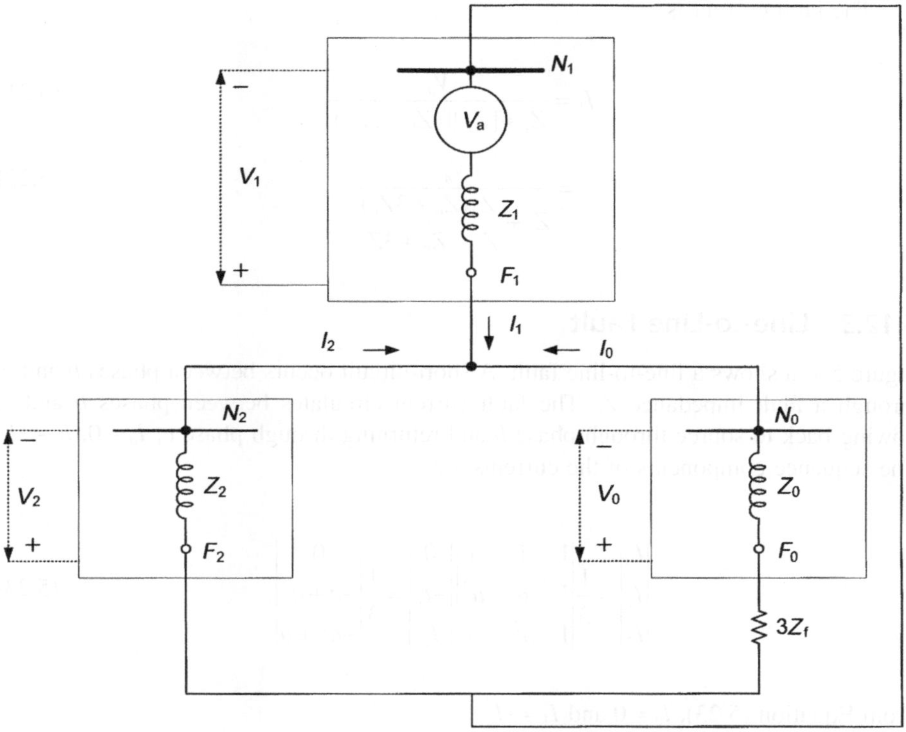


(c)

Figure 5.11. *(Continued)*

which gives $V_1 = V_2$ and

$$\begin{aligned} V_0 &= \frac{1}{3}(V_a + 2V_b) \\ &= \frac{1}{3}[(V_o + V_1 + V_2) + 2(I_b + I_c)Z_f] \\ &= \frac{1}{3}[(V_0 + 2V_1) + 2(3I_0)Z_f] \\ &= V_1 + 3Z_f I_0 \end{aligned} \quad (5.20)$$

This gives the equivalent circuit of Figure 5.11b,c.

The fault current is

$$I_1 = \frac{V_a}{Z_1 + [Z_2 \| (Z_0 + 3Z_f)]} \tag{5.21}$$

$$= \frac{V_a}{Z_1 + \dfrac{Z_2(Z_0 + 3Z_f)}{Z_2 + Z_0 + 3Z_f}} \tag{5.22}$$

## 5.12.3 Line-to-Line Fault

Figure 5.12a shows a line-to-line fault. A short-circuit occurs between phases $b$ and $c$, through a fault impedance $Z_f$. The fault current circulates between phases $b$ and $c$, flowing back to source through phase $b$ and returning through phase $c$; $I_a = 0$, $I_b = -I_c$. The sequence components of the currents are

$$\begin{vmatrix} I_0 \\ I_1 \\ I_2 \end{vmatrix} = \frac{1}{3}\begin{vmatrix} 1 & 1 & 1 \\ 1 & a & a^2 \\ 1 & a^2 & a \end{vmatrix}\begin{vmatrix} 0 \\ -I_c \\ I_c \end{vmatrix} = \frac{1}{3}\begin{vmatrix} 0 \\ -a+a^2 \\ -a^2+a \end{vmatrix}. \tag{5.23}$$

From Equation (5.23), $I_0 = 0$ and $I_1 = -I_2$.

$$\begin{aligned} V_b - V_c &= \begin{vmatrix} 0 & 1 & -1 \end{vmatrix}\begin{vmatrix} V_a \\ V_b \\ V_c \end{vmatrix} = \begin{vmatrix} 0 & 1 & -1 \end{vmatrix}\begin{vmatrix} 1 & 1 & 1 \\ 1 & a^2 & a \\ 1 & a & a^2 \end{vmatrix}\begin{vmatrix} V_0 \\ V_1 \\ V_2 \end{vmatrix} \\ &= \begin{vmatrix} 0 & a^2-a & a-a^2 \end{vmatrix}\begin{vmatrix} V_0 \\ V_1 \\ V_2 \end{vmatrix} \end{aligned}. \tag{5.24}$$

Therefore,

$$\begin{aligned} V_b - V_c &= (a^2 - a)(V_1 - V_2) \\ &= (a^2 I_1 + aI_2)Z_f \\ &= (a^2 - a)I_1 Z_f. \end{aligned} \tag{5.25}$$

This gives

$$(V_1 - V_2) = I_1 Z_f. \tag{5.26}$$

The equivalent circuit is shown in Figure 5.12b,c.

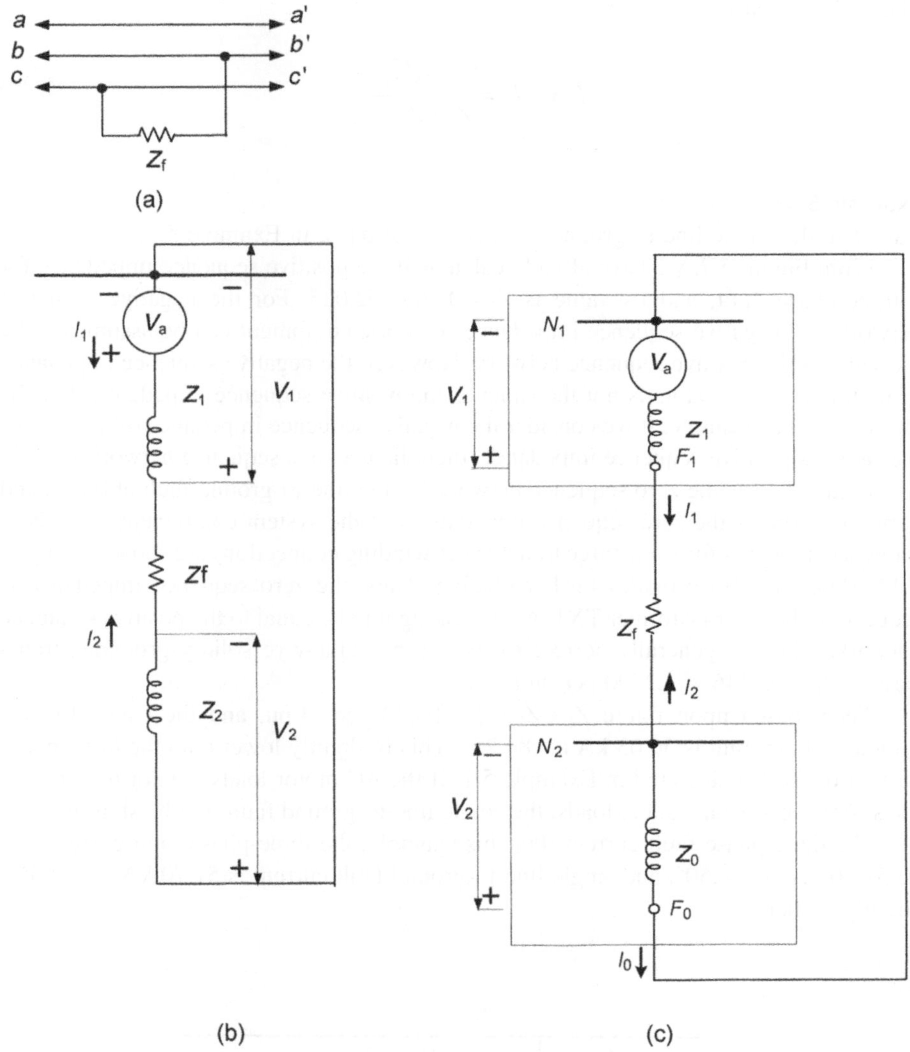


Figure 5.12. (a) Two-phase fault in a three-phase system, (b) Two-phase fault-equivalent circuit, and (c) sequence network interconnections.

Also

$$I_b = (a^2 - a)I_1 = -j\sqrt{3}I_1, \tag{5.27}$$

and

$$I_1 = \frac{V_a}{Z_1 + Z_2 + Z_f}. \tag{5.28}$$

The fault current is

$$I_b = -I_c = \frac{-j\sqrt{3}V_a}{Z_1 + Z_2 + Z_f}. \tag{5.29}$$

**Example 5.4**
Calculate the single line-to-ground fault current at bus 2 in Example 5.1.

From Figure 5.7, we have already calculated the positive sequence impedance for a three-phase fault, and its value is $Z_1 = 0.209 + j2.025$ .For the negative sequence network, the negative sequence impedances of static equipment can be assumed to be the same as the positive sequence network. However, the negative sequence impedance of motors and generators is not the same as the positive sequence impedance. For the purpose of this example, if we consider the negative sequence impedance of the motors the same as positive sequence impedance, then the negative sequence network $Z_2 = Z_1$.

Figure 5.13 is the zero sequence network for the line-to-ground fault at bus 2, and Table 5.17 shows the zero sequence impedances of the system components. The zero sequence networks for some three transformer winding connections are shown in Figure 5.14. Reference [9] provides further reading. Thus, the zero sequence impedance is reduced to that of transformer TX1. Again, taking it to be equal to the positive sequence network, which is generally correct for core type delta–wye solidly grounded transformer, $Z_0 = 0.2146 + j2.2900$ per unit.

Then, based upon figure $Z_1 + Z_2 + Z_0 = 0.633 = j6.34$ pu, and the single line-to-ground fault current is 56.63 kA at -84.25°. This is slightly lower than the three-phase current of 59.08 calculated in Example 5.1. If the AC motor loads are not present and the system serves only static loads, the single line-to-ground fault will be slightly more than the three-phase fault current. For this example, the three-phase fault current will be 51.40 kA at -84.50°, and single line-to-ground fault current is 51.70 kA at -84.55°, slightly higher.

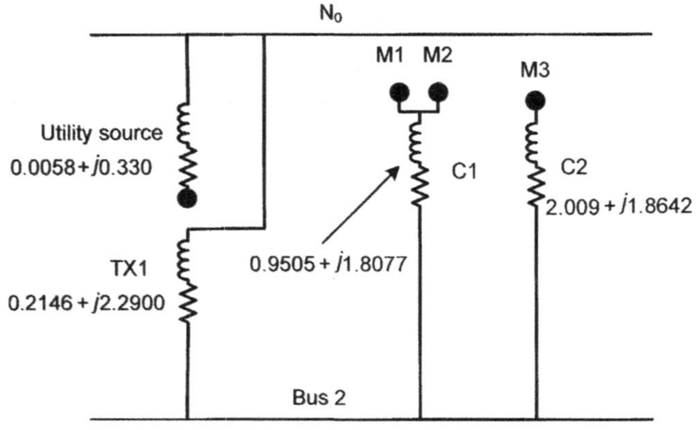


Figure 5.13. Equivalent zero sequence circuit for a fault on bus 2 (Example 5.4).

TABLE 5.17. Zero Sequence Impedance Data (Example 5.4)

| Description of Equipment | Per Unit Resistance on a 100-MVA Base | Per Unit Reactance on a 100-MVA Base |
|---|---|---|
| Utility's 13.8-kV source, three-phase fault level = 717.07 MVA, $X/R$ ratio = 15 | 0.0058 | 0.3390 |
| Single line-to-ground fault = 20 kA, $X/R$ = 8 | | |
| 2.5-MVA transformer TX1, 13.8 - 0.48 kV, delta-wye-connected solidly grounded, Z1 = Z0 = 5.75%, $X/R$ = 10.67 | 0.21462 | 2.2900 |
| Transformer TX2, 100 kVA, 0.48 - 0.24 kV, delta–wye-connected solidly grounded, Z1 = Z0 = 2.60%, $X/R$ = 1.92 | 12.0103 | 23.0598 |
| 600-V cable C1, 2-3/C, 500 KCMIL in nonmagnetic conduit, 100 ft. $R$ = 0.0276 Ω per 1000 ft and $X$ = 0.03110 Ω per 1000 ft | 0.9505 | 1.8077 |
| 600-V cable C2, 2-3/C, 500 KCMIL in magnetic conduit, 100 ft. $R$ = 0.0294 Ω per 1000 ft and $X$ = 0.03490 Ω per 1000 ft | 2.009 | 1.8642 |
| 600-V cable C3, 1-3/C, 4/0 in magnetic conduit, 50 ft. $R$ = 0.0640 Ω per 1000 ft and $X$ = 0.03810 ohms per 1000 ft | 6.8859 | 2.0355 |
| M1 and M3 7 of 100-hp motors, 460 V, each rated kVA = 98.79, locked rotor reactance = 16.7% rated kVA base, $X/R$ = 7.79 | infinite | Infinite |
| M2, 6 of 40-hp motors, rated kVA = 41.43, locked rotor reactance = 16.7%, rated kVA base, $X/R$ = 4.93 | Infinite | Infinite |

## 5.13 COMPUTER METHODS

The computer-based short-circuit calculations use matrix techniques [9]. Consider that the positive negative and zero sequence bus impedance matrices $Z^1_{ss}$, $Z^2_{ss}$ and $Z^0_{ss}$ are known. and a single line-to-ground fault occurs at the $r$th bus. The positive sequence current is then injected only at the $r$th bus, and all other currents in the positive sequence current vector are zero. The positive sequence voltage at bus $r$ is given by

$$V_r^1 = -Z_{rr}^1 I_r^1. \tag{5.30}$$

Similarly, the negative and zero sequence voltages are

$$\begin{aligned} V_r^2 &= -Z_{rr}^2 I_r^2 \\ V_r^0 &= -Z_{rr}^0 I_r^0 \end{aligned}. \tag{5.31}$$

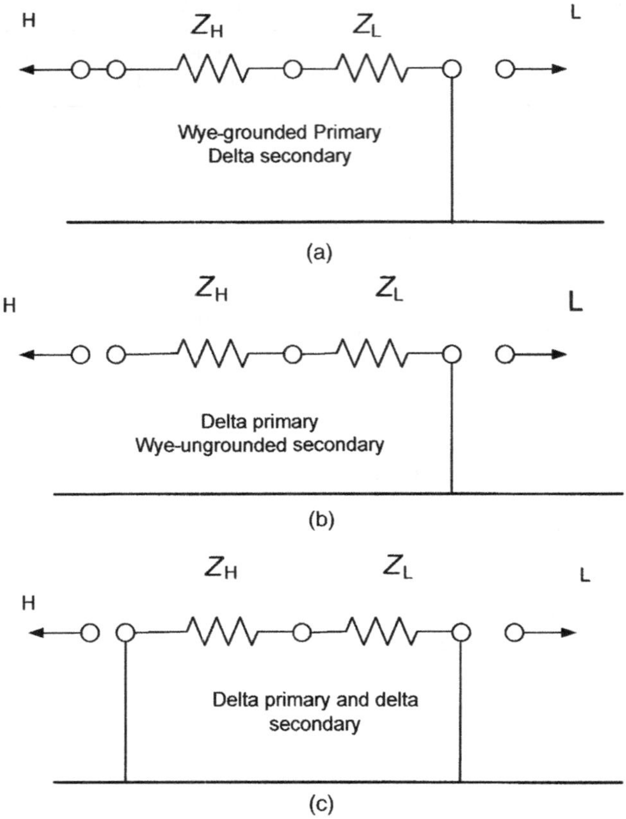


Figure 5.14. (a) Equivalent zero sequence circuit of a transformer with primary windings in wye connection and solidly grounded, and secondary windings in delta connection, (b) for a transformer with primary windings in delta connection and the secondary windings in wye connection and ungrounded, and (c) for a transformer with primary and secondary windings in delta connection.

From the sequence network connections for a line-to-ground fault:

$$I_r^1 = I_r^2 = I_r^0 = \frac{1.0}{Z_{rr}^1 + Z_{rr}^2 + Z_{rr}^0 + 3Z_f}. \tag{5.32}$$

This shows that the following equations can be written for a shorted bus $s$.

### 5.13.1 Line-to-Ground Fault

$$I_s^0 = I_s^1 = I_s^2 = \frac{1}{Z_{ss}^1 + Z_{ss}^2 + Z_{ss}^0 + 3Z_f}. \tag{5.33}$$

### 5.13.2 Line-to-Line Fault

$$I_s^1 = -I_s^2 = \frac{1}{Z_{ss}^1 + Z_{ss}^2 + 3Z_f}. \tag{5.34}$$

### 5.13.3 Double Line-to-Ground Fault

$$I_s^1 = \frac{1}{Z_{ss}^1 + \dfrac{Z_{ss}^2\left(Z_{ss}^0 + 3Z_{\mathrm{f}}\right)}{Z_{ss}^2 + \left(Z_{ss}^0 + 3Z_{\mathrm{f}}\right)}} \tag{5.35}$$

$$I_s^0 = -\frac{Z_{ss}^2}{Z_{ss}^2 + (Z_{ss}^0 + 3Z_{\mathrm{f}})} I_s^1 \tag{5.36}$$

$$I_s^2 = \frac{(-Z_{ss}^0 + 3Z_{\mathrm{f}})}{Z_{ss}^2 + (Z_{ss}^0 + 3Z_{\mathrm{f}})} I_s^1. \tag{5.37}$$

The phase currents are calculated by

$$I_s^{abc} = T_s I_s^{012}. \tag{5.38}$$

The voltage at bus $j$ of the system is

$$\begin{vmatrix} V_j^0 \\ V_j^1 \\ V_j^2 \end{vmatrix} = \begin{vmatrix} 0 \\ 1 \\ 0 \end{vmatrix} - \begin{vmatrix} Z_{js}^0 & 0 & 0 \\ 0 & Z_{js}^1 & 0 \\ 0 & 0 & Z_{js}^2 \end{vmatrix} \begin{vmatrix} I_s^0 \\ I_s^1 \\ I_s^2 \end{vmatrix}, \tag{5.39}$$

where $j = 1, 2, \ldots, s, \ldots, m$.

The fault current from bus $x$ to bus $y$ is given by

$$\begin{vmatrix} I_{xy}^0 \\ I_{xy}^1 \\ I_{xy}^2 \end{vmatrix} = \begin{vmatrix} Y_{xy}^0 & 0 & 0 \\ 0 & Y_{xy}^1 & 0 \\ 0 & 0 & Y_{xy}^2 \end{vmatrix} \begin{vmatrix} V_x^0 - V_y^0 \\ V_x^1 - V_y^1 \\ V_x^2 - V_y^2 \end{vmatrix}, \tag{5.40}$$

where

$$I_{xy}^0 = \begin{vmatrix} I_{12}^0 \\ I_{13}^0 \\ \vdots \\ I_{\mathrm{mm}}^0 \end{vmatrix} \tag{5.41}$$

and

$$\bar{Y}^0_{xy} = \begin{vmatrix} Y^0_{12,12} & Y^0_{12,13} & . & Y^0_{12,mn} \\ Y^0_{13,12} & Y^0_{13,13} & . & Y^0_{13,mn} \\ . & . & . & . \\ Y^0_{mn,12} & Y^0_{mn,13} & . & Y^0_{mn,mn} \end{vmatrix}, \tag{5.42}$$

where $Y^0_{xy}$ is the inverse of the primitive matrix of the system. Similar expressions apply to positive sequence and negative sequence quantities.

**Example 5.5**

Solve Example 5.1 by matrix method.

We can first form a Y matrix, as it can be easily formed by inspection of the circuit. Then it can be inverted to find the Z matrix. The diagonal elements of the z-matrix give the reduced short-circuit impedance directly on the bus.

From Figure 5.15, the y-matrix is:

$$\bar{Y} = \begin{vmatrix} y_{11} & y_{12} & y_{13} & y_{14} & y_{15} \\ y_{21} & y_{22} & y_{23} & y_{24} & y_{25} \\ y_{31} & y_{32} & y_{33} & y_{34} & y_{35} \\ y_{41} & y_{42} & y_{43} & y_{44} & y_{45} \\ y_{51} & y_{52} & y_{53} & y_{54} & y_{55} \end{vmatrix}. \tag{5.43}$$

Then, by inspection of Figure 5.15

$$\bar{Y} = \begin{vmatrix} 0.518 - j7.588 & -0.041 + j0.433 & 0 & 0 & 0 \\ -0.041 + j0.433 & 1.446 - j2.016 & -0.736 + j0.829 & -0.651 + j0.722 & -0.018 + j0.0321 \\ 0 & -0.736 + j0.829 & 0.7426 - j0.8751 & 0 & 0 \\ 0 & -0.651 + j0.722 & 0 & 0.656 - j0.759 & 0 \\ 0 & -0.018 + j0.0321 & 0 & 0 & 0.018 - j0.0321 \end{vmatrix}. \tag{5.44}$$

Its inversion gives Z bus matrix:

$$\bar{Z} = \begin{vmatrix} 0.0093 + j0.138 & 0.0091 + j0.116 & 0.0062 + j0.112 & 0.0064 + j0.113 & 0.0091 + j0.116 \\ 0.0091 + j0.116 & 0.211 + j2.026 & 0.161 + j1.961 & 0.164 + j1.967 & 0.211 + j2.026 \\ 0.0062 + j0.112 & 0.161 + j1.961 & 0.677 + j2.561 & 0.116 + j1.903 & 0.161 + j1.961 \\ 0.0064 + j0.113 & 0.164 + j1.967 & 0.116 + j1.903 & 0.77 + j2.663 & 0.164 + j1.967 \\ 0.0091 + j0.116 & 0.211 + j2.026 & 0.161 + j1.961 & 0.164 + j1.967 & 13.565 + j25.765 \end{vmatrix}. \tag{5.45}$$

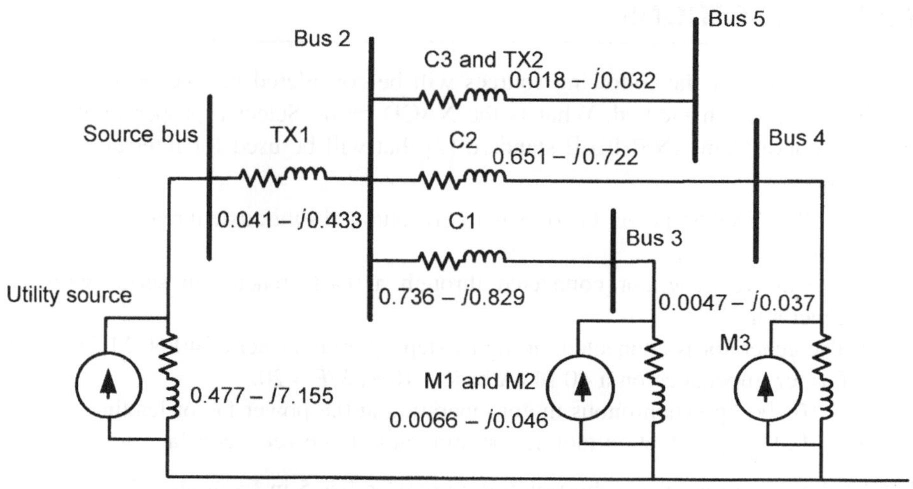


Figure 5.15. Equivalent admittance diagram of the system in Figure 5.5.

In Example 5.1, we calculated Z = 0.209 + *j*2.025 by hand calculations; the self-impedance of bus 2 in above matrix equation is. 0.211 + *j*2.026. Within the accuracy of calculations, this is acceptable. Similarly, for bus 3, the impedance calculated in Example 5.1 = 0.674 + *j*2.554 correlates well with 0.677 + *j*2.561 from the matrix. The short-circuit current at any bus can be calculated from the matrix, for example, at bus 5, 13.565 + *j*25.765 gives a three-phase short-circuit current of 8.26 kA at −62.23°. All the elements in Z matrix are populated, and the matrix is symmetrical. In computer application, direct inversion is not carried out, and other matrix methods are used. Here, we have only a five-bus system. In large systems, there can be more than 1000 buses.

The short-circuit calculations is a vast subject, and this chapter provides a synopsis of it. Though rigorous calculations of switching devices short-circuit duties are not required for arc flash hazard analyses, the concise description in this chapter does provide a conceptual base for an important and fundamental study in the power systems.

## 5.14 SHORT-CIRCUIT CURRENTS FOR ARC FLASH CALCULATIONS

Note that

- Only symmetrical three-phase short-circuit currents are considered for the arc flash calculations.
- Momentary or first cycle short-circuit currents need not be calculated.
- No post multiplying factors are required. The intent is to ascertain the symmetrical short-circuit currents and not to calculate the duties on the switching devices
- No unsymmetrical fault currents are required to be calculated; even if these can be greater than the three-phased bolted currents.

## REVIEW QUESTIONS

1. Indicate whether the following sources will be considered remote or local in the ANSI calculation method. What is the NACD ratio? Select a proper multiplying factor curve from ANSI/IEEE standard [2] that will be used for interrupting duty currents?
   (a) A 50-MVA 0.85 power factor generator; saturated subtransient reactance = 16%, *X/R* = 60.
   (b) The above generator connected through a 0.4-Ω reactor in series with the generator.
   (c) The generator is connected through a step-up transformer of 40/60 MVA; transformer impedance on a 40 MVA base = 10%, *X/R* = 20.
   (d) A 10,000-hp synchronous motor, operating at 0.8 power factor leading.
   (e) A 10,000-hp induction motor, operating at 0.93 power factor lagging.

2. Calculate the three-phase short-circuit current on bus 5 in Example 5.1.

3. In Example 5.2, construct a fault impedance network for the fault at bus 3. Calculate fault point impedance and the *X/R* ratio based on separate *R-X* networks. How does this impedance differ from the values arrived at in Table 5.8 from complex impedance reduction calculation?

4. Which of the following locations of circuit breakers in the same system and on the same bus will have the maximum fault duties: bus tie circuit breaker, incoming breaker, feeder breaker fully loaded, or feeder breaker lightly loaded?

## REFERENCES


1. IEEE Standard 551-Violet Book, *Recommended Practice for Calculating Short-Circuit Currents in Industrial and Commercial Power Systems*, 2006.
2. ANSI/IEEE Standard C37.010, *Application Guide for AC High-Voltage Circuit Breakers Rated on a Symmetrical Current Basis*, 1999 (revision of 1979 standard).
3. IEC Standard 60909-0, *Short Circuit Calculations in Three-Phase AC Systems. 0-Calculation of Currents*, 2001–2007.
4. IEEE Standard C37-013, *AC High-Voltage Generator Breakers Rated on Symmetrical Current Basis*, 1999.
5. ANSI Standard C37.06, *AC High-Voltage Circuit Breakers Rated on a Symmetrical Current Basis—Preferred Ratings and Related Capabilities*, 2000 (revision of 1987).
6. ANSI/IEEE Standard C37.13, *Standard for Low-Voltage AC Power Circuit Breakers in Enclosures*, 1997.
7. ANSI/IEEE Standard C37.04, *Rating Structure for AC High Voltage Circuit Breakers*, 1999 (Revision of 1979).
8. IEEE PC37.06/D11, *Draft Standard AC High Voltage Circuit Breakers Rated on a Symmetrical Current Basis—Preferred Ratings and Required Capabilities for Voltages above 1000 Volts*, 2008.

9. J.C. Das, *Power System Analysis-Short-Circuit Load Flow and Harmonics*, 2nd ed., CRC Press, Boca Raton, FL, 2012.
10. J.C. Das, "Reducing interrupting duties of high voltage circuit breakers by increasing contact parting time," IEEE Transactions on Industry Applications, vol. 44, pp. 1027–1034, 2008.
11. IEEE Standard 141, *Recommended Practice for Electrical Power Distribution for Industrial Plants*, 1993.
12. W.C. Huening, Jr., "Interpretation of new American standards for power circuit breaker applications," IEEE Transactions on Industry and General Applications, vol. IGA-5, pp. 121–143, 1969.
13. IEEE Standard 1110, *Guide for Synchronous Generator Modeling Practice and Applications in Power System Stability Analysis*, 2002.
14. J.R. Dunki-Jacobs, B.P. Lam, and R.P. Stafford, "A comparison of ANSI-based and dynamically rigorous short-circuit current calculation procedures," IEEE Transactions on Industry Applications, vol. 24, pp. 1180–1194, 1988.
15. A. Berizzi, S. Massucco, A. Silvestri, and D. Zaninelli, "Short-circuit current calculations-a comparison between methods of IEC and ANSI standards using dynamic simulation as reference," IEEE Transactions on Industry Applications, vol. 30, pp. 1099–1106, 1994.
16. A.J. Rodolakis, "A comparison of North American (ANSI) and European (IEC) fault calculation guidelines," IEEE Transactions on Industry Applications, vol. 29, pp. 515–521, 1993.
17. J.C. Das, "Limitations of fault current limiters for expansion of electrical distribution systems," IEEE Transactions on Industry Applications, vol. 33, pp. 1073–1082, 1997.

# 6

# ACCOUNTING FOR DECAYING SHORT-CIRCUIT CURRENTS IN ARC FLASH CALCULATIONS

This chapter illustrates that ANSI/IEEE methods for calculating the short-circuit duties on switching devices have limitations with respect to calculations of bolted three-phase currents for arc flash analysis *when decaying components of the short-circuit currents are involved*. As stated in Chapter 5, a continuous decay profile cannot be simulated in ANSI/IEEE empirical calculations. It should, however, be noted that over the course of many years, experience shows that ANSI/IEEE empirical calculations have given reliable results for applications of switching devices. ANSI/IEEE standards for short-circuit calculations are perhaps the oldest standard in the world, and recognized internationally.

We will discuss some theoretical aspects of decaying short-circuit currents and then demonstrate a comparison with ANSI/IEEE empirical calculations versus a new algorithm proposed in this chapter for arc flash calculations.

## 6.1 SHORT CIRCUIT OF A PASSIVE ELEMENT

Consider a sinusoidal time-invariant single-phase 60-Hz source of power, $E_m \sin \omega t$, connected to a single-phase short distribution line, $Z = (R + j\omega L)$, where $Z$ is the

*Arc Flash Hazard Analysis and Mitigation*, Second Edition. J.C. Das.

Published 2021 by John Wiley & Sons, Inc.

complex impedance, $R$ and $L$ are the resistance and inductance, $E_m$ is the peak source voltage, and ω is the angular frequency = $2\pi f$, $f$ being the frequency of the AC source. A differential equation of this single-phase circuit is written:

$$L\frac{di}{dt}+Ri=V\sin(\omega t+\theta). \tag{6.1}$$

This can be solved using differential equations or Laplace transforms [1–4]

Taking Laplace transform:

$$\begin{aligned} Ri(s)+Lsi(s)-LI(0) &= V(\sin\omega t\cos\theta+\cos\omega t\sin\theta) \\ &= V\left(\frac{\omega\cos\theta}{s^2+\omega^2}+\frac{s\sin\theta}{s^2+\omega^2}\right) \end{aligned} \tag{6.2}$$

$$i(s)=\frac{V}{L}\frac{1}{s+(R/L)}\left[\frac{\omega\cos\theta}{s^2+\omega^2}+\frac{s\sin\theta}{s^2+\omega^2}\right]. \tag{6.3}$$

This can be resolved into partial fractions.

$$\frac{1}{(s+\alpha)(s^2+\omega^2)}=\frac{1}{\alpha^2+\omega^2}\left[\frac{1}{s+\alpha}-\frac{s}{s^2+\omega^2}+\frac{\alpha}{s^2+\omega^2}\right] \tag{6.4}$$

where $\alpha = R/L$; therefore:

$$\begin{aligned} i(s) &= \frac{V\omega\cos\theta}{L(\alpha^2+\omega^2)}\left[\frac{1}{s+R/L}-\frac{s}{s^2+\omega^2}+\frac{\alpha}{s^2+\omega^2}\right] \\ &+ \frac{V\sin\theta}{L(\alpha^2+\omega^2)}\left[\frac{1}{s+R/L}-\frac{s}{s^2+\omega^2}+\frac{\alpha}{s^2+\omega^2}\right]. \end{aligned} \tag{6.5}$$

The inverse Laplace transform of function in Equation (6.4) is:

$$L^{-1}\frac{1}{(s+\alpha)(s^2+\omega^2)}=\frac{1}{(\alpha^2+\omega^2)}\left[e^{-\alpha t}-\cos\omega t+\frac{\alpha}{\omega}\sin\omega t\right]. \tag{6.6}$$

Therefore:

$$L^{-1}\frac{s}{(s+\alpha)(s^2+\omega^2)}=\frac{1}{(\alpha^2+\omega^2)}\left[-\alpha e^{-\alpha t}+\omega\sin\omega t+\cos\omega t\right]. \tag{6.7}$$

Substituting these results in Equation (6.5)

$$\begin{aligned} i &= \frac{V}{L(\alpha^2+\omega^2)}\left[\omega\cos\theta\left(e^{-\alpha t}-\cos\omega t+\frac{\alpha}{\omega}\sin\omega t\right)+\sin\theta\left(\alpha\cos\omega t+\omega\sin\omega t-\alpha e^{-\alpha t}\right)\right] \\ &= \frac{V}{L(\alpha^2+\omega^2)}[(\omega\cos\theta-\alpha\sin\theta)e^{-\alpha t}-(\omega\cos\theta-\alpha\sin\theta)\cos\omega t \\ &+(\alpha\cos\theta+\omega\sin\theta)\sin\omega t]. \end{aligned} \tag{6.8}$$

Remembering that:

$$\tan\phi = \omega L / R = \omega\alpha, \sin\phi = \omega / (\alpha^2 + \omega^2)^{1/2}, \cos\phi = \alpha / (\alpha^2 + \omega^2)^{1/2}$$

$$i = \frac{V}{\left(R^2 + \omega^2 L^2\right)^{1/2}}\left[\sin(\omega t + \theta - \phi) - \sin(\theta - \phi)e^{-\alpha t}\right]. \tag{6.9}$$

In power systems, the *X/R* ratio is high. A 100-MVA, 0.85 power factor generator may have an *X/R* of 110 and a transformer of the same rating, an *X/R* = 45. The *X/R* ratios for low voltage systems may be of the order of 2–8. Consider that $\phi \approx 90^0$.

If the short circuit occurs when the switch is closed at an instant at $t = 0$, $\theta = 0$, that is, when the voltage wave is crossing through zero amplitude on *x*-axis, the instantaneous value of the short-circuit current from Equation (6.9) is $2I_m$. This is sometimes called the *doubling effect*.

If the short circuit occurs when the switch is closed at an instant at $t = 0$, $\theta = \pi/2$, that is, when the voltage wave peaks, the second term in Equation (6.9) is zero, and there is no transient component.

A physical explanation of the DC transient is that the inductance component of the impedance is high. If the short circuit occurs at the peak of the voltage, the current is zero. No DC component is required to satisfy the physical law that the current in an inductance cannot change suddenly. When the fault occurs at an instant when $\theta - \phi = 0$, there has to be a transient current whose value is equal and opposite to the instantaneous value of the AC short-circuit current. This transient current can be called a DC component, and it decays at an exponential rate. Equation (6.9) can be simply written as:

$$i = I_m \sin\omega t + I_{DC}e^{-\frac{Rt}{L}}. \tag{6.10}$$

There are two distinct components of the short-circuit current: (1) an AC component and (2) a decaying DC component at an exponential rate, the initial value of the DC component being equal to the maximum value of the AC component.

Factor *L/R* can be termed as a time constant. The exponential then becomes $I_{DC}e^{-t/t'}$, where $t' = L/R$. In this equation, making $t = t'$ will result in approximately 62.3% decay from its original value, that is, the transient current is reduced to a value of 0.368 per unit after a elapsed time equal to the time constant (Figure 6.1). The DC component always decays to zero in a short time. Consider a modest *X/R* of 15; the DC component decays to 88% of its initial value in five cycles. This phenomenon is important for the rating structure of circuit breakers. The higher is the *X*/R ratio, the slower is the decay.

The presence of the DC component makes the total short-circuit current wave asymmetrical about zero line. The total asymmetrical rms current is given by:

$$i_t(\text{rms, asym}) = \sqrt{(\text{AC component})^2 + (\text{DC component})^2}. \tag{6.11}$$

In a three-phase system, the phases are displaced from each other by 120°. If a fault occurs when the DC component is phase *a* is zero, the phase *b* component will be

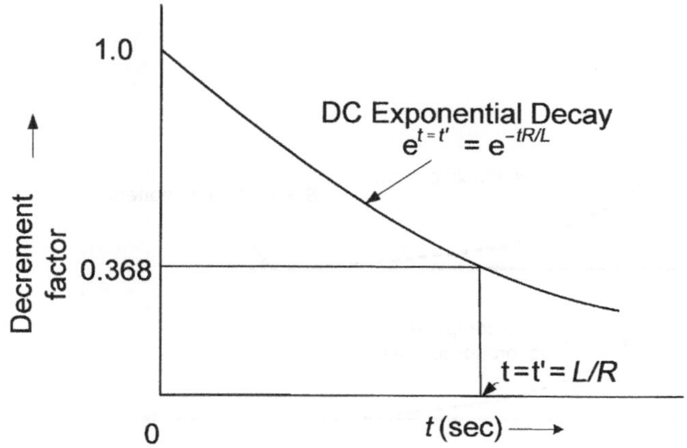


Figure 6.1. Time constant of exponential DC decaying component of the short-circuit current.

positive, and the phase component will be equal in magnitude but negative. As the fault is symmetrical, the identity $I_a + I_b + I_c = 0$, is approximately maintained.

## 6.2 SYSTEMS WITH NO AC DECAY

For the arc flash calculations, we do not consider the asymmetry, that is, ignore the DC component. Then for a passive element, the short-circuit current is the steady current, simply given by $E/Z$, that is, the first term of Equation (6.10). If we consider a distribution system consisting of only passive elements, and ignore the DC component, then the above condition of steady short-circuit current is fulfilled. Such a distribution system can be a utility source with transformers, static loads, loads served through static converters or rectifiers, shunt capacitors, series reactors, cables, and overhead lines—but no motors and generators. Utility sources are considered "remote" sources (Chapter 5). The short circuits in such a system can be diagrammatically represented as in Figure 6.2a.

When we consider the short circuit of a synchronous generator and rotating AC motor loads, $Z$ is not time invariant. Both AC and DC components decay. The short circuit in such a system can be represented by Figure 6.2b. More clearly, this decay of AC and DC components and total current wave shape is shown in Figure 5.1.

In Chapter 5, we discussed how this decay in the AC component is accounted for in the empirical calculations of short-circuit currents according to ANSI/IEEE standards. This decay is further discussed in this chapter for arc flash calculations. To appreciate the difference with respect to empirical calculations knowledge of modeling of rotating machines, especially from the short circuit, considerations is required.

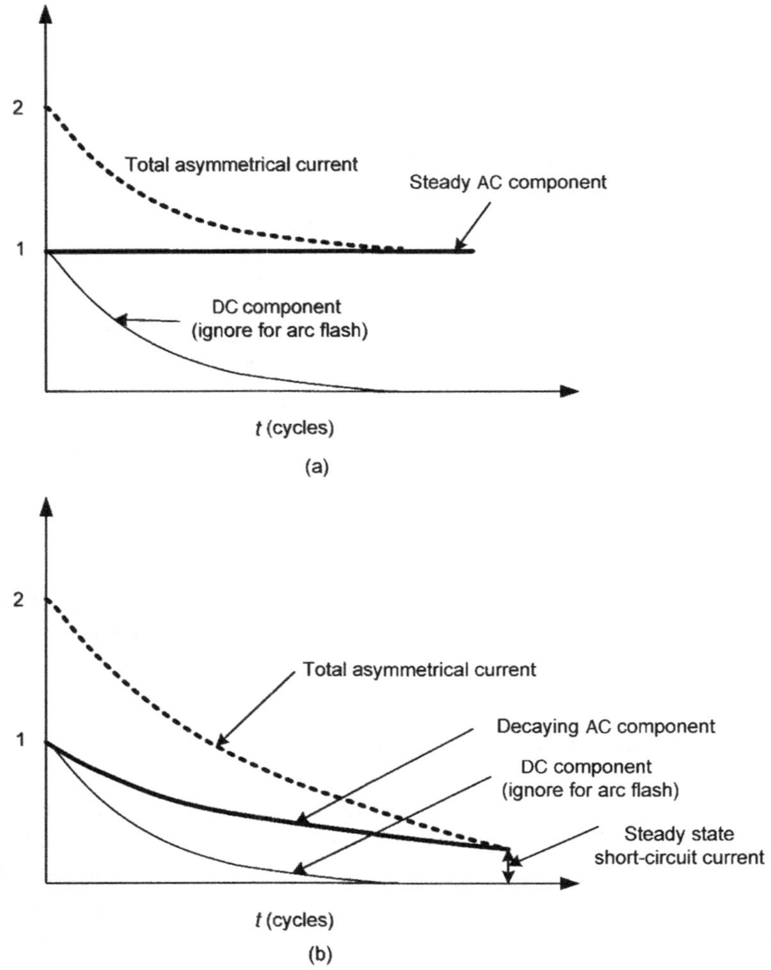


Figure 6.2. (a) Short-circuit current profile utility source only, no generators or motors or any other sources of short-circuit currents; (b) with decaying AC component when generators and motor loads are present.

## 6.3 REACTANCES OF A SYNCHRONOUS MACHINE

The following definitions are applicable.

### 6.3.1 Leakage Reactance

The leakage reactance, denoted by symbol $X_l$, can be defined but cannot be tested. It is the reactance due to flux setup by armature windings, but not crossing the air gap. It can be divided into end-winding leakage and slot leakage. A convenient way of picturing the reactances is to view these in terms of permeances of various magnetic paths in the machine, which are functions of dimensions of iron and copper circuits and independent of the flux density or the current loading. The permeances thus calculated can be multiplied by a factor to consider the flux density and current. For example, the leakage reactance is mainly given by the slot permeance and the end-coil permeance.

### 6.3.2 Subtransient Reactance

Subtransient reactance, denoted by symbol $X_d''$, equals the leakage reactance plus the reactance due to the flux setup by stator currents crossing the air gap and penetrating the rotor as far as the damper windings in a laminated pole machine or as far as the surface damping currents in a solid pole machine. The subtransient conditions last for one to five cycles on a 60 Hz basis.

### 6.3.3 Transient Reactance

Transient reactance, denoted by symbol $X_d'$, is the reactance after all damping currents in the rotor surface or amortisseur windings have decayed, but before the damping currents in the field winding have decayed. The transient reactance equals the leakage reactance plus the reactance due to flux setup by the armature that penetrates the rotor to the field windings. Transient conditions last for 5–200 cycles on a 60-Hz basis.

### 6.3.4 Synchronous Reactance

Synchronous reactance denoted by symbol $X_d$, is the steady-state reactance after all damping currents in the field windings have decayed. It is the sum of leakage reactance and a fictitious armature reaction reactance, which is much larger than the leakage reactance. Ignoring resistance, per unit synchronous reactance is the ratio of per unit voltage on an open circuit divided by per unit armature current on a short circuit for a given field excitation. This gives *saturated* synchronous reactance. The unsaturated value of the synchronous reactance is given by the per unit voltage on air-gap open circuit line divided by per unit armature current on short circuit. If 0.5 per unit field excitation produces full-load armature current on short circuit, the saturated synchronous reactance is 2.0 per unit. The saturated value may be only 60–80% of the unsaturated value.

### 6.3.5 Quadrature-Axis Reactances

Quadrature-axis reactances, denoted by symbols $X_q''$, $X_q'$, and $X_q$, are similar to direct-axis reactances, except that they involve the rotor permeance encountered when the stator flux enters one pole tip, crosses the pole, and leaves the other pole tip. The direct-axis permeance is encountered by the flux crossing the air gap to the center of one pole, then crossing from one pole to the other pole and entering the stator from that pole. The total armature reaction $F$ can be divided into two components, $F_{ad}$ and $F_{aq}$; $F_{ad}$ is directed across the direct axis and $F_{aq}$ across the quadrature axis [5–7]. As these MMFs (magnetomotive forces) act on circuits of different permeances, the flux produced varies. If damper windings across pole faces are connected, $X''_q$ is nearly equal to $X''_d$.

### 6.3.6 Negative Sequence Reactance

The negative sequence reactance $X_2$ is the reactance encountered by a voltage of reverse-phase sequence applied to the stator, with the machine running. Negative sequence flux revolves opposite to the rotor and is at twice the system frequency. Negative sequence reactance is practically equal to the subtransient reactance, as the damping currents in the damper windings or solid pole rotor surface prevent the flux from penetrating farther. The negative sequence reactance is generally taken as the average of subtransient direct-axis and quadrature-axis reactances.

### 6.3.7 Zero Sequence Reactance

The zero sequence reactance $X_0$ is the reactance effective when rated frequency currents enter all three terminals of the machine simultaneously and leave at the neutral of the machine. It is approximately equal to the leakage reactance of the machine with full-pitch coils. With two-thirds pitch stator coils, the zero sequence reactance will be a fraction of the leakage reactance.

## 6.4 SATURATION OF REACTANCES

Saturation varies with voltage, current, and power factor. *For short-circuit calculations, according to ANSI/IEEE, saturated subtransient reactance must be considered.* The saturation factor is usually applied to transient and synchronous reactances, though all other reactances change with saturation. In a typical machine, transient reactances may reduce from 5% to 25% on saturation. Saturated reactance is sometimes called the rated voltage reactance and is denoted by subscript "*v*" added to the "d" and "q" subscript axes, that is, $X''_{dv}$ and $X''_{qv}$ denote saturated subtransient reactances in direct and quadrature axes, respectively.

## 6.5 TIME CONSTANTS OF SYNCHRONOUS MACHINES

### 6.5.1 Open-Circuit Time Constant

The open-circuit time constant $T'_{do}$ expresses the rate of decay or buildup of field current when the stator is open circuited and there is zero resistance in the field circuit.

### 6.5.2 Subtransient Short-Circuit Time Constant

The subtransient short-circuit time constant $T''_d$ expresses the rate of decay of the subtransient component of current under a bolted (zero resistance), three-phase short circuit at the machine terminals.

### 6.5.3 Transient Short-Circuit Time Constant

The transient short-circuit time constant $T'_d$ expresses the rate of decay of the transient component of the current under a bolted (zero resistance), three-phase short circuit at the machine terminals.

### 6.5.4 Armature Time Constant

The armature time constant $T_a$ expresses the rate of decay of the DC component of the short-circuit current under the same conditions.

## 6.6 SYNCHRONOUS MACHINE BEHAVIOR ON TERMINAL SHORT CIRCUIT

The time immediately after a short circuit can be divided into three periods:

- The subtransient period lasting from one to five cycles
- The transient period, which may last up to 100 cycles or more
- The final or steady-state period.

Normally, the generator will be removed from service by protective relaying, much before the steady-state period is reached.

In the subtransient period, the conditions can be represented by the flux linking the stator and rotor windings. Any sudden change in the load or power factor of a generator produces changes in the MMFs, both in direct and quadrature axes. A terminal three-phase short circuit is a large disturbance. At the moment of short circuit, the flux linking the stator from the rotor is trapped to the stator, giving a stationary replica of the main pole flux. The rotor poles may be in a position of maximum or minimum flux linkage, and as these rotate, the flux linkages tend to change. This is counteracted by a current in the stator windings. The short-circuit current is, therefore, dependent on rotor angle. As the energy stored can be considered as a function of armature and field linkages, the torque fluctuates and reverses cyclically. The DC component giving rise to asymmetry is caused by the flux trapped in the stator windings at the instant of short circuit, which sets up a DC transient in the armature circuit. This DC component establishes a component field in the air gap that is stationary in space, and which, therefore, induces a fundamental frequency voltage and current in the synchronously revolving rotor circuits. Thus, an increase in the stator current is followed by an increase in the field current. The field flux has superimposed on it a new flux pulsating with respect to field windings at normal machine frequency. The single-phase induced current in the field can be resolved into two components, one stationary with respect to the stator that counteracts the DC component of the stator current, and the other component travels at twice the synchronous speed with respect to the stator and induces a second harmonic in it.

The armature and field are linked through the magnetic circuit, and the AC component of lagging current creates a demagnetizing effect. However, some time must elapse before it starts becoming effective in reducing the field current and the steady-state current is reached. The protective relays will normally operate to open the generator circuit breaker, and, simultaneously, the field circuit for suppression of generated voltage.

The above description is rather an oversimplification of the transient phenomena in the machine on short circuit. In practice, a generator will be connected in an interconnected system. The machine terminal voltage, rotor angle, and frequency all change depending on the location of the fault in the network, the network impedance, and the machine parameters. The machine output power will be affected by the change in the rotor winding EMF and the rotor position in addition to any changes in the impedance seen at the machine terminals. For a terminal fault, the voltage at the machine terminals is zero, and, therefore, power supplied by the machine to load reduces to zero, while the prime mover output cannot change suddenly. Thus, the generator accelerates. In a multimachine system with interconnecting impedances, the speeds of all machines change, so that these generate their share of synchronizing power in the overall impact, as these strive to reach a mean retardation through oscillations due to stored energy in the rotating masses.

In a dynamic simulation of a short circuit, the following may be considered:

- Network before, during, and after the short circuit
- Induction motors' dynamic modeling, with zero excitation
- Synchronous machine dynamic modeling, considering saturation
- Modeling of excitation systems
- Turbine and governor models.

## 6.6.1 Equivalent Circuits during Fault

An equivalent circuit of the terminal fault of a synchronous machine can be derived based on the above discussions. These are shown in Figure 6.3 in the subtransient, transient, and steady-state periods. As the flux penetrates into the rotor, and the currents die down, it is equivalent to opening a circuit element, that is, from subtransient to transient state, the damper circuits are opened.

The direct-axis subtransient reactance is given by

$$X_d'' = X_l + \frac{1}{1/X_{ad} + 1/X_f + 1/X_{kD}}, \tag{6.12}$$

where $X_{ad}$ is the reactance corresponding to the fundamental space wave of the armature in the direct axis, $X_f$ is the reactance of the field windings, $X_{kD}$ is that of the damper windings in the direct axis, and $X_l$ is the leakage reactance. $X_f$ *and* $X_{kD}$ *are also akin to leakage reactances.* Similarly, the quadrature-axis subtransient reactance is given by

$$X_q'' = X_l + \frac{1}{1/X_{aq} + 1/X_{kQ}}, \tag{6.13}$$

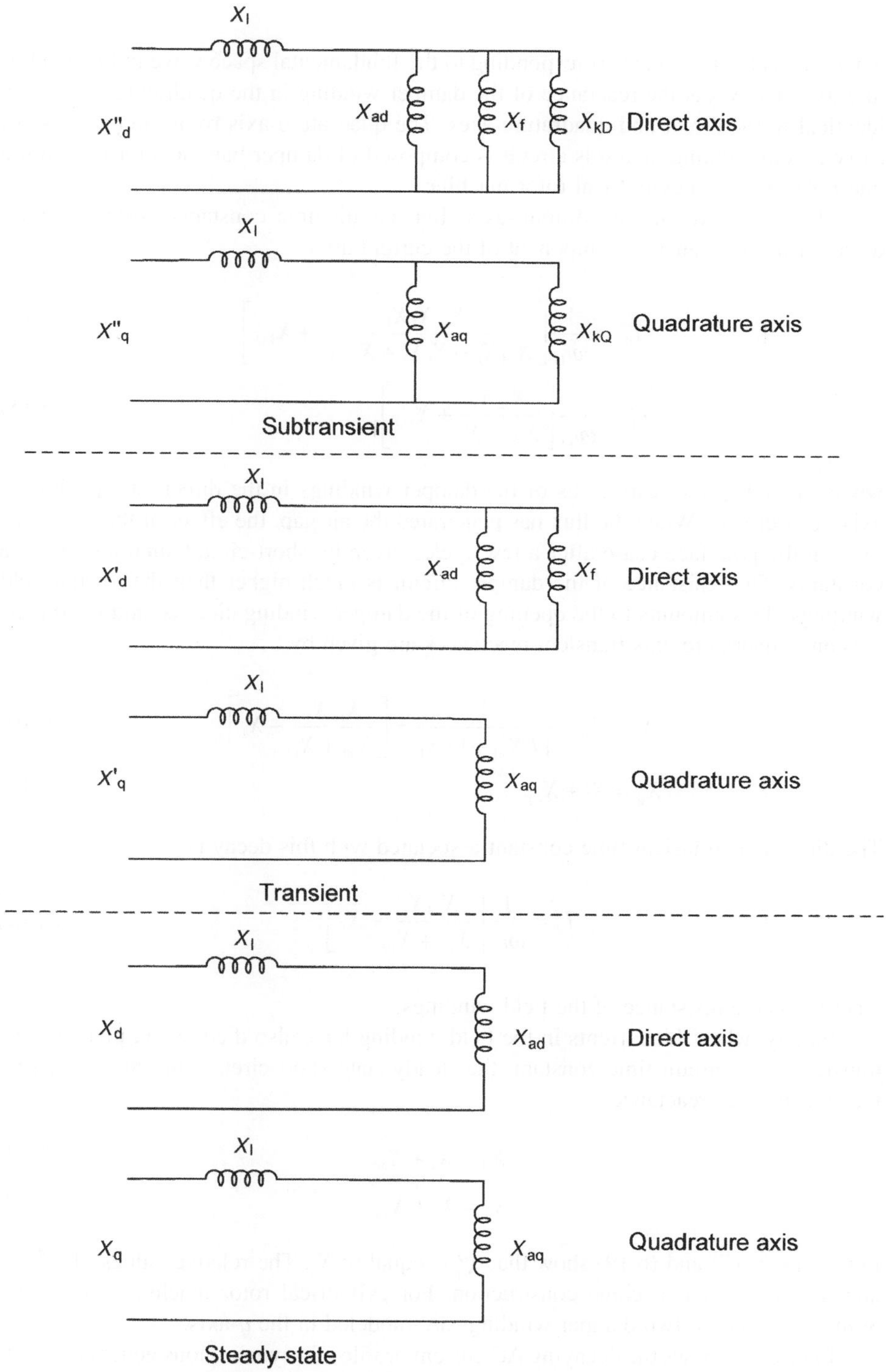


Figure 6.3. Equivalent transformer circuits of a synchronous generator during subtransient, transient, and steady-state periods after a terminal fault.

where $X_{aq}$ is the reactance corresponding to the fundamental space wave in the quadrature axis, the $X_{kQ}$ is the reactance of the damper winding in the quadrature axis; $X_l$ is identical in the direct and quadrature axes. The quadrature-axis rotor circuit does not carry a field winding, and this circuit is composed of damper bars or rotor iron in the interpolar axis of a cylindrical rotor machine.

The direct-axis and quadrature-axis short-circuit time constants associated with decay of the subtransient component of the current are

$$T_d'' = \frac{1}{\omega r_D}\left[\frac{X_{ad}X_f X_l}{X_{ad}X_f + X_f X_l + X_{ad}X_l} + X_{kD}\right] \tag{6.14}$$

$$T_q'' = \frac{1}{\omega r_Q}\left[\frac{X_{aq}X_l}{X_{aq} + X_l} + X_{kQ}\right], \tag{6.15}$$

where $r_D$ and $r_Q$ are resistances of the damper windings in the direct and quadrature axis, respectively. When the flux has penetrated the air gap, the effect of the eddy currents in the pole face cease after a few cycles given by short-circuit subtransient time constants. The resistance of the damper circuit is much higher than that of the field windings. This amounts to the opening of the damper winding circuits, and the direct-axis and quadrature-axis transient reactances are given by

$$X_d' = X_l + \frac{1}{1/X_{ad} + 1/X_f} = \left[\frac{X_{ad}X_f}{X_{ad} + X_f} + X_l\right] \tag{6.16}$$

$$X_q' = X_l + X_{aq}. \tag{6.17}$$

The direct-axis transient time constant associated with this decay is

$$T_d' = \frac{1}{\omega r_F}\left[\frac{X_{ad}X_l}{X_{ad} + X_l} + X_f\right], \tag{6.18}$$

where $r_F$ is the resistance of the field windings.

Finally, when the currents in the field winding have also died down, given by the transient short-circuit time constant, the steady-state short-circuit current is given by the synchronous reactance:

$$X_d = X_l + X_{ad} \tag{6.19}$$

$$X_q = X_l + X_{aq}. \tag{6.20}$$

Equations (6.17) and (6.19) show that $X_q'$ is equal to $X_q$. The relative values of $X_q''$, $X_q'$ and $X_q$, depend on machine construction. For cylindrical rotor machines, $X_q >> X_q'$. Sometimes, one or two damper windings are modeled in the $q$-axis.

Figure 6.4 shows the decaying AC current profile of a synchronous generator short-circuit current. Note the three distinct periods—subtransient, transient, and steady

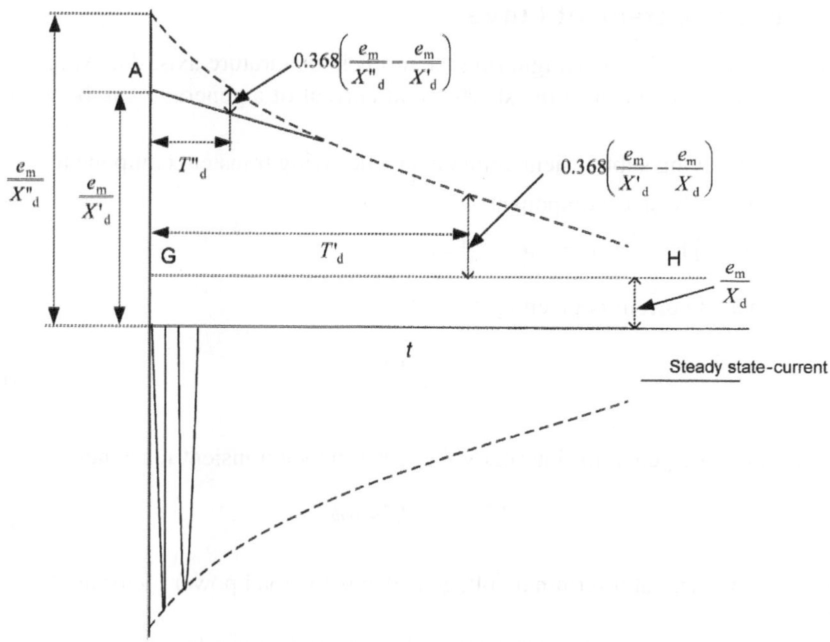


Figure 6.4. Decaying AC component of the short-circuit current showing subtransient, transient, and steady-state currents.

state—and the time constants associated with it. An analytical expression for the short-circuit current of a synchronous generator with damper windings can be written as [8–13]:

$$
\begin{aligned}
i_a = \sqrt{2}E\Bigg[&\left(\frac{1}{X_d}\right)\sin(\omega t+\delta)+\left(\frac{1}{X'_d}-\frac{1}{X_d}\right)e^{-t/T'_d}\sin(\omega t+\delta)\\
&+\left(\frac{1}{X''_d}-\frac{1}{X'_d}\right)e^{-t/T''_d}\sin(\omega t+\delta)\\
&-\frac{(X''_d+X''_q)}{2X''_dX''_q}e^{-t/T_a}\sin\delta-\frac{(X''_d-X''_q)}{2X''_dX''_q}e^{-t/T_a}\sin(2\omega t+\delta)\Bigg].
\end{aligned}
\tag{6.21}
$$

- The first term is final steady-state short-circuit current.
- The second term is normal-frequency decaying transient current.
- The third term is normal-frequency decaying subtransient current.
- The fourth term is decaying DC current.
- The fifth term is double-frequency decaying current.

### 6.6.2 Fault Decrement Curve

Based on Equation (6.21) and ignoring the terms in quadrature axis, the expression for a decaying AC component of the short-circuit current of a generator can be written as

$$\begin{aligned} i_{AC} &= \text{Decaying subtransient component} + \text{decaying transient component} \\ &\quad + \text{steady-state component} \\ &= (i''_d - i'_d)e^{-t/T''_d} + (i'_d - i_d)e^{-t/T'_d} + i_d. \end{aligned} \tag{6.22}$$

The subtransient current is given by

$$i''_d = \frac{E''}{X''_d}, \tag{6.23}$$

where $E''$ in is the generator internal voltage behind subtransient reactance:

$$E'' = V_a + X''_d \sin\phi, \tag{6.24}$$

where $V_a$ is the generator terminal voltage and $\phi$ is the load power factor angle prior to fault.

Similarly, the transient component of the current is given by

$$i'_d = \frac{E'}{X'_d}, \tag{6.25}$$

where $E'$ is the generator internal voltage behind transient reactance:

$$E' = V_a + X'_d \sin\phi. \tag{6.26}$$

The steady-state component is given by

$$i_d = \frac{V_a}{X_d}\left(\frac{i_F}{i_{Fg}}\right), \tag{6.27}$$

where $i_F$ is the field current at given load conditions (when regulator action is taken into account) and $i_{Fg}$ is the field current at no-load rated voltage.

The DC component is given by

$$i_{DC} = \sqrt{2}\, i''_d e^{-t/T_a}, \tag{6.28}$$

where $T_a$ is the armature short-circuit time constant, given by

$$T_a = \frac{1}{\omega r}\left[\frac{2X''_d X''_q}{X''_d + X''_q}\right], \tag{6.29}$$

where $r$ is the stator resistance.

The open-circuit time constant describes the decay of the field transient; the field circuit is closed, and the armature circuit is open:

$$T''_{do} = \frac{1}{\omega r_D}\left[\frac{X_{ad}X_f}{X_{ad}+X_f} + X_{kD}\right], \tag{6.30}$$

and the quadrature-axis subtransient open-circuit time constant is

$$T''_{qo} = \frac{1}{\omega r_Q}(X_{aq} + X_{kQ}). \tag{6.31}$$

The open-circuit direct-axis transient time constant is

$$T'_{do} = \frac{1}{\omega r_F}(X_{ad} + X_f). \tag{6.32}$$

The short-circuit direct-axis transient time constant can be expressed as

$$T'_d = T'_{do}\left[\frac{X'_d}{(X_{ad}+X_1)}\right] = T'_{do}\frac{X'_d}{X_d}. \tag{6.33}$$

It may be observed that the resistances have been neglected in the above expressions. These can be included, that is, the subtransient current is

$$i''_d = \frac{E''}{r_D + X''_d}, \tag{6.34}$$

where $r_D$ is defined as the resistance of the armortisseur windings on salient pole machines and analogous body of cylindrical rotor machines. Similarly, the transient current is

$$i'_d = \frac{E'}{r_f + X'_d}. \tag{6.35}$$

**Example 6.1**

Consider a 13.8-kV, 100-MVA 0.85 power factor generator. Its rated full load current is 4184 A. Other data are:

| | |
|---|---|
| Saturated subtransient reactance $X''_{dv}$ | = 0.15 per unit |
| Saturated transient reactance $X'_{dv}$ | = 0.2 per unit |
| Synchronous reactance $X_d$ | = 2.0 per unit |
| Field current at rated load $i_f$ | = 3.0 per unit |
| Subtransient short-circuit time constant $T''_d$ | = 0.012 second |
| Transient short-circuit time constant $T'_d$ | = 0.35 second |

Armature short-circuit time constant $T_a$ = 0.15 second

A three-phase short circuit occurs at the terminals of the generator when it is operating at its rated load and power factor. It is required to construct a fault decrement curve of the generator for: (1) the AC component, (2) DC component, and (3) total current.

From Equation (6.24), the voltage behind subtransient reactance at the generator-rated voltage, load, and power factor is

$$E'' = V + X''_d \sin\phi = 1 + (0.15)(0.527) = 1.079 \text{ PU}.$$

From Equation (6.23), the subtransient component of the current is

$$i''_d = \frac{E''}{X''_{dv}} = \frac{1.079}{0.15} \text{ per unit} = 30.10 \text{ kA}.$$

Similarly, from Equation (6.26), $E'$, the voltage behind transient reactance is 1.1054 per unit, and, from Equation (6.25), the transient component of the current is 23.12 kA.

From Equation (6.27), current $i_d$ at constant excitation is 2.09 kA rms. For a ratio of $i_f/i_{Fg}$ = 3, current $i_d$ = 6.28 kA rms. Therefore, the following equations can be written for the AC component of the current with constant excitation:

$$i_{AC} = 6.98^{-t/0.012} + 20.03e^{-t/0.35} + 2.09 \text{ kA}.$$

And with full load excitation:

$$i_{AC} = 6.98e^{-t/0.012} + 16.84e^{-t/0.35} + 6.28 \text{ kA}.$$

The AC decaying component of the current can be plotted from these two equations, with the lowest value of $t$ = 0.01 to $t$ = 1000 seconds. This is shown in Figure 6.5. The DC component is given by Equation (6.19):

$$i_{DC} = \sqrt{2} i''_d e^{-t/T_a} = 42.57e^{-t/0.15} \text{kA}.$$

This is not shown in Figure 6.5. At any instant, the total current is

$$\sqrt{i_{AC}^2 + i_{DC}^2} \text{ kA rms.} \tag{6.35}$$

The steady state short-circuit current with constant excitation is 50% of the generator full-load current. This can occur for a stuck voltage regulator condition. Though this current is lower than the generator full-load current, it cannot be allowed to be sustained. Voltage restraint or voltage-controlled overcurrent generator backup relays (ANSI/IEEE device number 51 V) or distance relays (device 21) are set to pick up on this current. The generator fault decrement curve is often required for appropriate setting and coordination of these relays with the system relays (Chapter 7).

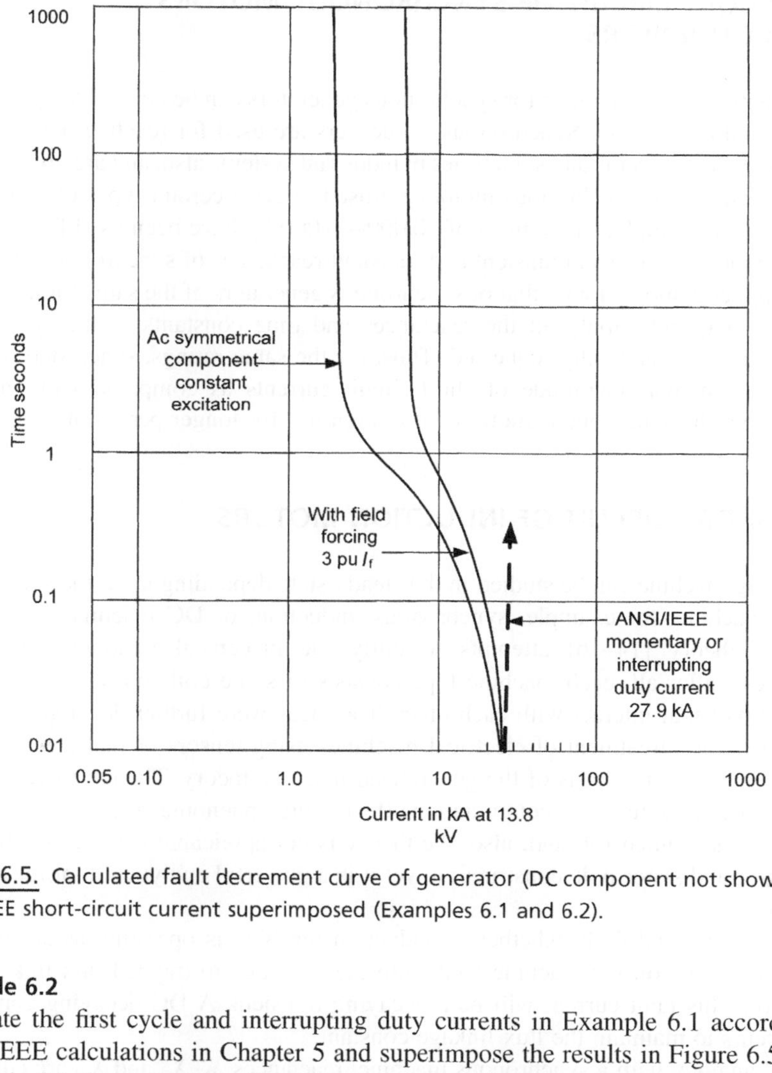


Figure 6.5. Calculated fault decrement curve of generator (DC component not shown) with ANSI/IEEE short-circuit current superimposed (Examples 6.1 and 6.2).

**Example 6.2**

Calculate the first cycle and interrupting duty currents in Example 6.1 according to ANSI/IEEE calculations in Chapter 5 and superimpose the results in Figure 6.5.

Table 5.1 shows that for all turbogenerators, all hydrogenerators with amortisseur windings, and all condensers, the multiplying factor for first cycle and interrupting duty networks is 1. This means that the first cycle (momentary) and interrupting duty short circuit currents are identical, given by:

$$i_d'' = i_d' = \frac{E}{0.15} = 27.89 \text{ kA}.$$

Here, we use rated system voltage of 13.8 kV. This is superimposed in Figure 6.5. Obviously, the ANSI/IEEE currents cannot provide the accurate arc flash calculation results.

## 6.7 SHORT CIRCUIT OF SYNCHRONOUS MOTORS AND CONDENSERS

The equations derived above for synchronous generators can be applied to synchronous motors and condensers. Synchronous condensers are used for reactive power support in transmission systems and sometimes in industrial systems also, and are not connected to any load. Large synchronous motors are used to drive certain types of loads in the industry: for example, single units of 30,000–40,000 hp have been used for refiners in the pulp industry. The subtransient and transient reactances of synchronous motors are much higher compared with that of synchronous generators of the same ratings. Manufacturers may not supply all the reactances and time constants of the synchronous motors, unless specifically requested. Thus, for the same ratings, synchronous motors contribute smaller magnitude of short-circuit currents as compared with induction motors, but the synchronous motor contribution last for longer period of time.

## 6.8 SHORT CIRCUIT OF INDUCTION MOTORS

A rotating machine can be studied in the steady state depending upon the specific type of the machine, for example, synchronous, induction, or DC machine. Generalized machine theory [14–16] attempts to unify the piecemeal treatment of rotating machines—after all, each machine type consists of some coils on the stationary and rotating part that interact with each other. The ideas were further developed by Kron and others, and treatment of electrical machines using tensors or matrices and linear transformation is the basis of the generalized machine theory. Though there are some limitations, that saturation, commutation effects, surge phenomena, eddy current losses cannot be accounted for, and, also, the theory is not applicable if there is saliency in both rotor and stator pole construction, yet it is a powerful analytical tool for machine modeling.

For a terminal fault, whether an induction machine is operating as an induction generator or motor, the machine feeds into the fault due to trapped flux linkage with the rotor. This fault current will be a decaying transient. A DC decaying component also occurs to maintain the flux linkage constant.

By analogy with a synchronous machine, reactances $X_a$, $X_f$ ,and $X_{ad}$ are equivalent to $X_1$, $X_2$, and $X_m$ in the induction machine. The transient reactance of the induction machine is

$$X' = X_1 + \frac{X_m X_2}{X_m + X_2}. \tag{6.36}$$

This is also the motor-locked rotor reactance. The open circuit transient time constant is

$$T'_o = \frac{X_2 + X_m}{\omega r_2}. \tag{6.37}$$

The short-circuit transient time constant is

$$T' = T_0' \frac{X'}{X_1 + X_m}. \tag{6.38}$$

This is approximately equal to

$$T' = \frac{X'}{\omega r_2}, \tag{6.39}$$

and the time constant for the decay of DC component is

$$T_{DC} = \frac{X'}{\omega r_1}. \tag{6.40}$$

The AC symmetrical short-circuit current is

$$i_{AC} = \frac{E}{X'} e^{-t/T'}, \tag{6.41}$$

and the DC current is:

$$i_{DC} = \sqrt{2} \frac{E}{X'} e^{-t/T_{DC}}, \tag{6.42}$$

where $E$ is the prefault voltage behind the transient reactance $X'$. At no load, $E$ is equal to the terminal voltage.

**Example 6.3**

Consider a 2.3 kV, 900 hp, four-pole motor. The manufacturer's data is: full load efficiency = 94.7%, full load power factor = 88%, full load current = 202 amperes, and locked rotor current = six times the full load current = 1212 A.

Using a computer subroutine, the following parameters are calculated:

Stator resistance $r_1 = 0.031$ pu, rotor resistance $r_2 = 0.015$ pu, magnetizing reactance
$X_m$ = 3.00 pu, $R_m$ = 100.0 pu, stator reactance $X_1$ = 0.0656 pu, rotor reactance $X_2$ = 0.0984 pu, motor transient reactance $X'$ = 0.165 pu, AC short-circuit time constant $T'$ = 0.0295 second, and DC short-circuit time constant $T_{DC}$ = 0.0142 second.

The AC component of the short-circuit current is

$$i_{AC} = 6.06e^{-t/0.0295}.$$

At $t = 0$, the AC symmetrical short-circuit current is 6.06 times the full load current.

The DC component of the short-circuit current is

$$i_{DC} = 8.56e^{-t/0.0142}.$$

Based upon these equations, the decay of the AC and DC component is plotted in Figure 6.6. Also, the ANSI/IEEE interrupting and first cycle (momentary) values are superimposed. Again it is obvious that ANSI/IEEE calculations cannot accurately simulate the real world decay of the short-circuit currents from the motors.

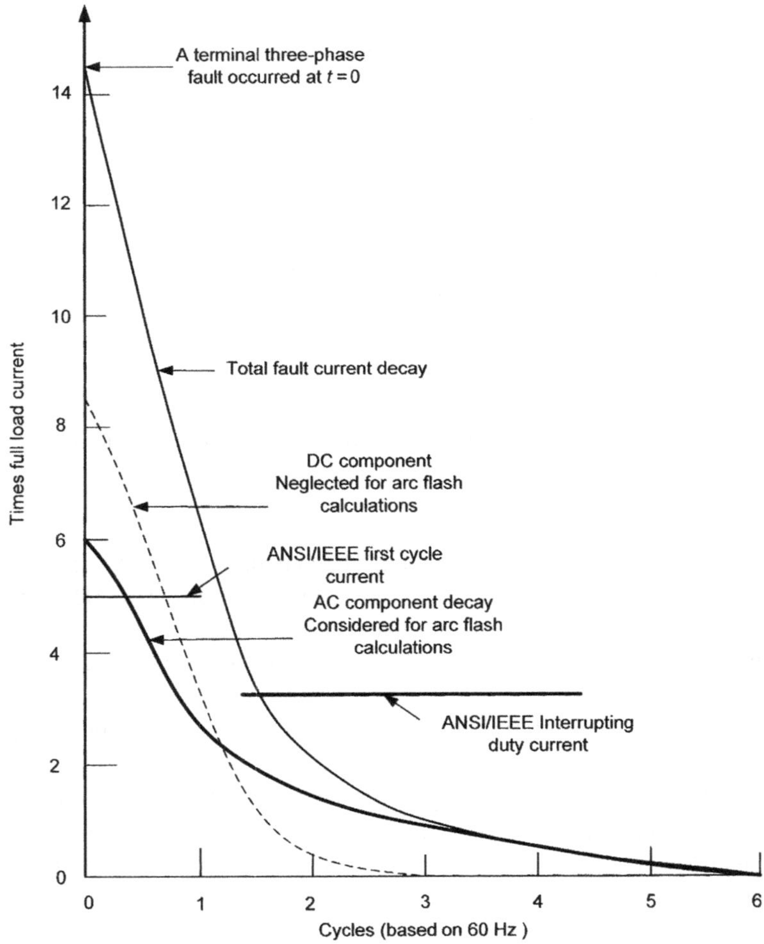


Figure 6.6. Calculated short-circuit currents for terminal fault of a 900 hp, 2.3 kV motor. The DC component is ignored for arc flash calculations.

Typically, the effect of short-circuit currents from induction machines can be ignored in couple of cycles depending upon the rating of the motor. For small motors less than 100 hp, this decay may be one to two cycles, and even for very large motors, it rarely exceeds seven to eight cycles. It depends upon the decay of trapped MMF in the motor windings after short circuit, and the decay will vary with motor and load inertia, the instant of short circuit on voltage wave, and the like.

## 6.9 A NEW ALGORITHM FOR ARC FLASH CALCULATIONS WITH DECAYING SHORT-CIRCUIT CURRENTS

The situation with respect to accurate calculations of short-circuit currents for arc flash is demonstrated with respect to a simple system configuration in Figure 6.7. This shows a generator of 40 MVA operating in parallel with a 50-MVA utility transformer. The motor loads in the system are lumped together on equivalent transformers. The

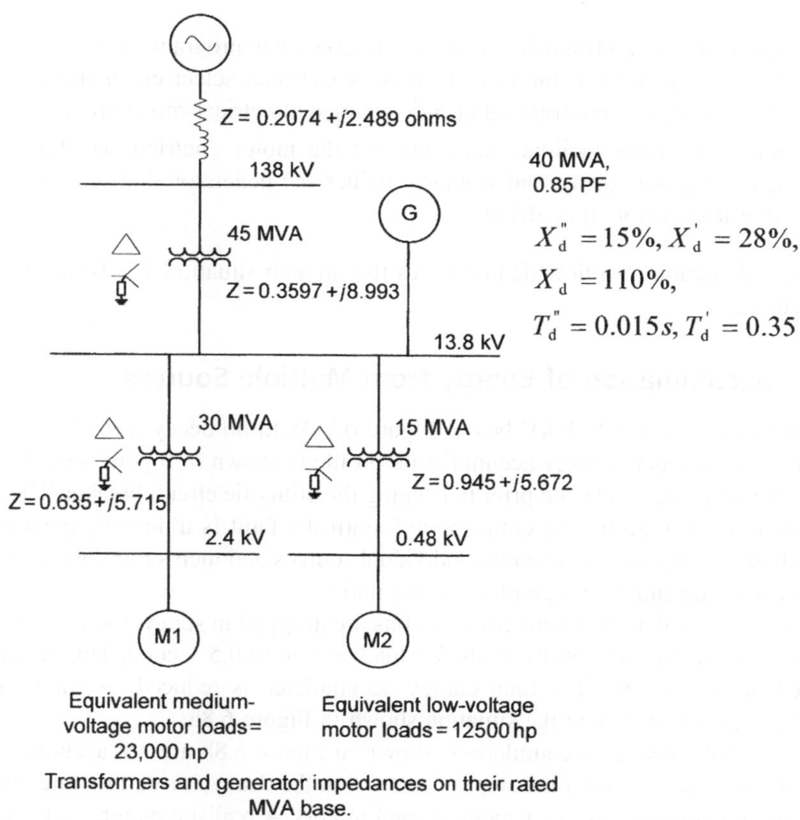


Figure 6.7. A 13.8 kV bus with multiple sources of short-circuit currents.

impedance data and utility source impedance is shown in this figure, based on which the three-phase short-circuit current at 13.8-kV bus is 37.9 kA symmetrical.

It is interesting to note that in the industrial distribution environment, much larger transformers and generators have been interconnected, and the application of 13.8-kV circuit breakers is limited to an interrupting rating of 40.2 kA (earlier 15 kV, $K = 1.3$, 37 kA breakers). Following is a real-world example of a large distribution system:

> Four utility tie transformers each of 30/50 MVA, 110–13.8 kV, six plant generators having a total installed capacity of 270 MVA, total running load 190 MVA—the excess power generated is supplied into the utility system, full stream production can be maintained on forced or maintenance outage of one or more generators and one utility tie transformer, yet, 13.8 kV circuit breakers with interrupting rating of 40.2 kA at 13.8 kV have been applied.

### 6.9.1 Available Computer-Based Calculations

The commercially available software programs use the short-circuit calculations as the base to which the arc flash routines are added. With respect to the calculations of short-circuit currents for arc flash analysis, there are two situations:

1. There is no algorithm in the available computer program to account for the decay from the generators or motors. A user can select either the first cycle (momentary) or interrupting (1.5–5 cycle) symmetrical rms currents.
2. Some programs facilitate knocking out the motor contribution after a user-selectable time delay, and similarly reduce the generator short-circuit current, after a selectable time delay.

The calculation to follow demonstrates that though situation 2 is better, it is still not accurate.

### 6.9.2 Accumulation of Energy from Multiple Sources

Consider a fault on the 13.8 kV bus of Figure 6.7. With no decay from the generator or motors, the incident energy accumulation profile is shown in Figure 6.8a. The generator circuit breaker is tripped prior to tripping the utility tie circuit breaker. The motor contributions continue for the entire period, until the fault is ultimately cleared. The figure shows energy release from the individual sources and then sums these to give an overall energy accumulation graphic representation.

Now, assume that the motor contributions are dropped in six cycles, and generator contribution is reduced to 300% of its full load current in 0.5 second. This situation is depicted in Figure 6.8b. The total energy accumulated is reduced, as shown in the shaded area, compared with the situation shown in Figure 6.8a.

Yet the total energy accumulations shown in Figure 6.8b are not accurate. In the real world situation, it is not the step reduction, but the decay at a certain rate given by the transient parameters of the generators and motors. A realistic picture of the energy accumulation from the decaying short-circuit currents of the motors and generator and their accumulation is shown in Figure 6.8c.

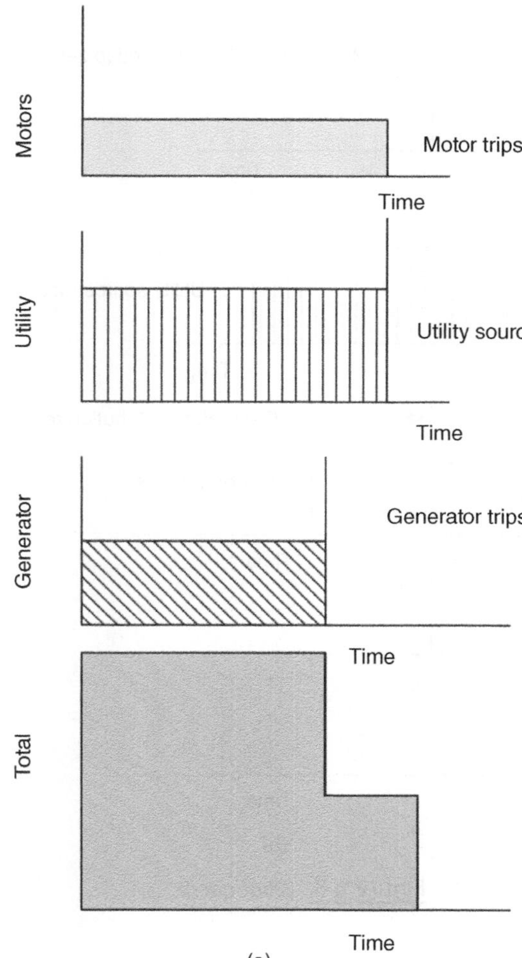


Figure 6.8. (a–c) Accumulation of incident energy profiles (see text).

Therefore, incident energy calculated using the three figures is:

$$E_{\text{fig8a}} > E_{\text{fig8b}} > E_{\text{fig8c}}.$$

In other words, the calculations are overly conservative in Figure 6.8a. The results shown in Figure 6.8b can vary depending upon when the contributions are reduced. In Figure 6.8b, it is only guesswork when the step change should be made.

*There is no commercial computer software that can simulate the results of Figure 6.8c.* Theoretically, the trapezoidal rule of integration or other step-by-step numerical techniques can be used.

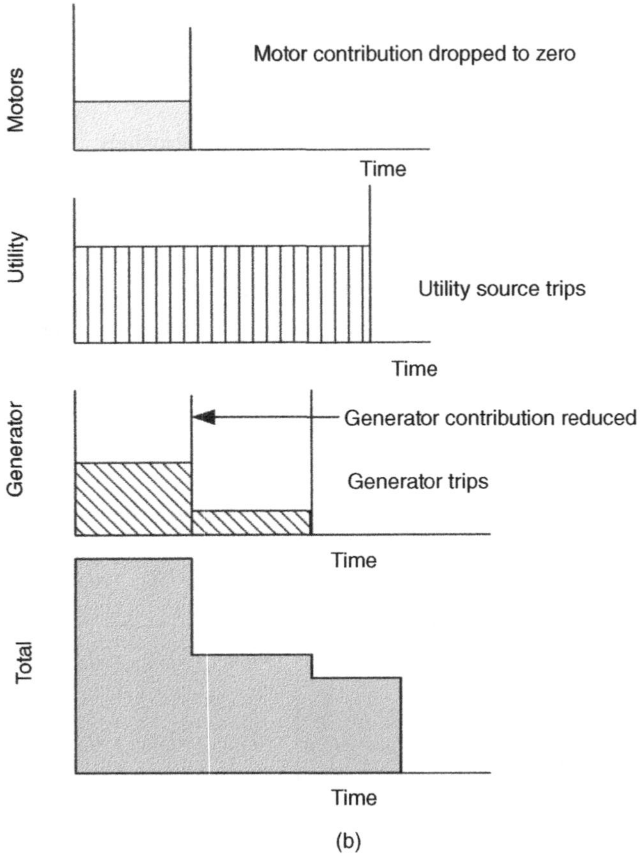


Figure 6.8. (*Continued*)

### 6.9.3 Comparative Calculations

Consider that in Figure 6.7, the fault on a 13.8 kV bus fed from the generator and utility tie is removed simultaneously in 0.5 second. This is arbitrary, to show the difference in calculations using the methodology shown in Figure 6.8a–c.

***Calculation 1.*** No decay of short-circuit current from the generator or motors. The calculated results are shown in Table 6.1, row 1. Incident energy = 36 cal/cm$^2$ and hazard risk category (HRC) 4; Figure 6.8a.

***Calculation 2.*** Motor contribution knocked out in eight cycles and generator contribution reduced to 350% at 15 cycles. Incident energy release = 30 cal/cm$^2$ and HRC 4 (Table 6.1, row 2); Figure 6.8b.

***Calculation 3.*** The concept is to first plot/calculate the overall current-time decrement curve of the short-circuit currents. Utility source is considered nondecaying,

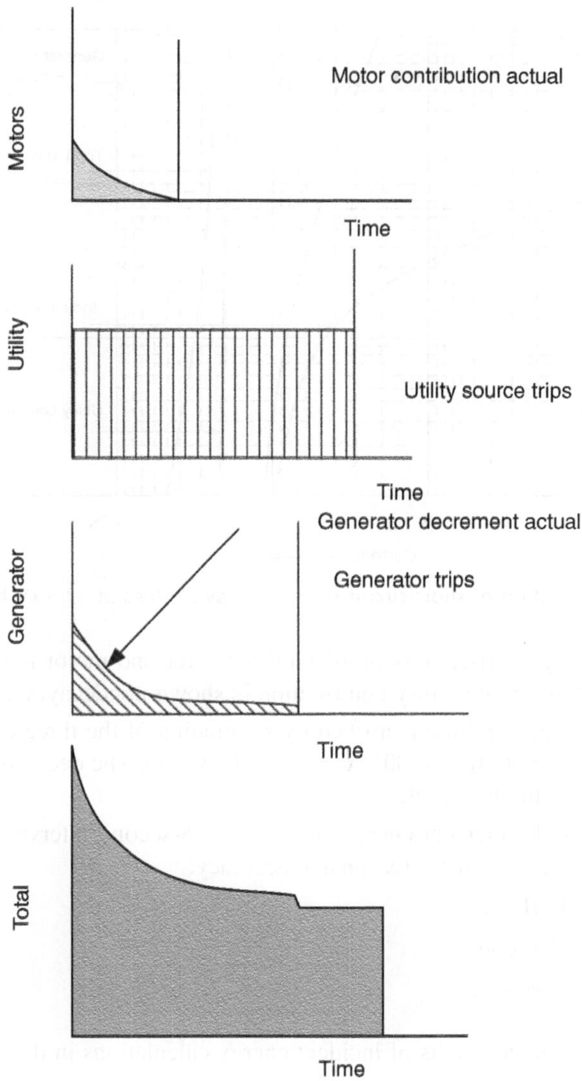


(c)

Figure 6.8. *(Continued)*

and motor and generator short-circuit contributions decay, as shown in Figure 6.8c. A step-by step procedure and explanations of the calculations are provided:

- Plot decrement curves of the generator.
- Plot the decrement curve of the motor loads; motors considered lumped together. The equivalent motor parameters can be derived.

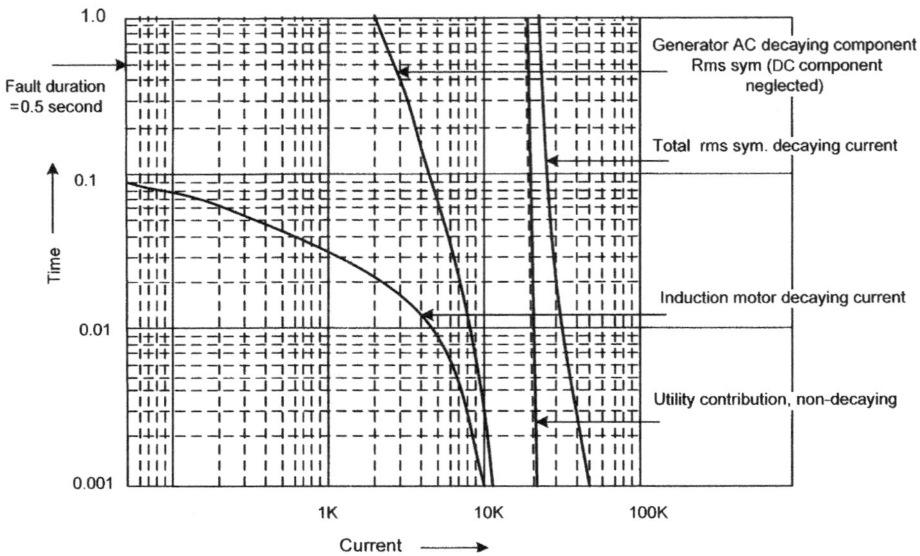


Figure 6.9. Calculation of short-circuit current decay profiles at 13.8 kV bus (Figure 6.7).

- The decaying short-circuits profiles of generator and motor loads are shown in Figure 6.9. Also, the utility contribution is shown nondecaying.
- An overall decay profile is arrived by summation of the three components. The plot is extended to time 0.001 second (0.06 cycles). The decay of current is very prominent in the first cycle.
- To calculate the incident energy, divide the 0.5-second interval into three intervals, arbitrary chosen for reasonable accuracy:

  0.001–0.01 second

  0.01–0.1 second

  0.1–0.5 second.

Table 6.1 shows the results of incident energy calculations in three cases depicted in Figure 6.8.

TABLE 6.1. Incident Energy Calculations in Figures 6.5, 6.6 and 6.7

| Source | Figure 6.8a Incident Energy (cal/cm$^2$) | Figure 6.8b Incident Energy (cal/cm$^2$) | Figure 6.8c Incident Energy (cal/cm$^2$) |
|---|---|---|---|
| From motors | 16 | 5.02 | 3.16 |
| From generator | 13.3 | 10.47 | 7.97 |
| From utility | 32.4 | 32.4 | 32.4 |
| Total | 61.7 | 47.89 | 42.63 |

Fortunately, commercially available software for power system analysis and arc flash calculations do consider the decay in the short-circuit currents; either with step change of the time period is divided into couple of time steps.

To summarize:

- Plot decrement curves of the generators and motors. Static impedance, for example, that of a transformer can be simply added to the equations above.
- Adjust the plot for the operating time of the protective devices.
- Divide into number of segments. Initially, the decay will be faster, so closer time interval will be appropriate.
- Read the average current in each interval.
- Using computer-based program and IEEE 1584 equations, calculate energy release in each segment.
- Summate the energy in each time interval

Practically, it will be time consuming to plot the decay at each bus; divide it into a number of segments and then calculate energy release in each segment and summate. However, the algorithm can be computerized [17].

Again, for simplicity, consider three sources connected to a bus: $S_1$, $S_2$, and $S_3$. These contribute incident energy $E_1$, $E_2$, and $E_3$ at times $t_1$, $t_2$ and $t_3$.

$t_1 < t_2 < t_3$.

Then the total energy is $E_1 + E_2 + E_3$.

This may be conservative, but strictly speaking not correct. It implies that energy $E_1$ remains unchanged for the time $t_3 - t_1$ and energy $E_2$ remains constant for the time $t_3 - t_2$; though the sources contributing to these energies have been removed from service.

Currently there is no publication on this aspect.

## 6.10 CROWBAR METHODS

For three-phase bolted fault, the arc flash does not occur. A radically different method of dealing with arcing faults is what is known as crowbar method. This method is well known in Europe and is recognized as a viable method for the medium-voltage switchgear by IEC Standard 62771-200. It is not popular in USA and as yet no ANSI/IEEE Standard exists.

The crowbar method consists of high-speed detection of arcing and intentional creation of a three-phase bolted fault with a high-speed grounding switch. The circuit voltage is brought to zero and the arc collapses. Arcing is detected by an optical sensor. An electronic relay energizes an actuator which closes three-phase grounding switch. The bolted fault is sensed by a relay and the circuit breaker tripped. See Figure 6.10.

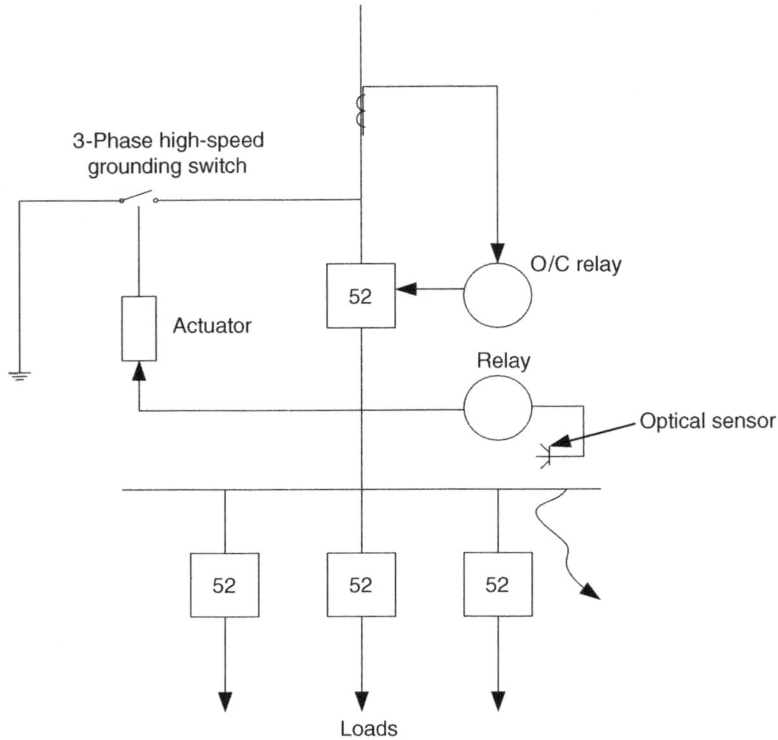


Figure 6.10. Crowbar arrangement.

## REVIEW QUESTIONS

1. Calculate the AC fault decrement curves of a 120 MVA generator, 13.8-kV, 0.85 power factor, synchronous reactance = 1.95 pu, saturated transient reactance = 0.278 pu, saturated subtransient reactance = 0.164 pu, short-circuit subtransient time constant = 0.015 second, short-circuit transient time constant = 0.597 second. Consider a field current of 1 per unit at no load voltage. Plot the results to 1000 seconds and superimpose ANSI/IEEE momentary and interrupting duty currents, similar to Figure 6.5.

2. A 4-kV, 5000-hp four-pole induction motor, full load kVA = 4200, has: $r_1 = r_2 = 0.0075$, $X_1 = 0.0656$, $X_2 = 0.0984$, $X_m = 3.0$, $R_m = 100$, all specified in per unit on motor full load kVA base. Calculate the equations for a sudden terminal fault and plot the results similar to Figure 6.6. Superimpose ANSI/IEEE momentary and interrupting duty currents.

## REFERENCES


1. R. Bronson and G. Costa, *Differential Equations*, 3rd edition, Schaum's Outline Series, McGraw-Hill, New York, 2006.
2. J. Cronin, *Ordinary Differential Equations: Introduction to Qualitative Theory*, Chapman and Hall/CRC, Boca Raton, FL, 2008.
3. S. Goldman, *Laplace Transform Theory and Electrical Transients*, Dover Publications, New York, 1966.
4. M.G. Smith, *Laplace Transform Theory*, Van Nostrand, London/New York, 1966.
5. A.E. Fitzgerald, Jr., S.D. Umans, and C. Kingsley, *Electrical Machinery*, McGraw-Hill Higher Education, New York, 2002.
6. R.H. Park, "Two reaction theory of synchronous machines, part I," AIEE Trans., vol. 48, pp. 716–730, 1929.
7. R.H. Park, "Two reaction theory of synchronous machines, part II," AIEE Trans., vol. 52, pp. 352–355, 1933.
8. N.N. Hancock, *Matrix Analysis of Electrical Machinery*, Pergamon Press, Oxford, UK, 1964.
9. C. Concordia, *Synchronous Machines*, John Wiley, New York, 1951.
10. I. Boldea, *Synchronous Generators*, CRC Press, Boca Raton, FL, 2005.
11. P.M. Anderson and A. Fouad, *Power System Control and Stability*, IEEE Press, New York, 1991.
12. P.M. Anderson, *Analysis of Faulted Power Systems*, Iowa State University Press, Ames, 1973.
13. J.C. Das, *Power System Analysis—Short-Circuit Load Flow and Harmonics*, 2nd edition, CRC Press, Boca Raton, FL, 2011.
14. B. Adkins, *The General Theory of Electrical Machines*, Chapman and Hall, London, 1964.
15. C.V. Jones, *The Unified Theory of Electrical Machines*, Pergamon Press, Oxford, UK, 1964.
16. A.T. Morgan, *General Theory of Electrical Machines*, Heyden & Sons Ltd., London, 1979.
17. J.C. Das, "Protection planning and system design to reduce arc flash incident energy in a multi-voltage level distribution system to 8 cal/cm$^2$ (HRC 2) or less—Part 1: Methodology," IEEE Trans. Ind. Appl., vol. 47, no. 1, pp. 398–407, Jan./Feb. 2011.

# 7

# PROTECTIVE RELAYING

The system protection has a profound effect on arc flash hazard analysis and reduction. The protective relaying has been called an "art" and also a "science." This is so because there is a judgment involved in making selections, which require compromises between conflicting objectives, such as maximizing reliability, fast fault clearance times, economics, and selectivity. A fault in the system should be detected fast, and only the faulty section isolated without impacting the unfaulted system. Protective relaying is an essential feature of the electrical system and is considered concurrently with the system design. Protection is not a substitute for poorly designed systems—that is, protecting a poorly designed system will be more complex and less satisfactory than a properly designed system.

In many continuous-processes industrial plant distribution systems, a single nuisance trip can result in colossal loss of revenue, and it may take many hours to days to restore the processes to full stream production.

## 7.1 PROTECTION AND COORDINATION FROM ARC FLASH CONSIDERATIONS

This chapter and the following Chapters 8–10 discuss protective relaying from arc flash considerations. Protective relaying can be distinctively classified into two categories:

*Arc Flash Hazard Analysis and Mitigation*, Second Edition. J.C. Das.

Published 2021 by John Wiley & Sons, Inc.

- equipment protection
- system protection.

Equipment protection narrows down the protection to individual equipment, that is, generator, transformer, bus, cable, transmission line, and motor protection. System protection considers a group of elements in a certain configuration. A large generator may be provided with following protective functions:

1. generator backup overcurrent, voltage restraint, or voltage controlled or distance, ANSI device types, 51V and 21
2. negative sequence current (46)
3. loss of field (40)
4. reverse power or antimotoring (32)
5. out-of-step (loss of synchronism), (78)
6. phase overcurrent differential, (87)
7. ground overcurrent differential, (for low-resistance grounded generators), (87G)
8. 100% stator winding ground fault protection through third harmonic biasing, for high resistance grounded unit connected generators (27TN third harmonic neutral undervoltage, 27X and 59X, auxiliary undervoltage and overvoltage)
9. overfrequency and underfrequency protection and rate of change of frequency (81O, 81U, 81R)
10. auto-synchronizing with manual synchronizing check (25)
11. volts per hertz protection, (24)
12. phase undervoltage and overvoltage and neutral overvoltage (27P, 59P, 59N)
13. negative sequence overvoltage (59-2)
14. negative sequence directional overcurrent (67-2)
15. phase and neutral directional overcurrent (67, 67N)
16. phase instantaneous and neutral overcurrent (50P, 50N)
17. phase and neutral time overcurrent (51P, 51N)
18. backup ground fault protection, low resistance grounded generators, ground instantaneous, and time delay (50G, 51G)
19. generator stator windings and bearings temperature monitoring
20. inadvertent energization of generator breaker (27/50)
21. generator breaker failure monitoring and upstream trips.

The numbers in parentheses denote ANSI/IEEE protective devices nomenclature [1].

Figure 7.1 shows some of above protections for a generator, using a modern microprocessor-based multifunction relay (MMPR). This term is synonymous with digital relays and numerical relays. The numerical relays can be even more versatile,

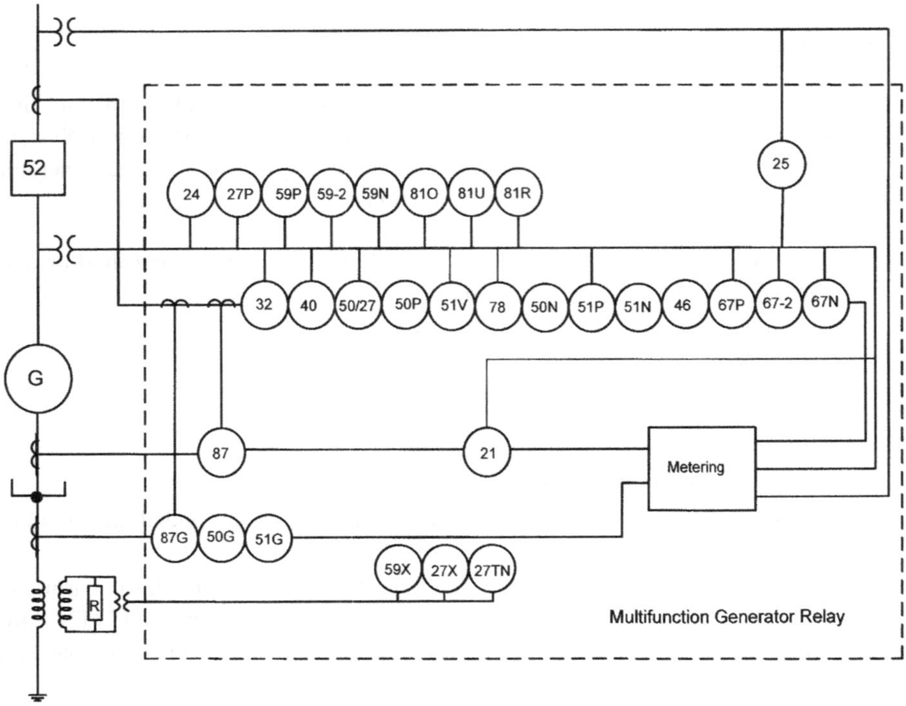


Figure 7.1. Functionality of a MMPR (multifunction microprocessor-based relay) for generator protection.

and the same relay with varied functions can be applied, say, for the protection of a generator, motor, or transformer. All the above protective features and some additional features are shown in this figure. The relay can be applied to generators with split-phase windings, pumped storage generators, bus-connected generators, and generators and their step-up transformers protected as a unit. The relay has communication facilities: IEC61850 [2], DNP 3.0, Ethernet Global Data (EGD), IEC60870-5-104, Modbus™ RTU, and Modbus TCP/IP. It has GPS time stamp, a host of digital and analogue inputs, and programmable outputs.

Many features and reports like oscillography, setting files, password protection, event log, fault reports, data logger, and health reports for problems in hardware and software are standard features of any MMPR. Even the front-panel LEDs are programmable. The settings are generated using the manufacturer's software. These settings may run into 30–40 pages and the setting file can be loaded directly into the relay. It can be safely said that 20 years back, the implementation of all the functionality in this single MMPR would have taken more than two 6-ft × 4-ft panels of individual discrete devices with much interconnecting wiring between them, and yet all the versatility would not be available.

Though MMPRs have self-diagnostic capabilities for hardware and software alarms or communication link failures, yet for large equipments, it is usual to duplicate these with full redundancy.

Looking from the prospective of arc flash considerations, it is enough to know that phase faults will be cleared in the differential zone of protection. It will be necessary to correctly ascertain the operating time of the differential element and the interrupting time of the breaker, so that arcing time is calculated. None of the above-stated generator protection functions, their applicability to a particular generator, and protection philosophy will be of interest for the arc flash calculations, though very much required for the protection of the generator. Similar situations exist with respect to protections of other electrical equipment.

These qualifications make it abundantly clear that:

1. A protection engineer must understand and apply all the protective features demanded by an equipment and system protection for an efficient and effective protection. For arc flash considerations, only the clearing of three-phase bolted faults enter into the picture.
2. Protective relaying is a vast subject. The short-circuit currents in a system cannot be easily reduced, especially in an existing distribution system. It is the manipulation, coordination, selection, and application of overcurrent protection devices that can impact the results of arc flash hazard calculations. It is recognized that the descriptive literature can only lay down rules and guidelines with specimen examples, yet it requires a good deal of experience and practice to apply these to real-world situations.
3. Special knowledge and experience is required to apply proper protection philosophies depending upon the power system; for example, utility systems, industrial systems, commercial systems each have their specific characteristics and requirements. Standards and industry practices have been established over the course of years and coupled with that much innovations are occurring in the protective relaying on account of microprocessor technology.
4. System Integrity Protection Schemes (SIPS), and Adaptive protection can be cited as two modern trends. According to the IEEE Power Systems Relaying Committee (PSCR), SIPS embraces a wide range of measures, like underfrequency and undervoltage load shedding, adaptive load mitigation, out-of-step tripping, voltage and angular instability, advanced warning schemes, overload and congestion mitigation, system separation, shunt capacitor switching, tap changer control, SVC/STATCOM control, HVDC controls, and the like. A discussion of these is not required for arc flash hazard analysis and mitigation.

Thus, it can be qualified that the chapters on system protection and coordination in this book concentrate on arc flash analysis and its mitigation strategies in an industrial distribution system environment.

## 7.2 CLASSIFICATION OF RELAY TYPES

The protective relays can be: Regulating, controlling, alarming, restoring, synchronizing, load shedding, frequency or voltage sensing, out of step, rate of pressure rise, breaker failure sensing, inadvertent energization, temperature, vibration sensing, and the like. None of these are discussed. Also, all ground fault relaying, other than that described in Chapters 4 and 11, is not discussed.

The protective relays have also been classified based upon:

**Input**

- Current
- Voltage
- Power
- Temperature
- Vibration
- Frequency

**Operating Principle**

- Percentage
- Product
- Electromecahnical
- Thermal

**Performance**

- Distance
- Mho
- Ground or phase
- Undervoltage
- Directional comparison

For the arc flash analysis, only the phase overcurrent protection is of interest.

## 7.3 DESIGN CRITERIA OF PROTECTIVE SYSTEMS

The logic of protective relaying looks at a complex distribution system as an integration of subsystems. In all cases, some common criteria are applicable. These are:

- selectivity
- speed
- reliability

- simplicity
- economics

and sometimes a sixth criterion is added of

- maintainability.

### 7.3.1 Selectivity

A protection system must operate so as to isolate the faulty section only. In a radial distribution system, which is a common system configuration in the industrial power distribution systems, inverse time overcurrent relays are used as the primary protection. The desired selectivity is obtained by coordinating upstream relays with the downstream relays, so that the upstream relay is slower than the downstream relay. A proper time delay should be selected between two overcurrent relays in series by either (1) providing a certain appropriate time-delay, called coordinating time interval (CTI) or variations of the inverse time–current characteristics; not forgetting the definite time–current characteristics. This coordination is discussed in Chapter 10. This increases the time delay for fault clearance toward the source, which is not desirable from arc flash hazard limitation and equipment damage. Separate zones of protection can be established around each equipment, unit protection systems. This is discussed in Chapters 8 and 9.

### 7.3.2 Speed

Fault damage to the system components and the stability between synchronous machines and interconnected systems are related to the speed of operation of the protective systems. In case all faults could be cleared instantaneously, the *equipment damage, as well as the arc flash hazard,* will be a minimum. Thus, there is a direct relation between limiting the arc flash hazard and equipment damage, though as yet there is no documented data on the subject. Unit protection systems, with overlapping zones of protection can mitigate this problem, see Chapter 8. Practically, unit protection systems are not applied throughout an industrial distribution.

From transient stability considerations, there is a critical fault clearing time and even a slight delay of 1/4 of cycle exceeding this time can result in system separation. Single-pole closing, fast load shedding, bundle conductors, fast excitation systems, power system stabilizers, series and shunt compensation of transmission lines, SVC and STATCOM and FACT controllers can enhance the stability limits of a power system. In industrial plants having cogeneration facilities, fast fault clearance times and system separation for a fault close to the generator become of importance.

### 7.3.3 Reliability

Dependability and security are the measures of reliability. The protection must be dependable and operate in response to system faults within its protected zone and be secure against incorrect trips from all other conditions, for example, voltage regulation due to load demand changes, high magnitude of through fault currents, inrush currents, and the like. Thus, these two objectives of reliability mutually oppose each other.

Designing more flexibility into system designs, for example, double-ended substations, duplicate feeders, auto-switching, and bus transfer schemes will increase the complexity and hence reduce the security of the protective systems.

### 7.3.4 Backup Protection

In a protective system design, the protection system is backed up in the sense that if the primary protection fails to trip, the second protective device in line must trip. The back-up protection considers failure of the relaying scheme, a breaker, or control supply failure. Relaying for a mesh or ring-connected bus configuration will be different from that for a radial system, even though these systems may interconnect same size of transformers, feeders, and generators. In a time–current coordinated system, the back-up protection is inherent. If the intended relay or circuit breaker fails to trip, the next upstream breaker will trip with a greater time delay, which will increase the fault and arc flash damage. As a general practice, a unit protection system, for example, differential relaying, is backed up with time overcurrent protection.

While a back-up protection is an important safeguard and feature of all protective systems, for the purpose of arc flash hazard and incident energy release calculations, the back-up protection is not considered. The industry practice is to consider the primary protective relay only for fault clearance times. This qualification is important.

## 7.4 OVERCURRENT PROTECTION

We can divide overcurrent protection in three categories:

1. Relayed high voltage circuit breakers are the fault interrupting devices and protective relays, separate from the breakers, are the fault sensing devices that trip the breakers through shunt trip coils. A breaker can be reclosed after it trips and the relay can be reset to restore the operation. For reliability, the high voltage circuit breakers can be provided with duplicate trip coils. The continuity of the breaker trip coils and closing coil can be continuously monitored and alarmed.
2. Low voltage circuit breakers are provided with direct acting trips, which can be microprocessor based, discussed further. These can also be equipped with shunt trip coils, actuated by relays, much alike high voltage circuit breakers. Also, the low voltage circuit breakers can be provided with undervoltage trip coils.
3. Fuses are fault sensing and tripping devices. These will open up the circuit for a fault, but destroy themselves in the process. Thus, to restore operation, these must be replaced. Further comparison is presented in a following section.

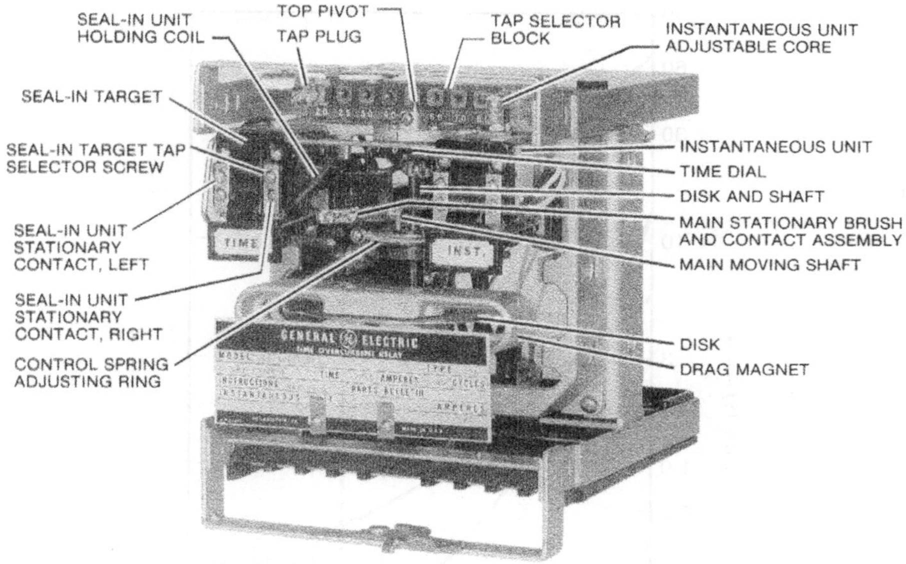


Figure 7.2. An electromechanical overcurrent relay. *Source*: IEEE standard 141-1993.

### 7.4.1 Overcurrent Relays

The most commonly used overcurrent relays are time delay and instantaneous relays, used as primary and backup devices.

***Electromechanical Relays.*** The electromechanical induction pattern relays have held their field in the industry since long, but are no longer being used. The construction of an induction pattern relay, out of its draw-out case is shown in Figure 7.2. The pulling out of the relay element from the case short-circuits the CT secondary winding circuit, and prevents open-circuit of the CT secondary. The disk is mounted on a rotating shaft, restrained by a spring. The moving contact is fastened to the shaft, and the operating torque is produced by an electromagnet having a main and lag coil, which produce out of phase magnetic flux. A damping magnet provides the restraint after the disk starts to move and this results in the desired time–current characteristics. There are discrete taps that determine the current pickup setting, and the time dial setting determines the initial position of the moving contact. Different current characteristics are obtained by modification of the electromagnetic design. Figure 7.3 shows comparison of typical ANSI curve shapes. An instantaneous unit is mounted in the same case. The relays are single phase type, that is, three units are required for three-phase, three-wire overcurrent service.

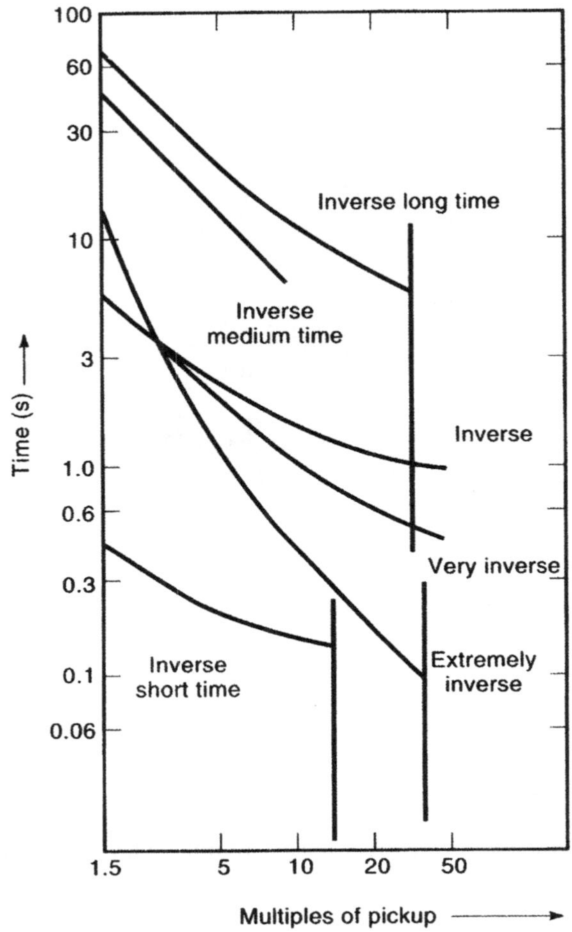


Figure 7.3. Time–current inverse characteristics of various overcurrent relay types, ANSI/IEEE curve shapes.

Mathematically, ANSI curve shapes [3] are given by the following equation:

$$
\begin{aligned}
T &= M \times \left[ A + \frac{B}{(1.03 - C)} + \frac{D}{(1.03 - C)^2} + \frac{E}{(1.03 - C)^3} \right] \quad 1 \le \frac{I}{I_{\mathrm{PKP}}} < 1.03 \\
&= M \times \left[ A + \frac{B}{(1/I_{\mathrm{PKP}} - C)} + \frac{D}{(1/I_{PKP} - C)^2} + \frac{E}{(1/I_{\mathrm{PKP}} - C)^3} \right] \quad 1.03 \le \frac{I}{I_{\mathrm{PKP}}} < 20, \qquad (7.1) \\
&= M \times \left[ A + \frac{B}{(20 - C)} + \frac{D}{(20 - C)^2} + \frac{E}{(20 - C)^3} \right] \quad 20.0 \ge \frac{I}{I_{PKP}}
\end{aligned}
$$

TABLE 7.1. Constants for Overcurrent Relay ANSI/IEEE Curves

| | Constants | | | | |
|---|---|---|---|---|---|
| ANSI Curve | A | B | C | D | E |
| Extremely inverse | 0.0399 | 0.2294 | 0.5000 | 3.0094 | 0.7222 |
| Very inverse | 0.0615 | 0.7989 | 0.3400 | -0.2840 | 4.0505 |
| Normally inverse | 0.0274 | 2.2614 | 0.3000 | -4.1899 | 9.1272 |
| Moderately inverse | 0.1735 | 0.6791 | 0.8000 | -0.0800 | 0.1271 |

where:

$T$ = operate time (second)
$M$ = multiplier set point, commonly called time dial
$I$ = input current
$I_{PKP}$ = pickup current set point, commonly called "pickup."

$A$, $B$, $C$, and $D$ are constants given in Table 7.1.

A typical setting range of an electromechanical relay for phase fault protection, of a certain characteristics can be:

- Time overcurrent taps setting, range 1–12 A, tap adjustments 1, 1.5, 2, 3, 4, 5, 6, 7, 8, 10, 12 A
- Overcurrent instantaneous settings: 40–160 A, in terms of CT secondary current
- Time dial: 0.5–10
- Indicator contactor switch settings are 0.2 or 2 A.

The indicator contactor switch indicates the operation of the relay and is hand reset. Two indicator contactor switches, one for time delay and the other for instantaneous function are provided, Figure 7.2. These indicator contactor switches have target coils in series with the main trip circuit and when these drop out for indication, the coils and springs are bypassed. The time–current characteristic is preset and cannot be changed. Thus, a protection engineer has to decide in advance what type of characteristics will be appropriate in a certain application. More often than not, it is rather difficult to decide the time–current characteristics unless a rigorous coordination study is undertaken in the design stage. This is one major limitation. This simple and rugged construction held its field for the past 50 years, but is not recommended for modern projects and arc flash reduction strategies.

## 7.4.2 Multifunction Overcurrent Relays

In today's protective relaying environment, it will be hard to buy an MMPR having only phase overcurrent and time delay protections. The simplest multifunction relay will have host of other protection features, like ground fault, directional phase overcurrent and ground fault, voltage, frequency, synchronizing and metering functions, self-diagnostics, oscillography, event logger, prefault and postfault data capture, programmable analog and digital inputs, and communication features as described for generator protection. Sometimes, a user can select the required functionality by inserting or removing required function cards in the spare relay slots, much like memory upgrades in digital computers. Concentrating on phase overcurrent protections:

- Any ANSI or IEC time–current characteristics can be programmed or changed afterwards. Compare this with electromechanical relays, which have single fixed characteristics.
- The current pickup setting can be in increments of 0.1 amps, and the time settings in increment of 0.01 seconds. This is of much importance for close coordination for arc flash mitigation.
- Load shedding, auto-closing, loss of loads can be programmed.
- Metering functions and waveform capture is available. Harmonic spectrum analysis can be carried out.
- For the electromechanical relays, manufacturers do not even project the time–current curve in the range of 1–1.5 times the pickup current. This is so because the time for operation at these low multiples of pickup can have many variations. A MMPR has following accuracies:

  Steady state pickup ±0.05 A and ±3% of setting (5 A CT secondary)

- The settings can be arranged in groups, for example, group 1–4, with each group having different settings depending upon the system conditions. The setting in each group can be brought into action and triggered by a system change as programmed in the relay.
- The "backlash" of electromechanical relays is eliminated (see Chapter 10). The CTI between series-connected devices for coordination can be reduced.
- The application flexibility is enhanced as each group of settings permits much closer time delay and instantaneous settings.
- The trip outputs can be directed and programmed to different output relays for indications, controls, alarms, or trip. Sometimes, as many as five to six trip outputs are available with programmable logic.
- Much routine testing with electromechanical relays due to wear and tear of moving parts is eliminated.
- In electromechanical relays, very short-time dial settings can result in nuisance trips. For example, at a setting of 0.5 time dial, the moving contacts are too close to the trip contacts, the gap distance is small, and even a jerk or vibration, say on opening the switchgear door on which the relay is mounted, can bridge the contacts and trip the circuit. Practically, these nuisance trips have been experienced. This is not the case with MMPRs.
- Sophisticated remote communication facilities, event log, oscillography, and fault capture data makes the pre- and postfault diagnostics reliable and dependable.

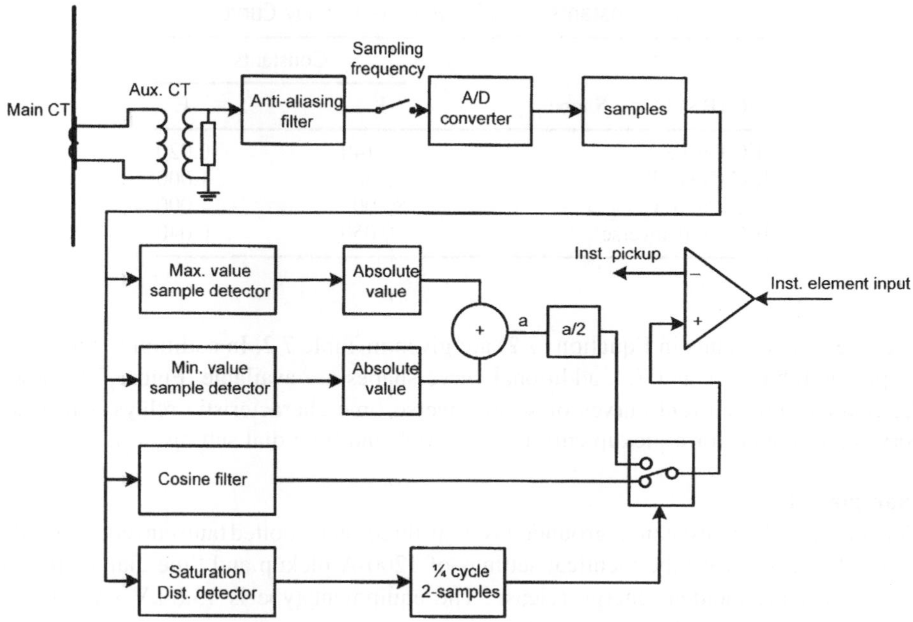


Figure 7.4. Instantaneous overcurrent function logic in an MMPR.

As many as 15 records can be stored and retrieved at a later date. The current and voltage phasors can be plotted, and the sequence components can be displayed.

- Sometimes, in electromechanical systems, it becomes difficult to ascertain the event that initiated a series of trip functions. Not so with MMPRs. The data can be captured with a time stamp on multiple trips.

Figure 7.4 shows the instantaneous overcurrent logic of a modern MMPR. In the examples of time–current coordination in this book, the MMPR relays are used.

### 7.4.3 IEC Curves

The equations for IEC curves are:

$$
\begin{aligned}
T &= M \times \left[\frac{K}{(1.03)^E - 1}\right] \quad 1 \le \frac{I}{I_{\text{PKP}}} < 1.03 \\
&= M \times \left[A + \frac{K}{(1/I_{\text{PKP}})^E - 1}\right] \quad 1.03 \le \frac{I}{I_{\text{PKP}}} < 20. \\
&= M \times \left[A + \frac{K}{(20)^E}\right] \quad 20.0 \le \frac{I}{I_{PKP}}
\end{aligned}
\tag{7.2}
$$

TABLE 7.2. Constants for IEC Overcurrent Relay Curves

| IEC (BS) Curve Shape | Constants K | Constants E |
|---|---|---|
| IEC Curve A | 0.140 | 0.020 |
| IEC Curve B | 13.50 | 1.000 |
| IEC Curve C | 80.00 | 2.000 |
| IEC short-inverse | 0.050 | 0.040 |

The relevant constants in Equation (7.2) are given in Table 7.2. In addition to the curve shapes in Tables 7.1 and 7.2, additional curve shapes are available. Figure 7.5 shows the relative time–current curves of seven inverse time characteristic relays. All these curves are plotted for a pickup current of 1200 A and time dial setting of 3.

**Example 7.1**

Consider a 13.8-kV resistance-grounded system, three-phase bolted fault current = 30 kA. Using the curves with the identical settings of 1200-A pickup and time dial setting of 3, calculate the incident energy release. The equipment type is 13.8 kV switchgear, provided with five-cycle circuit breakers.

The results of the calculation are shown in Table 7.3. This considers an IEEE 1584 Guide gap of 153 mm, working distance of 36 in, and 0.85 $I_a$ to find second arc duration time. Note how the trip time varies with the type of characteristics, which results in incident energy release from 6.5 cal/cm$^2$ (minimum) for IEC extremely inverse characteristics to 37.6 cal/cm$^2$ (maximum) for IEC standard inverse characteristics. The selection of proper characteristics in a certain application has a profound effect on the arc flash incident energy release. In a practical coordination exercise, the selection of overcurrent characteristics must consider the devices downstream and upstream with which coordination is required.

**Example 7.2**

A 1.5 MVA , 13.8 -0.48 kV substation, with protective devices, is illustrated in Figure 7.6. It is required to select characteristics for relay R1 for the primary breaker of the transformer, which gives the *minimum* incident energy for a fault at the low voltage switchgear bus. There is no main secondary breaker.

The coordination can be tried with various time–current characteristics shown in Figure 7.5, and it is seen that the lowest arc flash incident energy release is obtained with IEC extremely inverse characteristics. Even then the incident energy release is high = 45.9 cal/cm$^2$, *extreme* danger. The IEC extremely inverse characteristics and coordination is shown in Figure 7.7.

Figure 7.7 also shows a three-step curve in dotted line that can be obtained by proper settings in a MMPR. The incident energy is reduced from 45.9 to 10.5 cal/cm$^2$. The results of these calculations are shown in Table 7.4

This example shows that an effective coordination in a given case can also simultaneously lower the incident energy release and arc flash hazard.

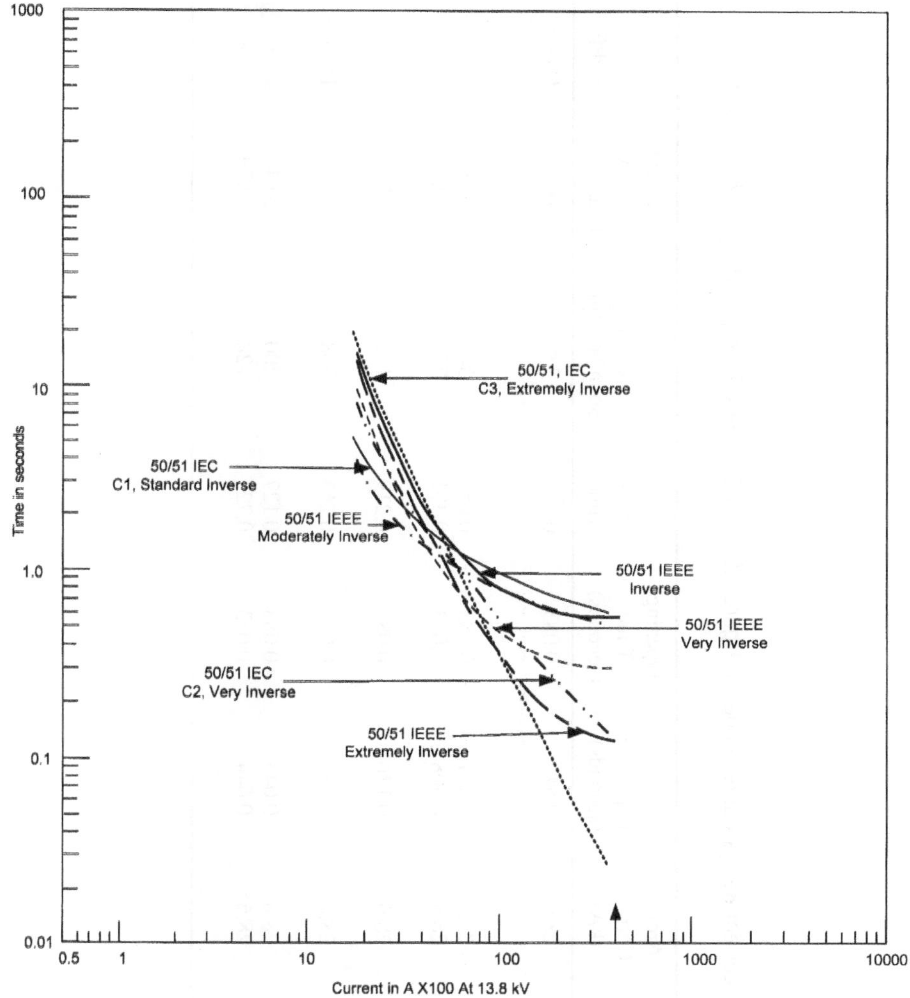


Figure 7.5. Time–overcurrent characteristics of various overcurrent relay types drawn for the same pickup setting of 1200-A pickup and time dial setting of 3.

## 7.5 LOW VOLTAGE CIRCUIT BREAKERS

The three classifications of low voltage circuit breakers are: (1) molded case circuit breakers, (2) insulated case circuit breakers, and (3) low voltage power circuit breakers [4–9].

### 7.5.1 Molded Case Circuit Breakers (MCCBs)

In MCCBs, the current carrying parts, mechanism, and trip devices are completely contained in a molded-case insulating material, and these breakers are not maintainable.

TABLE 7.3. Incident Energy for Various Overcurrent Relays Curve Shapes in Figure 7.5, All Relays Have Pickup Setting = 1200 A, and Time Dial = 3, 13.8 kV, EC = HCB, Example 7.1

| Relay Characteristics | Arc Gap (mm) | Working Distance (in) | Arcing Current (kA) | Trip Time (seconds) | Opening Time (seconds) | Arcing Time | Arc Flash Boundary (in) | Incident Energy (cal/cm$^2$) | PPE |
|---|---|---|---|---|---|---|---|---|---|
| IEEE, moderately inverse | 152 | 36 | 28.6 | 0.546 | 0.083 | 0.63 | 338 | 48.9 | Danger |
| IEEE, inverse | 152 | 36 | 28.6 | 0.571 | 0.083 | 0.654 | 346 | 50.8 | Danger |
| IEEE, very inverse | 152 | 36 | 28.6 | 0.308 | 0.083 | 0.391 | 254 | 30.4 | 4 |
| IEEE, extremely inverse | 152 | 36 | 28.6 | 0.136 | 0.083 | 0.219 | 179 | 17 | 3 |
| IEC, standard inverse | 152 | 36 | 28.6 | 0.18 | 0.083 | 0.263 | 368 | 56.5 | Danger |
| IEC, very inverse | 152 | 36 | 28.6 | 0.043 | 0.083 | 0.127 | 200 | 20.4 | 3 |
| IEC, extremely inverse | 152 | 36 | 28.6 | 0.644 | 0.083 | 0.727 | 128 | 9.79 | 3 |

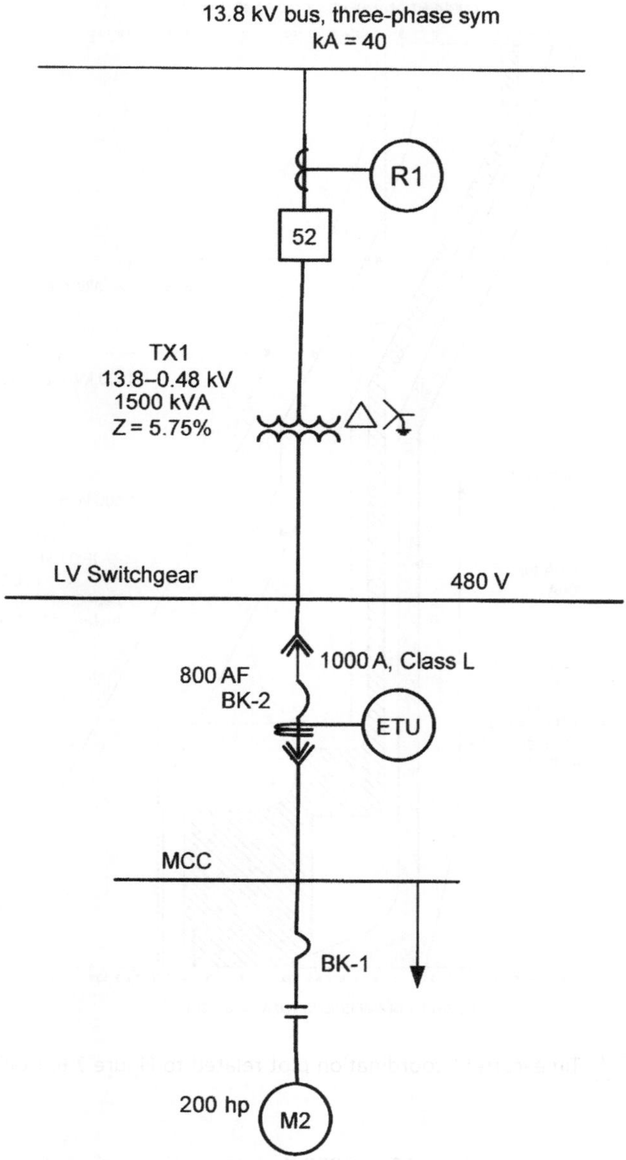


Figure 7.6. A low voltage system for illustration of selection of transformer primary relay characteristics for minimum arc flash incident energy release on low voltage switchgear bus (see Example 7.2).

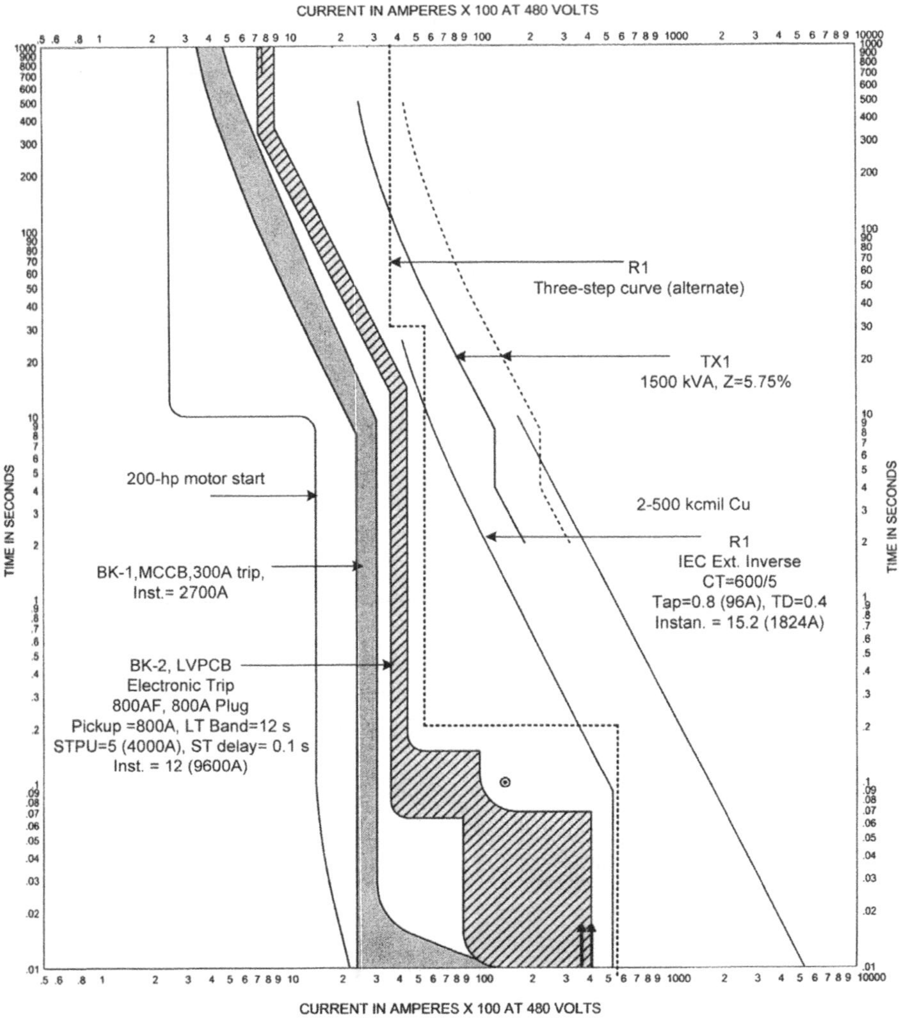


Figure 7.7. Time–current coordination plot related to Figure 7.6, Example 7.2.

Available frame sizes range from 15 to 6000 A, interrupting ratings from 10 to 100 kA symmetrical without integral current limiting fuses and to 200 kA symmetrical with current limiting fuses. These can be provided with electronic- and microprocessor-based trip units, and have limited short-time delay and ground fault sensing capability. When provided with thermal magnetic trips, the trips may be adjustable or nonadjustable, and are instantaneous in nature. Motor circuit protectors (MCPs) may be classed as a special category of MCCBs and are provided with instantaneous trips only. MCPs do not have an interrupting rating by themselves and are tested in conjunction with motor starters. An MCCB can be applied for a variety of applications, in residential and industrial

TABLE 7.4. Arc Flash Hazard Calculations, Example 7.2, LV switchgear, gap = 32mm, working distance 24″, Dimensions 36″W x 90″H x 48″D, LV MCC, gap = 25mm, working distance 18″, Dimensions 30″W x 90″H x25″D, EC = HCB

| Bus Identification | Arc Gap (mm) | Working Distance (in) | Bolted Current (kA) | Arcing Current (kA) | Trip Time (seconds) | Opening Time (seconds) | Arcing Time | Arc Flash Boundary (in) | Incident Energy (cal/cm$^2$) | PPE |
|---|---|---|---|---|---|---|---|---|---|---|
| LV switchgear R1 IEC Ext. Inverse | 32 | 24 | 1.052 | 0.465[a] | 1.444 | 0.083 | 1.528 | 154 | 49.4 | Extreme danger |
| LV switchgear Three-step curve | 32 | 24 | 1.052 | 0.465[a] | 0.21 | 0.083 | 0.293 | 74 | 11.6 | 3 |
| MCC | 25 | 18 | 28.26 | 15.83 | 0.071 | 0 | 0.071 | 41 | 6.62 | 2 |

[a] Secondary current reflected at 13.8 kV, as seen by Relay R1.

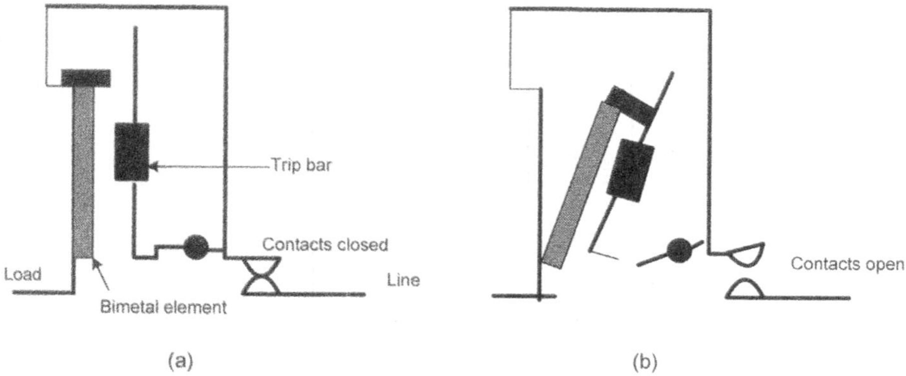


Figure 7.8. A bimetallic thermal trip device. (a) Normal state and (b) tripped state.

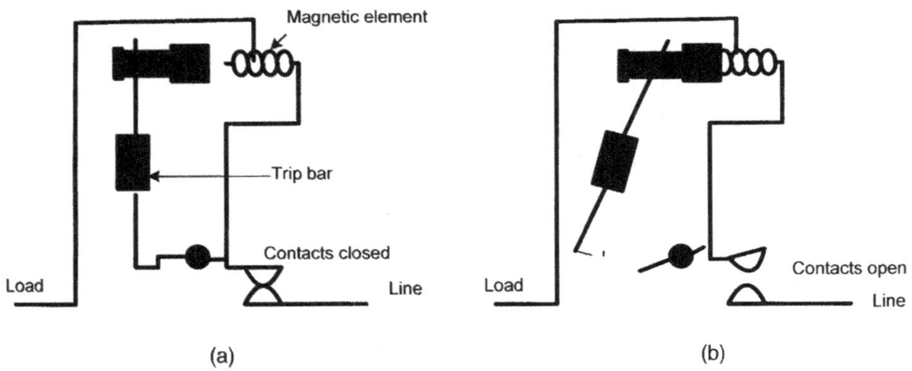


Figure 7.9. A magnetic instantaneous trip device. (a) Normal state and (b) tripped state.

distribution panels, in main power feed panels, for controlling low voltage motor starters, and many other commercial applications.

The predominant standard is UL 489. For UL listing, the MCCB must undergo well-defined rigorous series of test sequences. These include temperature rise measurements at MCCB terminals while carrying rated current after the MCCB has interrupted an overload current of 600% of the full load current 50 times.

Figure 7.8 shows thermal overload trip action. The bimetallic design parameters controlling the thermal action are: type of material, its resistance, thermal capacity, and overall length of the element. Figure 7.9 shows magnetic trip action and adjustable characteristics are obtained by varying air-gap length or adjustment of spring force. In Figure 7.7, the thermal magnetic curve is shown for a 400 AF breaker, (with 300 A current rating plug) with magnetic adjustable from 5 to 10 times the plug rating. The electronic trip units permit different rating plugs in the same frame size, and may use a flux transfer device requiring very small energy of 3 mj to shift the flux and trip. For high continuous currents of 1000 A or more, there are parallel main and arcing contacts; the main contacts make last and break first. The current commutation from mains to arcing is of the order of 300–600 μs. MCCBs use deion plates to form the arc chute. Much effort has gone into the optimization of arc runners for rapid arc motion.

### 7.5.2 Current-Limiting MCCBs

The 1970s saw the development of current-limiting MCCBs. The arc in a MCCB serves the additional function of suddenly injecting a resistive element into the circuit to limit the fault current. The current limiting MCCBs have rapid contact motion after fault initiation and rapid arc voltage development that is achieved by arc runners or blowout effect of deion plates, and fast gap recovery voltage. The effectiveness is given by both peak let-through current and also $\int i^2 dt$ values. Figure 7.10 is a representative arc voltage and current waveform for a current-limiting MCCB.

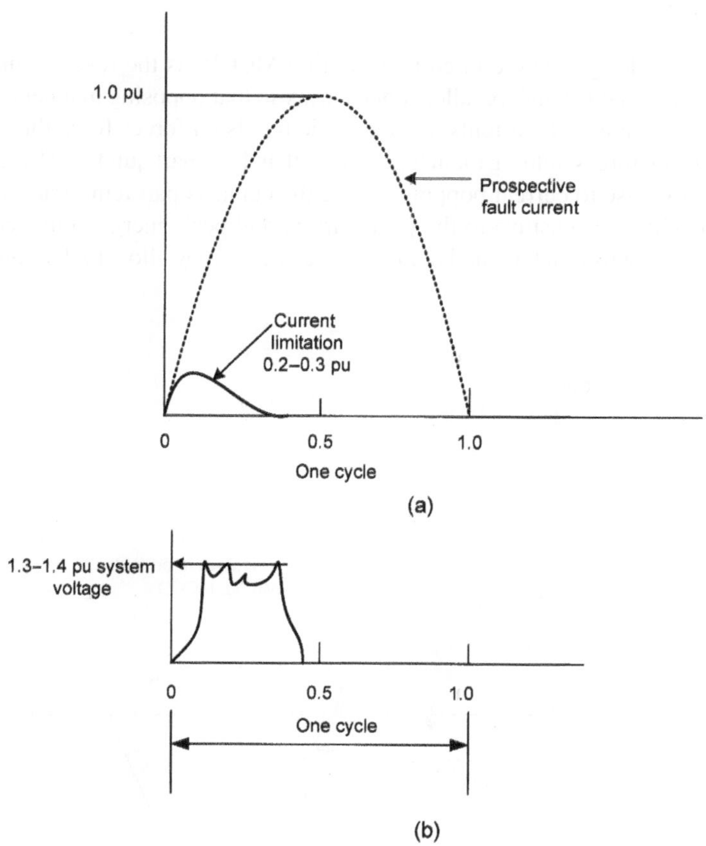


Figure 7.10. (a) Current limitation, operation of a current limiting MCCB; (b) arc voltage generated during operation.

The operation of MCCBs not *classified as current limiting* is illustrated in Figure 7.11. This figure clearly depicts that to some extent the MCCBs are current limiting. Figure 7.12 illustrates the typical characteristics of a current limiting MCCB. The current limiting MCCBs can be thermal magnetic or provided with electronic trips. These trip units, much alike the trip units of LVPCBs (Figure 7.13), can have adjustable current pickups, adjustable LT delays, short time pickup and delays, and must be provided with instantaneous trips. The MCCBs can be true rms sensing (for harmonic loads) and also be provided with shunt trip coils for operation from separate relaying devices, undervoltage trips, zone interlocking, and host of other features.

For arc flash reduction, the development of MCCBs, faster in operation with much higher short-circuit ratings, has attracted manufacturers. The current limiting action has three distinct benefits.

- Lesser let-through energy and reduction in arc flash hazard
- Current limitation can help series ratings (Section 7.7)
- Better coordination can be obtained with coordination on instantaneous basis (Chapter 10).

A common design feature in current limiting MCCBs is the reverse current loop. The current is routed through parallel contact arms so that opposing magnetic forces are formed. During high fault currents, the magnetic repulsion forces force the contacts to overcome spring forces holding them together, so that these part quickly. These magnetic forces may give rise to current popping, where the contacts part temporarily.

The MCCBs are sensitive to the peak current and peak energy delivered over the first few milliseconds of a fault and then limit the energy they allow to flow on complete interruption.

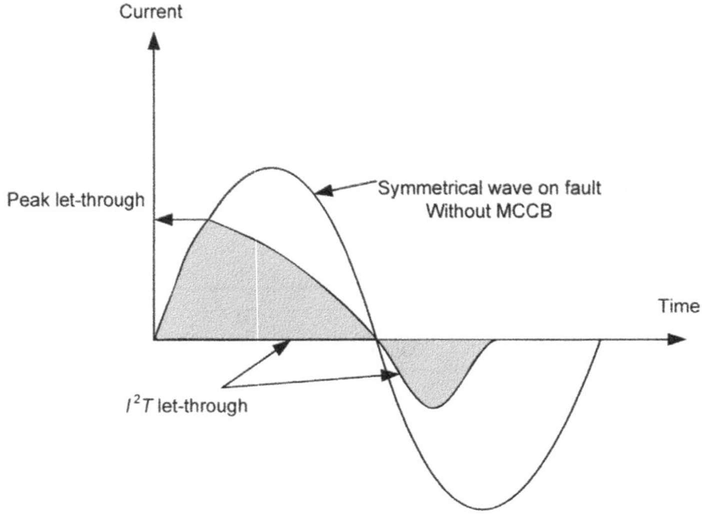


Figure 7.11. Operation of a MCCB, not specifically designated as current limiting.

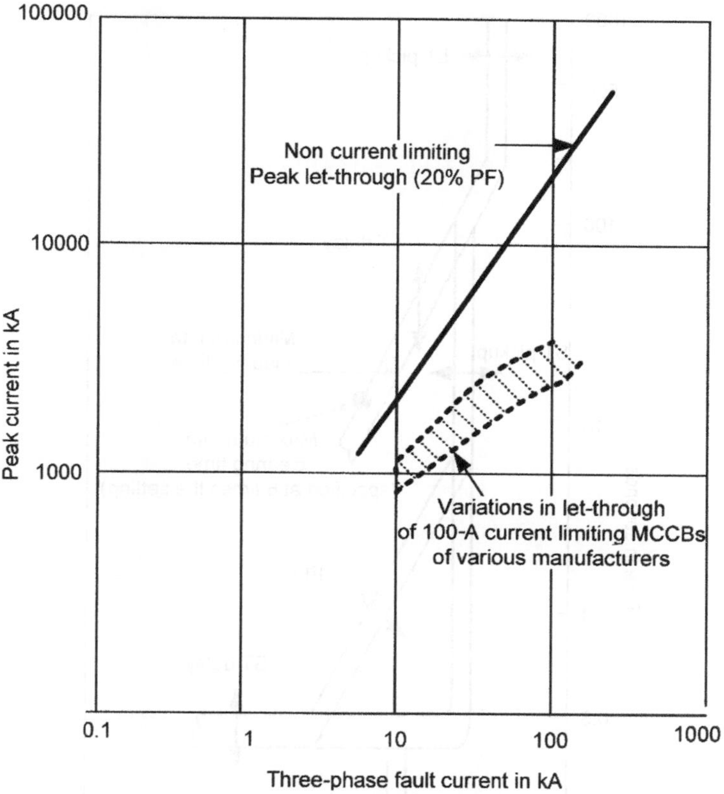


Figure 7.12. Let-through characteristics of a current limiting MCCB.

For arc flash considerations, many efforts are concentrated on development of lighter mechanisms and faster operation. Figure 7.14 shows low arc-flash circuit breaker design [10].

### 7.5.3 Insulated Case Circuit Breakers (ICCBs)

Insulated case circuit breakers utilize characteristics of design from the power circuit breakers and MCCBs. These are not fast enough to qualify as current-limiting type, and are partially field maintainable. These can be provided with electronic trip units and have short-time ratings and ground fault sensing capabilities. These are available in ratings up to 5000 A and 85 kA interrupting. These utilize stored energy mechanisms similar to low voltage power circuit breakers.

MCCBs and ICCBs are rated and tested according to UL 489 standard [7]. Both MCCBs and ICCBs are tested in the open air without enclosure and are designed to carry 100% of their current rating in open air. When housed in an enclosure, there is 20% derating, though some models and frame sizes may be listed for application at 100% of their continuous current rating in an enclosure. MCCBs are fixed mounted in switchboards and bolted to bus bars. ICCBs can be fixed mounted or provided in draw-out designs.

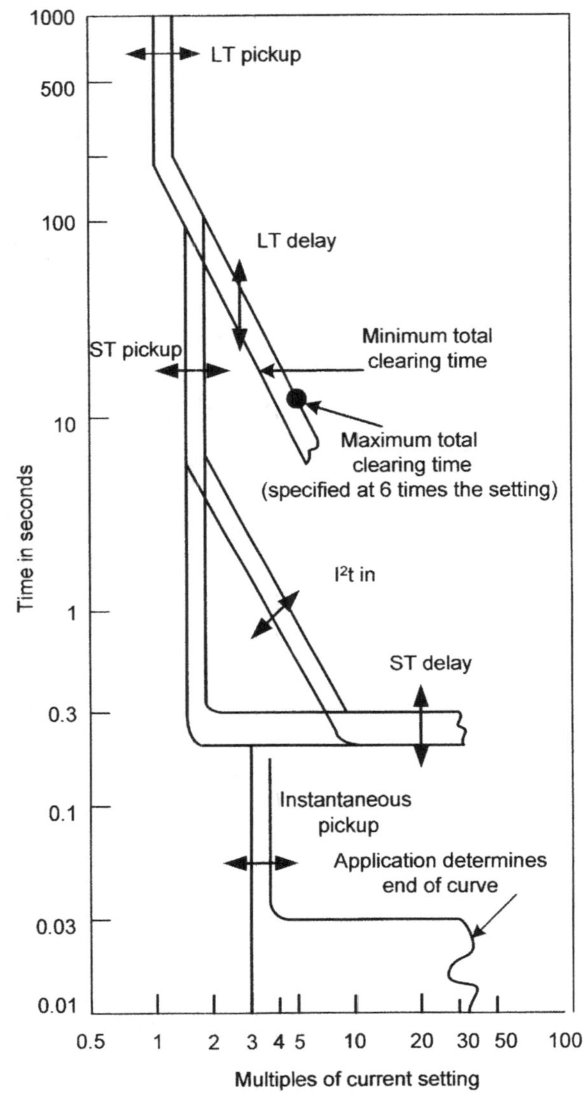


Figure 7.13. Time–current characteristics of a modern low voltage electronic trip programmer. See text.

### 7.5.4 Low Voltage Power Circuit Breakers (LVPCBs)

Low voltage power circuit breakers are rated and tested according to ANSI C37.13 [4–5] and are used primarily in draw-out switchgear. These are the largest in physical size and are field maintainable. Electronic trip units are now almost standard with these circuit breakers and these are available in frame sizes from 800 to 6000 A, interrupting ratings, 40–100 kA symmetrical without integral current-limiting fuses.

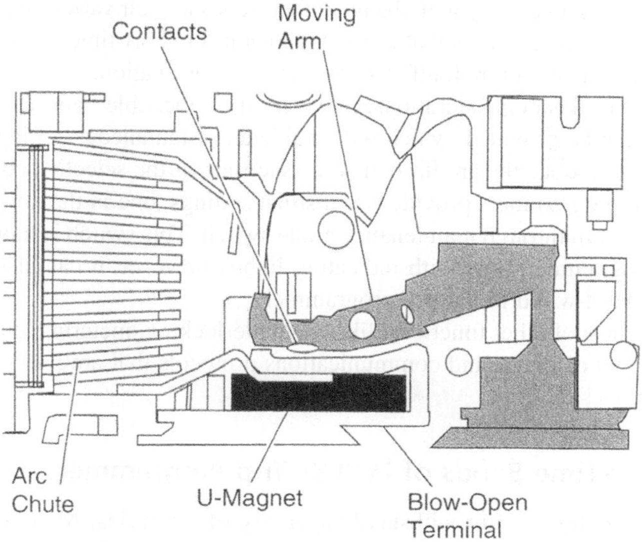


Figure 7.14. Low arc flash circuit breaker design. *Source*: Arc-Flash Protection with Masterpact NW and NT Circuit Breakers, Data Bulletin 0613DB0202R603. Austin, TX: Square D/Schneider Electric Engineering Services, 2003.

Figure 7.13 is a typical phase overcurrent time–current characteristics of a LVPCB. It is provided with an electronic trip programmer, designated as LSIG. Here, the letter L stands for long time, S for short-time, I for instantaneous, and G for ground fault.

- Consider that the breaker is 1600 A frame, is provided with sensors 1600 A; call the sensor current rating "*s*"
- Plug ratings of 600, 800, 1000, 1100, 1200, and 1600 A can be provided; call this setting "*x*." This virtually changes the current rating of the circuit breaker.
- The long-time pickup can be adjusted at 0.5, 0.6, 0.7, 0.8, 0.9, 0.95, and 1.1 times the plug setting *x*. Call this setting "c." Say for a 1600 A plug, $x = 0.6$, $c = 960$ A.
- Long time delay band can be selected at 2, 3, 4, 5, 6, 8, 12, 20, 24, and 32 seconds.
- Short-time pick up is adjustable from 1.5 to 9 times the *c* setting in increments of 0.5.
- Short-time delay band can be adjusted; there can be three to seven time delay bands.
- $I^2t$ function of short-time can be set in or out. When the $I^2t$ function is switched in, the shape of the curve slopes as shown in Figure 7.13. This slope is particularly helpful for coordination with the fuse characteristics.
- Instantaneous pickup is adjustable from, say from 1.5 to 12 times or more of the setting *x*.

These setting ranges vary with the manufacturers and their various trip programmer types. A recent advancement is that the instantaneous or short-time pickup settings can be switched off when required, affording selective coordination.

From the arc flash considerations, a two-position lockable "normal/maintenance" mode switch can be provided, which will enable an instantaneous setting in the maintenance mode to reduce the arc flash at some sacrifice of the selective coordination. In fact, some trip programmers provide two distinct settings, one in the normal mode and the other brought in through maintenance mode switch. This switch is mounted directly on the trip programmer itself with indicating lights; however, it can also be remotely mounted and hard-wired to the trip programmer.

There are host of other functions, like zone interlocking, metering, energy management, front panel displays, and communications protocols that are provided in the trip programmers.

### 7.5.5 Short-Time Bands of LVPCBs Trip Programmers

LVPCBs have a short-circuit withstand capability of 30 cycles; MCCBs do not have any short-circuit withstand capability and must be provided with instantaneous protection; ICCBs may have short-circuit withstand capability of 15 cycles, yet these are provided with high set instantaneous override. Thus, selection of appropriate low voltage breaker types can be an important criteria in selective coordination.

There has been an attempt to split the 500 ms (30 cycles on 60 Hz. basis) time withstand of LVPCBs in to much smaller short-time delay bands. Figure 7.15 shows three short-time time delay bands. Coordination is obtained between the bands, though

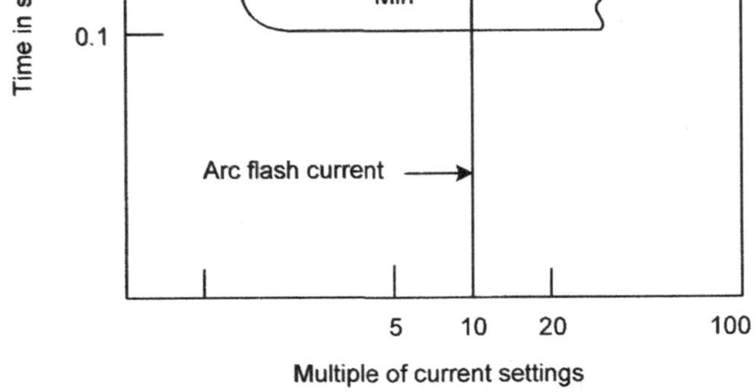


Figure 7.15. Short-time bands of a low voltage trip programmer, to show coordination between bands. The operating time is considered on the top of the band for conservatism.

TABLE 7.5. Short Time Bands of Old versus New LVPCB Trip Devices

| | Breaker A | Breaker B |
|---|---|---|
| Band | Max. trip time, ms | Max. trip time, ms |
| 1 | 200 | 92 |
| 2 | 300 | 158 |
| 3 | 500 | 200 |
| 4 | | 267 |
| 5 | | 317 |
| 6 | | 383 |
| 7 | | 500 |

the gap between the bands is small and sometimes the bands may be even overlapping. The device can trip anywhere between the time zone of the band, but for conservatism, it is the maximum operating time that is considered for arc flash analysis and coordination. This figure shows only three ST delay bands, minimum, intermediate, and maximum. The time associated with these bands is 0.2, 0.3, and 0.5 seconds, respectively, shown with bold dots.

Table 7.5 shows a modern LVPCB, provided with an electronic trip device having seven short-time delay bands.

**Example 7.3**

A low voltage double-ended substation is shown in Figure 7.16. Three-step coordination is considered, between the main breaker, bus section breaker and the feeder breaker. Breaker-type A, provided with three short-time bands (Table 7.5) will clear the fault in 200, 300, and 500 ms respectively, while breaker type B, provided with seven short-time bands, will clear it in 92, 158, and 200 ms, respectively. The end result can be two levels of higher HRC with breaker type A, as compared with breaker type B. Calculation results for some transformer sizes with these two types of breakers are shown in Table 7.6. The reduction in the incident energy obtained by using a low voltage trip programmer provided with seven bands (breaker type B) is noteworthy.

The example is illustrative of the possibility of arc flash reductions by proper selection of the protective devices in the design stage of the project.

## 7.6 SHORT-CIRCUIT RATINGS OF LOW VOLTAGE CIRCUIT BREAKERS

All the three types of circuit breakers have different ratings, short-circuit test requirements, and applications. As discussed in Chapter 5, the symmetrical interrupting rating of the circuit breaker takes into account the initial current offset due to circuit *X/R* ratio.

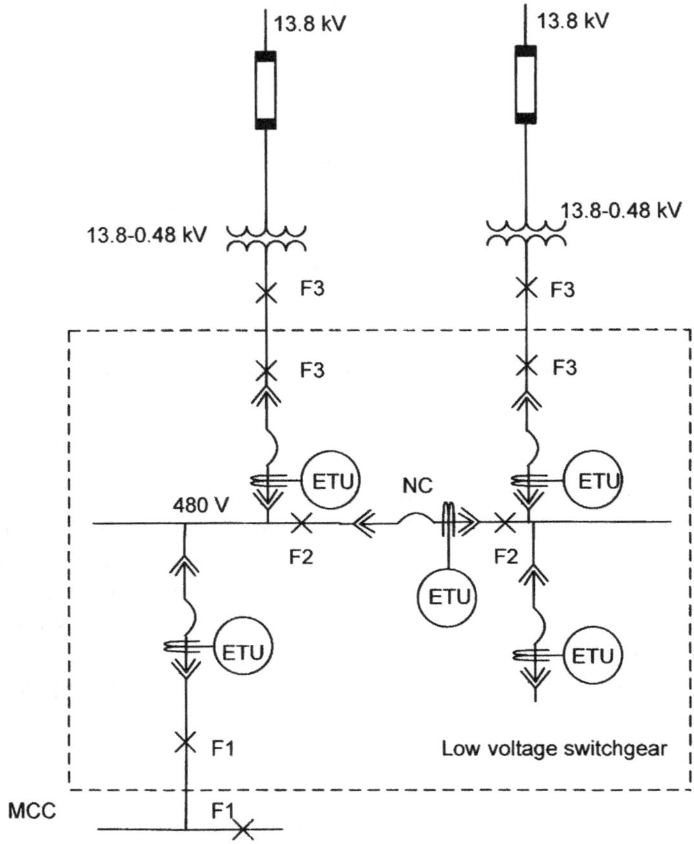


Figure 7.16. A low voltage double-ended substation for calculations of arc flash hazard and time–current coordination, Example 7.3.

The value of the standard *X/R* ratio is that used in the test circuit. For LVPCBs, this *X/R* = 6.6, corresponding to a 15% power factor (ANSI/IEEE standard. C37.13). Table 7.7 shows the multiplying factor (MF) for other *X/R* ratios. The recommended MFs for unfused circuit breakers are based on highest peak current and can be calculated from

$$\text{MF} = \frac{\sqrt{2}\left[1 + e^{-\pi/(X/R)}\right]}{2.29}. \tag{7.3}$$

The MF for the fused breaker is based on the total rms current (asymmetrical) and is calculated from:

$$\text{MF} = \frac{\sqrt{1 + 2e^{-2\pi/(X/R)}}}{1.25}. \tag{7.4}$$

TABLE 7.6. Arc Flash Hazard Calculations in Low Voltage Systems, Various Transformer Sizes, LV switchgear, gap = 32 mm, working distance 24", Dimensions 36"W x 90"H x48"D, LV MCC, gap = 25 mm, working distance 18", Dimensions 30"W x 90"H x25"D, EC = HCB

| Tran. KVA (Total) | Primary Fuse | Fault At | Bolted Fault Current (kA rms) | Breaker Type A Three ST Delay Bands, 200, 300, and 500 ms | | Breaker Type B ST Delay Bands, Table 6 92, 158, and 200 ms | |
|---|---|---|---|---|---|---|---|
| | | | | Incident Energy (cal/cm$^2$) | HRC | Incident Energy (cal/cm$^2$) | HRC |
| 2500 | 175E | F1(FB) | 49.44 | 16.7 | Danger | 7.06 | 2 |
| | | F2 (BS) | 62.77 | 28.3 | 4 | 15 | 3 |
| | | F2 (M) | 62.77 | 78.8 | Danger | 31.5 | 4 |
| 2000 | 150E | F1 (FB) | 41.40 | 14.8 | 3 | 13.1 | 3 |
| | | F2 (BS) | 50.01 | 25.2 | 4 | 14.8 | 3 |
| | | F2 (M) | 50.01 | 60.1 | Danger | 25 | 3 |
| 1500 | 125E | F1 (FB) | 33.0 | 12.3 | 3 | 5.08 | 2 |
| | | F2 (BS) | 38.48 | 24 | 3 | 10.9 | 3 |
| | | F2 (M) | 38.48 | 50.1 | Danger | 23.6 | 3 |
| 1000 | 80E | F1 (FB) | 23.55 | 9.04 | 3 | 4.16 | 2 |
| | | F2 (BS) | 26.21 | 16 | 3 | 7.82 | 2 |
| | | F2 (M) | 26.21 | 42 | Danger | 16.8 | 3 |

F1 (FB), feeder breaker tripped; F2 (BS), bus section breaker tripped; F2 (M), main secondary breaker tripped. Calculations for fault F3 in Figure 7.16 are not shown.

TABLE 7.7. Multiplying Factors for Low Voltage LVPCBs

| System Short-Circuit Power Factor (%) | System *X/R* Ratio | Multiplying Factors for the Calculated Current | |
|---|---|---|---|
| | | Unfused Circuit Breakers | Fused Circuit Breakers |
| 20 | 4.9 | 1.00 | 1.00 |
| 15 | 6.6 | 1.00 | 1.07 |
| 12 | 8.27 | 1.04 | 1.12 |
| 10 | 9.95 | 1.07 | 1.15 |
| 8.5 | 11.72 | 1.09 | 1.18 |
| 7 | 14.25 | 1.11 | 1.21 |
| 5 | 20.0 | 1.14 | 1.26 |

*Source*: Data from Reference [10].

In general, when *X/R* differs from the test power factor, the MF can be approximated by

$$\mathrm{MF} = \frac{1+e^{-\pi(X/R)}}{1+e^{-\pi/\tan\phi}} \tag{7.5}$$

where $\phi$ is the test power factor.

TABLE 7.8. Test Power Factors of MCCBs

| Interrupting Rating (kA, rms sym) | Test Power Factor Range | *X/R* |
|---|---|---|
| 10 or less | 0.45–0.50 | 1.98–1.73 |
| 10–20 | 0.25–0.30 | 3.87–3.18 |
| Over 20 | 0.15–0.20 | 6.6–4.9 |

TABLE 7.9. Short-Circuit Multiplying Factors for MCCBs and ICCBs

| | | Interrupting Rating Multiplying Factor | | |
|---|---|---|---|---|
| Power Factor (%) | *X/R* Ratio | 10 kA or Less | 10–20 kA | >20 kA |
| 5 | 19.97 | 1.59 | 1.35 | 1.22 |
| 6 | 16.64 | 1.57 | 1.33 | 1.20 |
| 7 | 14.25 | 1.55 | 1.31 | 1.18 |
| 8 | 12.46 | 1.53 | 1.29 | 1.16 |
| 9 | 11.07 | 1.51 | 1.28 | 1.15 |
| 10 | 9.95 | 1.49 | 1.26 | 1.13 |
| 13 | 7.63 | 1.43 | 1.21 | 1.09 |
| 15 | 6.59 | 1.39 | 1.18 | 1.06 |
| 17 | 5.80 | 1.36 | 1.15 | 1.04 |
| 20 | 4.90 | 1.31 | 1.11 | 1.00 |
| 25 | 3.87 | 1.24 | 1.05 | 1.00 |
| 30 | 3.18 | 1.18 | 1.00 | 1.00 |
| 35 | 2.68 | 1.13 | 1.00 | 1.00 |
| 40 | 2.29 | 1.08 | 1.00 | 1.00 |
| 50 | 1.98 | 1.04 | 1.00 | 1.00 |

MCCBs and ICCBs are tested in the prospective fault test circuit according to UL 489 [7]. Power factor values for the test circuit are different from LVPCBs and are given in Table 7.8. If a circuit has an *X/R* ratio that is equal to or lower than the test circuit, no corrections to the interrupting rating are required. If the *X/R* ratio is higher than the test circuit *X/R* ratio, the interrupting duty requirement for that application is increased by a MF from Table 7.9. The MF can be interpreted as a ratio of the offset peak of the calculated system peak (based on *X/R* ratio) to the test circuit offset peak.

While testing the breakers, the actual trip unit type installed during testing should be the one represented by referenced specifications and time–current curves. The short-circuit ratings may vary with different trip units, that is, a short-time trip only (no instantaneous) may result in reduced short-circuit interrupting rating compared with testing with instantaneous trips. The trip units may be rms sensing or peak sensing, electronic or electromagnetic, and may include ground fault trips.

IEC standards do not directly correspond to the practices and standards in use in North America for single-pole duty, thermal response, and grounding. A direct comparison is not possible.

### 7.6.1 Single-Pole Interrupting Capability

A single-pole interruption connects two breaker poles in series, and the maximum fault current interrupted is 87% of the full three-phase fault current. The interrupting duty is less severe as compared with a three-phase interruption test, where the first-pole-to-clear factor can be 1.5. Therefore, the three-phase tests indirectly prove the single-pole interrupting capability of three-pole circuit breakers. For the rated *X/R*, every three-pole circuit breaker intended for operation on a three-phase circuit can interrupt a bolted single-phase fault. LVPCBs are single-pole tested with maximum line-to-line voltage impressed across the single pole, and at the theoretical maximum single-phase fault current level of 87% of maximum three-phase bolted fault current. Generally, single-pole interrupting is not a consideration. Nevertheless, all MCCBs and ICCBs do not receive the same 87% test at full line-to-line voltage. In a corner-grounded delta system (not much used in the industry), a single line-to-ground fault on the load side of the circuit breaker will result in single-phase fault current flowing through only one pole of the circuit breaker, but full line-to-line voltage impressed across that pole. A rare fault situation in ungrounded or high-resistance grounded systems can occur with two simultaneous bolted faults on the line side and load side of a circuit breaker and may require additional considerations. Some manufacturers market circuit breakers rated for a corner-grounded systems.

Thus, normally, the three-phase faults, calculated at the point of application, gives the maximum short-circuit currents on which the circuit breaker rating can be based, adjusted for fault point *X/R*. But in certain cases, a line-to-ground fault in solidly grounded system can slightly exceed-phase symmetrical fault, and care has to be exercised in selecting the short-circuit rating.

### 7.6.2 Short-Time Ratings

The short-time ratings (Section 7.5.5) impact the applications. Consider that in a practical application, if an ICCB is applied where the available short-circuit current is 40 kA, and the 15-cycle short-time withstand rating of the ICCB is 20 kA for 15 cycles only, and then a high set override protects the breaker. Coordination at higher levels of short-circuit current >20 kA will therefore be sacrificed with the downstream instantaneous devices. LVPCBs are designed to have short-time capabilities, typically 30 cycles, and can withstand short-time duty cycle tests. Akin to ICCBs, LVPCBs may also be provided with high set instantaneous overrides, as these may not have the 30-cycle withstand capability throughout the specified short-circuit rating range.

Short-time rating becomes of concern when two devices are to be coordinated in series and these see the same magnitude of fault current. If an upstream device has a short-time withstand capability, a slight delay in the settings can ensure selective coordination. This is an important concept from time–current coordination point of view and arc flash hazard reduction.

For an unfused LVPCB, the rated short-time current is the designated limit of prospective current at which it will be required to perform its short-time duty cycle of two periods of 0.5-second current flow separated by 15 seconds intervals of zero current at rated maximum voltage under prescribed test conditions. This current is expressed in rms symmetrical ampères. The unfused breakers will be capable of performing the short-time current duty cycle with all degrees of asymmetry produced by three-phase or single-phase circuits having a short-circuit power factor of 15% or greater. Fused circuit breakers do not have a short-time current rating.

## 7.7 SERIES-CONNECTED RATINGS

Series connection of MCCBs or MCCBs and fuses permits a downstream circuit breaker to have an interrupting rating less than the calculated fault duty, and the current limiting characteristics of the upstream device "protects" the downstream lower-rated devices. Series combination is recognized for application by *testing* only. The upstream device is fully rated for the available short-circuit current and protects a downstream device, which is not fully rated for the available short-circuit current by virtue of its current-limiting characteristics. The series rating of the two circuit breakers makes it possible to apply the combination as a single device, the interrupting rating of the combination being that of the higher rated device. As an example, a single upstream incoming breaker of 65 kA interrupting may protect a number of downstream feeder breakers of 25 kA interrupting, and the complete assembly will be rated for 65 kA interrupting. The series rating should not be confused with cascading arrangement. IEC also uses this term for their series-rated breakers [9]. A method of cascading that is erroneous and has been in use in the past is shown in Figure 7.17.

Consider a series combination of an upstream current limiting fuse of 1200 A, and a downstream MCCB. The available short-circuit current is 50 kA symmetrical, while the MCCB is rated for 25 kA. Figure 7.17 shows the let-through characteristics of the fuse. The required interrupting capability of the system, that is, 50 kA is entered at point A, and moving upwards the vertical line is terminated at the 1200-A fuse let-through characteristics. Moving horizontally, point C is intercepted, and then moving vertically down, point D is located. The symmetrical current given by D is read off, which in Figure 7.17 is 19 kA. As this current is less than the interrupting rating of the downstream device to be protected, the combination is considered safe. *This method can lead to erroneous results*, as the combination may not be able to withstand the peak let-through current given by point E in Figure 7.17 on the $y$-axis. Calculations of series ratings are not permissible, and these can only be established by testing.

A disadvantage of series combination is lack of selective coordination. On a high fault current magnitude, both the line side and load side circuit breakers will trip. A series combination should not be applied if motors or other loads that contribute to short-circuit current are connected between the line-side and load-side MCCBs. NEC (240.86 (C)) [11] specifies that series rating will not be used where:

- Motors are connected on load side of higher-rated overcurrent device and on the line side of the lower-rated overcurrent device.
- The sum of motor full load currents exceeds 1% of the interrupting rating of the lower rated circuit breaker.

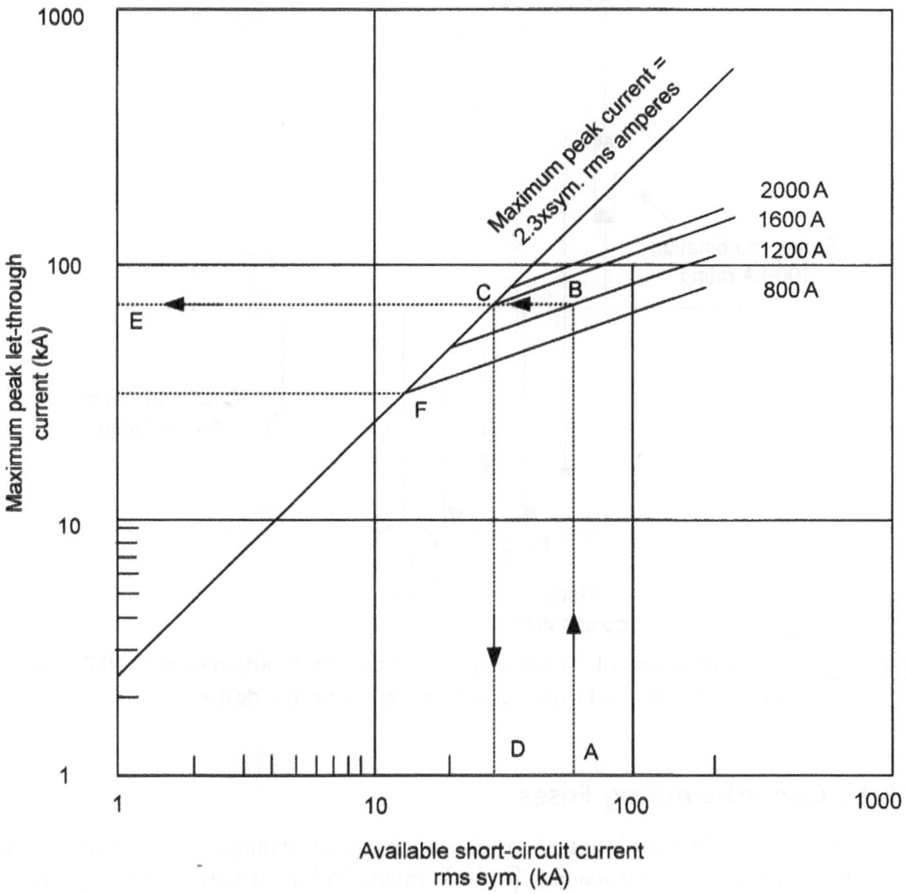


Figure 7.17. Let-through characteristics of a current-limiting fuse.

Figure 7.18 illustrates this situation. This does not apply to integrally fused circuit breakers. There is no motor contribution at the common junction of the fuse and the circuit breaker, with fuses mounted on the draw-out stabs of the circuit breaker. Integrally fused circuit breakers have been used extensively in the industry for many years.

## 7.8 FUSES

Fuses are fault sensing and interrupting devices, while circuit breakers must have protective relays as sensing devices before these can operate to clear short-circuit faults. Fuses are direct acting, single-phase devices, which respond to magnitude and duration of current. The relevant standards are References [12–17].

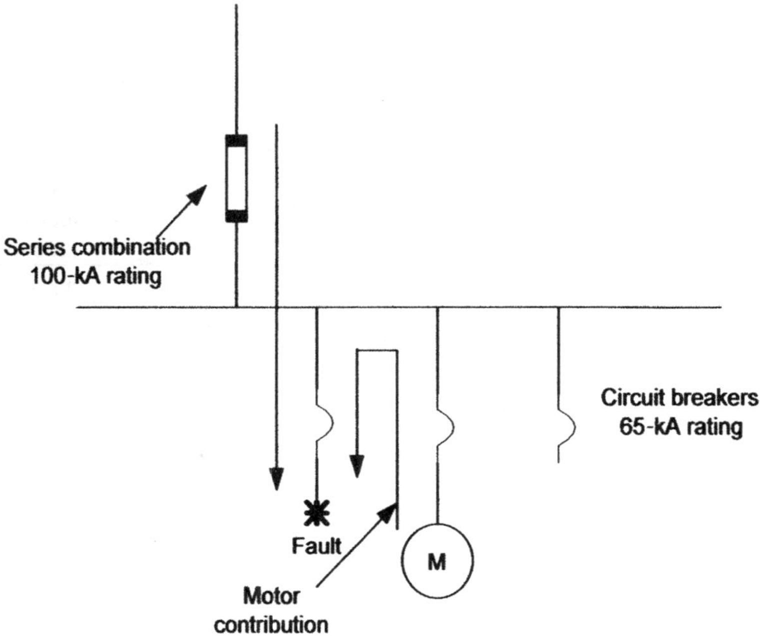


Figure 7.18. Example of an installation where the motor contribution exceeds NEC requirements (<1%) for the lowest rated circuit breaker series-connected devices.

### 7.8.1 Current-Limiting Fuses

A current-limiting fuse is designed to reduce equipment damage by interrupting the rising fault current before it reaches its peak value. Within its current limiting range, the fuse operates within 1/4 to 1/2 cycle. The total clearing time consists of melting time, sometimes called the prearcing time, and the arcing time. This is shown in Figure 7.19. The let-through current can be much lower than the prospective fault current peak, and the rms symmetrical available current can be lower than the let-through current peak. The prospective fault current can be defined as the current that will be obtained if the fuse was replaced with a bolted link of zero impedance. By limiting the rising fault current, the $I^2t$ let-through to the fault is reduced because of two counts: (1) high speed of fault clearance in 1/4 cycle typically in the current limiting range, and (2) fault current limitation. This reduces the fault damage.

Current-limiting fuses have a fusible element of nonhomogeneous cross-section. It may be perforated or notched, and, while operating, it first melts at the notches because of reduced cross-sectional area. Each melted notch forms an arc that lengthens and disperses the element material into the surrounding medium. Generally, a silver element is placed in a sand medium. When it is melted by current in the specified current-limiting range, it abruptly introduces a high resistance to reduce the current magnitude and duration. It generates an internal arc voltage, much greater than the system voltage, to force the current to zero, before the natural current zero crossing.

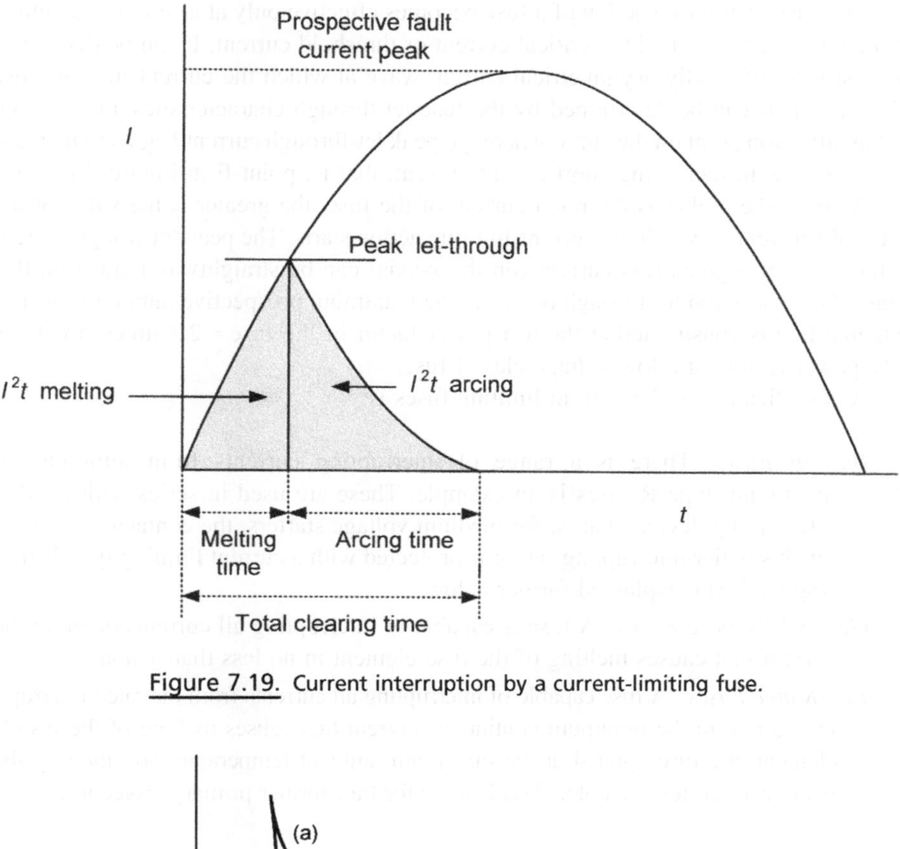


Figure 7.19. Current interruption by a current-limiting fuse.

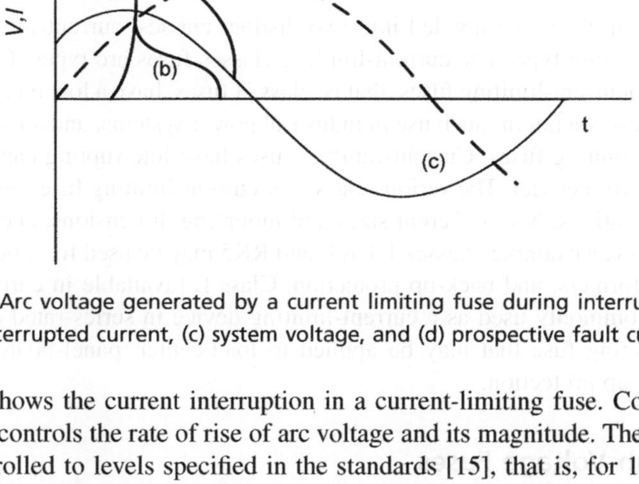


Figure 7.20. Arc voltage generated by a current limiting fuse during interruption: (a) arc voltage, (b) interrupted current, (c) system voltage, and (d) prospective fault current.

Figure 7.20 shows the current interruption in a current-limiting fuse. Controlling the arcs in series controls the rate of rise of arc voltage and its magnitude. The arc voltages must be controlled to levels specified in the standards [15], that is, for 15.5-kV fuses of 0.5–12 A, the maximum arc voltage is 70 kV peak, and for fuses >12 A, the arc voltage is 49 kV peak.

The current-limiting action of a fuse becomes effective only at a certain magnitude of the fault current, called the critical current or threshold current. It can be defined as the first peak of a fully asymmetrical current wave at which the current-limiting fuse will melt. This can be determined by the fuse let-through characteristics and is given by the inflection point on the curve where the peak let-through current begins to increase less steeply with increasing short-circuit current, that is, point F in Figure 7.17 for a 800-A fuse. The higher is the rated current of the fuse, the greater is the value of the threshold current at which the current-limiting action starts. The peak let-trough current of the fuse for a given rms current (on the $x$-axis) can be straightway read from this figure. The maximum let-through occurs at the maximum prospective fault current. The diagonal line is constructed at the test power factor of the fuse = 2.3 times rms for a 15% power factor for a low voltage class J fuse.

A classification of the current limiting fuses is:

*Backup Fuses.* There is a range of interrupting currents from minimum to maximum: type R fuses is an example. These are used in series with another interrupting device, that is, for medium voltage starters, the contactor that is of much smaller interrupting rating is protected with a current limiting type R fuse (Figure 7.31), explained further below.

*General Purpose Fuses.* A fuse is capable of interrupting all currents down to the current that causes melting of the fuse element in no less than 1 hour.

*Full-Range Fuses.* A fuse capable of interrupting all currents from the rated interrupting current to the minimum continuous current that causes melting of the fusible element, the fuse applied at the maximum ambient temperature specified by the manufacturer, for example, class E fuses for transformer primary protection.

### 7.8.2 Low Voltage Fuses

Low voltage fuses can be divided into two distinct classes, current-limiting type and noncurrent-limiting type. The current-limiting classes fuses are types: CC, T, K, G, J, L, and R. Noncurrent-limiting fuses, that is, class H fuses, have a low interrupting rating of 10 kA. These are not in much use in industrial power systems, and are being replaced with current-limiting fuses. Current-limiting fuses have interrupting capabilities up to 200 kA rms symmetrical. The various classes of current-limiting fuses are designed for specific applications, have different sizes and mounting dimensions, and are not interchangeable. As an example, classes J, RK1, and RK5 may be used for motor controllers, control transformers, and back-up protection. Class L (available in current ratings up to 6 kA) is commonly used as a current-limiting device in series-rated circuits. Class T is a fast-acting fuse that may be applied to load-center, panel-board, and circuit-breaker back-up protection.

### 7.8.3 High Voltage Fuses

High voltage fuses can be divided into two distinct categories: distribution fuse cutouts and power fuses. Distribution cutouts are meant for outdoor pole or cross-arm mounting (except

TABLE 7.10. Short-Circuit Ratings of Various Fuse Types

| Fuse Type | Current Ratings | Nominal Voltage Rating in kV-Maximum Short-Circuit Interrupting Rating(kA rms Symmetrical) |
|---|---|---|
| Distributions fuse cutouts | Up to 200 A | 4.8–12.5, 7.2–15, 14.4–13.2, 25–8, 34.5–5 |
| Solid-material boric acid fuses | Up to 300 A | 17.0–14.0, 38–33.5, 48.3–31.5, 72.5–25, 121–10.5, 145–8.75 |
| Current-limiting fuses | Up to 1350 A for 5.5 kV, up to 300 A for 15.5 kV, and 100 A for 25.8 and 38 kV | 5.5–50, 15.5–50 (85 sometimes), 25.8–35, 38.0–35 |

distribution oil cutouts), have basic insulation levels (BILs) at distribution levels, and are primarily meant for distribution feeders and circuits. These are available in voltage ratings up to 34.5 kV. The interrupting ratings are relatively low, 5.00 kA rms sym. at 34.5 kV. The power fuses are adapted to station and substation mounting, have BILs at power levels and are meant primarily for applications in stations and substations. These are of two types: expulsion-type fuses and current-limiting fuses. Expulsion-type fuses can again be of two types: (1) fiber-lined fuses having voltage ratings up to 169 kV and (2) solid boric acid fuses that have voltage ratings up to 145 kV. The solid boric acid fuse can operate without objectionable noise or emission of flame and gases. High voltage current-limiting fuses are available up to 38 kV, and these have comparatively much higher interrupting ratings. Table 7.10 shows comparative interrupting ratings of distribution cutouts, solid boric acid, and current-limiting fuses. While the operating time of the current-limiting fuses is typically one-quarter of a cycle in the current-limiting range, the expulsion-type fuses will allow the maximum peak current to pass through and interrupt in more than one cycle. This can be a major consideration in some applications where a choice exists between the current-limiting and expulsion-type fuses.

Class E fuses are suitable for protection of voltage transformers, power transformers, and capacitor banks, while class R fuses are applied for medium voltage motor starters. All class E-rated fuses are not current limiting; E rating merely signifies that class E rated power fuses in ratings of 100E or less will open in 300 seconds at currents between 200 and 240% of their E rating. Fuses rated above 100E open in 600 seconds at currents between 220% and 264% of their E ratings.

### 7.8.4 Electronic Fuses

Electronically actuated fuses are a recent addition, and these incorporate a control module that provides current sensing, electronically derived time–current characteristics, energy to initiate tripping, and an interrupting module that interrupts the current. These are available in ratings of up to 1200 A and 25 kV. The time–current characteristics are determined by the electronic module and have a sensing current transformer for actuating the electronics and setting the time–current characteristics. No external power supply is required. The interrupting module and control module are mechanically attached through a threaded connection. The interrupting module passes the main current to the load without its flowing through the fusible elements. This has an advantage that repeated inrush currents, say on energizing a transformer, do not deteriorate the fusible element. After the main current circuit is opened, the fault current is shunted into the fusible elements. Only the fusible element needs replacement after a fault.

### 7.8.5 Interrupting Ratings

The interrupting ratings relate to the maximum rms asymmetrical current available in the first half-cycle after fault, which the fuse must interrupt under the specified conditions. The interrupting rating itself has no direct bearing on the current-limiting effect of the fuse. Currently, the rating is expressed in maximum rms symmetrical current, and, thus, the fault current calculation based on an *E/Z* basis can be directly used to compare the calculated fault duties with the short-circuit ratings. Many power fuses and distribution cutouts were earlier rated on the basis of maximum rms asymmetrical currents, which represents the maximum current that the fuse has to interrupt because of its fast-acting characteristics. For power fuses, the rated asymmetrical capability is 1.6 times the symmetrical current rating. The asymmetrical rms factor can exceed 1.6 for high *X/R* ratios or a low power factor short-circuit currents. Figure 7.21 from Reference [13] relates rms multiplying factors and peak multiplying factors.

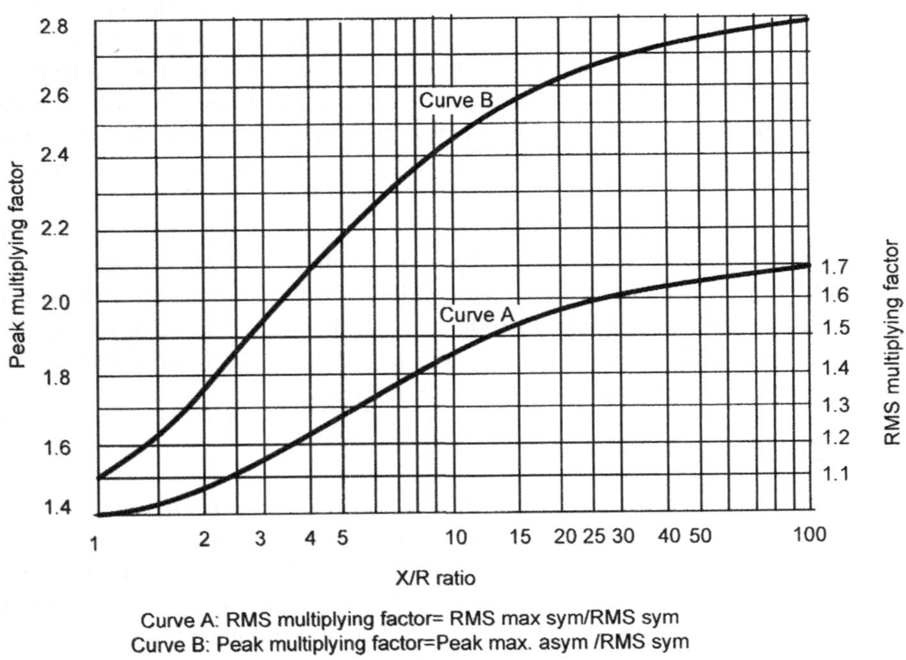


Figure 7.21. Relation of *X/R* to rms and peak multiplying factors. *Source*: Based on IEEE Standard C37.41 [13].

The test *X/R* ratio is 25 only for expulsion type and current limiting type fuses [16]. For distribution class fuse cutouts interrupting tests (except current limiting and open-link cutouts), the minimum *X/R* ratio varies, 1.5–15 [13]. It is important to calculate the interrupting duty based upon the actual system X/R and apply proper adjustment factors.

## 7.9 APPLICATION OF FUSES FOR ARC FLASH REDUCTION

### 7.9.1 Low Voltage Motor Starters

The application of fuses in low voltage MCCs serving motor loads is sometimes advocated for arc flash reduction. However, an analysis of fuses versus MCCBs for motor starters shows that properly selected and applied MCCBs give identical arc flash incident energy release. The magnetic trip in the MCCBs is set at eight times the current rating, which is adequate for motor starting inrush currents.

Fuses at motor starter level can give rise to single phasing, say due to a single line-to-ground fault in a solidly grounded low voltage system, which must be cleared fast to prevent motor damage. The thermal magnetic trip elements in low voltage motor starters, which are very common in the industry, are not sensitive enough to operate fast under these conditions. On a complete loss of a phase, a three-phase induction motor will continue rotating, (though it cannot be started on two-phase power supply); the negative sequence currents are produced, which have a heating effect of approximately six times the positive sequence currents, and the motor can be damaged. A fuse must be replaced on operation, while a MCCB can be reset. This requires maintaining an inventory of the proper fuse sizes and types.

### 7.9.2 Medium Voltage Motor Starters

NEMA E2 starters for medium voltage motors [18] are provided with type R fuses for the short-circuit protection. The incident energy release for a fault on the load side of the fuse in the medium voltage motor starter is low.

Unlike low voltage fuses, the medium voltage type R fuses are provided with a trigger that opens all the three-poles of the motor contactor. Also, MPPR motor protection relays provide sensitive negative sequence protection, generally set as:

- Alarm, 10% negative sequence current (of the motor full load current) time delay 5 seconds
- Trip, 15% negative sequence current (of the motor full load current) time delay 10 seconds.

Further, the medium voltage systems are, generally, low resistance grounded. A pickup setting of 3–5 A for the ground faults with a slight time delay is adequate.

### 7.9.3 Low Voltage Switchgear

A low voltage system configuration with integrally fused breakers is illustrated in Figure 7.22. The main 4000 AF breaker BK3 is provided, with 4000 A class L limiter, and the 800 AF feeder breaker BK2 is provided with 1000-A class L limiter. The coordination with this arrangement is shown in Figures 7.23 and 7.24. Observe that the limiters do not coordinate well with the settings. Say, for a fault F2, exceeding 10 kA, 1000 A limiters will operate faster that the ST or

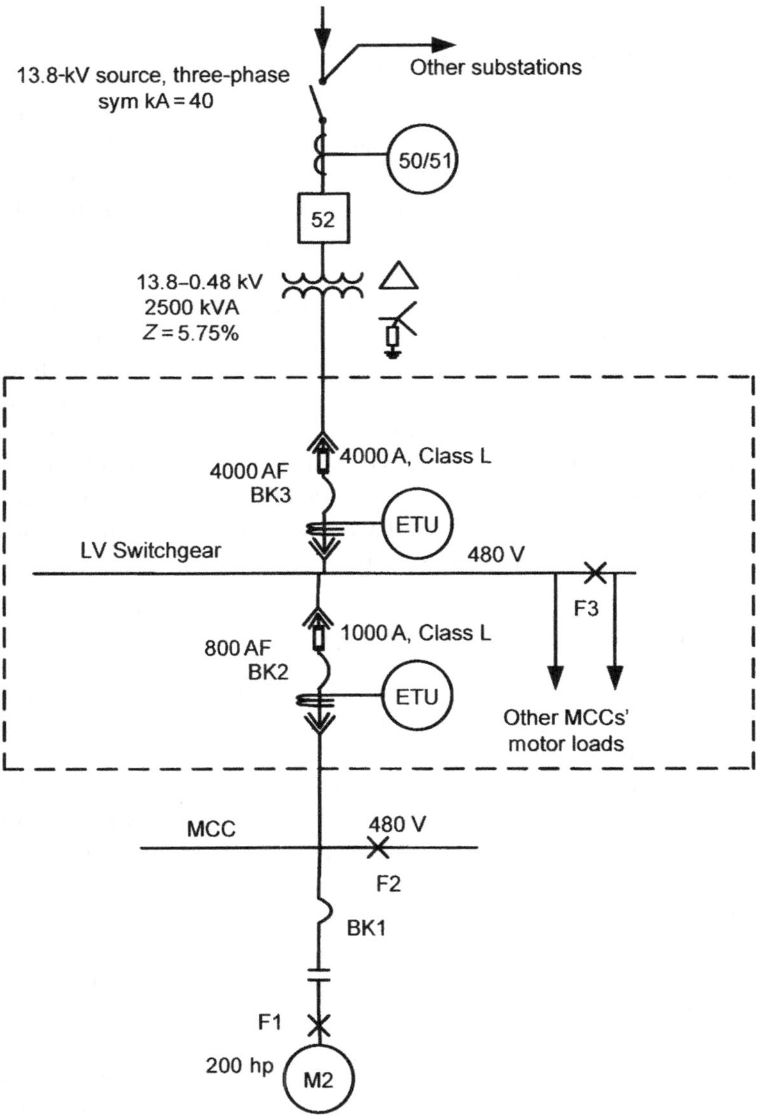


Figure 7.22. Low voltage distribution system with integrally fused main secondary and feeder breakers.

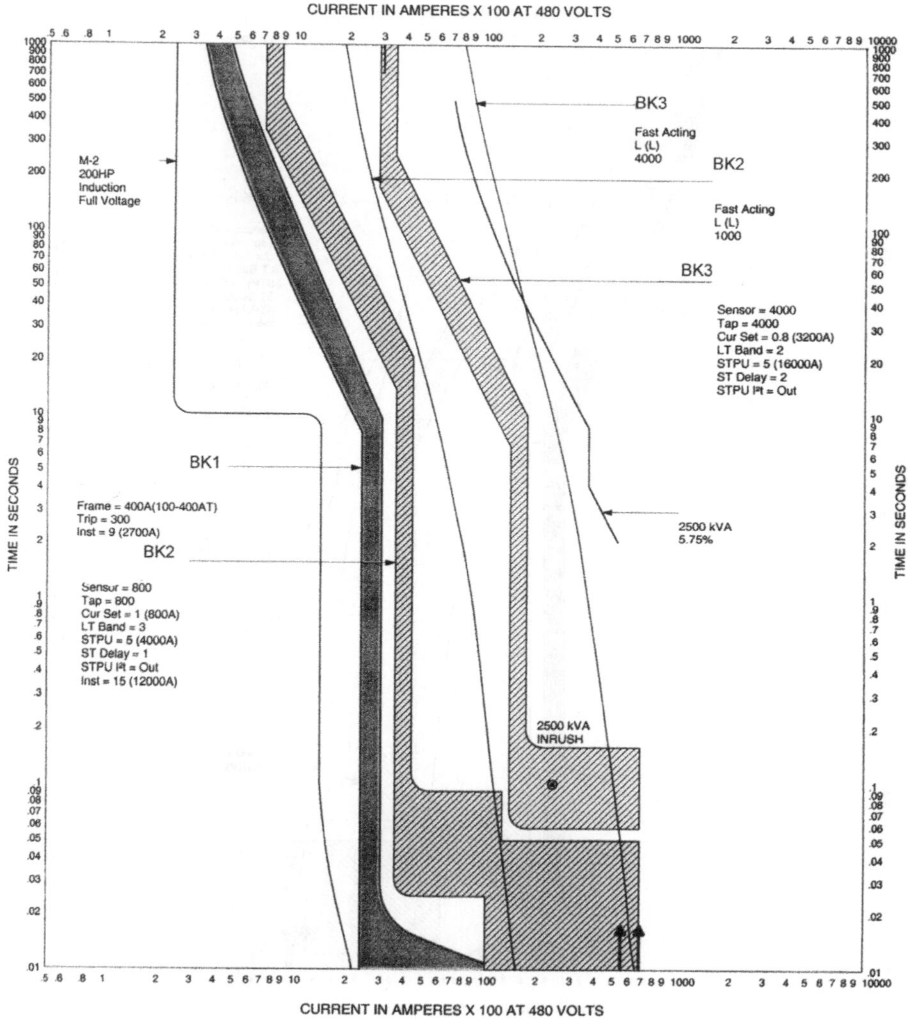


Figure 7.23. Coordination in the distribution system of Figure 7.22. Current limiters on main secondary breaker do not reduce arc flash hazard (see text).

instantaneous settings on the feeder breaker. This limits the incident energy, and HRC on MCC is reduced to zero.

The coordination in Figure 7.23 shows that 4000 A limiter does not help in limiting the HRC at LV switchgear, which is at category 3. This is so because the arc flash current of 26.87 kA is much lower than the threshold current level of 4000-A limiter, and for this value of arc flash current, the short-time setting operating in 0.158 second, gives HRC 3.

If some compromise is made or an instantaneous setting on main 4000AF breaker is brought into action through a maintenance switch, the coordination can be altered with the addition of instantaneous settings as shown in Figure 7.24. This reduces the incident energy to 5.3 cal/cm$^2$, and HRC is 2.

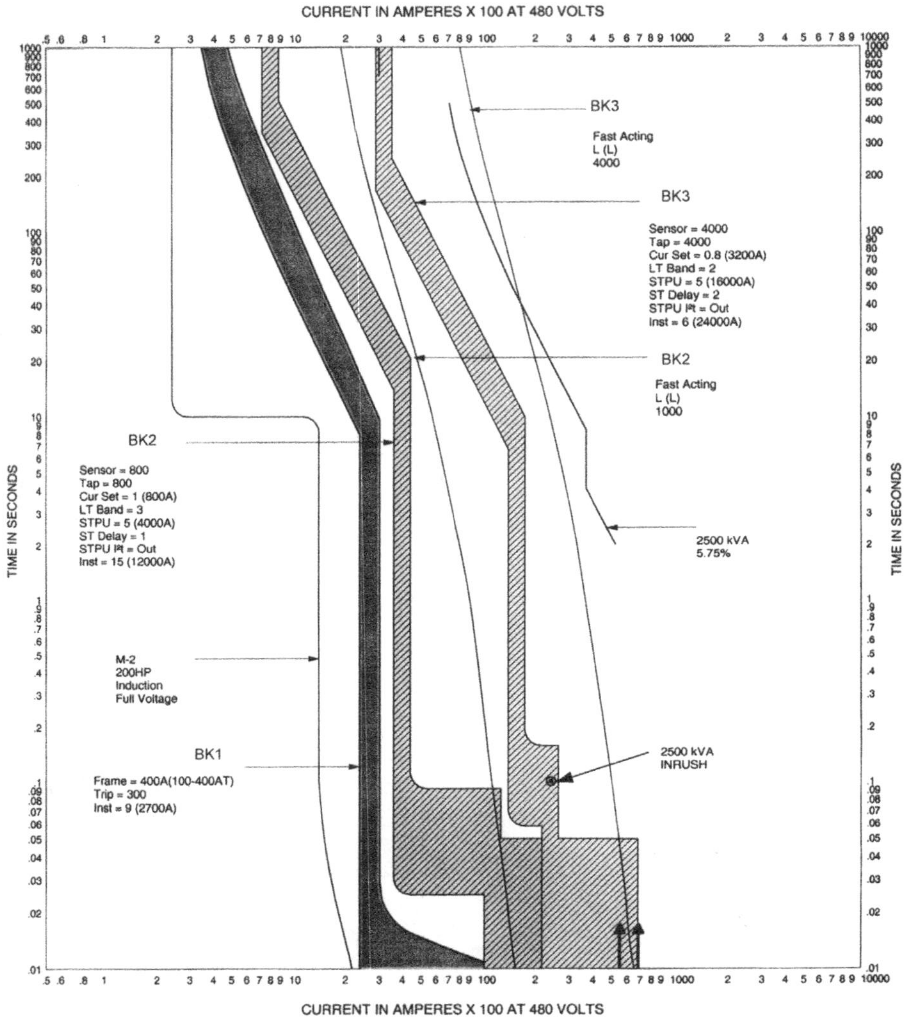


Figure 7.24. Coordination in the distribution system of Figure 7.22, modified by adding instantaneous setting on main secondary breaker, which reduces arc flash hazard.

The time–current characteristics of fuses can vary, depending upon the manufacturer. Figure 7.25 illustrates this. It shows four characteristics, all of 150-A, 15.5-kV fuses. Two characteristics are for 150-A current limiting class E fuses and two characteristics are for boric acid expulsion type fuses, one for the standard operating time delay and other for delayed time. Note the much wider spread at shorter operating times with expulsion type fuses; which makes the selective coordination with other devices more difficult. This is unlike ANSI/IEEE overcurrent relay characteristics. There is practically no variation in the relay time–current characteristics of different manufacturers for the same inverse characteristics.

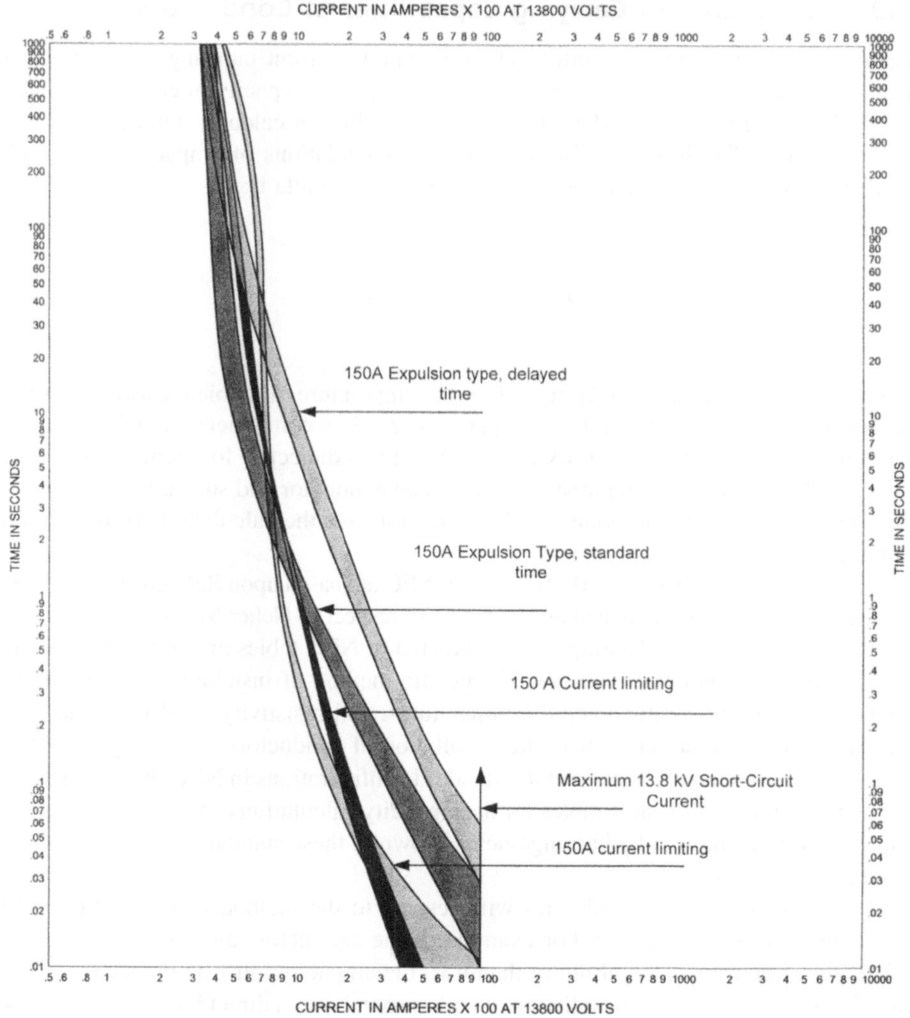


Figure 7.25. Variations in the time–current characteristics of 150-A class E fuses; two fuses are current limiting type and two are expulsion type.

## 7.10 CONDUCTOR PROTECTION

The conductors in a power distribution system are sized from three distinct conditions:

1. These must be suitable for carrying the continuous and short-time overloads.
2. These must be large enough to arrest unacceptable voltage dips due to long lengths and inrush currents; say on account of starting of motors.
3. These must not be damaged due to system short-circuit currents.

Much can be said about these three aspects. Here an overview is provided.

### 7.10.1 Load Current Carrying Capabilities of Conductors

Conductors can be bare or insulted. The continuous current carrying capabilities of insulated conductors are provided in tables in NEC [11]. Ampacity of conductors rated 0–2000 V are permitted to be determined from the tables or calculated under engineering supervision. The Neher McGrath method of calculations of ampacities is recognized. This is based upon Reference [19]. The basic formula is:

$$I = \sqrt{\frac{T_c - (T_e + \Delta T_d)}{R_{DC}(1 + Y_c)R_{ca}}}, \tag{7.6}$$

where $T_c$ is the conductor temperature, $T_e$ = the temperature of ambient earth, $R_{DC}$ is the DC resistance of the conductor at temperature $T_c$, $Y_c$ = component of AC resistance resulting from skin effect and proximity effect, $\Delta T_d$ = dielectric loss temperature rise, and $R_{ca}$ is the effective thermal resistance between conductor and surrounding ambient. The paper contains 62 main mathematical equations for the calculations of delineated parameters.

Tables 310.15(B)(16) and 310.15(B)(17) in NEC are based upon Reference [20]. There is some difference in the tabulated ampacities with respect to Neher McGrath method.

It is noteworthy that the ampacities tabulated in NEC tables are applicable for the specific insulted conductor type, for the specific method of installation, air or underground, ambient temperatures, earth temperatures, soil resistivity, load factor and the like. There can be situations where the installation of conductors, especially in underground duct banks, is different from the standard configurations in NEC. ICEA standard [21] contains more exhaustive tables for the ampacity calculations. When the configurations do not meet the standard arrangements shown in these standards, computer-based calculations are required.

NEC contains further guidelines with respect to the method of installations and their impact on the ampacities. For example, if the conductors are installed in a tray, which are covered for a length more than 6 ft, the ampacity must be reduced to 95%. Guidelines are provided for installation in underground ducts, direct buried, metal trays, flexible metal conduit (FMC), lightweight flexible metal conduit (LFMC), PVC conduit, electrical metallic tubing (EMT), electrical nonmetallic tubing (ENT), bus ways and gutters, and cellular floor raceways. Yet all possible installation methods are not addressed, for example de-rating of cables in semi-enclosed trenches, with ventilated covers.

The methods of installation, conductor type, and insulation type impact the ampacities. It is noteworthy that ampacities in underground duct banks are much lower than that in the open air or trays, especially if a large number of cables have to be laid in underground ducts. For example, NEC table 310.60(C)(77) shows that ampacity of six 500 kcmil single insulated conductors in UG electrical ducts, MV-90, 2001–5000 V, is 300 A/cable with earth ambient temperature of 20°C, and RHO = 90 Ω-m, LF = 100%. (Approximately 90% of the soil in United States has a soil resistivity of 90 Ω-m.) Compare this with MV-90 single conductor ampacities given in Table 310-60(C)(73) of NEC, which shows 485 A for 40°C ambient.

### 7.10.2 Conductor Terminations

Most terminations are designed for 60°C or 75°C maximum temperatures. The higher rated ampacities of conductors at 90°C and 105°C cannot be used unless the terminals to which these are connected have comparable ratings.

While this overview is relevant with respect to selection of proper conductor ampacities. It is only indicative and not exhaustive.

### 7.10.3 Considerations of Voltage Drops

When long lengths of cables are involved, a voltage dip at the load end will occur due to flow of current. This should be limited, generally, not more than 2–3% by increasing the conductor size. Starting currents of induction motors can be six times the full load current or more and at a low power factor, of the order of 12–20%. This becomes of considerations for starting of large motors connected through long cables. An excessive starting voltage dip can result in:

- Dropout of motor contactors
- Motor-starting torque, which in case of an induction motor, approximately varies as the square of the voltage, may drop below the load torque and the motor cannot be started at all.
- It is also possible that the starting voltage dip is large enough to cause the running motors on the same bus to increase their slip, and lose speed to a point where on the return of the voltage these will not accelerate and will lockout.
- The starting voltage dip lasts approximately 90% of the time of the motor starting. It is possible that after a successful start, the motors that lost speed during the starting reaccelerate and take an increased current, which causes further voltage dip and a possible lockout.

These scenarios are practical possibilities. It is not intended to discuss the strategies for controlling the motor starting voltage dips and ensure system stability. Load flow and motor starting studies are conducted to properly size the conductors. For the arc flash evaluations, we will assume that the engineering aspects and system designs have been properly implemented for conductor sizing.

### 7.10.4 Short-Circuit Considerations

Power cables should be designed to withstand short-circuit currents so that these are not damaged within the total fault clearing time of the protective devices. During short-circuit, approximately, all heat generated is absorbed by the conductor metal, and the heat transfer to insulation and surrounding medium can be ignored. An expression relating the size of copper conductor, magnitude of fault current, and duration of current flow is

$$\left(\frac{I}{\text{CM}}\right)^2 tF_{\text{AC}} = 0.0297\log_{10}\frac{T_{\text{f}}+234}{\text{T}_0+234}, \tag{7.7}$$

where $I$ is the magnitude of fault current in amperes, CM is the conductor size in circular mils, $F_{AC}$ is the skin effect ratio or AC resistance/DC resistance ratio of the conductor, $T_f$ is the final permissible short-circuit conductor temperature, depending on the type of insulation, and $T_0$ is the initial temperature prior to current change. For aluminum conductors, this expression is

$$\left(\frac{I}{\text{CM}}\right)^2 tF_{AC} = 0.00125\log_{10}\frac{T_f + 228}{T_0 + 228}, \tag{7.8}$$

where $F_{ac}$ is given in Table 7.11 [22]. The short-circuit withstand capability of 4/0 (211600 CM) copper conductor cable of 13.8-kV, breaker 2F4, Figure 5.8, is 0.238 second. This is based on an initial conductor temperature of 90°C, a final short-circuit temperature for XLPE (cross-linked polyethylene) insulation of 250°C, and a fault current of 31.18 kA sym ($F_{AC}$ from Table 7.11 = 1.02). The breaker interrupting time is five cycles, which means that the protective relays must operate in less than nine cycles to clear the fault. Major cable circuits in industrial distribution systems are sized so that these are not damaged even if the first zone of protective relays (instantaneous) fail to operate and the fault has to be cleared in the time-delay zone of the backup device. From these criteria, the cable may be undersized. Instead of Equation (7.8), conductor short-circuit withstand curves from IEEE standard 242 [23] can be used. The maximum short-circuit temperatures for insulation types are specified in ICEA P-32-382 [24]. The adjustment factors for initial conductor tem-

TABLE 7.11. AC/DC Resistance Ratios: Copper and Aluminum Conductors at 60 Hz and 65°C

| Conductor Size (KCMIL or AWG) | 5–15-kV Nonleaded Shielded Power Cable, Three Single Concentric Conductors in Same Metallic Conduit | |
|---|---|---|
| | Copper | Aluminum |
| 1000 | 1.36 | 1.17 |
| 900 | 1.30 | 1.14 |
| 800 | 1.24 | 1.11 |
| 750 | 1.22 | 1.10 |
| 700 | 1.19 | 1.09 |
| 600 | 1.14 | 1.07 |
| 500 | 1.10 | 1.05 |
| 400 | 1.07 | 1.03 |
| 350 | 1.05 | 1.03 |
| 300 | 1.04 | 1.02 |
| 250 | 1.03 | 1.01 |
| 4/0 | 1.02 | 1.01 |
| 3/0 | 1.01 | <1% |
| 2/0 | 1.01 | <1% |

perature and final conductor temperature, factor *Kt*, are also graphically illustrated in IEEE Standard 242.

***Overhead Line Conductors.*** Calculations of load current and short-circuit withstand ratings for overhead line conductors must also receive similar considerations as cables, that is, these should be sized not only for load current and voltage drop consideration, but also from short-circuit considerations. For ACSR (aluminum conductor steel reinforced) conductors, a temperature of 100°C (60°C rise over 40°C ambient) is frequently used for normal loading conditions, as the strands retain approximately 90% of rated strength after 10,000 hours of operation. Under short circuit, 340°C may be selected as the maximum temperature for all aluminum conductors and 645°C for ACSR, with a sizable steel content. An expression for safe time duration based on this criterion and no heat loss during short-circuit for ACSR is [22]:

$$t = \left(0.0862 \frac{\text{CM}}{I}\right)^2, \tag{7.9}$$

where $t$ is the duration is seconds, CM is the area of conductor in circular mils, and $I$ is the current in amperes, rms.

From Equation (7.9), no. 4 (41740 CM) ACSR of the transmission line connected to breaker 2F2, Figure 5.8 has a short-circuit withstand capability of 0.013 second for a symmetrical short-circuit current of 31.28 kA, close to bus 2. The conductors, though adequately sized for the load current of a 1-MVA transformer, are grossly undersized from short-circuit considerations.

## 7.10.5 Overcurrent Protection of Conductors

***Protective Devices Not More Than 600 V.*** The overcurrent protection of conductors is addressed in Article 240 of NEC. This addresses a number of issues, like current limiting devices, tap conductors, transformer secondary conductors, ungrounded conductors, grounding requirements, and the like, which cannot be completely addressed in this book. It is imperative that a protection engineer is conversant with these NEC provisions. Special provisions apply to establishments like fire pumps, where the overcurrent protection is sacrificed altogether. The following brief synopsis is of interest:

For devices rated 800 A or less, protection of conductors 600 V or less, NEC permits next higher standard device overcurrent rating above the ampacity of the conductor being protected. For example, a 500-kcmil THWN conductor has an allowable ampacity of 380 A from NEC table 310.15(B)(16). It can be protected by a 400-A overcurrent device. There are qualifications to this rule:

- That conductors being protected are not a part of a multibranch circuit supplying receptacles.
- That ampacity does not correspond with the standard ampere rating of a fuse or circuit breaker.

- That next higher standard rating selected does not exceed 800 A.

For devices rated 800 A and more, the ampacity of the conductors it protects shall be equal to or greater than the rating of the overcurrent device defined in Article 240.6. Here, the intent is that the next higher overcurrent setting or fuse cannot be used. For example, consider that we need to protect 1100-A conductors, the overcurrent setting cannot exceed 1100 A, that is, a 1200-A overcurrent device is not permissible. For practical proposes, a 1000-A standard setting will be applied.

***Over 600-V Nominal.*** For the systems with voltages >600 V nominal and for the feeder overcurrent protection, it is permissible to select a fuse with continuous rating not exceeding three times the ampacity of conductors. The long-time trip setting of a breaker or the minimum trip setting of an electronically operated fuse shall not exceed six times the ampacity of conductor. The operating time of the protective device, the available short-circuit current and the conductors shall be coordinated to prevent damaging or dangerous temperatures in the conductors or conductor insulations under short-circuit.

Practically, this relaxation in the overcurrent protection of conductors at voltages >600 V nominal is of significance. The overcurrent coordination in these systems shows that it is difficult to protect the conductors within their ampacities. Yet, every attempt should be made to protect these as close to their ampacities as possible.

## 7.11 MOTOR PROTECTION

A MMPR for protection of medium voltage motors may have the following functionality:

- motor thermal model, which will account for overload curves, unbalance biasing, hot/cold safe stall ratio, motor cooling time constants, start inhibit and emergency start, and RTD biasing
- motor start supervision
- mechanical jam and acceleration times
- phase differential protection
- ground fault protection
- voltage and frequency protection
- protection system blocks, say from repeated starts or starting before the set time has elapsed
- metering functions
- zero speed switch inputs
- RTD (resistance temperature detectors) protection of stator windings and bearings
- power elements, current and voltage inputs, digital and analogue inputs and outputs, monitoring and metering, event recorder, advanced motor diagnostics, and communications. Synchronous motors will have in addition synchronizing

and resynchronizing protection, pull-out protection, incomplete starting sequence protection, and loss of field protection.

The older bimetallic thermal elements are no longer in use for the medium voltage motors. Also, for low voltage motors, all the major features as described above for the medium voltage motor protection are now available in digital relays.

From the arc flash point of view, selection of motor overcurrent/thermal protection, and short-circuit protection indirectly enter into the picture for coordination of the upstream devices. A motor coordination study will start by:

- First plotting the motor starting current time curve. Boiler ID (Induced Draft) fan motors driving large inertia may take 45 seconds or more to start. The calculation of the starting time of the motor for a given load requires computer simulation and dynamic motor starting studies, not discussed here. This data should be accurate, based upon motor starting studies.
- Next, the motor thermal withstand curve is required. Special care is required when the locked rotor time of the motor is much less than the starting time. Though during starting, some heat dissipation will take place due to rotation of the rotor, as compared with the full-voltage locked rotor withstand time, yet to be safe, the motor should be designed with locked rotor hot withstand time slightly greater than the starting time. If this cannot be achieved, additional protection features are required. One solution is to use impedance-type relay for protection; another method is to use zero speed switch, which senses the rotation of the motor and its acceleration up to the speed. This bypasses the protection for a predetermined time.
- A proper standard curve built in the relay or custom designing a curve to protect the motor thermal withstand is chosen.
- The short-circuit protection, with type R fuses for the medium voltage motors and their coordination with contactors for NEMA E2 starters [18], is considered.
- Low voltage high efficiency motors may draw a first cycle starting current exceeding 15 times the rated full load current. NEC allows setting magnetic only (MCPs) up to 17 times the motor full load current.
- Selection of appropriate motor starter fuses, MCPs or MCCBs for low voltage motors comes into picture. NEC table 430.52 specifies the maximum settings on nondelay fuse, dual element time delay fuse, instantaneous trip breaker (MCP), and inverse time breaker for various types of motors. This table is not reproduced. Practically, much lower settings for coordination will be adequate.

### 7.11.1 Coordination with Motor Thermal Damage Curve

IEEE Standard C37.96-2000, Guide for AC Motor Protection [25] recommends overcurrent relays for overload and locked rotor protection and IEEE standard 620-1996, Guide for Presentation of Thermal Withstand Curves for Squirrel Cage Induction Motors [26] lays down guidelines that a manufacturer must follow to supply the thermal damage curves. The allowable locked rotor thermal limit is given for rated locked rotor current. It can also be given as accelerating thermal limit curves both for cold and hot starts at various voltages,

100%, 90% and 80%. Induction and synchronous motors starts are specified in NEMA MG-1 [27]. This provides for two starts in succession coasting to rest with the motor initially at ambient temperature and one start when the motor is at a temperature not exceeding its rated load operating temperature. The motors may be specifically designed for a higher number of starts, inching and jogging. Figure 6a of Standard C37.96, not reproduced here may be seen for a thermal withstand curve. As thermal conditions are protected by overcurrent protection, a question of correlation between the two arises.

Figure 7.26 shows the thermal withstand curve supplied by a manufacturer for 400-hp, 2.4-kV motor, SF = 1.15. The basic thermal protection model is given by:

$$t_{\text{H-Curve}} = T_{\text{th}} \ln\left(\frac{I^2 - I_{\text{H}}^2}{I^2 - I_{\text{SF}}^2}\right)$$
$$t_{\text{C-Curve}} = T_{\text{th}} \ln\left(\frac{I^2 - I_{\text{C}}^2}{I^2 - I_{\text{SF}}^2}\right), \tag{7.10}$$

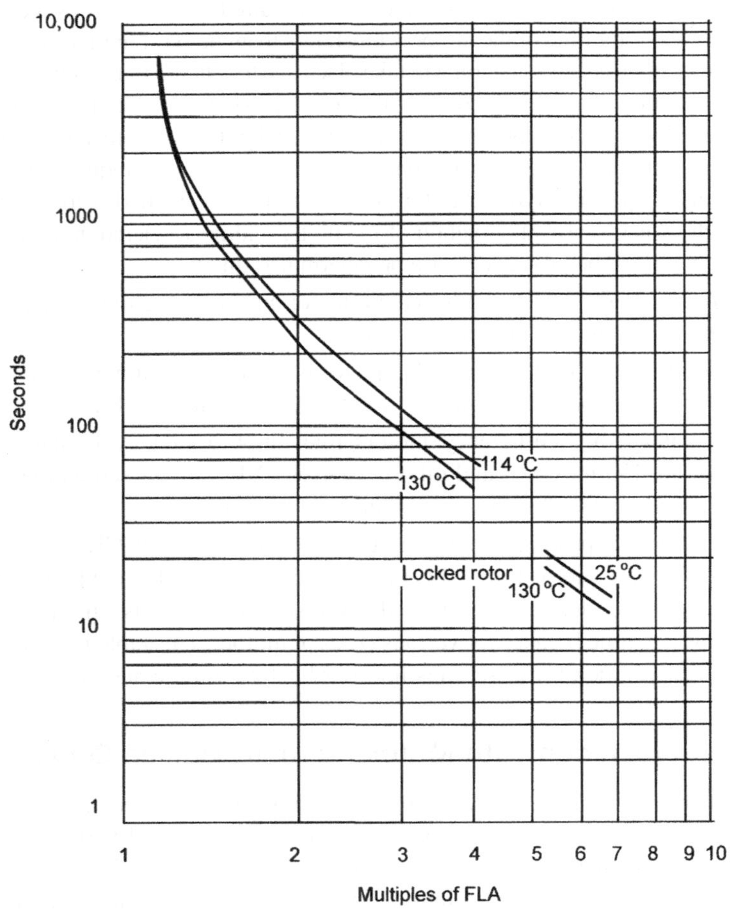


Figure 7.26. Thermal withstand curve of a 2.4-kV, 400-hp, SF = 1.15 motor.

where $T_{th}$ is thermal time constant, $I$ = motor current in pu of full load current, and $I_{SF}$ is the current at service factor. $I_H$ is current that raised temperature to 130°C, and $I_C$ is the current that raised the temperature to 114°C.

We can write the constraint that:

$$\frac{I_H^2}{I_C^2} = \left(\frac{130-25}{114-25}\right) = 1.179. \tag{7.11}$$

By solving the following equations in terms of service factor, a fit can be obtained:

$$\begin{aligned} I^2\left[1-e^{\frac{t_{\text{H-curve}}}{T_{th}}}\right] + I_H^2 e^{\frac{t_{\text{H-curve}}}{T_{th}}} &= 1.15^2 \\ I^2\left[1-e^{\frac{t_{\text{C-Curve}}}{T_{th}}}\right] + \frac{I_H^2}{1.179} e^{\frac{t_{\text{C-Curve}}}{T_{th}}} &= 1.15^2 \end{aligned}. \tag{7.12}$$

Consider a point on thermal limit curve, say at 2.0 per unit $I$, then from Figure 7.26, time is 223 seconds (hot) and 279 seconds (cold). Considering $T_{th}$ = 1370 seconds, this gives $I_H^2 = 0.846$ and $I_C^2 = 0.717$.

Then any point on the thermal curve can be calculated:

$$\begin{aligned} t_H &= 1370 \ln\left(\frac{I^2 - 0.846}{I^2 - 1.15^2}\right) \\ t_C &= 1370 \ln\left(\frac{I^2 - 0.717}{I^2 - 1.15^2}\right) \end{aligned}. \tag{7.13}$$

The overcurrent model is implemented by integrating the reciprocal of hot thermal limit curve as specified in IEEE Standard C37.112-1996 [3]. The incremental equations for this process are:

For $I > 1.15$

$$\theta_n = \theta_{n-1} + \frac{\Delta t}{t_H}. \tag{7.14}$$

For $I < 1.15$

$$\theta_n = \left(1 - \frac{\Delta t}{1370}\right)\theta_{n-1}. \tag{7.15}$$

Equation (7.14) is used to calculate response of overcurrent relay above the pickup current. $\theta_n$ and $\theta_{n-1}$ are consecutive samples displaced by one time step. An overcurrent relay can trip, say on a cyclic load, even before the thermal limit is reached as the relay does not have a thermal memory.

When selecting an overcurrent relay curve to protect a given thermal withstand characteristics of the motor, it is important that overcurrent characteristic is matched closely to the thermal characteristics. The use of microprocessor-based relays provides more accurate means of determining the coordination under starting conditions. In modern MPPRs for motor protection, there are two options:

- A standard built-in curve can be selected to match the thermal curve.
- A user can create a curve to match the thermal curve.

The test of a thermal model is its ability to adequately protect the motor from overheating during cyclic overloads. To this end, see the two settings provided to protect the same motor, settings A and B in Figure 7.27a,b. The response of the two settings to cyclic overloads is shown in Figure 7.28. The setting A prematurely trips the motor on cyclic overloads.

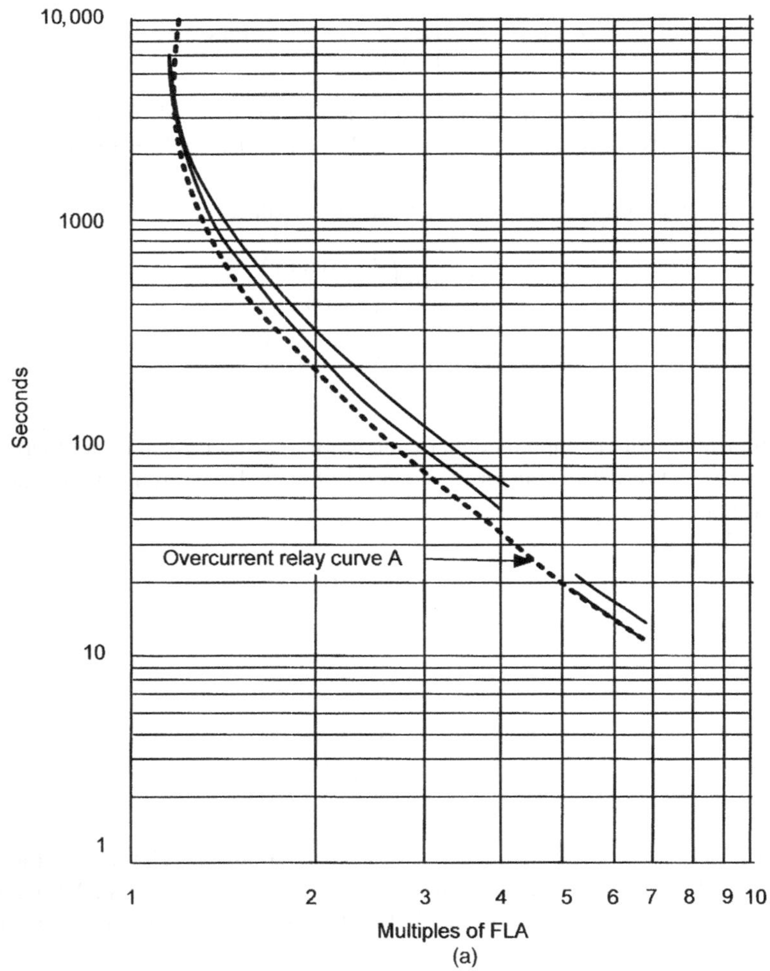


Figure 7.27. (a) and (b) Protection of thermal withstand curve of the motor in Figure 7.26 with overcurrent relay curves "A" and "B," respectively.

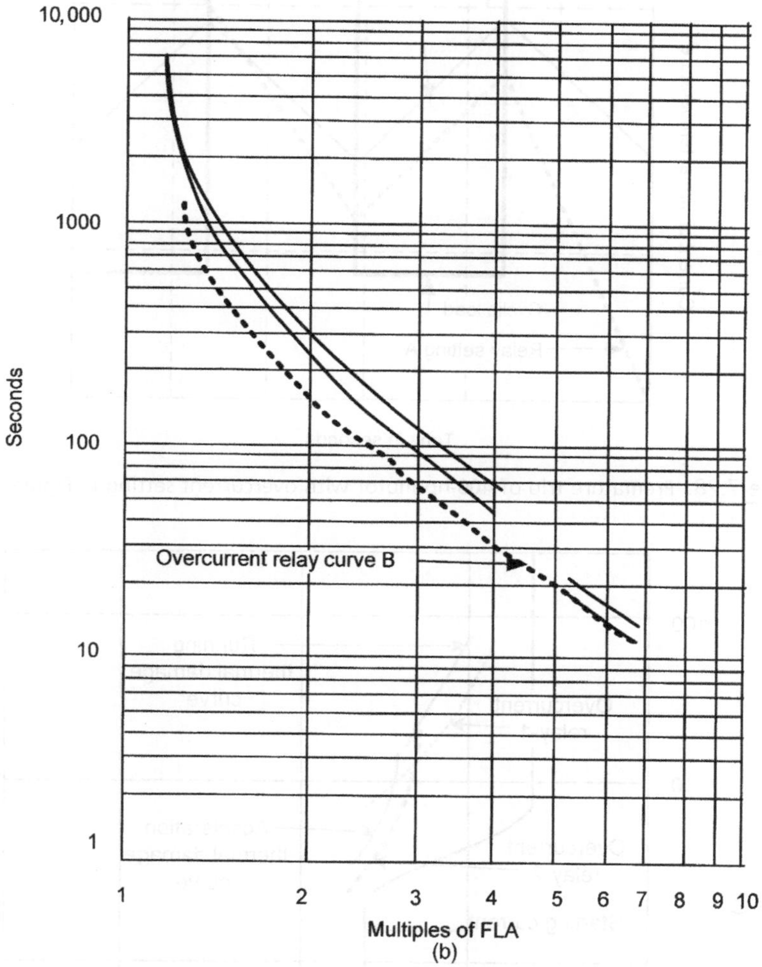


Figure 7.27. (*Continued*)

Though we talk of thermal time constant as a single number, practically, the thermal model of a machine is fairly complex. The slots embedded in the iron core; the overhangs in air, the frame, the end rings, the shaft, and the rotor structure—all have different materials, mass and conductivity varying over large limits.

The overcurrent model and thermal model may not be in step, and it is possible that the motor may be prematurely tripped or subjected to overloads; though attempts have been made that the two models correlate as much as possible.

Figure 7.29 shows a two-step overcurrent relay for coordinating with motor thermal withstand curve and also the starting characteristics. As stated before, a problem of starting can arise when the locked rotor withstand time is much lower than the motor starting time. With a properly set thermal protection system, it will not allow starting the motor. One solution is that the zero speed switch (device 12 in Figure 7.30) acts in conjunction with an overcurrent element. As soon as the motor starts to accelerate, the 51 start is disabled, leaving the overcurrent protection to longer time overcurrent relay 51.

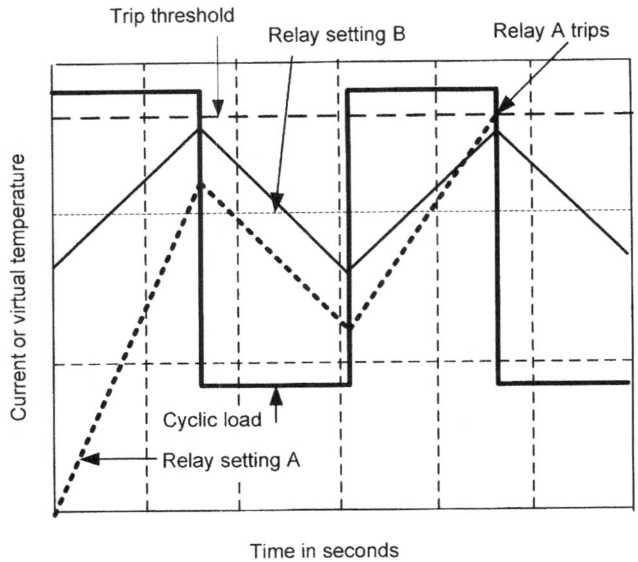


Figure 7.28. Premature trip of 400-hp motor with overcurrent setting in Figure 7.27a.

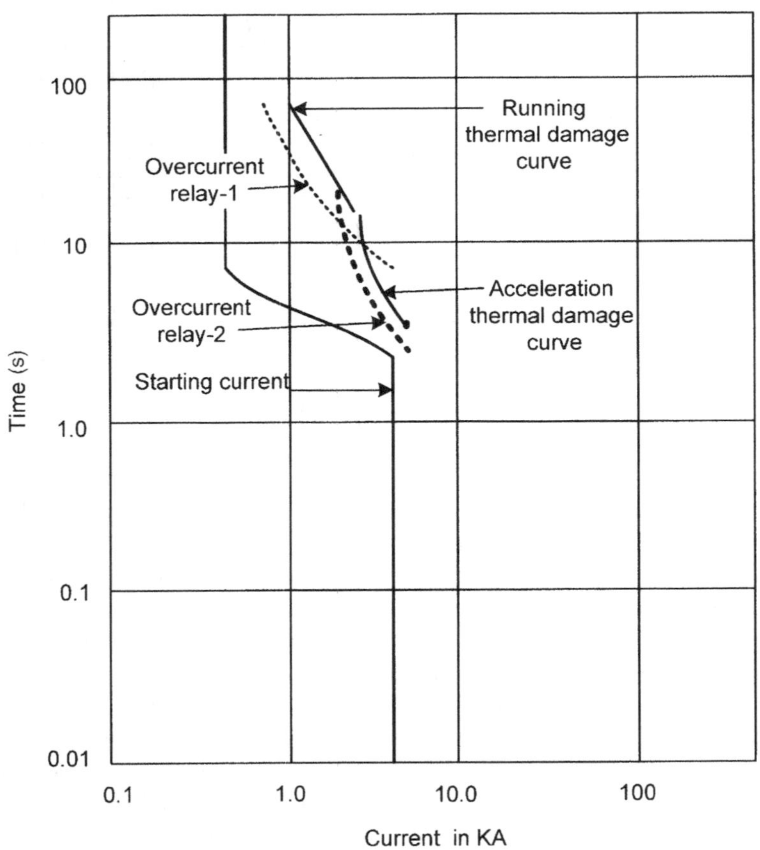


Figure 7.29. A two-step overcurrent relay characteristics and motor starting curve for coordination with motor thermal withstand curve.

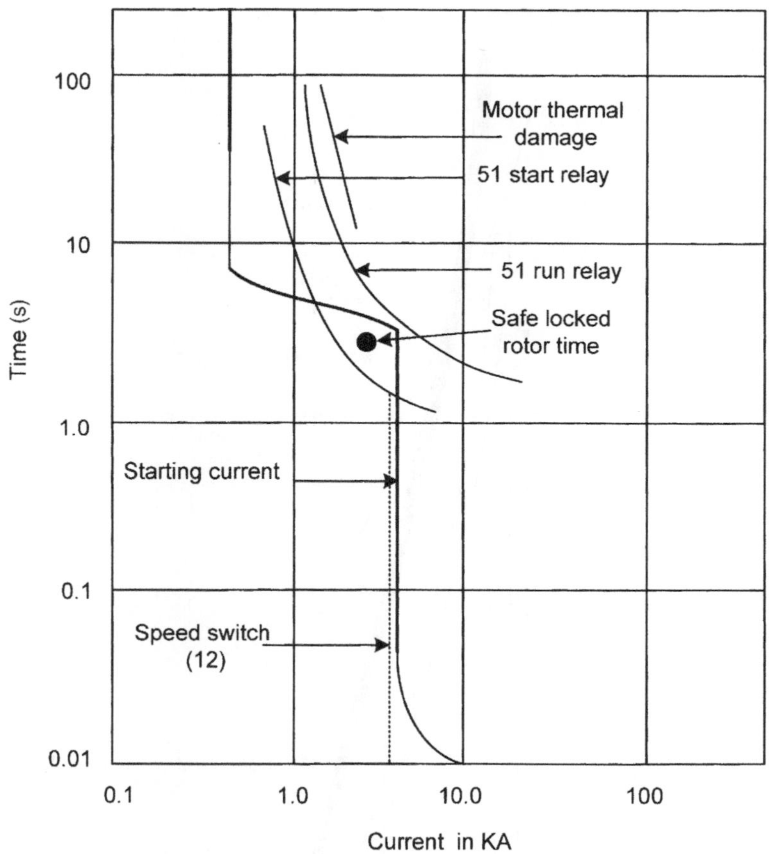


Figure 7.30. Application of a zero-speed switch and an overcurrent element for protection of motor where the accelerating time exceeds the locked rotor time (hot).

**Example 7.4**

This is a practical coordination of protection of a 1000-hp, 2.3-kV motor, connected to a 2.5 MVA transformer. The selection of appropriate motor fuse size and motor protective relay settings is illustrated. Consider that the motor starting curve has been calculated and the thermal damage curve is supplied by the manufacturer. As a first step, these two curves can be plotted (Figure 7.31). The motor is controlled through a NEMA E2 starter, with a 400-A vacuum contactor.

Increase the starting current (ignoring the increased inrush during in the first cycle or so) by 10%, the dotted line in this figure, and select R-type fuse rating, so that it clears this dotted line as well as the locked rotor withstand of the motor. A fuse should not operate for the locked rotor condition of the motor, and the vacuum contactor should clear this condition. An R-type fuse of 18R is selected, and it meets these criteria. The symmetrical interrupting current of 8.5 kA for the 400-A vacuum contactor based on manufacturer data is shown in this figure. It is also pertinent to draw the contactor

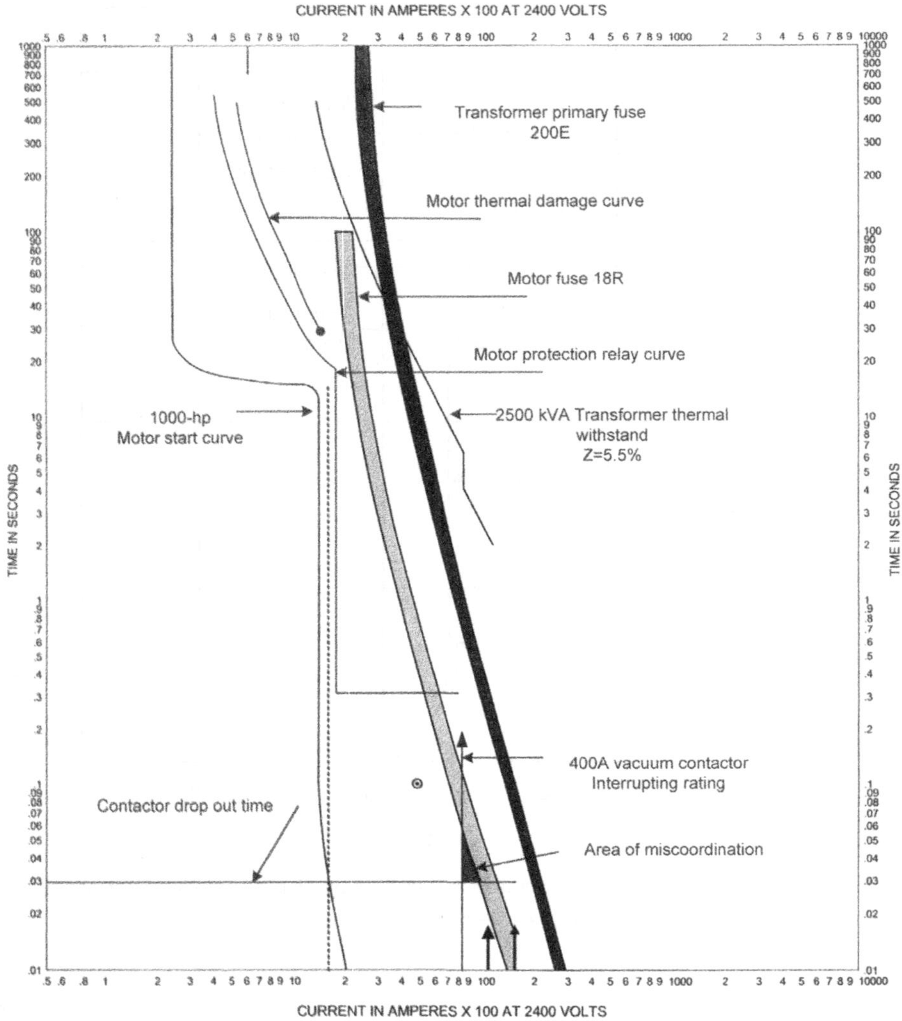


Figure 7.31. A practical time–current coordination of a 2.3-kV, 1000-hp motor connected to the secondary of a 2.5-MVA transformer, Example 7.4.

dropout line. It is recognized that the dropout time varies with the residual magnetism and is not a fixed number. A dropout time of 0.03 second is shown. In the area of lack of coordination between the fuse and the contactor interrupting rating, marked in this figure, the vacuum contactor will clear a fault beyond its interrupting rating. Therefore, it is necessary to provide a vacuum contactor of 800 A rating, which has an interrupting capability of 12.5 kA.

The thermal damage curve of the motor is protected and the instantaneous setting with a time delay of 0.3 second protects the fuse and trips the contactor. This delay of 0.3 second can be further reduced.

The motor damage curve for the cold condition is plotted. The relay allows reducing the thermal capacity of the motor based upon the ratio of the locked rotor time in the hot and cold conditions. The manufacturer's recommendation for the particular relay type to be used should be followed.

The starting time is 15 seconds, and with the coordination shown, the motor is capable of two consecutive starts per hour from cold, with motor coasting to rest between starts, or one start with the motor at the operating temperature according to NEMA standards [27].

The transformer is protected with a primary fuse of 200E. It is assumed that there is no main secondary breaker secondary breaker for the 2500-kVA transformer. From arc flash point of view, the lack of secondary protection will give high incident energy for a fault on the motor starter bus; see Chapters 11 and 13.

## 7.12 GENERATOR 51-V PROTECTION

IEEE Guide for AC Generator Protection [28] provides reference to many publications on generator protection. Here we are interested in backup protection provided by 51 V or 21 devices. The function of 51 V is to disconnect a generator form service if other generator relays have failed to clear the fault. It protects the distribution system components against excessive damage and its auxiliaries from exceeding their thermal limits. For unit-connected generators, a distance relay device 21 is used. Overcurrent relays without voltage control are difficult to coordinate with downstream protection and also be sensitive to generator decaying fault currents discussed in Chapter 6. There are two versions of 51-V relays:

**Voltage Restraint**

Restraining torque is proportional to voltage. Say at 100% voltage, the pickup setting, as a percentage of tap setting, is 100%. If the voltage dips to zero, the pickup setting is reduced to 25%. At 48% voltage, the pickup setting is reduced to 52%. Manufacturers publish the restraint characteristics with varying voltage.

For a microprocessor relay, an algorithm is:

$$T_{op} = \frac{K}{[(I/I_{pu})/(V/V_{nom})]^{0.5} - 1}, \tag{7.16}$$

where $T_{op}$ is operating time is seconds, $V_{Nom}$ = nominal voltage, and $V$ = restraint voltage, = 1.732 phase-to-ground voltage for wye-connected transformer and = phase-to-phase voltage for delta connected transformers.

**Voltage Controlled**

In voltage controlled relay, the torque is adjusted over a range of 65–83% of rated voltage. When the applied voltage is above the pickup settings, no operating torque is produced regardless of current magnitude.

### 7.12.1 Arc Flash Considerations

The calculations of incident energy on bus fault F2 incident energy release of 45 cal/cm$^2$: dangerous. Both relays $R_2$ and $R_3$ must operate to clear the bus fault. A bus differential scheme will be invariably provided, this bus fault will be removed quickly, and all the breakers on this bus tripped. This will reduce the incident energy to 6.6 cal/cm$^2$. Generally, the bus, utility tie breaker, and generator will have overlapping zones of differential protection.

A fault at location F3, say in the incoming section of generator breaker 52G sees two sources of fault current, one from the utility source through 7.5 MVA transformer and the other contributed by the generator. The generator will be tripped in a short time, approximately six cycles, assuming five-cycle rated breakers and one cycle operating time of generator differential relay. But due to rotating inertia of the generator and turbine, the fault at F3 continues to be fed by the generator. NEMA standard requires that a generator should be capable of withstanding a three-phase bolted fault at its terminals for 30 seconds, without injury, when operating at its rated kVA and power factor, at 5% overvoltage with fixed excitation [27].

An EMTP simulation will show that the fault will continue to be fed with decaying magnitude for many seconds, even though the field circuit breaker is tripped and the generator excitation is removed. *The generator time constants described in Chapter 6 are not valid for this situation by their definitions.*

The available computer-based programs do not account for this decay after the generator breaker is opened. The generator side terminal compartment of the circuit breaker continues to be fed from this fault current releasing additional incident energy.

**Example 7.5**

This example demonstrates the application of a 51-V relay for generator protection and its coordination with downstream relays. Consider a system configuration as shown in Figure 7.32. A 10-MVA 2.4-kV generator operates in synchronism with a transformer of 7.5 MVA connected to a 13.8-kV source. The load has 2000 Hp of AC motors, and all other loads served from this 2.4-kV bus are static in nature.

Coordination with primary and secondary protective relays and generator 51-V relay settings is shown in Figure 7.33. In practical installations, the generator may be relatively small compared with the utility source or vice-versa. This means that load shedding is adopted to reduce the load if the utility source goes out of service. For a fault location F1, which will be fed both from the utility source and generator, 7.5 MVA transformer is taken out of service, leaving the generator to supply the load. For a fault at F3, only generator is taken out. For a fault F2 on the bus itself, both sources will be tripped and there will be a complete loss of power. This selective tripping is obtained with differential zones, Chapter 8 and by directional overcurrent relays, not shown in this figure. Overcurrent relays are retained as backup.

Observe the settings on 51-V relay in Figure 7.33. For a fault at location F1 or F2, the voltage will be practically reduced to zero and the zero voltage restraint characteristics of 51 V relay should be considered. A short-circuit study to calculate the fault

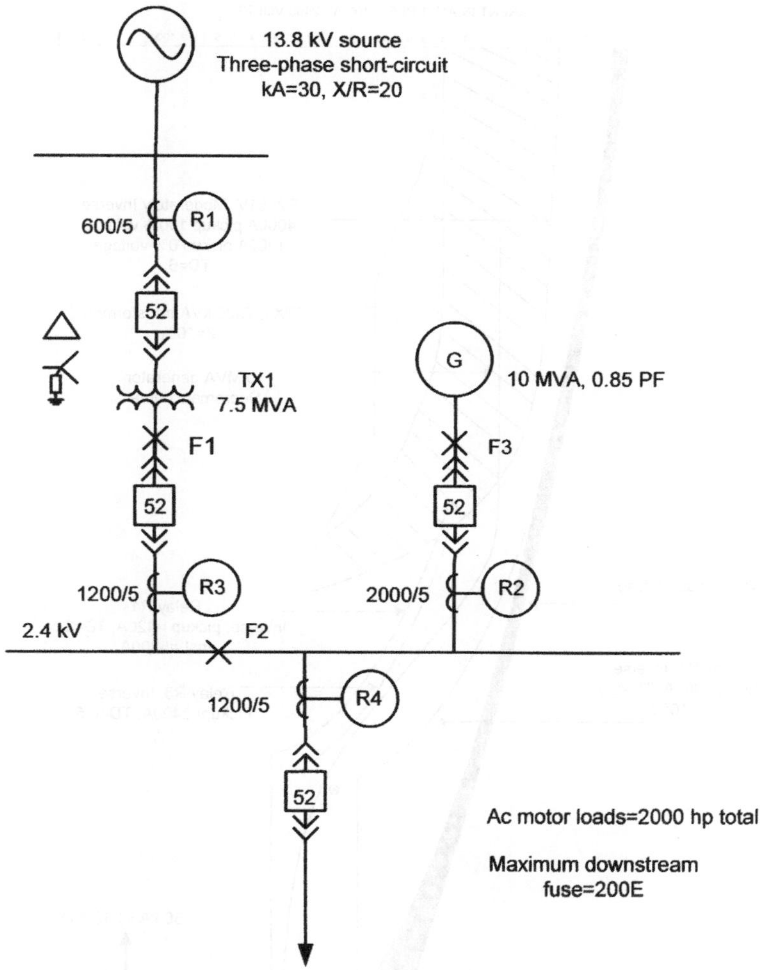


Figure 7.32. A parallel running 10 MVA generator with a utility source for application of a voltage restraint relay on the generator.

voltages, as the fault is removed away from the generator can be conducted and the 51 V overcurrent relay characteristics varies between the 100% restraint and 0% restraint with respect to voltage at the fault. The zero restraint characteristic protects the generator for a stuck regulator condition. The steady state fault current of the generator can even be lower than the generator load current.

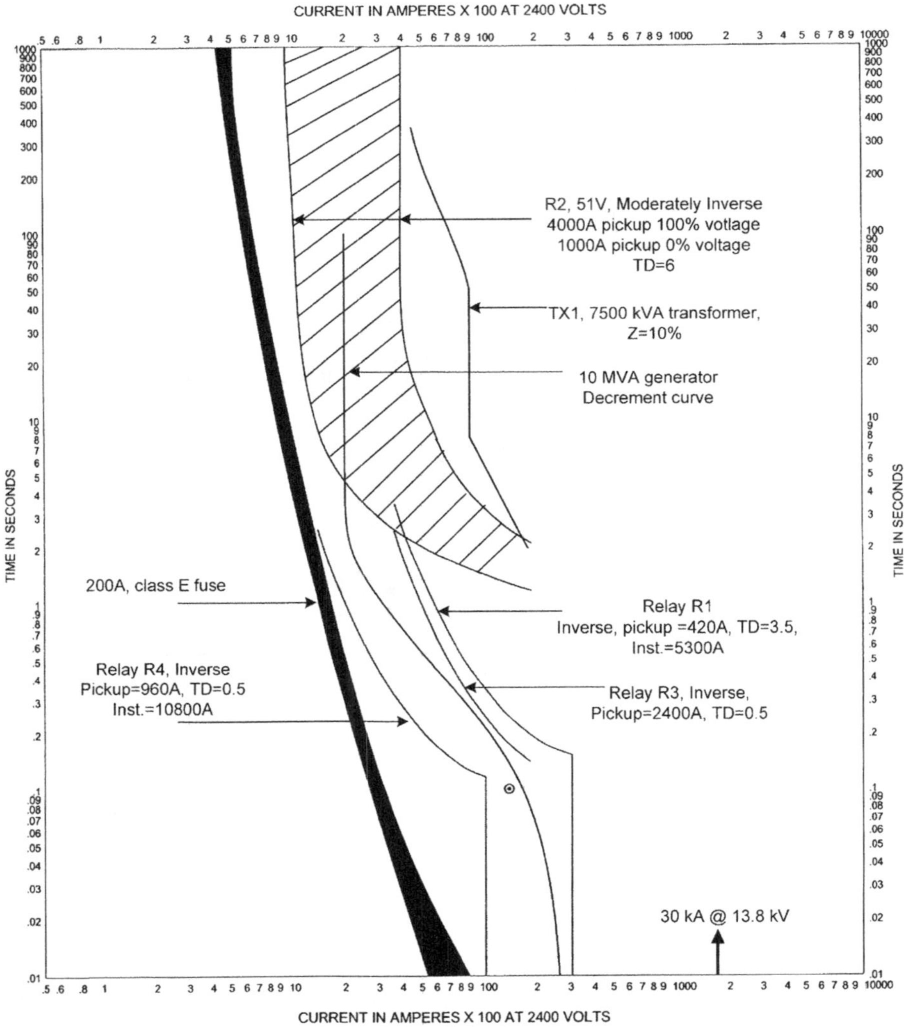


Figure 7.33. 51-V relay settings and coordination of the generator for the configuration shown in Figure 7.32, Example 7.5.

It is prudent to calculate the additional incident energy released by hand for a period of 2 seconds. The generator transient fault current for this duration will give conservative results. This calculation, considering generator transient reactance of 23%, will give 10.45 kA for 2 seconds, which adds 34 cal/cm$^2$.

It is recommended that generator circuit breaker should not be maintained in the energized condition.

## REVIEW QUESTIONS

1. A LVPCB, ICCB, and MCCB are similarly rated at 65 kA symmetrical interrupting. What other short-circuit rating is important for their application and protection coordination?
2. Each type of breaker in Problem 1 is subjected to a fault current of 50 kA, $X/R = 7.0$. Calculate the interrupting duty multiplying factors from the tables in this chapter.
3. What are the advantages and disadvantages of current-limiting fuses as compared with relayed circuit breakers for short-circuit interruption? How do these compare with expulsion-type fuses?
4. Explain the series interrupting ratings of two devices. What are the relative advantages and disadvantages of this configuration? Why should the series rating of two devices not be calculated? What is the impact on coordination and selectivity?
5. Recommend pickup settings for a 51-V relay for protection of a 13.8-kV, 30-MW, and 0.85 power factor generator. Why do we not use overcurrent relays without voltage control for generator backup protection?

## REFERENCES


1. ANSI, C37.2. *IEEE Standard for Electrical Power System Device Function Numbers, Acronyms and Contact Designations*, 2008.
2. IEC, 61850. *Communication Networks and Systems*, 2002–2012.
3. IEEE Std., C37.112. *IEEE Standard Inverse-Time Characteristic Equations for Overcurrent Relays*, 1996.
4. ANSI/IEEE Standard, C37.13. *Standard for Low-Voltage AC Power Circuit Breakers Used in Enclosures*, 2008.
5. IEEE Standard, C37.13.1. *IEEE Standard for Definite-Purpose Switching Devices for use in Metal-Enclosed Low-Voltage Power Circuit Breaker Switchgear*, 2006.
6. NEMA Standard, AB-1. *Molded Case Circuit Breakers and Molded Case Switches*, 1993.
7. UL Standard, 489. *Molded Case Circuit Breakers and Circuit-Breaker Enclosures*, 1991.
8. IEC Standard, 60947-2. *Low-Voltage Switchgear and Control Gear-Part 2: Circuit Breakers*, 2009.
9. IEEE Standard, 1015. *Applying Low-Voltage Circuit Breakers Used in Industrial and Commercial Power Systems*, 1997.
10. W.A. Brown and R. Shapiro, "Incident energy reduction techniques," IEEE Industry Applications Magazine, vol. 15, no. 3, pp. 53–61, 2009.
11. NEC, NFPA 70. *National Electric Code*, 2011.
12. NEMA Standard, SG2. *High-Voltage Fuses*, 1981.
13. IEEE Standard, C37.41. *IEEE Standard Design Tests for High-Voltage (>1000 V) Fuses, Fuse and Disconnecting Cutouts, Distribution Enclosed Single-Pole Air Switches, Fuse Disconnecting Switches, and Fuse Links and Accessories Used with These Devices*, 2008.

14. IEEE Standard, C37.42. *IEEE Standard Specifications for High-Voltage( >1000 V) Expulsion-Type Distribution-Class Fuses, Fuse and Disconnecting Cutouts, Fuse Disconnecting Switches, and Fuse Links, and Accessories Used with These Devices*, 2008.
15. ANSI Standard, C37.46. *American National Standard for High-Voltage Expulsion and Current Limiting Type Power Class Fuses and Fuse Disconnecting Switches*, 2000.
16. ANSI Standard, C37.47. *American National Standard for High-Voltage Current-Limiting Type Distribution Class Fuses and Fuse Disconnecting Switches*, 2000.
17. IEEE, C37.48. *IEEE Guide for Application, Operation and Maintenance of High Voltage Fuses, Distribution Enclosed Single Pole Air Switches, Fuse Disconnect Switches, and Accessories*, 1997.
18. NEMA, ICS2-234. *AC General Purpose HV Contactor and Class E Controllers, 50 and 60 Hz*, 1974.
19. N. McGrath, "The calculations of the temperature rise and load capability of cable systems," paper presented at AIEEE general meeting, Montréal, Quebec, June 24–28, 1956.
20. NEMA, "Report of determination of maximum permissible current carrying capacity of code insulated wires and cables for building purposes," June 27, 1938.
21. IEEE, S-135/ICEA P-46-426. *Power Cable Ampacities for Copper and Aluminum Conductors*, 2009.
22. D.G. Fink and J.M. Carroll (eds.), *Standard Handbook for Electrical Engineers*, 10th edition, McGraw-Hill, New York (see section 4, Table 4-8), 1968.
23. ANSI/IEEE Std., 242. *IEEE Recommended Practice for Protection and Coordination of Industrial and Commercial Power Systems*, 1986.
24. ICEA, P-32-382. *Short-Circuit Characteristics of Insulated Cable*, 1969.
25. IEEE Standard, 620. *Guide for the Presentation of Thermal Limit Curves for Squirrel Cage Induction Machines-R2000*, revised 2000.
26. IEEE Standard, C37.96. *IEEE Guide for AC Motor Protection*, 2000.
27. NEMA, MG-1. *Motors and Generators*, 1993.
28. IEEE, C37.102. *IEEE Guide for AC Generator Protection*, 2006.


## Further Reading on Protective Relaying


C.R. Mason, *Art and Science of Protective Relaying*, John Wiley, New York, 1956.

A.F. Silvia, *Protective Relaying Principles*, CRC Press, Boca Raton, FL, 2009.

Westinghouse Electric Corporation, *Applied Protective Relaying*, Westinghouse, Coral Springs, FL, 1982.

J.L. Blackburn and T.J. Domin, *Protective Relaying: Principles and Applications*, CRC Press, Boca Raton, FL, 2006.

D. Reimert, *Protective Relaying for Power Generation Systems*, CRC Press, Boca Raton, FL, 2005.

W.A. Elmore, *Protective Relaying: Theory and Applications*, Marcel Dekker, New York, 2004.

P.M. Anderson, *Power System Protection*. IEEE Press Series on Power Engineering, IEEE Press, NJ, 1998.

# 8

# UNIT PROTECTION SYSTEMS

Unit protection systems are of special significance for arc flash reduction. In Chapter 7, we studied the time–current coordination systems and their limitations; that starting from downstream, as higher sources of power upstream are coordinated, the fault clearance times go on increasing. From an arc flash reduction point of view, this is not desirable. The higher short-circuit currents upstream, coupled with higher fault clearance times, will give rise to increased arc flash energy release and equipment damage.

A separate zone of protection can be established around each system element so that any fault within that zone will cause tripping of the circuit breakers to isolate the fault quickly, without looking at the coordination or protective devices in the rest of the system. If a fault occurs outside the protective zone, the protective system will not operate, that is, it is stable for all faults outside the protective zone. Such zones of protection constitute unit protection systems [1]. Differential relaying is one form of unit protection system and is discussed first in this chapter.

Unit protection systems (differential relays) can be applied to any individual system element or group of system elements in power distribution systems. That is, separate zones of protection can be created around:

- generators
- transformers

*Arc Flash Hazard Analysis and Mitigation*, Second Edition. J.C. Das.

Published 2021 by John Wiley & Sons, Inc.

- motors
- bus bars
- cables
- overhead lines.

Sometimes, more than one equipment to be protected is covered in a single zone of protection. An example is unit-connected generator, where the generator and transformer are protected as one unit and there is no generator circuit breaker.

Figure 8.1 shows a single-line diagram of two interconnected 13.8-kV buses for primary distribution of power in an industrial plant with cogeneration facility. Bus 1 receives utility source power through a 40/64-MVA transformer, and a 50-MVA 13.8-kV generator is connected to bus 2; it operates in synchronism with the utility source. The plant loads are served from both the buses 1 and 2.

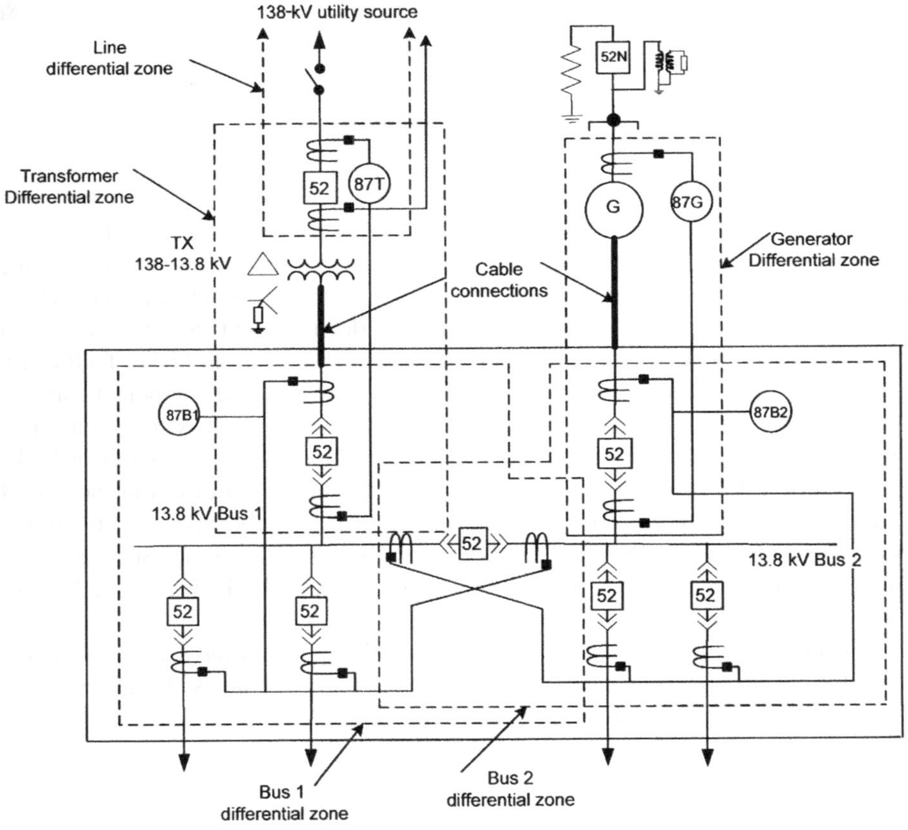


Figure 8.1. A 13.8-kV sectionalized bus with overlapping zones of differential protection. *Note*: All protections, for example, ground fault, directional, frequency, time overcurrent, and so on, are not shown.

This figure shows distinct zones of protection created to protect the utility tie transformer TX, bus tie reactors, and 13.8-kV buses 1 and 2. The secondary cables from transformer TX and generator G are included in the transformer and generator differential zones. For the utility incoming line to the transformer, a separate zone of protection is provided by the utility company. This figure also shows the location of current transformers and the circuit breakers. The cables from the feeder circuit breakers are not in any differential zone of protection.

## 8.1 OVERLAPPING THE ZONES OF PROTECTION

It is noteworthy that the overlapping of the differential zones of protection is achieved by proper location of the current transformers, and constructing a zone of protection so that one zone overlaps the other and no area is left unprotected. Consider that a differential protection is provided only for a 13.8-kV bus in Figure 8.2. It is metal-clad

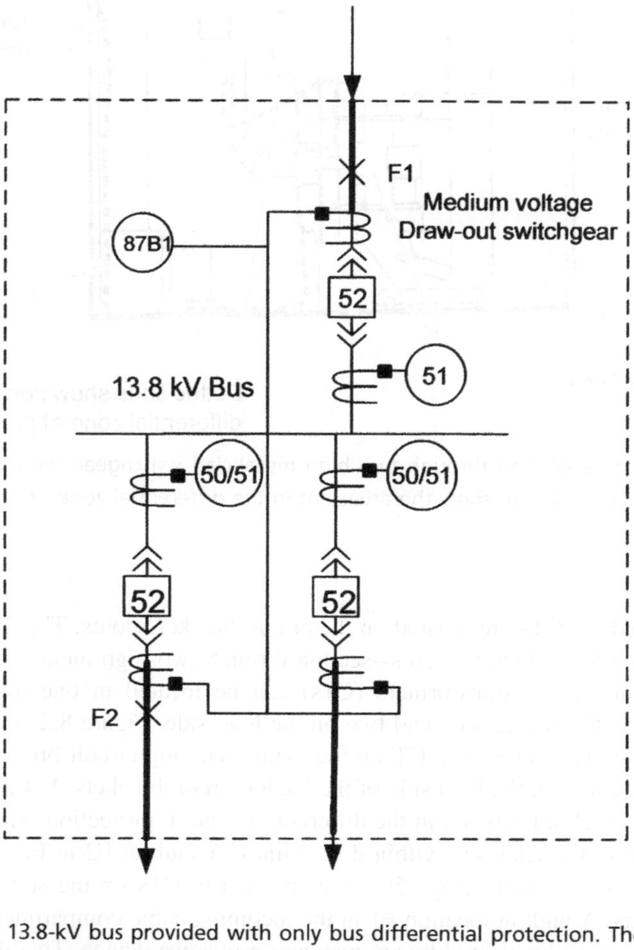


Figure 8.2. A 13.8-kV bus provided with only bus differential protection. The areas shown in thick lines are outside the differential zone of protection.

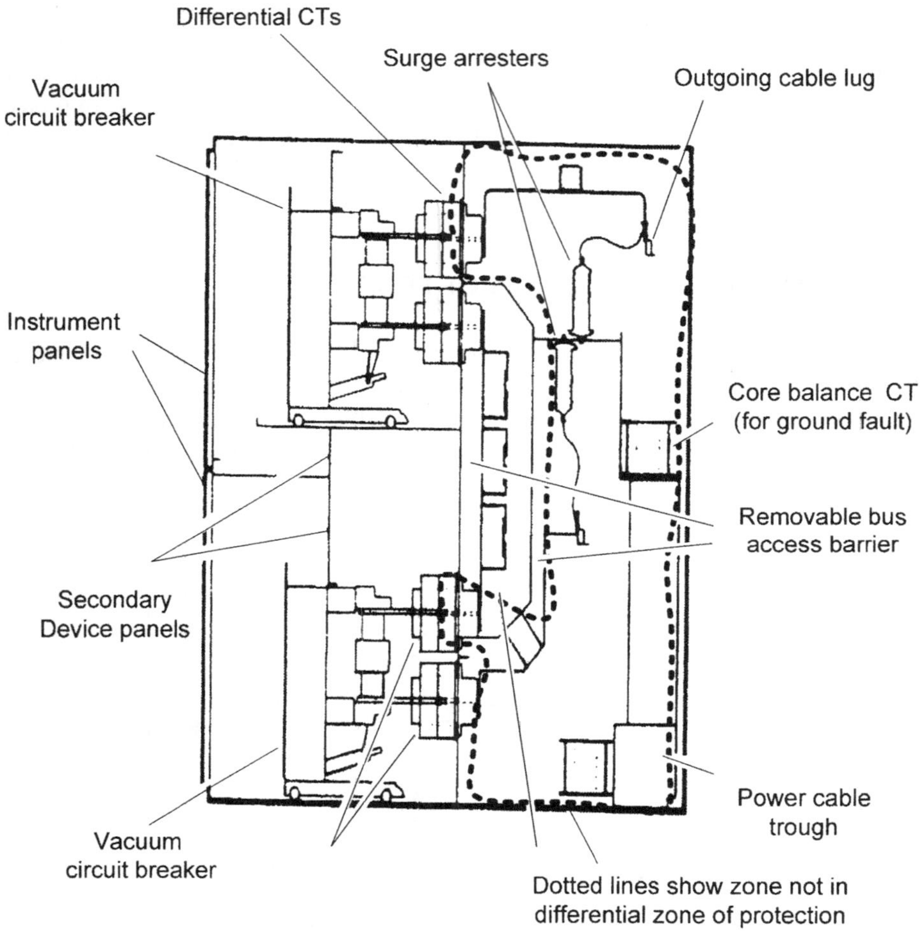


Figure 8.3. A cross-section through two high metal-clad switchgear, with bus differential protection. The dotted lines show the areas not in the differential zone of protection.

switchgear, and the CTs are located on the circuit breaker spouts. This is further illustrated in Figure 8.3, which is a cross-section through two-high metal-clad switchgear. Generally, four current transformers (CTs) can be located in one draw-out circuit breaker, two on the source side and two on the load side. Figure 8.2 shows that a bus differential zone is created by a CT on the main incoming circuit breaker source side and the CTs located on the load side of the feeder circuit breakers. In this diagram, the area shown in thick lines is not in the differential zone of protection. The same area is shown in Figure 8.3, enclosed within dotted lines. A fault at F2 in Figure 8.2 will be cleared by the overcurrent relays 50/51, connected to CTs on the source side of the circuit breakers. A fault at location F1 in the incoming cable compartment of the main circuit breaker must be cleared by an upstream protective device. Therefore, the cable

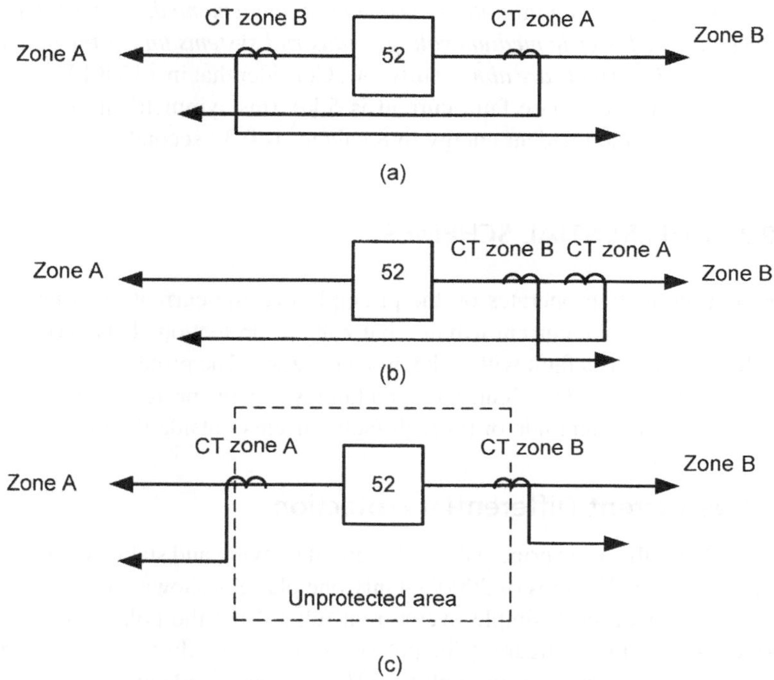


Figure 8.4. (a) Overlapping zones of differential protection provided by proper connections of CTs on the source and load side of a circuit breaker, (b) overleaping zones provided when the CTs are located on the load side only, and (c) defective connections of CTs showing unprotected area.

terminations and cable compartments remain outside the zone of differential protection. This concept is important that the differential zone of protection is dictated by the location of the CTs. A single bus differential zone in a metal-clad switchgear leaves the areas shown unprotected.

In some earlier vintage of switchgear, the CTs could not be provided on the source side, and were located only on the load side of the circuit breaker. Still, overlapping zones of protection can be achieved by proper CT connections as shown in Figure 8.4b. The connections of CTs shown in Figure 8.4c will be inappropriate, the zones do not overlap, and the area shown dotted remains unprotected. The CT connections shown in Figure 8.4a are currently used in metal-clad switchgear.

## 8.2 IMPORTANCE OF DIFFERENTIAL SYSTEMS FOR ARC FLASH REDUCTION

The importance of arc flash reduction with differential protections is evident. This type of protection operates fast within 0.75–1.5 cycles. Comparatively, an overcurrent relay in the instantaneous zone will also operate fast within one to two cycles.

*Thus, time–current coordination, howsoever implemented, cannot reduce the hazard level to 2 or lower in medium voltage industrial systems unless the bolted three-phase short-circuit currents are abnormally low*. Consider that in a 13.8-kV ungrounded system, the bolted three-phase fault current is 5 kA rms symmetrical. The calculated arcing time to limit the incident energy to 8 cal/cm$^2$ is 0.32 seconds.

## 8.3 BUS DIFFERENTIAL SCHEMES

A differential protection operates on the principle that the current entering a zone of protection is equal to the current leaving that zone of protection. This current balance does not hold if there is a fault within the protected zone. The protection should operate fast even for low magnitudes of currents for a fault within the protected zone and should be stable for a large magnitude of through fault currents outside the protected zone.

### 8.3.1 Overcurrent Differential Protection

Figure 8.5a shows the basic principal of differential relaying and stability for an external fault of 20 kA, with CT ratios of 2000/5. Only one phase is shown, and the differential protection is provided by a simple overcurrent relay. Note the polarity of the current transformers shown in this figure. (The polarities of CTs are discussed in Chapter 12.) Fifty-ampere current circulates through the CT secondary leads and none flows in the relay. This is an ideal situation.

Figure 8.5b shows that for an internal fault of 500-A, 1.25-A secondary current should flow in the overcurrent relay, and if this relay is set to pick up at this level, it will trip for an internal fault.

This situation is not realized in practice because:

- The CT leads may be of different length, imposing different burdens on the CTs.
- Even if the CTs are exactly of the same ratio, there can be variations and CTs do not perform exactly according to their ratios. This difference is caused by variations in manufacture, difference in secondary loading, and magnetic history. Residual magnetic flux of varying magnitude may be trapped in the CT core (see Chapter 12).
- Though CTs may be selected to avoid saturation, under high magnitudes of through fault currents, these may have different saturation characteristics.
- The CT accuracies and errors are discussed in Chapter 12.

Yet simple overcurrent relays have been applied for differential schemes. Figure 8.6a shows such a scheme for two buses with a bus section switch. This will have poor sensitivity; the overcurrent element must be set high to override the CT spill currents. Also, it will be more susceptible to nuisance operation due to CT saturation [2].

### 8.3.2 Partial Differential Schemes

Figure 8.6b shows a further simplification of the differential protection called a *partial differential scheme*. This is the slowest and least sensitive of any differential scheme, though sometimes used. There are no differential CTs on the feeder circuit breakers. The relay must be set high enough to coordinate with feeder relaying.

From the arc flash considerations, the schemes shown in this figure are not recommended.

### 8.3.3 Percent Differential Relays

This has the advantage of high stability for external faults, when errors are more likely to produce erroneous differential currents, and good sensitivity to faults in the protected zone is required. These are commonly applied to generator, transformer, and motor differential protection. There are three types of restraint characteristics [3]:

- fixed percentage

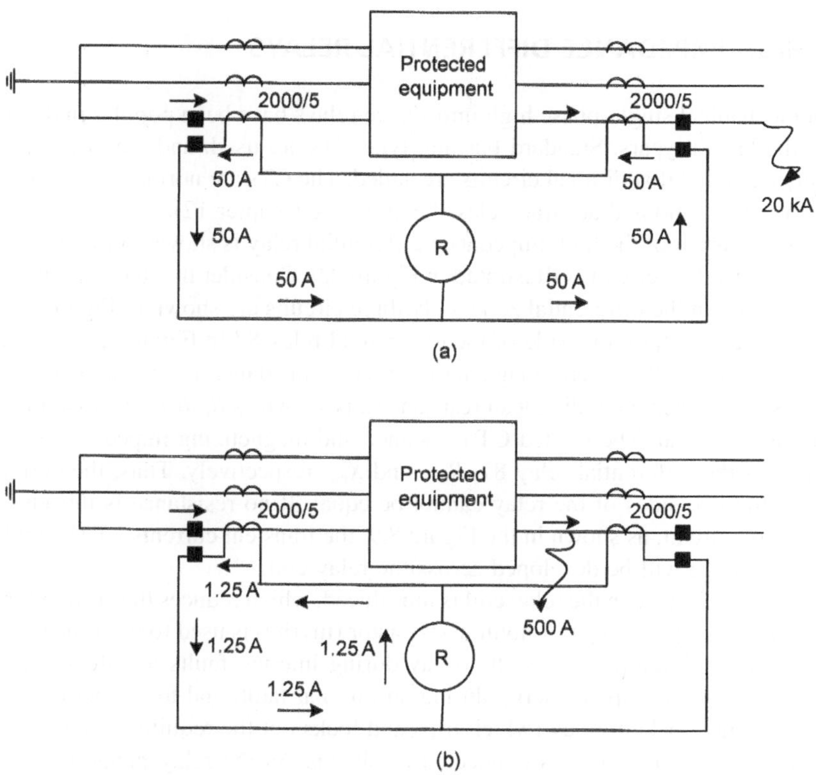


Figure 8.5. Concepts of differential protection. (a) A fault external to the protected equipment; (b) a fault within the differential protected zone.

- variable percentage
- harmonic-restraint percentage.

Variable percentage relays are used for generator protection and are less likely to have nuisance tripping. Harmonic restraint blocks tripping on transformer inrush currents.

The principle of an electromechanical percentage differential relay can be explained with reference to Figure 8.7a. The relay uses an induction principle. The current in the restraining coils, $R_1$ and $R_2$, produce a restraining or contact opening torque. An internal fault in the protected zone will unbalance the secondary currents, forcing an operating current $I_o$ in the relay operating coil, O. For a fixed-percentage differential relay, the operating current required to overcome restraining torque is a fixed percentage of restraining current (see Figure 8.7b). For example, for a fixed 10% percentage differential current, the relay will trip if the operating current was greater than 10% of restraint current.

In a variable percentage differential relay, the amount of differential current required to overcome restraining toque is a variable percentage of the restraining current, having a higher percentage at higher currents, Figure 8.7c.

## 8.4 HIGH IMPEDANCE DIFFERENTIAL RELAYS

Electromechanical single-phase high impedance relays have been popular in the industry for the last 50 years. Standard bushing type CTs are used, and the protection can be easily extended if additional circuits are added. The CTs are normally dedicated and are of the same ratio and accuracy classification (see Chapter 12).

A basic circuit of the high impedance differential relay is shown in Figure 8.8, and a general equivalent circuit is illustrated in Figure 8.9. Consider that there are $n$ circuits to be protected in the differential zone, only three circuits are shown in Figure 8.8. The circulating current $I_1$ on one side of the differential relay 87 in Figure 8.9 is driven by all the nonfaulted CTs. There, magnetizing circuit impedance is shown as $X_m/(n-1)$; and series impedance, including lead resistances, is shown as $R_{CT}/(n-1)$, assuming that all CTs are identical. The faulted CT resistance and magnetizing impedance is on the right side of the differential relay 87, $R_{CTF}$ and $X_{mF}$, respectively. Thus, the currents $I_1$ and $I_2$ on the two sides of the relay cannot be equal. If no resistance is introduced in the relay coil circuit, as shown in the Figure 8.9, the transient currents will be high and not much voltage will be developed across the relay coil.

A series resistance in the relay coil is introduced, which reduces the current through the coil to milliampere range. A nonlinear resistor (thyrite) is used to limit the overvoltage that can be developed across the relay during internal faults to safe values. The thyrite blocks will dissipate energy during an internal fault, and to protect these from thermal damage, lockout relay (which trips and locks out the required circuit breakers to remove the fault) contact is connected as shown. As the relay picks up, it short-circuits the thyrite blocks. The relay also has a high set element, 87H.

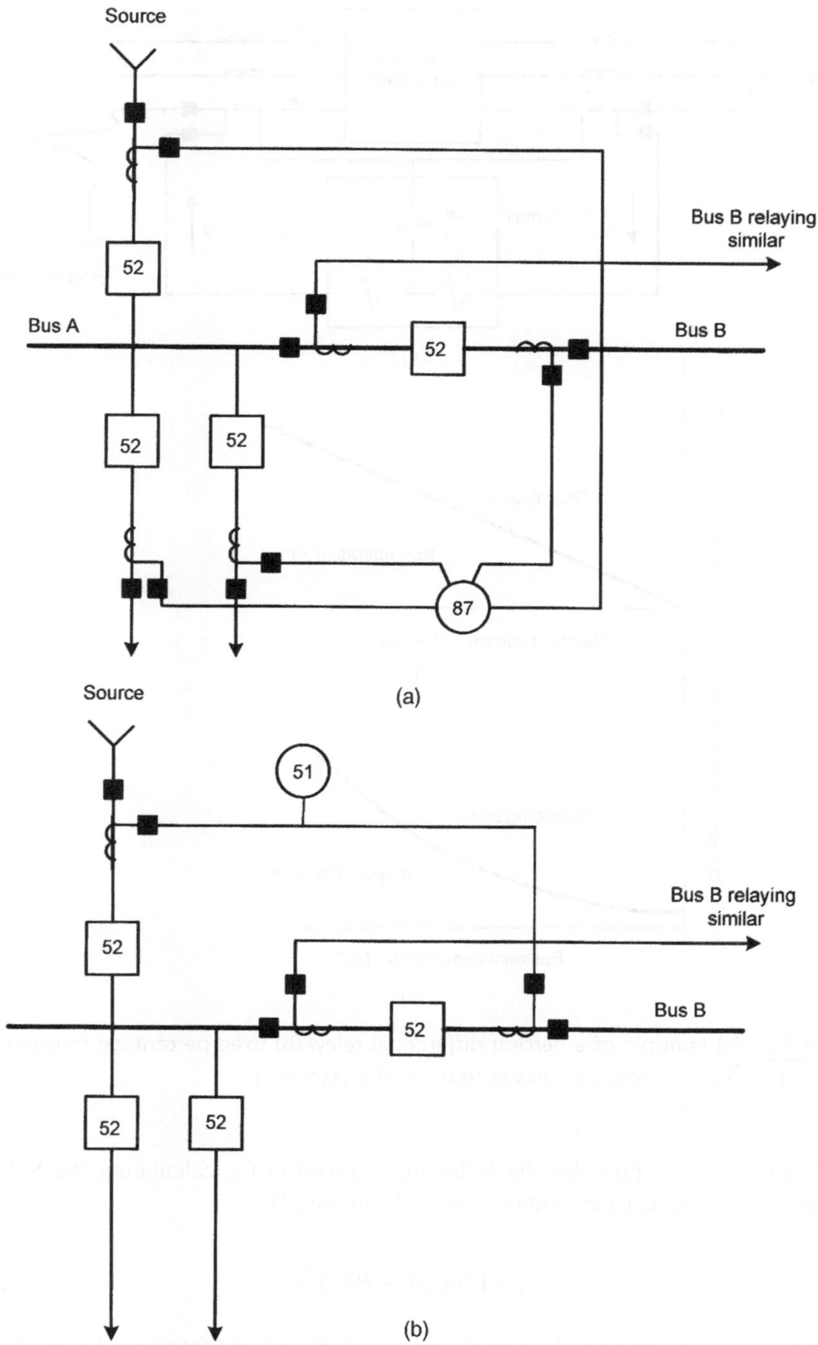


Figure 8.6. (a) A differential protection using a simple overcurrent relay; (b) partial differential protection. See text.

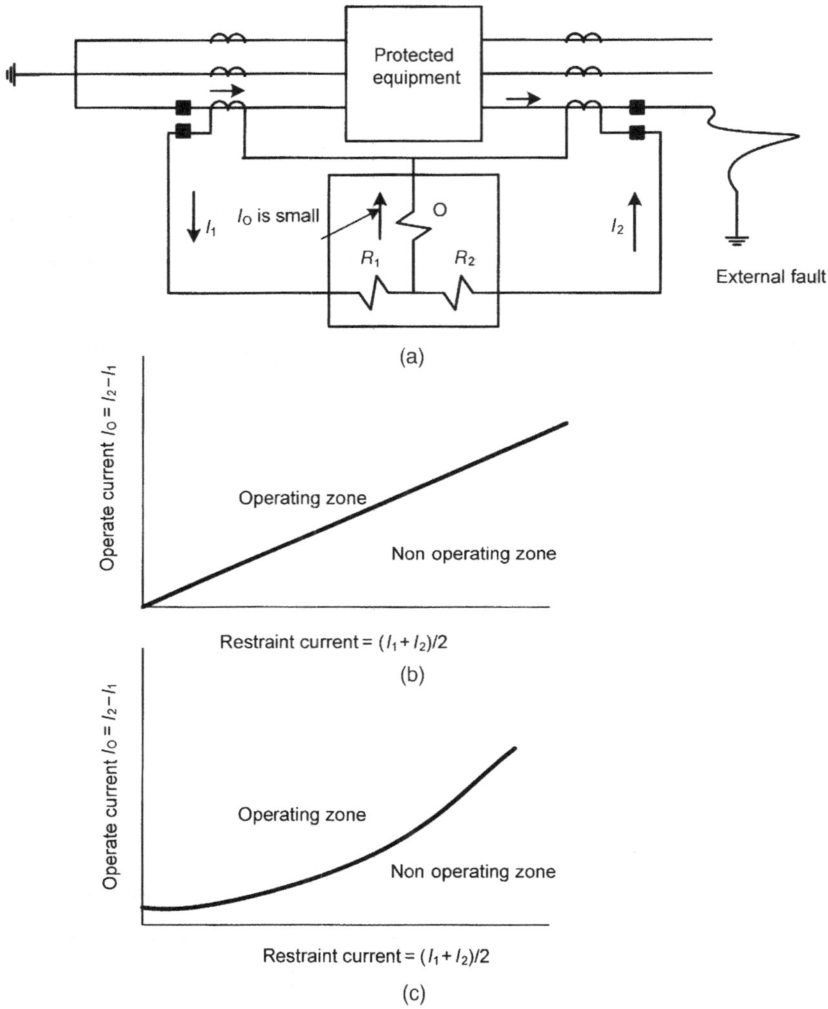


Figure 8.7. (a) Principle of a percent differential relay, (b) fixed percentage restraint characteristics, and (c) variable percentage restraint characteristics.

A manufacturer provides the following expression for calculating the 87L unit setting, considering CT performance and CT burden [4]:

$$V_R = 1.6K(R_s + PR_L)\frac{I_F}{N}, \tag{8.1}$$

where $V_R$ = voltage developed across the relay coil, $K$ = CT performance factor, $R_s$ = CT secondary winding and lead resistance, $R_L$ = one way cable resistance from the junction point to CT, $I_F$ = primary fault current, $N$ = CT ratio and $P$ = 1 for three-phase faults, and = 2 for single line-to-ground faults. If the calculated $V_R$ remains below the knee point voltage of the CT, the application is acceptable. A value higher than the knee point voltage is acceptable if the calculated $V_R$ is below 0.67 times the excitation voltage at 10 A excitation of the CT. A CT magnetization curve is required for the application. A reader may like to familiarize with CT characteristics before proceeding with this section (Chapter 12).

The CT performance factor is given by a graph provided by the manufacturer, not reproduced here. To ascertain K-factor, first calculate:

$$\frac{(R_S + PR_L)I_F}{E_S N}, \tag{8.2}$$

where $E_s$ is the CT saturation voltage. For numerical values between 0.8 and 2.0 given by Equation (8.2), K = 0.7. Between the numerical values 0 to 0.8, it increases linearly to 1.0.

### 8.4.1 Sensitivity for Internal Faults

The sensitivity of the relay for internal faults is of consideration. Again, a manufacturer provides the following relation:

$$I_{min} = \left[\sum_{X=1}^{n}(I)X + I_R + I_1\right]N, \tag{8.3}$$

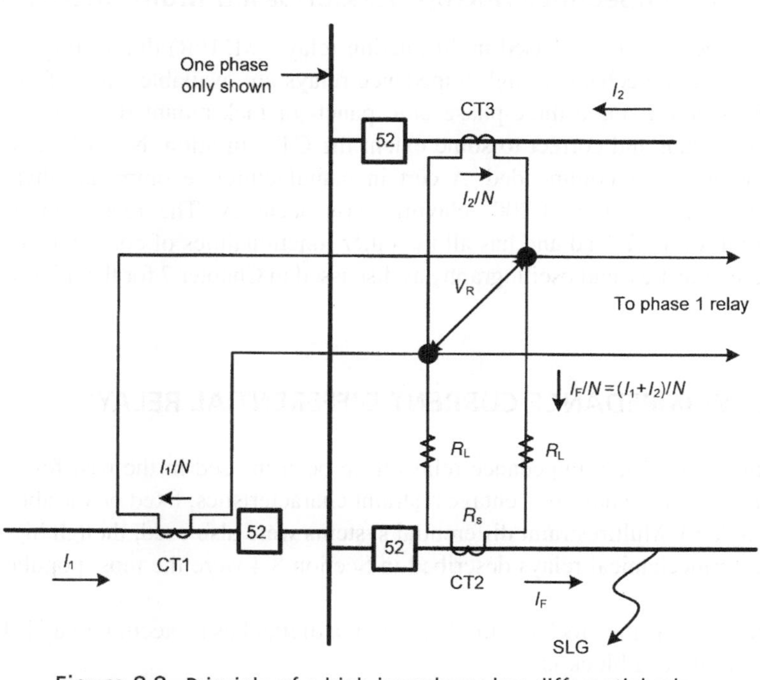


Figure 8.8. Principle of a high impedance bus differential relay.

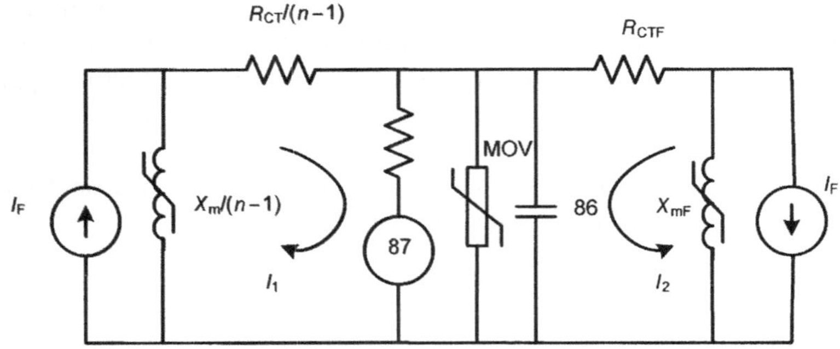


Figure 8.9. Equivalent circuit of a bus differential relay for an external fault (see text).

where $I_R$ = current in the 87L unit at pickup voltage, $I_1$ = current in thyrite unit at pickup voltage, $I_{min}$ = minimum internal fault current to trip 87L, and $I$ = secondary excitation current of the individual CT at the pickup of 87L.

The calculation will demonstrate that the sensitivity is dependent upon the voltage setting and the number of circuit breakers being protected. It can be of the order of 200–400 A. *Thus, for resistance grounded system where the ground fault current can be limited to 200–400 A, separate ground fault differential protection is required.*

Figure 8.10 shows the three-phase circuit connections of the relays.

### 8.4.2 High Impedance Microprocessor-Based Multifunction Relays

Recently, microprocessor-based multifunction relays (MMPR) duplicating the performance of electromechanical high impedance relays are available. Apart from smaller dimensions and a single three-phase unit, panel- or rack-mounted, these incorporate filters to account and correct to some extent the CT saturation. Nevertheless, varying CT ratios are not recommended. A certain manufacturer recommends that the CTs should have a minimum C200 relaying class accuracy. The relay can detect the open circuit of a CT lead and has all the other functionalities of communication, fault capture, diagnostics, and oscillography as discussed in Chapter 7 for the microprocessor relays.

## 8.5 LOW IMPEDANCE CURRENT DIFFERENTIAL RELAYS

Electormechanical low-impedance relays have been in used in the past for protection of generators. These have percentage restraint characteristics, fixed or variable restraint (Figure 8.7b,c). Multirestraint differential systems were also used, though high impedance electormechanical relays described in Section 8.4 were the most popular and are still in use.

Figure 8.11 is the block circuit diagram of a digital bus protection relay [5]. A brief description of each block is:

**Example 8.1**

Consider a system fault current of 40 kA, five-cycle symmetrical breakers, CT ratio = 1200/5, $R_s$ = 0.61 Ω, and the CT lead resistance corrected for temperature at 50°C = 0.548 Ω. For this example, consider that the CT characteristics shown in Figure 12.1 for 1200/5 CT are applicable.

For each circuit breaker being protected, the $V_R$ is separately calculated and then the maximum value used. Alternatively, $V_R$ is calculated using $P$ = 2, even for three-phase faults.

To apply Equation (8.1), first the factor $K$ given by Equation (8.2) is calculated. The CT saturation voltage from Figure 12.1 = 300 V. Then:

$$\frac{(0.61+2\times0.548)\times40\times10^3}{300\times240}=0.94.$$

Therefore, $K$ = 0.7. Calculate $V_R$ from (8.1)

$$V_R=\frac{0.7\times1.6\times(0.61+2\times0.548)\times40\times10^3}{240}=318\text{ V}.$$

This exceeds the CT saturation voltage of 300 V. Apply the second criteria of application and calculate the CT voltage at 10 A excitation from Figure 12.1; it is 495 V. Then, 0.67 multiplied by 495 gives 331.7 V, which is higher than the calculated value of 318 V. Therefore, the application is acceptable.

From Equation (8.1), it is obvious that the $V_R$ can be reduced by increasing the CT ratio. With higher CT ratios, higher relaying C-class accuracy is obtained, but as the CT ratio rises so will its secondary resistance (see Figure 12.1). If the voltage setting is above the knee-point voltage, a more accurate calculation of the excitation current is required. The settings above knee-point voltage up to the C rating of the CT are generally acceptable (see Chapter 12).

- *Block 1*: Currents are digitally filtered to remove DC components and distortions.
- *Block 2*: A dynamic bus replica is created. The filtered input signals are brought to a common scale, taking into account the transformation ratios of the connected CTs.
- *Blocks 3*: The phasors of differential zone currents are estimated digitally.
- *Blocks 4 and 5*: Restraining and differential signals are calculated.
- *Block 6*: The magnitudes of the differential signals are compared with a threshold and an appropriate flag indicating operation of the unbiased bus differential protection is produced.
- *Blocks 7 and 8*: The magnitudes of differential and restraining currents are compared and two auxiliary flags that correspond to two separately shaped portions of differential characteristics, DIF1 and DIF2, are produced.

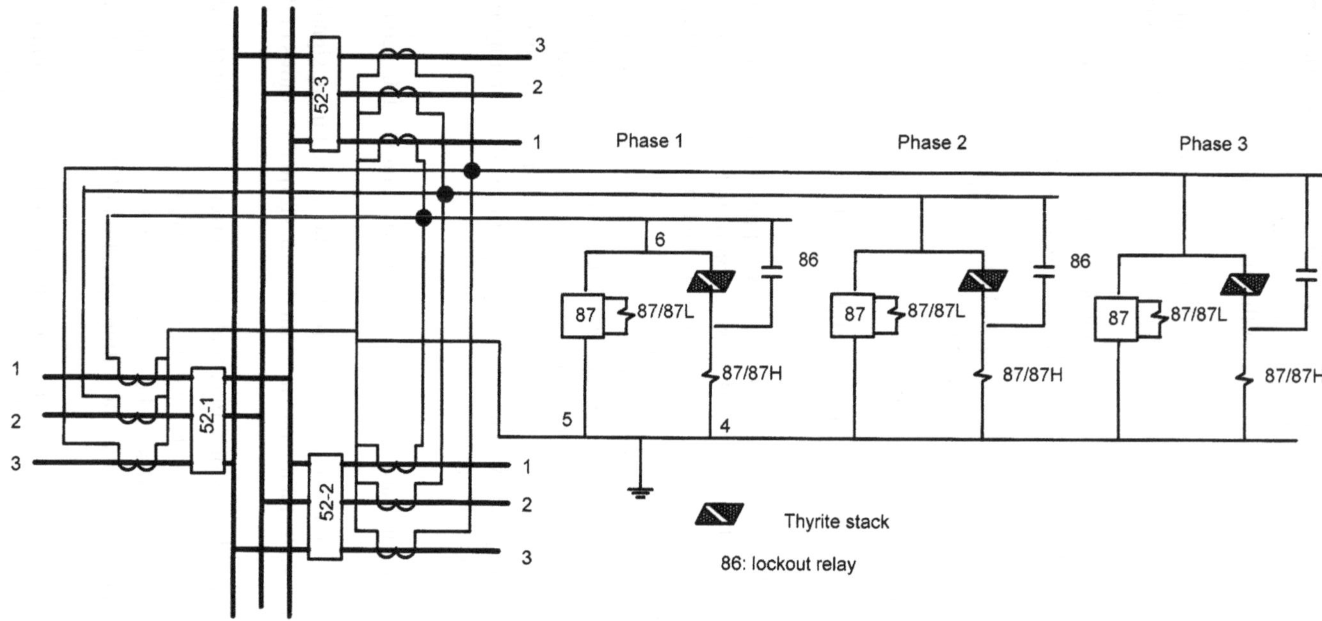


Figure 8.10. Connections of a three-phase differential relay. *Source*: Reference [4]. © General Electric Company.

- *Block 9*: The saturation detector analyzes the differential and restraining currents, as well as samples of the input currents. This block sets up an output flag upon detecting CT saturation.
- *Block 10*: Directional element block 10 supervises the biased differential characteristics as necessary. The current directional principal is used.
- *Block 11*: The output block 11 combines the differential, directional, and saturation flags into biased differential operation flag.

The relay can accommodate CTs of different ratios, and uses a dual-breakpoint operating characteristics as shown in Figure 8.12. A low pickup setting is provided to cope with the spurious signals when the bus carries a light load, and there is no effective restraining signal. The first breakpoint (low breakpoint) specifies the linear operation of the CTs in the most unfavorable conditions of residual magnetism left in the cores or multiple autoclosure shots. The second breakpoint is provided to specify the limits of operation of CTs without any substantial saturation. The higher slope acts as a percentage bias regardless of the value of restraining signal.

The directional comparison principal is that:

- If all the fault currents flow in one direction, the fault is internal.
- If at least one fault current flows in an opposite direction compared with the sum of remaining currents, the fault is external.

At first, based upon the magnitude of fault current, it is determined whether it is fault current or load current, for example, if its magnitude is >2 times the CT rating or

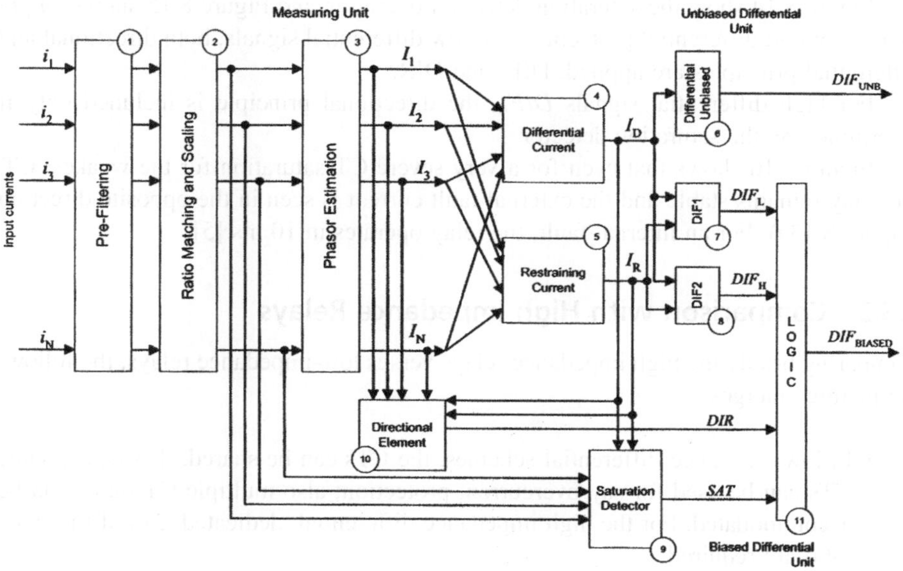


Figure 8.11. Block circuit diagram of a modern multi-function, microprocessor based relay. *Source*: Courtesy of General Electric Company.

>$K$ times the restraint current, where $K$ varies with the number of feeders, it can be flagged as fault current.

Second, for the selected fault current, the phase angle between a given current and the sum of all the other currents is checked. The phase angle check is not initiated for the load current, as the direction will be out of the bus even during internal faults. Ideally, for external faults this angle is 180°, and for internal fault, it is 0°. The phase angle between a given current and the sum of all the remaining currents is checked. The sum of all the remaining currents is the differential current minus the current under consideration, say for a $p$th path, the angle between $I_P$ and $I_D$–$I_P$ phasors is checked (see Figure 8.13a,b).

## 8.5.1 CT Saturation

A CT has a certain finite time constant to saturate. Thus, even under high primary currents, it will take a finite time for the CT to saturate. As a result, for an external fault, the differential current will be low during initial period of the operation of the CT, while the restraint signal develops rapidly. Once the CT saturate, the differential current will increase, but restraint current does not change at least for a few milliseconds. For internal faults, both the differential and restraining currents develop simultaneously. The relay declares CT saturation if the restraining signal is higher than the second breakpoint, and at the same time, differential current is below the low slope. In order to cope with fast saturation, another condition is checked, which uses signals at the waveform level. The sample-based stage of the saturation detector uses a time derivative of the restraining signal *di/dt* to trace the saturation pattern, and the saturation detector is capable of detecting saturation occurring in approximately 2 ms into a fault.

Figure 8.14 show the saturation detection operation and Figure 8.15 shows output logic of biased differential protection. For low differential signals, both directional and differential principles are applied, $DIF_L$ and DIR.

For high differential signals $DIF_H$, the directional principle is included only if determined by the saturation detector

Figure 8.16 shows that even for a very severe CT saturation for the weakest CT, the relay remains stable and the external fault current is seen in the opposite direction (Figure 8.13a). For an internal fault, the relay operates in 10 ms [5].

## 8.5.2 Comparison with High Impedance Relays

Comparing briefly the high impedance relays versus low-impedance relays, the following picture emerges:

- In low-resistance differential schemes, the CTs can be shared, that is, the same CTs can be used for say overcurrent protection; also multiple CT ratios can be accommodated. For the high impedance differential, dedicated CTs of the same ratio are required.
- It is possible to detect a shorted CT in low-resistance differential, but not so in high-resistance differential. On an open circuit CT, the high impedance

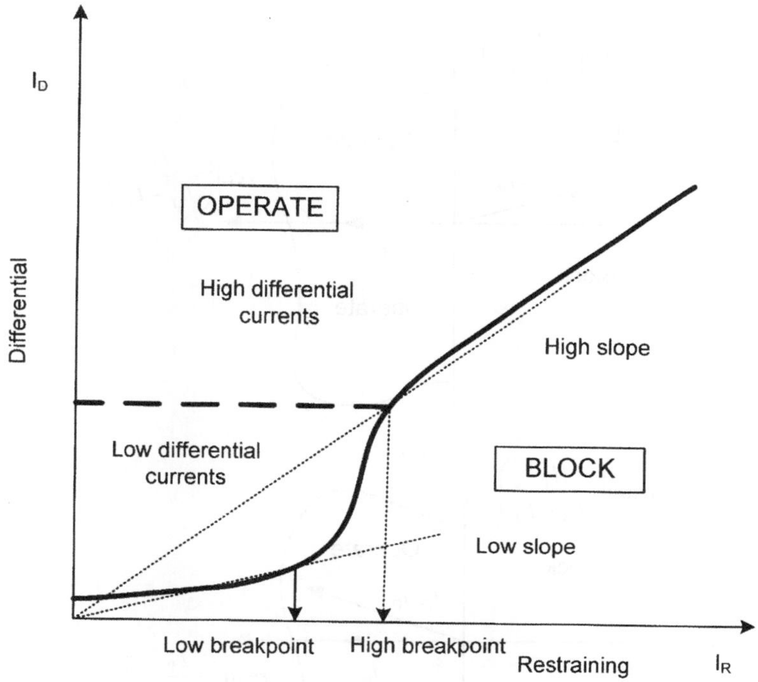


Figure 8.12. Two-slope characteristics of the relay in Figure 8.11, with low and high break points, as shown. *Source*: Courtesy of General Electric Company.

differential relay will trip, and a low-resistance differential relay can alarm the situation on unbalance.

- CT polarity compensation can be provided in low-resistance differential schemes, but not in high resistance differential relays.
- The low-resistance differential schemes are somewhat faster as compared with high resistance differential schemes, one cycle typical, versus approximately 1.5 cycle.
- A low-resistance differential scheme, generally, has selective circuit breaker failure protection, selective end-zone protection, direct circuit breaker tripping, and individual circuit metering facilities

See also References [6, 7].

## 8.6 ELECTROMECHANICAL TRANSFORMER DIFFERENTIAL RELAYS

When applying differential protection to transformers, additional considerations are:

- There is a phase shift in three-phase transformer windings (see Chapter 11).

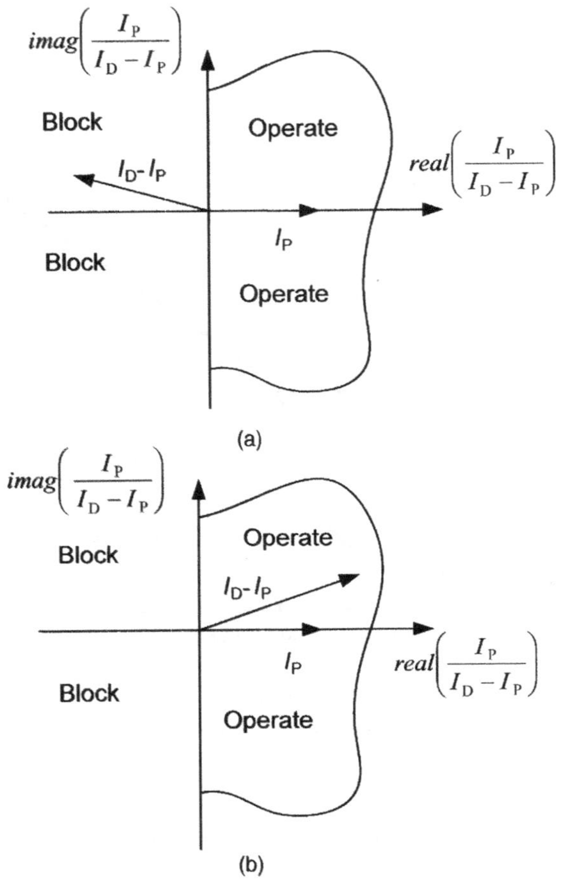


Figure 8.13 (a) Directional operating principle during external faults; (b) directional operating principle during internal faults.

- The CT ratios on the primary and secondary sides of the transformers cannot be exactly matched with respect to primary and secondary currents.
- The transformer inrush currents contain harmonics, particularly second and fifth harmonics.

Figure 8.17 illustrates the circuit of a harmonic restraint electromechanical relay. Differential relays without harmonic restraint have also been used for protection of transformers, but are no longer in use. The figure shows the CT connections on the high and low voltage side for a delta-wye-connected transformer. The CTs on the wye side are connected in delta, and those at the delta side are connected in wye for proper phasing. In case the CTs on the wye-grounded side are connected in wye, then for an external ground fault, the current will flow in the relay operating coil and the relay will operate. Delta-connected CTs block this zero sequence current. The phasing can be

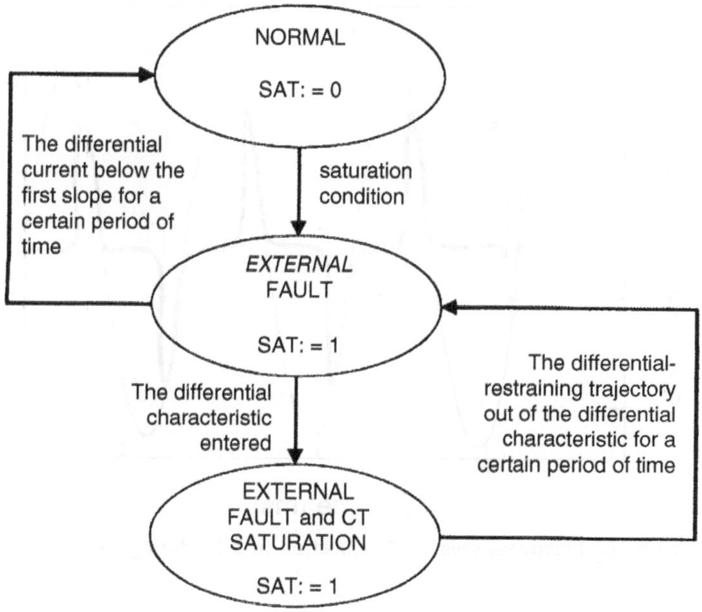


**Figure 8.14.** Principle of CT saturation detection in Figure 8.11. *Source*: Courtesy of General Electric Company.

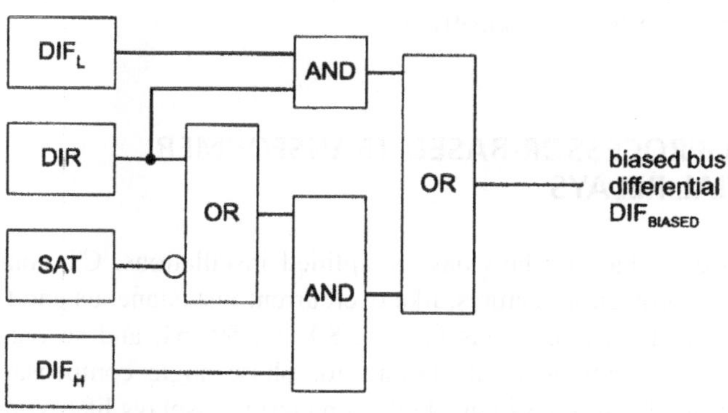


**Figure 8.15.** Output logic of biased differential protection, Figure 8.11. *Source*: Courtesy of General Electric Company.

checked for current flows in the transformer and CT windings, as shown in this figure. A three-phase external fault on the high side (138 kV) is shown in this figure, 13.8-kV side may be connected to a source like a generator. Next, a ratio check is made, and the mismatch should be reduced to acceptable limits.

### 8.6.1 Harmonic Restraint

The harmonic restraint in the relay has a second harmonic blocking filter in the operating and restraint coils; the circuit is designed to hold the contacts open with second

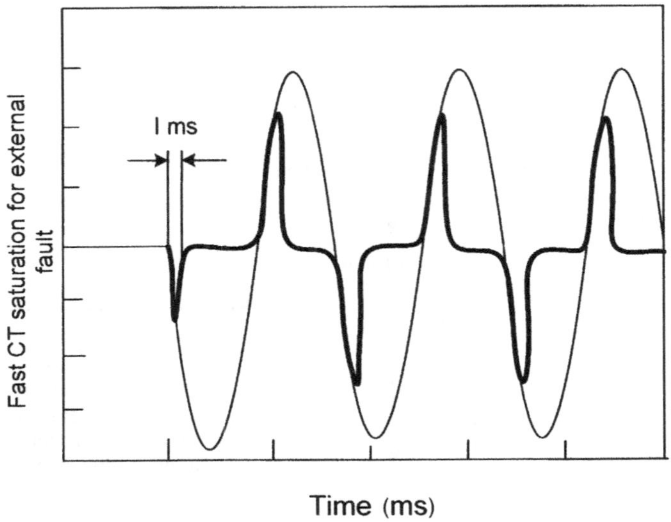


Figure 8.16. Illustrates severe CT saturation of a CT during external fault.

harmonic of 15% of the fundamental. The percentage characteristic varies from 20% on light loads to approximately 60% for heavy faults. The minimum pickup current is 30% of the tap value for 30% sensitivity.

## 8.7 MICROPROCESSOR-BASED TRANSFORMER DIFFERENTIAL RELAYS

The microprocessor-based relays have simplified installations, CT connections, and added additional protection features, like overcurrent instantaneous and time delay for phase and ground faults, functions like 24, 80, 27, 59, 64, and so on. These relays provide internal mathematical calculations for phase angle compensation and zero sequence filtering. These relays have built-in metering displays like secondary current magnitudes and phase angles, restraint and operate quantities, which are very useful during commissioning and fault diagnosis.

The dual slope characteristic is shown in Figure 8.18. The lowest portion of the trip threshold represents the minimum sensitivity of the relay. Some relays provide options to set as low as 5%; however, this may give rise to nuisance trips. Settings of 10% have held without nuisance trips with matched CTs. Some manufacturers recommend a minimum setting of no less than 30%. Also, the minimum settings are sometimes specified in terms of the calculated primary and secondary tap currents. The CT errors due to saturation are likely to be a problem at higher currents, and the steeper slope 2 characteristics add security for heavy external faults.

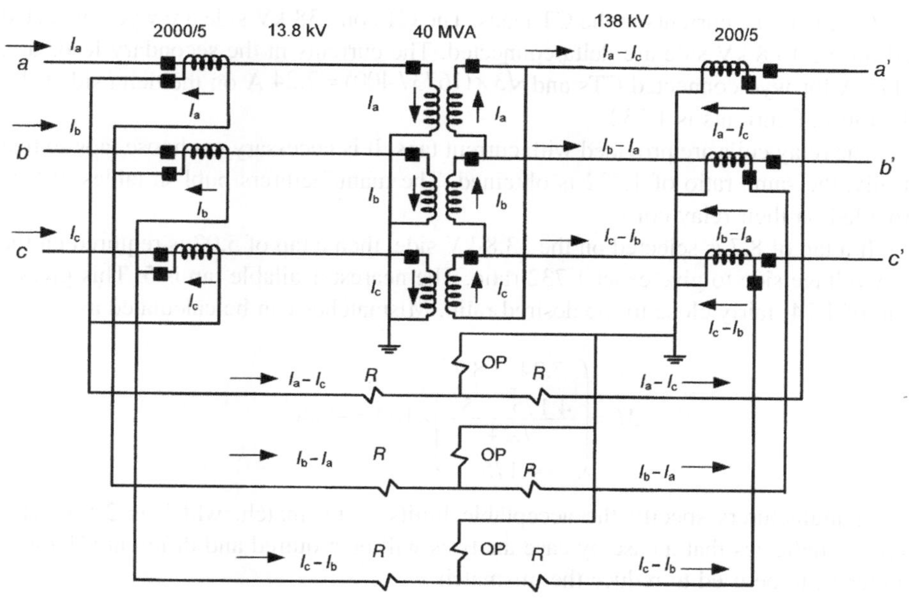


Figure 8.17. Differential protection of a 40-MVA, wye–delta connected, 138–13.8-kV transformer, showing CT connections and current flows for an external three-phase fault.

## 8.7.1 CT Connections and Phase Angle Compensation

Considering a delta–wye transformer, the CTs on the delta and wye can be both wye connected. This requires that the phase shift in the transformer windings should be corrected, and also the zero sequence current that can flow for an external ground fault should be blocked.

Some advantages of connecting CTs in wye connection are that these present a lower burden than the delta connected CTs, as much as three times less depending on fault type. With wye connected CTs, it is possible to use zero-sequence elements, like ground fault protection.

However, for the differential protection, external ground fault results in flow of currents through CTs without corresponding current flows through the transformer delta winding side CTs. To prevent relay operation, this zero sequence current

**Example 8.2**

A 40-MVA transformer, 138–13.8-kV, delta primary (138 kV), and wye-grounded secondary windings (13.8 kV) transformer is assumed for this example.

Calculate the full load currents on the primary and secondary sides. The 138-kV side full load current is 167 A, and on 13.8-kV side, it is 1673 A.

A CT ratio close to these currents is selected. Select a 200/5 CT on the 138 kV windings and 2000/5 CT on the 13.8 kV windings.

Calculate the currents in the CT leads. The CTs on 138 kV side are wye connected and on the 13.8 kV side are delta connected. The currents in the secondary leads are, 4.175 A for wye-connected CTs and $\sqrt{3}\times(1673/400)=7.24$ A on the delta side CTs. The ratio of currents is 1.732.

The relay coils are provided with current taps. It is necessary to choose taps so that ideally, the same ratio of 1.732 is obtained. The manufacturers publish tables of taps provided on their relay coils.

If a tap of 8.7 is selected on the 13.8-kV side, then a tap of 5.02 is required on the high voltage side to give exact 1.732 ratio. The nearest available tap is 5. This gives a ratio of 1.74, fairly close to the desired ratio. Mismatches can be calculated as:

$$M=\left(\frac{\dfrac{7.24}{4.175}-\dfrac{8.7}{5}}{\dfrac{7.24}{4.175}}\right)\times 100=-0.46. \tag{8.4}$$

The manufacturers specify the acceptable limits of mismatch, which is 2.5%. This exercise indicates that a case by case analysis will be required and different CT ratios and taps can be tried to reduce the mismatch.

*Considerations of CT accuracy*

With respect to CT accuracy, the following equation is provided by one manufacturer:

$$\left[\frac{N_p V_{CL}-(I_{ext}-100)R_s}{I_{ext}}\right]>Z_t, \tag{8.5}$$

where

$N_p$ = proportion of total CT turns in use for multiratio CTs, as the relaying accuracy is specified for the maximum turns ratio and should be proportionate if a tap on the CT winding is used (see Chapter 12)

$V_{CL}$ = CT C-class accuracy

$I_{ext}$ = maximum external fault current in CT secondary in rms, not <100

$R_s$ = CT secondary winding resistance.

$Z_t$ = total transformer secondary burden.

This can be calculated for both primary and secondary side faults.

The expressions of $Z_t$, assuming that the CTs, are dedicated for the differential protection are:

$$\begin{aligned}
&Z_t=1.13R_L+\text{relay burden}+Z_A && \text{Three-phase fault}\\
&Z_t=1.13(2R_L)+\text{relay burden}+Z_A && \text{Phase-to-ground fault}\\
&Z_t=2(1.13R_L+\text{relay burden}+Z_A) && \text{Phase-to-ground fault, delta-connected CTs.}
\end{aligned} \tag{8.6}$$

$R_L$ is one-way lead resistance, and $Z_A$ is burden of any other device connected to the CT.

**Example 8.3**

In Example 8.2, the CTs have a C-class accuracy of 100. Calculate if this is adequate for the application.

Continuing with Example 8.2, if the transformer has a percentage impedance of 9% on 40 MVA base, and the 138 kV source has a short-circuit current of 40 kA, then the through fault current on 13.8-kV side can be calculated = 15 kA approximately. Reflected in the CT leads, it is = 64.95 A (CT ratio 400, delta connected).

As this is less than 100, we consider $I_{ext}$ = 100. CT Class C accuracy = 100, CT secondary resistance = 1 5Ω. $N_p$ = 1, as it is the single ratio transformer of 2000/5. Substituting the values, $Z_t$ should be <1 Ω.

should be removed. This can be done by selecting correct phase angle compensation, which simultaneously corrects phase angle as well as compensates for zero sequence currents.

With the modern transformer differential relays, this can be done by two methods:

1. A differential relay will accept the actual transformer winding connection phasors directly. Figure 8.19 illustrates some phasor diagrams and vector groups for two-winding and three-winding transformers. Depending upon the transformer vector group, it is adequate to input that group.

   In a three-phase two winding transformer manufactured according to ANSI/IEEE standards the low-voltage side, whether in wye or delta connection, has a phase shift of 30° lagging with respect to phase-to- neutral voltage vectors on the high side, or the high side phasors lead the low-side phasors by 30°. These transformers are called Dy1 or Yd1. The letter 1 indicates that the low side lags the high side by 1 hour on the clock. See Chapter 11 for further discussions. A user can directly program the right connection into the relay.
2. For some other differential relays, a connection matrix is used, which corrects the phase shift and compensates for the zero sequence. Some connection matrices are shown in Figure 8.20. Phase rotation must be considered while selecting the appropriate matrix for the compensation.

   Consider that we have a Dy1 transformer, phase rotation ABC. First, the phasors with proper phase rotation on the high side of the transformer to represent high side currents are drawn. Also, consider phase to bushing connections. When we choose appropriate matrix from Figure 8.20, the compensated currents produce zero operate current for external faults. If the primary winding phase current is taken as a reference, the compensation matrix is a unity matrix, because the transformer primary winding is in delta, and there is no zero sequence source current between the CT and transformer windings. Next, move the low side vector in counterclockwise direction till it coincides with the vector on the primary side. A Dy1 or Yd1 connection requires a 30° movement in the counterclockwise direction. Thus, matrix $M_1$ is selected.

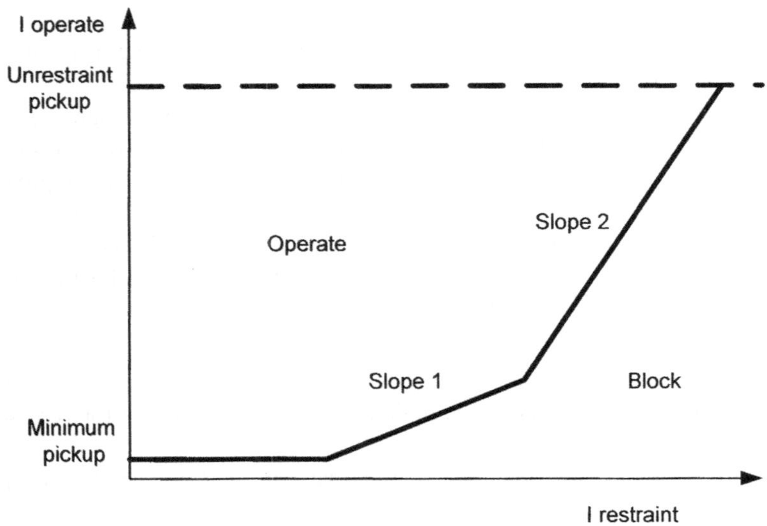


Figure 8.18. A two-slope restraint characteristics of a modern transformer differential relay.

$$\overline{I}_{abc\text{-comp}} = \frac{1}{\sqrt{3}}\begin{vmatrix} 1 & -1 & 0 \\ 0 & 1 & -1 \\ -1 & 0 & 1 \end{vmatrix}\begin{vmatrix} I_a \\ I_b \\ I_c \end{vmatrix}(\ )$$

or

$$I_{a\text{-comp}} = \frac{(I_a - I_b)}{\sqrt{3}} \tag{8.7}$$

$$I_{b\text{-comp}} = \frac{(I_b - I_c)}{\sqrt{3}}$$

$$I_{c\text{-comp}} = \frac{(I_c - I_a)}{\sqrt{3}}.$$

This produces compensated secondary currents which are 180° out of phase with the primary currents. Operate or difference current is therefore zero. Matrix $M_{12}$ filters the zero-sequence currents without any phase corrections. See also References [8–11].

## 8.7.2 Dynamic CT Ratio Corrections

We noted in Section 8.6 for electromechanical relays that the CTs on primary and secondary side must be carefully selected, and also proper taps on the relay must be selected to minimize mismatch. With the MMPR, this is not required. A differential relay will accommodate automatically a certain mismatch in the ratio, and only the CT ratios on the primary and secondary sides of the transformer are input to the relay. A

| Transformer Type | Wdg # | Connection | Voltage Phasors | Phase Shift |
|---|---|---|---|---|
| D/d300° | 1 | DELTA | 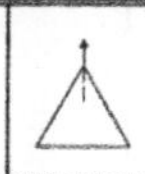 | 300° lag |
| | 2 | DELTA 300° lag |  | 0° |
| D/y30° | 1 | DELTA | 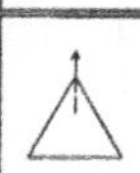 | 0° |
| | 2 | WYE (gnd 1/2) 30° lag | 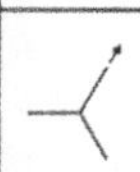 | 330° lag |
| D/y150° | 1 | DELTA | 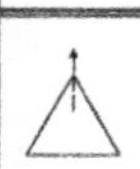 | 0° |
| | 2 | WYE (gnd 1/2) 150° lag | 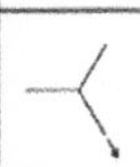 | 210° lag |
| D/y210° | 1 | DELTA | 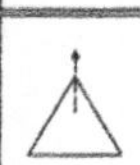 | 0° |
| | 2 | WYE (gnd 1/2) 210° lag | 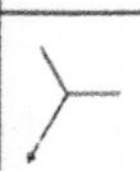 | 150° lag |
| D/y330° | 1 | DELTA |  | 0° |
| | 2 | WYE (gnd 1/2) 330° lag | 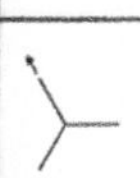 | 30° lag |
| Y/z30° | 1 | WYE (gnd 1/2) | 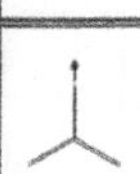 | 30° lag |
| | 2 | ZIG ZAG (gnd 2/3) 30° lag | 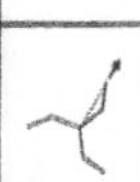 | 0° |

| Transformer Type | Wdg # | Connection | Voltage Phasors | Phase Shift |
|---|---|---|---|---|
| Y/z150° | 1 | WYE (gnd 1/2) | 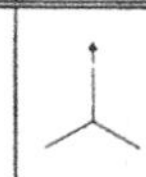 | 150° lag |
| | 2 | ZIG ZAG (gnd 2/3) 150° lag | 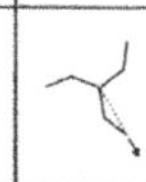 | 0° |
| Y/z210° | 1 | WYE (gnd 1/2) | 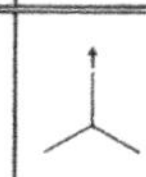 | 210° lag |
| | 2 | ZIG ZAG (gnd 2/3) 210° lag | 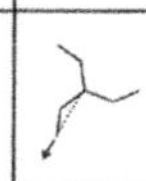 | 0° |
| Y/z330° | 1 | WYE (gnd 1/2) | 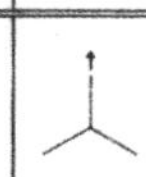 | 330° lag |
| | 2 | ZIG ZAG (gnd 2/3) 330° lag | 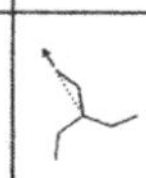 | 0° |
| D/z0° | 1 | DELTA | 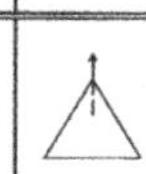 | 0° |
| | 2 | ZIG ZAG (gnd 1/2) 0° lag | 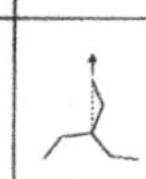 | 0° |
| D/z60° | 1 | DELTA | 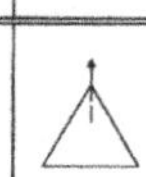 | 60° lag |
| | 2 | ZIG ZAG (gnd 1/2) 60° lag | 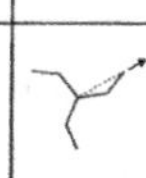 | 0° |
| D/z120° | 1 | DELTA |  | 120° lag |
| | 2 | ZIG ZAG (gnd 1/2) 120° lag | 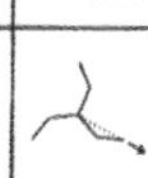 | 0° |

(a)

Figure 8.19. (a) Phase shifts in two-winding transformers; (b) phase shifts in three-winding transformers.

| Transformer Type | Wdg # | Connection | Voltage Phasors | Phase Shift |
|---|---|---|---|---|
| Y/y180°/d150° | 1 | WYE (gnd 1/2) | 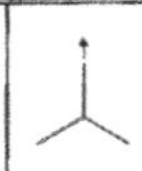 | 150° lag |
| | 2 | WYE (gnd 2/3) 180° lag | 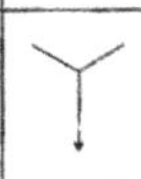 | 330° lag |
| | 3 | DELTA 150° lag | 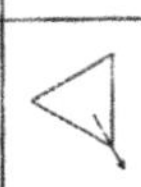 | 0° |
| Y/y180°/d210° | 1 | WYE (gnd 1/2) | 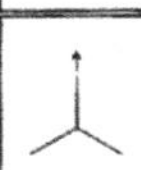 | 210° lag |
| | 2 | WYE (gnd 2/3) 180° lag | 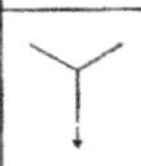 | 30° lag |
| | 3 | DELTA 210° lag | 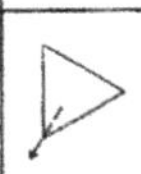 | 0° |
| Y/y180°/d330° | 1 | WYE (gnd 1/2) | 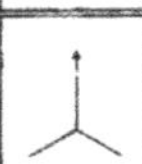 | 330° lag |
| | 2 | WYE (gnd 2/3) 180° lag | 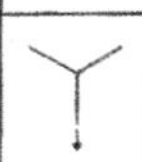 | 150° lag |
| | 3 | DELTA 330° lag | 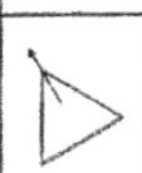 | 0° |
| Y/d30°/y0° | 1 | WYE (gnd 1/2) | 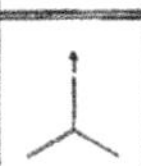 | 30° lag |
| | 2 | DELTA 30° lag |  | 0° |
| | 3 | WYE (gnd 2/3) 0° | 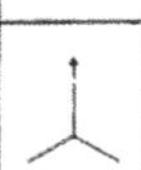 | 30° lag |

| Transformer Type | Wdg # | Connection | Voltage Phasors | Phase Shift |
|---|---|---|---|---|
| Y/d30°/y180° | 1 | WYE (gnd 1/2) | 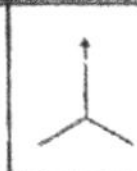 | 30° lag |
| | 2 | DELTA 30° lag | 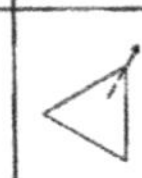 | 0° |
| | 3 | WYE (gnd 2/3) 180° lag | 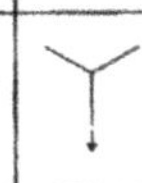 | 210° lag |
| Y/d30°/d30° | 1 | WYE (gnd 1/2) | 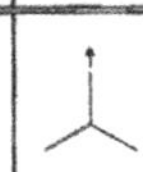 | 30° lag |
| | 2 | DELTA 30° lag | 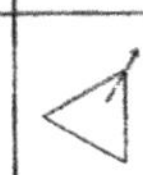 | 0° |
| | 3 | DELTA 30° lag | 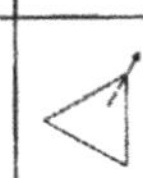 | 0° |
| Y/d30°/d150° | 1 | WYE (gnd 1/2) | 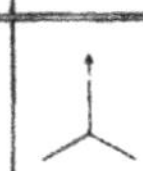 | 30° lag |
| | 2 | DELTA 30° lag | 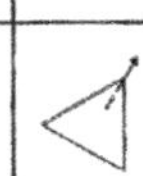 | 0° |
| | 3 | DELTA 150° lag |  | 240° lag |
| Y/d30°/d210° | 1 | WYE (gnd 1/2) | 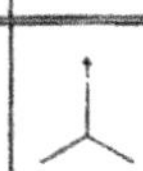 | 30° lag |
| | 2 | DELTA 30° lag |  | 0° |
| | 3 | DELTA 210° lag | 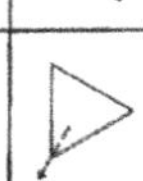 | 180° lag |

(b)

Figure 8.19. (*Continued*)

$$\overline{M}_1 = \frac{1}{\sqrt{3}}\begin{vmatrix} 1 & -1 & 0 \\ 0 & 1 & -1 \\ -1 & 0 & 1 \end{vmatrix}$$

$$\overline{M}_2 = \frac{1}{3}\begin{vmatrix} 1 & -2 & 1 \\ 1 & 1 & -2 \\ -2 & 1 & 1 \end{vmatrix}$$

$$\overline{M}_3 = \frac{1}{\sqrt{3}}\begin{vmatrix} 0 & -1 & 1 \\ 1 & 0 & -1 \\ -1 & 1 & 0 \end{vmatrix}$$

$$\overline{M}_4 = \frac{1}{3}\begin{vmatrix} -1 & -1 & 2 \\ 2 & -1 & -1 \\ -1 & 2 & -1 \end{vmatrix}$$

$$\overline{M}_5 = \frac{1}{\sqrt{3}}\begin{vmatrix} -1 & 0 & 1 \\ 1 & -1 & 0 \\ 0 & 1 & -1 \end{vmatrix}$$

$$\overline{M}_6 = \frac{1}{3}\begin{vmatrix} -2 & 1 & 1 \\ 1 & -2 & 1 \\ 1 & 1 & -2 \end{vmatrix}$$

$$\overline{M}_7 = \frac{1}{\sqrt{3}}\begin{vmatrix} -1 & 1 & 0 \\ 0 & -1 & 1 \\ 1 & 0 & -1 \end{vmatrix}$$

$$\overline{M}_8 = \frac{1}{3}\begin{vmatrix} -1 & 2 & -1 \\ -1 & -1 & 2 \\ 2 & -1 & -1 \end{vmatrix}$$

$$\overline{M}_9 = \frac{1}{\sqrt{3}}\begin{vmatrix} 0 & 1 & -1 \\ -1 & 0 & 1 \\ 1 & -1 & 0 \end{vmatrix}$$

$$\overline{M}_{10} = \frac{1}{3}\begin{vmatrix} 1 & 1 & -2 \\ -2 & 1 & 1 \\ 1 & -2 & 1 \end{vmatrix}$$

$$\overline{M}_{11} = \frac{1}{\sqrt{3}}\begin{vmatrix} 1 & 0 & -1 \\ -1 & 1 & 0 \\ 0 & -1 & 1 \end{vmatrix}$$

$$\overline{M}_{12} = \frac{1}{3}\begin{vmatrix} 2 & -1 & -1 \\ -1 & 2 & -1 \\ -1 & -1 & 2 \end{vmatrix}$$

Figure 8.20. Transformer winding phase-shift correction matrices, CTs on both sides of the transformer connected in wye-configuration.

particular relay can accommodate a mismatch in the primary and secondary CT ratios, up to a mismatch factor of 16. The CT currents are scaled with respect to the rating of the transformer to reflect the load currents in the windings.

### 8.7.3 Security under Transformer Magnetizing Currents

The relay takes advantage of harmonics and the DC component of the current waveform during inrush. The inrush current of a transformer is rich in second harmonics, which is of the order of 60% of the fundamental. The overexcitation of the transformer produces 3rd, 5th, and 7th harmonics, of the order of 49, 23, and 8% of the fundamental, respectively. The third harmonic is eliminated by the delta connection of the transformer windings. Thus, overexcitation can be detected with fifth harmonic.

Figure 8.21 depicts that the input currents from the CT secondary are converted into voltage signals. Low-pass filters remove high frequency components from the voltage signals. Digital filters extract the fundamental, second, and fifth harmonics from the digital signals. These are then scaled and the zero sequence components are removed. The currents are also compensated for transformer phase shift. The winding two compensated currents are similarly obtained. These are inputted to differential elements and the second and fifth harmonic blocking. The magnitude of the compensated fundamental frequency currents determines the operating quantity of the restrained differential element. The relay compares the operating quantity to the restraint quantity. 87R trip is actuated if the operating quantity is greater than restraint quantity. The relay also calculates the second and fifth harmonic content of the differential current. It compares the user-settable percentage with the operating quantity. If the harmonic percentage is greater than the operating quantity, the harmonic blocking element asserts to block the percentage restraint differential element. The unrestraint element compares the operating quantity against a settable threshold. And 87U element is asserted if operating quantity is bigger than the unrestrained element threshold. Some typical settings are:

| | |
|---|---|
| 87P restraint element pickup | = 0.50 of per unit tap setting. |
| Slope 1 | = 25% |
| Slope 2 | = 50% |
| Second harmonic blocking | = 15% |
| Fifth harmonic blocking | = 8% |

## 8.8 PILOT WIRE PROTECTION

With hard-wired differential schemes, a limitation of the distance to which the protection zone can be covered is soon reached depending upon the fault currents, CT burdens, and the long CT leads. Also, intertripping is required. Generally, the schemes can be used for distances no more than approximately 200–350 ft.

The earlier pilot wire scheme using electromechanical relays and metallic pilot wires for short lines up to approximately 20–30 mi is shown in Figure 8.22. It is a current-only system requiring no PT connections. These provide fast fault clearing for the 100% length of the protected line.

The three-phase currents at each terminal are converted into single-phase voltage, and this single phase voltage is compared with the remote terminal via a pilot wire, whether the fault is within the protected zone, or it is outside the protected zone. For the through fault conditions, the filter output voltages $V_A$ and $V_B$ have the polarity as shown. This allows them to support a circulating current flow through the restraint coils and the pilot wires, and little current flows through the operating coils. When an internal fault occurs, one of the sequence filter output voltages reverses. For the connections shown, $V_B$ will reverse its polarity, and most of the current flows through the operating coils, and the relays at both the terminals will trip. The pickup sensitivity for phase faults is of the order of 4 A for phase-to-ground faults, 0.5 A for phase-to-ground faults, and 2.5–7 A for phase-to-phase faults.

A shielded no. 19 or larger twisted pair pilot wire is required. There is a limitation of the series resistance and capacitance of the pilot wires. Extraneous voltages are of concern. These voltages arise from a rise in station ground potential by induction from power circuits, which can be reduced by properly shielding the pilot wires. Extraneous voltages will be at maximum during a ground fault, as the zero sequence currents return in the earth; mutual impedance $Z_M$ exists between the power line and the pilot wire. For the protection of the

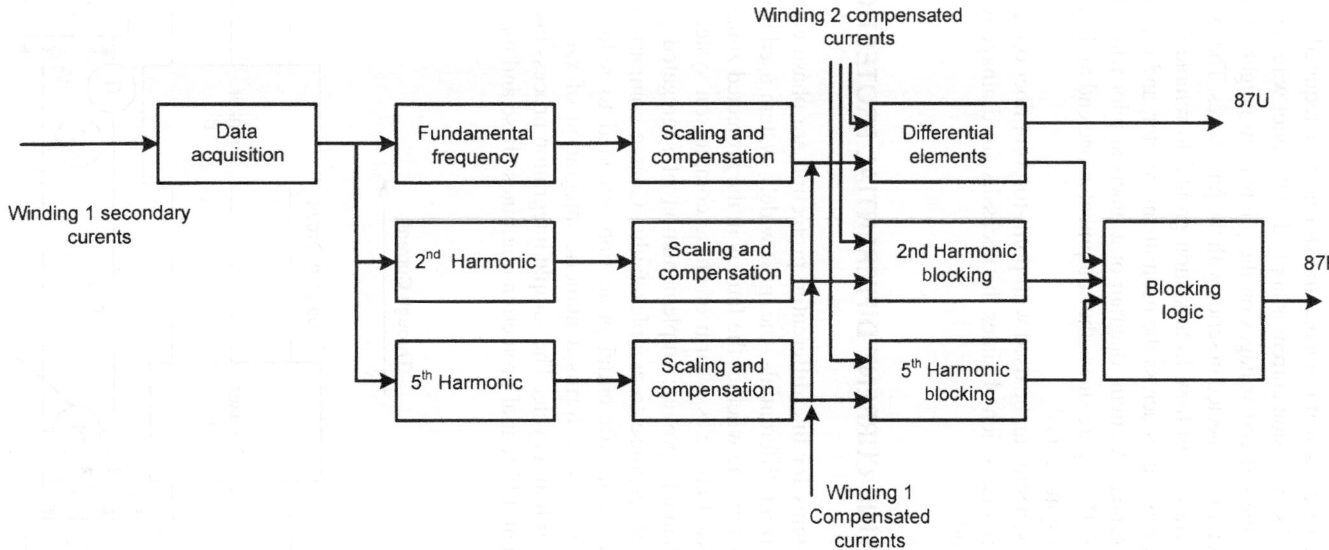


Figure 8.21. A block circuit diagram of a modern transformer differential relay, illustrating harmonic blocking.

pilot wires gaps/arresters, carbon blocks and protector tubes are employed. Also, drainage reactors, neutralizing reactors, and insulating transformers are required. To detect pilot wire faults, continuous DC supervision current is applied to the pilot wire. The drainage reactor and gas tube control the induced voltages on the pilots from neighboring power stations. The neutralizing reactor is connected in series with the pilot wires. The two small capacitors permit magnetizing currents to flow from station ground to remote ground. Most of the station rise voltage drop occurs across the neutralizing reactor, and the pilot wires remain at remote ground potential. A circuit diagram of a practical pilot wire relaying scheme is shown in Figure 8.23. This figure shows the equipment at one end of the protected line; the other terminal end is symmetrical.

This pilot wire scheme can be used with point-to-point fiber optic interface at each end and fiber optic pilots, which obviates the necessity for drainage reactors and surge arresters at either end.

## 8.9 MODERN LINE CURRENT DIFFERENTIAL PROTECTION

Unit protection systems for line differential protection are phase comparison, charge comparison, and current differential—the most widely system used is directional comparison. All these determine whether the fault is in the protected zone. The earlier pilot wire scheme discussed above is a form of phase comparison system. Modern digital communications channels provide implementation of segregated phase comparison systems that enhance response to complex faults. Charge comparison systems are an alternate form of current differential protection intended to reduce communication channel band width. This performs a numeric integration of samples of phase and residual currents over half a cycle. The sample integration process takes place between current zeros. Current differential protection combines phase and magnitude of current

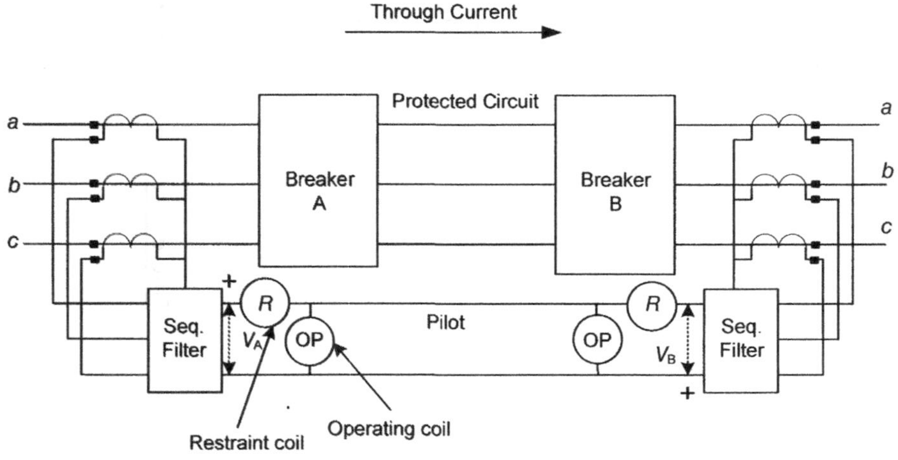


Figure 8.22. A pilot wire scheme using metallic pilots and electromechanical relays.

information in a single comparison. Microprocessor-based line current differential schemes utilize digital communication channels. These can be:

- digital microwave
- direct fiber connections
- synchronous optical network (SONET) or
- synchronous digital hierarchy (SDH).

The areas that impact the security of a line current differential system are:

- the robustness of relay hardware and firmware
- the robustness of applied algorithms and logic
- the ability to deal with channel impairment
- self-monitoring.

Figure 8.24 shows the interconnection of relays at two terminals through direct fiber and also DTT (direct transfer trip). Each relay samples its analog input current (A/D converter); the sampling rate varies, and in this respect, various line current differential relays differ considerably. The channels traditionally used for line current differential protection are limited in bandwidth; 64 kbps is a typical value [12, 13].

For the communication of phasors, twice the bandwidth is needed to send real and imaginary parts of the current required. The phasor exchange rate is not high enough to facilitate phasor interpolation and the relay sampling clocks must be synchronized (through GPS signals). The implementations may use communication channels to force synchronization of the sampling clocks. This is one of the key elements of line current differential system.

The local current data is communicated to remote terminals $i_{TX}$–$i_{RX}$ in Figure 8.24. Each relay then aligns the data and runs its differential trip equations. If a relay does not receive all the data from its remote terminal, say because of lack of communication, it is said to operate in a slave mode and DTT from master terminal allows slave relay to issue trip commands to their circuit breakers. Dedicated point-to-point channels remove any third-party devices and the failure modes associated with these.

## 8.9.1 The Alpha Plane

The alpha plane demonstrates the complex ratio $I_R$ / $I_L$. It compares individual magnitudes and phase angles of the currents. For a two-terminal line, and underbalanced conditions, $I_L$ (local end) and $I_R$ (remote end) are equal in magnitude and opposite in phase. This gives an operating point of $k$ = 1 < 180° = −1. This neglects the line-charging current. Practically all the following parameters must be considered:

- line charging current
- CT saturation
- channel time delay compensation errors
- tapped loads.

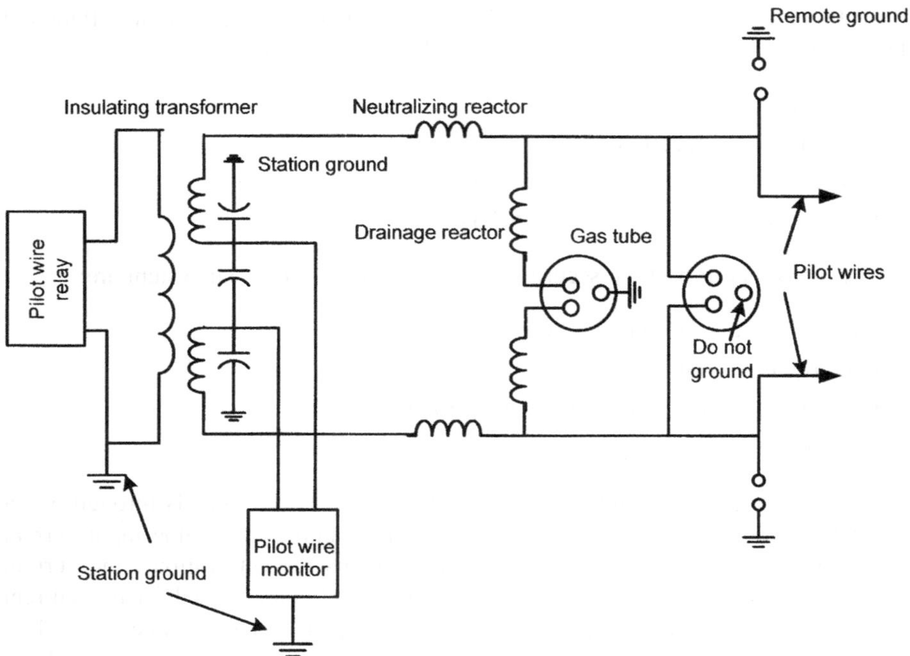


Figure 8.23. The details of the equipment at one end of a pilot wire protection, using electromechanical relays (see text).

If we define the operating current and restraint current as:

$$I_{op} = |I_L + I_R| \qquad (8.8)$$
$$I_{RT} = k|I_L - I_R|$$

and consider

$$I_{OP} > kI_{RT} \qquad (8.9)$$

with the relation

$$\left|\frac{I_R}{I_L}\right| = a + jb, \qquad (8.10)$$

Equation (8.9) can be written as:

$$|1 + a + jb| \geq |1 - a - jb|. \qquad (8.11)$$

Expanding,

$$a^2 + b^2 + 2\frac{1+k^2}{1-k^2}a + 1 \geq 0. \qquad (8.12)$$

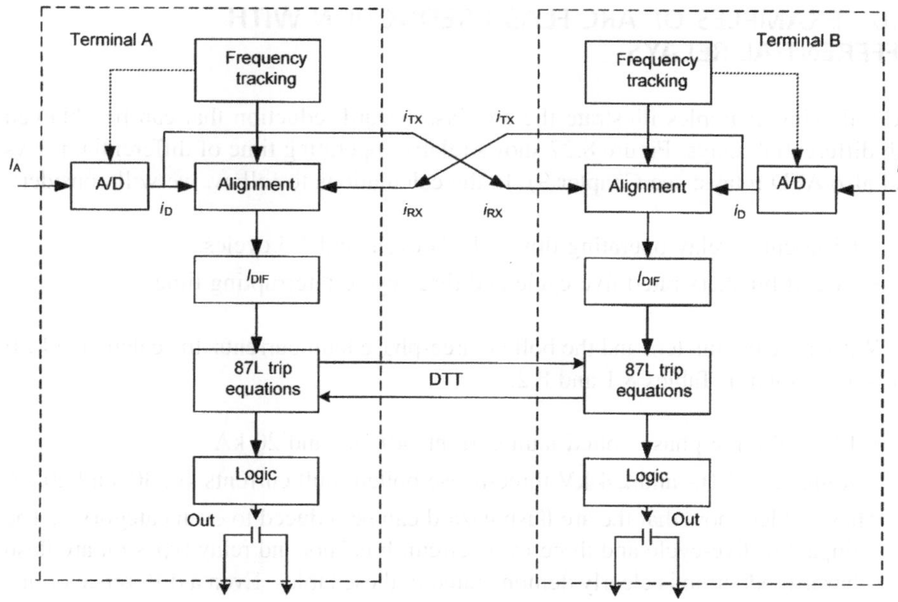


Figure 8.24. Simplified architecture of modern line current differential system.

This is the equation of a circle in the alpha plane, the operating characteristics of the differential relay (Figure 8.25).

## 8.9.2 Enhanced Current Differential Characteristics

The key factors in defining the required shape of the differential relay in alpha plane are: power system impedance nonhomogeneity, CT saturation, and low frequency oscillations in compensated lines. The effect of charging currents and system power angle can be eliminated by using negative or zero sequence currents. Blocking and operating functions are shaped as in Figure 8.26, the new differential element characteristics for protection of transmission lines. The relay restraining region is the area between two circular arcs and two straight lines and includes $a = -1$ point. Amplitude comparison provides the circular part of the characteristics with independent settings $R$ and $1/R$ circle radii. Phase comparison provides the linear part of the characteristics and defines the angular setting $\alpha$. The characteristics are symmetrical with reference to $a$-axis.

The traditional circular characteristics of the differential relay have low tolerance to the channel asymmetry. The new characteristics provide better stability, reliability, and sensitivity. The relays have the capability of single-phase reclosing and tripping, self-monitoring and communication channel loss. A host of ancillary protective functions, such as backup distance protection, circuit breaker failure protection, synchronization check, and directional overcurrent, can be provided. The robustness of relay hardware and firmware are of importance.

An operating time in subcycle range is obtained, and low settings of 0.5-A pickup for differential current, two-phase and line-to-ground fault are possible.

## 8.10 EXAMPLES OF ARC FLASH REDUCTION WITH DIFFERENTIAL RELAYS

The following examples illustrate the arc flash hazard reduction that can be obtained with differential relays. Figure 8.27 shows relative operating time of differential relays (and also AFD relays; see Chapter 9). In the calculations to follow, we will consider:

- Differential relay operating times of 3/4 cycle and 1.5 cycles.
- Circuit breakers rated five-cycle and three-cycle interrupting time.

With these parameters and the bolted three-phase fault currents, the calculated PPE levels are shown in Tables 8.1 and 8.2.

- 13.8-kV three-phase bolted fault currents 40, 30, and 20 kA
- *Table 8.2*: 4.16- and 2.4-kV three-phase bolted fault currents 40, 30, and 20 kA

These tables show that the arc flash hazard can be reduced to even category 1. The impact of five-cycle and three-cycle circuit breakers and relay types on arc flash hazard reduction is clearly demonstrated in these tables. Even a difference of one cycle in the arc flash time matters at high currents.

Though differential protection can be applied on all the medium voltage buses, the cost factor can be a deterrent.

TABLE 8.1. Arc Flash Hazard Reduction in Medium Voltage Systems through Differential Relays, 13.8-kV System, Gap = 152 mm, Working Distance = 36 in, Dimensions 26″W x 95″H, 96″D, EC = HCB

| Bus Bolted Fault, kA rms sym | Arc Fault, kA rms sym | Breaker Interrupting Time Cycles | Device Operating Time Cycles | Arc Flash Boundary, in | Incident Energy, cal/cm$^2$ | PPE | Remarks |
|---|---|---|---|---|---|---|---|
| 40 | 35.95 | 5 | 1.5 | 144 | 12 | 3 | 1.5 cycles diff |
| | | | 0.75 | 134 | 10.6 | 3 | 3/4 cycle diff |
| | | 3 | 1.5 | 116 | 8.3 | 3 | 1.5 cycles diff |
| | | | 0.75 | 104 | 6.9 | 2 | 3/4 cycle diff |
| 30 | 27.20 | 5 | 1.5 | 120 | 8.84 | 3 | 1.5 cycles diff |
| | | | 0.75 | 112 | 7.81 | 2 | 3/4 cycle diff |
| | | 3 | 1.5 | 97 | 6.14 | 2 | 1.5 cycles diff |
| | | | 0.75 | 87 | 5.11 | 2 | 3/4 cycle diff |
| 20 | 18.11 | 5 | 1.5 | 72 | 5.69 | 2 | 1.5 cycles diff |
| | | | 0.75 | 86 | 5.03 | 1 | 3/4 cycle diff |
| | | 3 | 1.5 | 74 | 3.95 | 1 | 1.5 cycles diff |
| | | | 0.75 | 66 | 3.29 | 1 | 3/4 cycle diff |

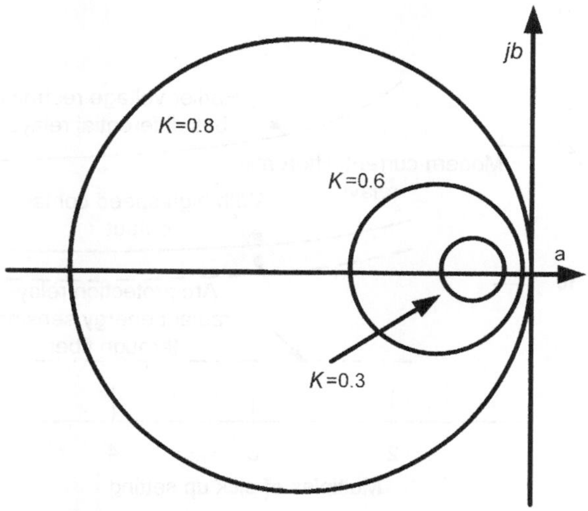


Figure 8.25. Alpha plane characteristics of a differential relay described by Equation (8.12).

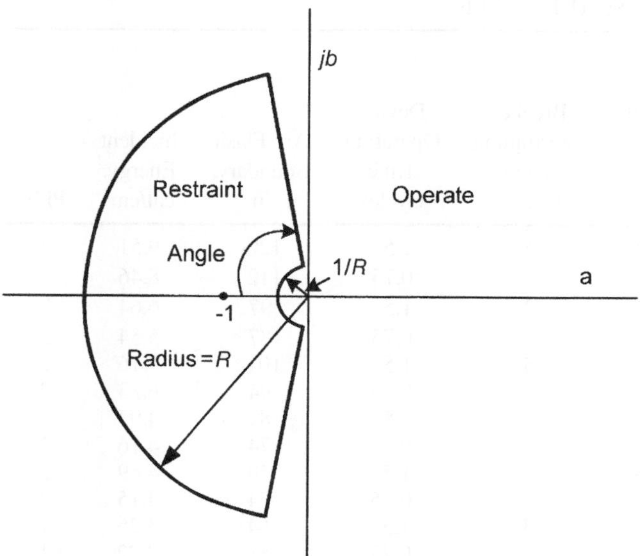


Figure 8.26. Characteristics of a new differential scheme in the alpha-plane.

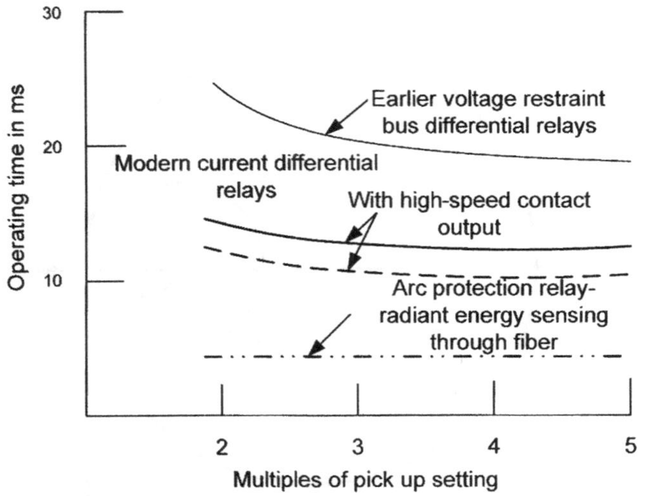


Figure 8.27. Typical operating times of differential and arc flash detection (AFD) relays.

TABLE 8.2. Arc Flash Hazard Reduction in Medium Voltage Systems through Differential Relays, 4.16- and 2.4-kV Systems, Gap = 104 mm, Working Distance = 36 in, Dimensions 26"W x 95"H, 96"D, EC = HCB

| Bus Bolted Fault, kA rms sym | Arc Fault, kA rms sym | Breaker Interrupting Time Cycles | Device Operating Time Cycles | Arc Flash Boundary, in | Incident Energy, cal/cm$^2$ | PPE | Remarks |
|---|---|---|---|---|---|---|---|
| 40 | 33.55 | 5 | 1.5 | 120 | 9.51 | 3 | 1.5 cycles diff |
| | | | 0.75 | 112 | 8.46 | 3 | 3/4 cycle diff |
| | | 3 | 1.5 | 97 | 6.64 | 2 | 1.5 cycles diff |
| | | | 0.75 | 87 | 5.54 | 2 | 3/4 cycle diff |
| 30 | 25.68 | 5 | 1.5 | 101 | 7.16 | 2 | 1.5 cycles diff |
| | | | 0.75 | 94 | 6.33 | 2 | 3/4 cycle diff |
| | | 3 | 1.5 | 82 | 4.97 | 2 | 1.5 cycles diff |
| | | | 0.75 | 74 | 4.16 | 2 | 3/4 cycle diff |
| 20 | 17.86 | 5 | 1.5 | 79 | 4.69 | 2 | 1.5 cycles diff |
| | | | 0.75 | 74 | 4.15 | 2 | 3/4 cycle diff |
| | | 3 | 1.5 | 64 | 3.26 | 1 | 1.5 cycles diff |
| | | | 0.75 | 58 | 2.72 | 1 | 3/4 cycle diff |

## REVIEW QUESTIONS

1. Write three advantages of using wye-connected CTs on both sides of a three-phase delta-wye transformer protected with MMPR transformer differential relays.
2. How are the zero sequence currents in question 1 eliminated?
3. Describe the limitations of hard-wire high impedance bus differential protection with respect to distance and CT lead length.
4. Compare high impedance bus differential protection with low-resistance current differential schemes for bus protection.
5. Write five limitations of pilot wire protection with metallic wires.
6. Compare the sensitivity for phase and ground fault of electromechanical relays with digital direction comparison relays.
7. A fault occurs on the wye side of a three-phase delta-wye transformer external to the protected zone. Describe the phase relations of the primary and secondary currents.
8. Simple overcurrent relays are sometimes used in partial differential schemes. What is their limitation with respect to arc flash reduction?
9. Check one specimen calculation of arc flash in Table 8.5 by hand calculations using the equations in IEEE 1584 Guide (see Chapter 1).

## REFERENCES


1. Westinghouse Electric Corporation, *Applied Protective Relaying*, Chapters 8, 9, and 14, Coral Springs, FL, 1982.
2. Westinghouse and ABB, "Bus Protection Guide," Publication No. PRSC-9, Nov. 1989.
3. IEEE Standard, C37.103. *Guide for Differential and Polarizing Relay Circuit Testing*, 2004.
4. GE, "PVD Differential Voltage Relay," Manual GEK-45405, 1992.
5. GE, "B30 Bus Differential Relay," Manual GEK-113371, 2011.
6. A. Guzman, S. Zocholl, and H. Altuv, "Performance analysis of traditional and improved transformer differential protective relays," in 27th Annual Western Protective Relay Conference Proceedings, Spokane, WA, 2000.
7. IEEE Standard, C37.97. *Guide for Protective Relay Applications to Power System Buses*, 1979.
8. IEEE Standard, C37.91. *Guide for Protective Relay Application to Transformers*, 2000.
9. D. Costello, "Lessons learned through commissioning and analyzing data from transformer differential installations—Part 1," in Conf. Record. IEEE Technical Conference on Pulp and Paper Industry, pp. 93–102, June 2007.
10. D. Costello, "Lessons learned through commissioning and analyzing data from transformer differential installations—Part 2," in Conf. Record. IEEE Technical Conference on Pulp and Paper Industry, pp. 103–118, June 2007.

11. Westinghouse, "Transformer Protection Guide," Publication No. PRSC-3D, November 1985.
12. G. Benmouyal, "Trajectories of line current differential faults in the Alpha plane," Proceedings of the 32nd Annual Western Protective Relay Conference, Spokane, WA, October 2005.
13. K. Behrendt, N. Fischer, and C. Labuschagne, "Considerations for using harmonic blocking and harmonic restraint techniques on transformer differential relays," Proceedings of 33rd Annual Western Protective Relay Conference, Spokane, October 2006.

# 9

# ARC FAULT DETECTION RELAYS

Arc flash detection relays provide even faster fault detection times as compared to the differential relays discussed in Chapter 8. With modern arc fault detection (AFD) technology, this AFD time is of the order of 1/4 cycles or even lower, based on 60 Hz. High-speed insulated gate bipolar transistors (IGBTs) are used to provide fully rated trip outputs rather than the relatively slow conventional dry contacts. Thus, AFDs form a technically acceptable and cost effective way to reduce the arc flash hazard. As stated in Chapter 1, when we reduce the arc flash hazard, simultaneously:

- The worker safety is enhanced.
- The equipment damage is reduced.
- The downtime to repair the equipment is reduced. This can result in savings in lost production.
- The equipment useful life is increased as it is subjected to short-circuit stresses for a very short time.

References [1–4] provide further reading.

*Arc Flash Hazard Analysis and Mitigation*, Second Edition. J.C. Das.

Published 2021 by John Wiley & Sons, Inc.

## 9.1 PRINCIPLE OF OPERATION

The light emitted during an arc flash event is significantly brighter than the normal lighting background. This radiant energy can be easily detected using light sensors.

Though relatively new in North America, the arc flash relays have been around since early 1990s. These earlier relays used only single point light receptacles called lens sensors. One or more lens sensors are located in a compartment where a potential arc hazard can occur. Each sensor is connected radially to electronics through clad fibers. Because of opaque cladding, only the extreme ends of the radial fibers are light sensitive. A limited number of sensors may be attached to an electronic package. Multiple electronic packages are then interconnected to provide the required coverage.

During the year 2000, a second generation of AFD relays was introduced. In addition to the lens sensors, a long unclad fiber optic that could absorb light throughout its length was developed. This had several advantages, as it reduced the cost of installation. A single optic fiber sensor can be as long as 200 ft and can cover the same zone of protection as bus differential protection (Chapter 8) but at a reduced cost. The concern for any shadows from the internal structures that could block the light is eliminated. Also, if the fiber sensor is configured in a loop, which is normally done in the current installation it can provide regular self-checking features of integrity and breakage of fiber loop. Any problem in the integrity of the sensor can be alarmed.

Thousands of arc flash detection systems have been installed in 36 countries worldwide over the past 6 years. Optically based arc flash relaying is the fastest protection available for arc flash hazard reduction.

## 9.2 LIGHT INTENSITY

Arc flash produces a bright light. The light produced can be in the range of 100,000–1,000,000 lx at a distance of 3 m. A camera flash can produce 234,000 lx at 18 in, and direct sunlight at any distance is 100,000 lx. High-intensity LED flash light can produce 28,000 lx. The intensity of normal substation lighting is 200–300 lx. Table 9.1 shows some typical illumination levels.

TABLE 9.1. Typical Ambient Illumination Levels

| Light Level | Example |
|---|---|
| 50 lx | Living room |
| 80 lx | Brightly lit room |
| 500 lx | Brightly lit office |
| 1000 lx | TV studio |
| 100,000 lx | Direct sunlight |
| >234,000 lx | Camera flash at 18 in |
| 100,000 to >1,000,000 lx | Arc flash event |

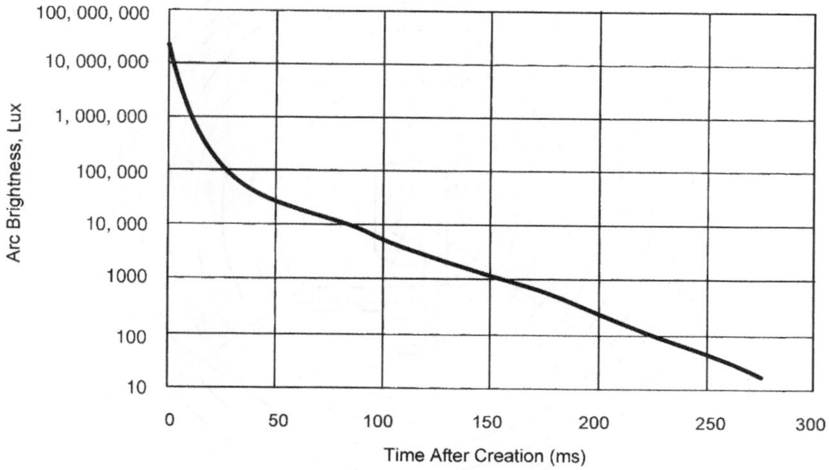


Figure 9.1. Arc brightness verses time after a 20-kA arc flash event.

Visible light consists of light spectrum ranging from 400 to 700 nm wavelengths. The arc flash tests show that most of the radiated energy has a wavelength of 200–600 nm. Consequently, AFDs are designed to operate in the lower end of the visible light spectrum, including ultraviolet light.

Figure 9.1 shows photometric data for an arc current of 20 kA, based upon the test results. As the intensity of light decreases very fast after an arc flash event, the response time of the AFD system must be small.

The light captured by the sensor is amplified and compared with a preselected light reference level. Once it exceeds the set level, a light signal is activated. AFD relays may be provided with a selector switch to select auto or manual light reference level. In the manual mode, the light reference level can be set through a front potentiometer. Figure 9.2 shows the sensitivity of lens sensors of a manufacturer at various backlight compensation settings [5, 6].

## 9.3 LIGHT SENSOR TYPES

The common types of sensors are lens-point sensors and bare fiber optic sensors. Figure 9.3 is a depiction of sensitivity of sensors. The normal operating sector is 130°. Practically, the light is also reflected from the compartment walls. For fiber sensors, the incident angle of lighting is not applicable. The bare fiber sensor consists of a high-quality plastic fiber-optic cable without a jacket. Figure 9.4a,b show a lens (point sensor), and Figure 9.4c depicts a fiber sensor. Figure 9.5 shows a fiber optic sensor running in a switchgear enclosure, the minimum bending radius is 2 in. The figure also shows a loop of spare sensor properly anchored during installation. Special termination and splice kits are required, which are not discussed. A bare fiber sensor makes possible the detection in large areas. The detecting reach depends upon several factors:

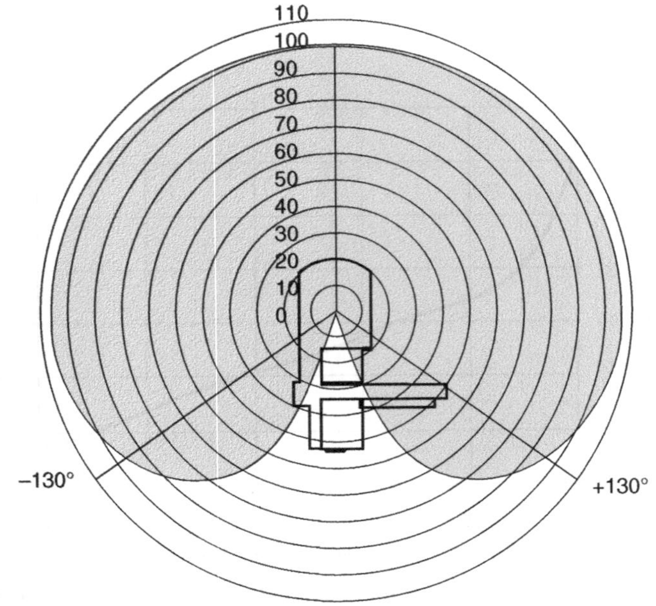


Figure 9.2. Relative sensitivity of a lens sensor from different angles of lighting. *Source*: Courtesy of ABB.

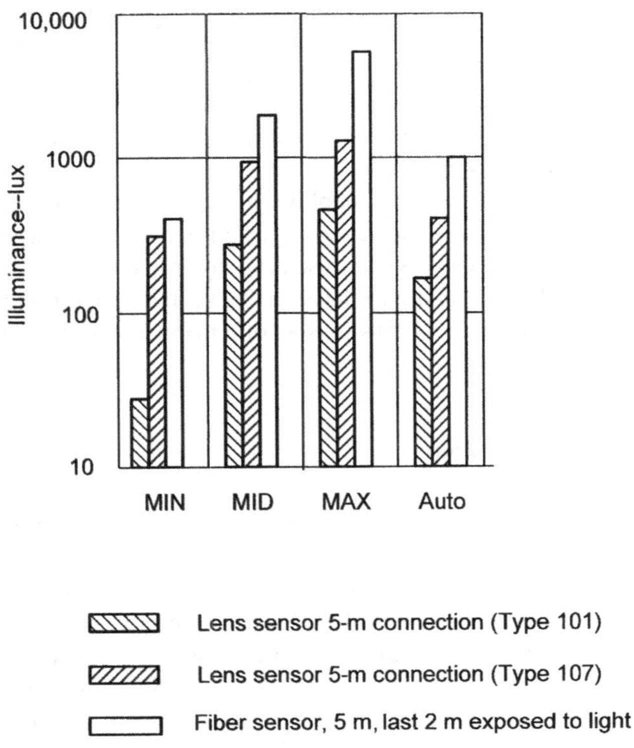


Figure 9.3. Sensitivity of sensors at various compensation settings. *Source*: Courtesy of ABB.

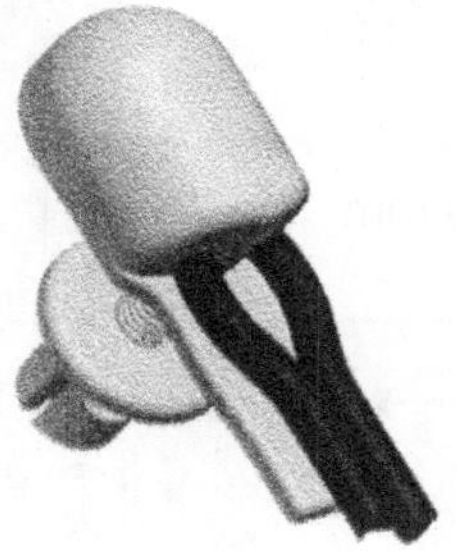

(a)

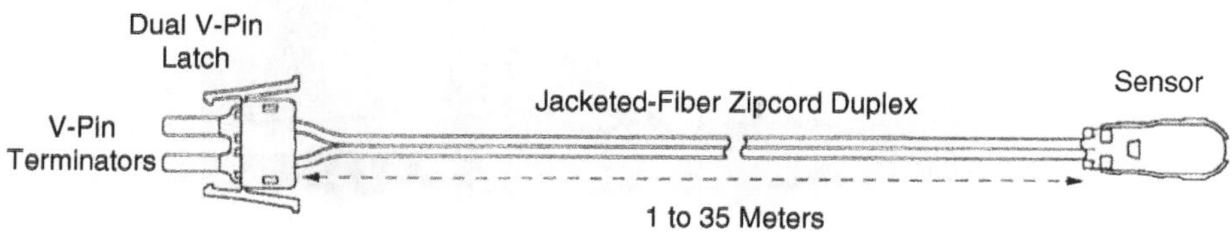


(b)

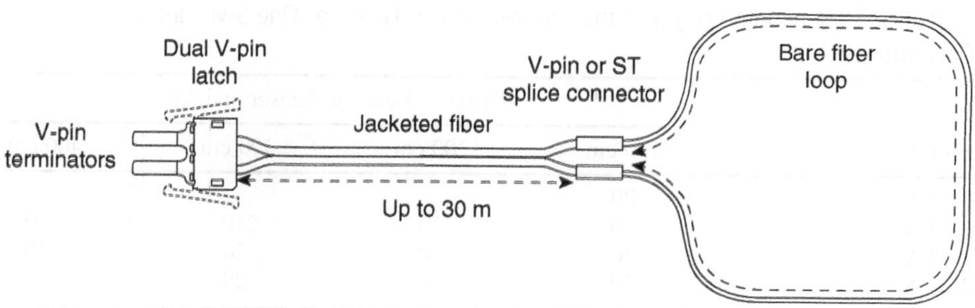


(c)

Figure 9.4. (a) and (b) A lens or point sensor; (c) a fiber sensor.

- light source energy
- fiber length
- reflectances
- backlight compensation settings.

The length of the sensor fiber for a switchgear compartment is selected according to available short-circuit or ground fault currents. Table 9.2 lists these selection criteria.

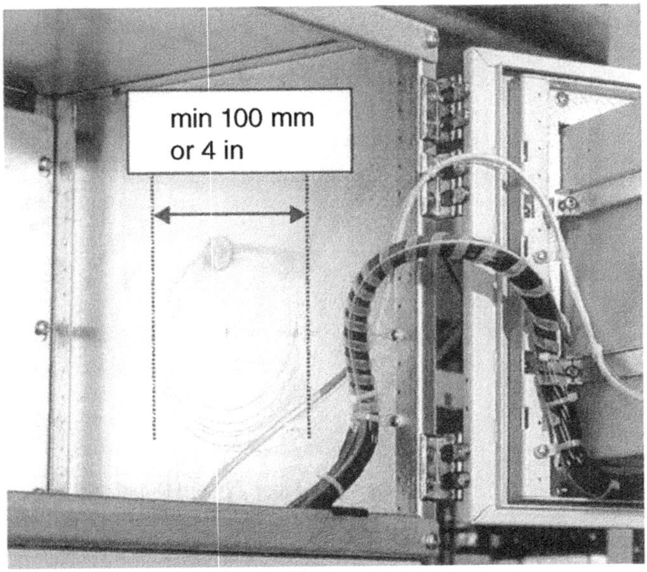


Figure 9.5. Layout of fiber sensor in switchgear, showing minimum bending radius of 2 in, that is, 4-in diameter. A loop of the excess fiber is also shown. From Reference [5].

TABLE 9.2. Minimum Length of the Exposed Sensor Fiber per One Switchgear Compartment

| | Distance Between Sensor and Arc | | | |
|---|---|---|---|---|
| Fault Current (rms) | 100 cm | 200 cm | 300 cm | 400 cm |
| 0.5 kA | 20 | | | |
| 0.7 kA | 20 | 70 | 210 | 280 |
| 1.4 kA | 20 | 20 | 20 | 140 |
| 2.2 kA | 20 | 20 | 20 | 20 |

*Note:* Blank cells indicate "not operational."

The table has been compiled based upon:

- copper bus bars
- arc length = 10 cm
- surrounding light = 400 lx
- no reflecting surfaces
- light reference level is set one scale mark to the right from the minimum.

According to one manufacturer, the detection distance of a lens sensor is 3 m. Thus, for protection of the bus bars lens, sensors should be placed at 6-m intervals. Arc detection systems typically use a combination of lens and bare fiber sensors. Proper location for AFD is important. Arc detection for a specific sensor will trip the upstream circuit breaker. Multiple sensor inputs provide coverage and redundancy that can be built in the layouts.

The sensors are exposed to harsh environments. Because the arc flash event can occur anywhere in the protected zone, the light sensor will be exposed to arc plasma and ejected particles, high temperature, pressure, and intense radiation. The light sensor must communicate with the electronics in AFD relay, without damage to the environmental conditions occurring due to an arc flash event.

A manufacturer will place the sensors to detect arc flash radiant light properly in an air-insulated switchgear to be protected. Some guidelines for the installation of the sensors are:

- Each sensor fiber is routed to cover the specific protection zone, that is, bus bars, cable, circuit breaker and PT compartments, and the like.
- The best location is near the top and rear of a circuit breaker cell and away from the breaker. For bus bars, the best location is usually along the back wall, centrally located with respect to bus bars.
- The high temperature surfaces are to be avoided. Temperatures above 60°C can decrease the performance over a period of time. The fibers are not secured directly to the bus.
- The fiber is not installed in conduit or in a raceway that will shield it from the light emitted by an arc flash event. Where exposure to the light is not required, protective tubing is installed.
- The fiber is secured using nylon ratchet clamps or similar fiber management products. The purpose is to keep the fiber away from moving parts and prevent snagging the fiber when racking the breakers in and out.
- Protective rubber grommets are used when routing sensors through metal walls. The hole-size required is small, of the order of 10 mm.
- Sharp bends are avoided. The minimum permanent bending radius of the sensor fiber is 2 in. The fiber can be broken or damaged if this minimum bending radius is not maintained.
- Correct installation tools are required. There is limit to the number of splices that can be installed in a given length of sensor fiber to maintain reliable operation. The following rules are advocated by one manufacturer:

  Maximum total length without splices or with one splice: 160 m

  With two splices: 50 m

  With three splices: 40 m
- When the existing equipment is to be retrofitted with AFD protection, additional care in the installation and testing is required.

In a radial unit installation system, the breaks cannot be detected.

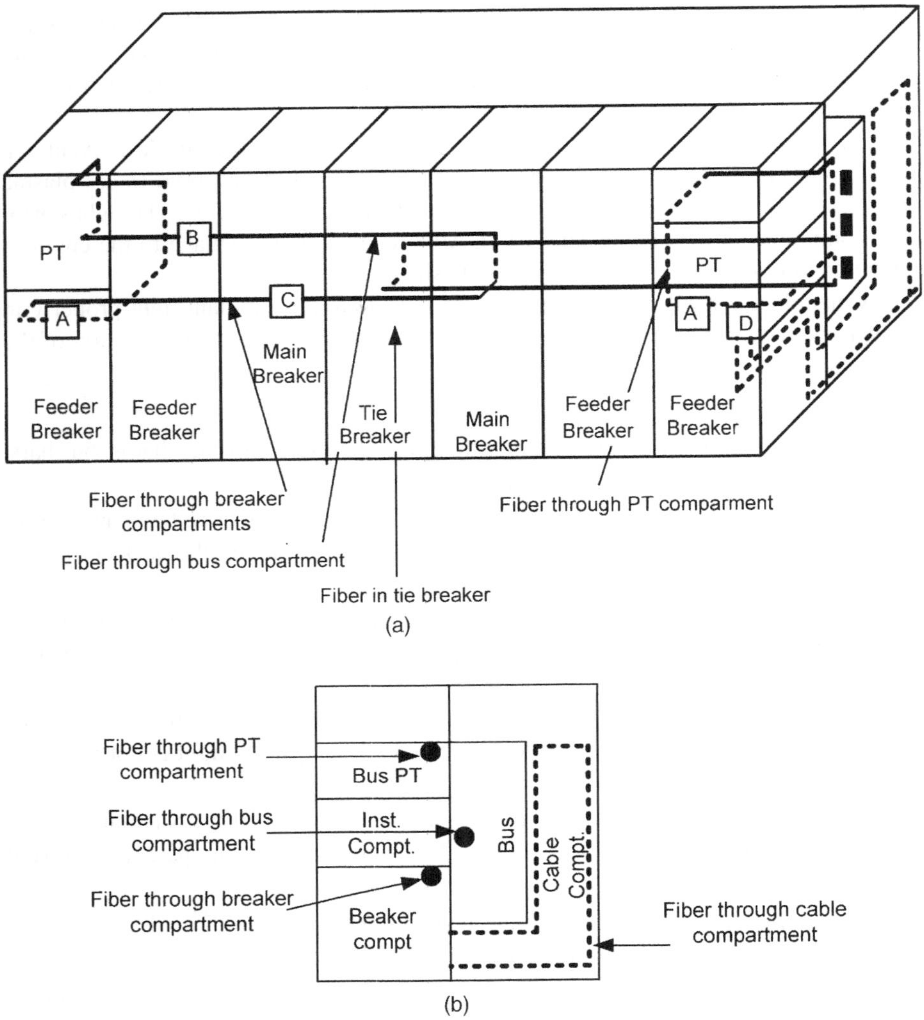


Figure 9.6. (a and b) Routing of fibers through bus compartment, breaker compartment, PT and cable compartment, and bus tie breaker compartment. *Source*: From Reference [5].

## 9.4 OTHER HARDWARE

The routing of fibers in a 13.8-kV, single high metal-clad switchgear lineup is shown in Figure 9.6a,b. This shows two feeder breakers, one tie breaker, and two main breakers. The sensing fibers are run through the main bus compartment, circuit breaker compartments, PT compartment, cable compartment, and bus section breaker compartment. In the bus section breaker compartment, the fibers overlap much like differential zones of protection. The fiber run through lower main and feeder breakers and returns through the bus compartments and through the PT compartment to device *A*. The right and left sides are symmetrical (details not completely shown in Figures 9.6a,b). The fiber in the cable compartment is routed through device *B*; only one relay is shown. *One B device per circuit breaker is used.* The devices functionality is described below:

*A*: central unit

*B*: high-speed trip unit, or AFD relay

*C*: fiber extension unit

*D*: lens sensor extension unit [5, 6].

## 9.5 SELECTIVE TRIPPING

With the arrangement shown in Figure 9.6a, selective trip arrangement is obtained. For a fault in the cable compartment, only a feeder circuit breaker can be tripped through device *B*. The main fiber loop does not enter the cable compartments. This ensures that only the respective feeder circuit breaker is tripped for a fault that is downstream of the feeder circuit breaker. Similarly for the faults on the right and left sides of the tie circuit breaker, only the circuit breakers on that side plus the tie circuit breaker are tripped. Regardless of where the flash occurs along the fiber, the same circuit breakers will be tripped. Thus, in order to achieve selectivity, separate fibers are used, as shown in this figure. Compare this with Figure 8.3, where for a bus differential protection, the cable compartments faults remain outside the differential zone of protection. With AFDs, the faults in the cable or PT compartments can be easily detected.

The light is transmitted from the sensor to the detector located remotely in the relay. For reliability, this is done through a fiber optic loop. For lens sensors, each lens has an input and output connection. The input is connected to the transmitter in the relay and output to the detector in the relay. The loop connection allows periodic testing of the system by injecting light from the transmitter through the loop and back to the detector. This can work with either the lens sensor or the bare fiber sensor; see Section 9.7.3.

About 200 ft of optical fiber loop is adequate to cover most applications. Extension units can be daisy chained, and up to approximately 20 additional fiber loops of 200 ft each can be connected for a total effective sensor length of 4000 ft.

A simple tripping scheme is shown in Figure 9.7a. For a fault in any one phase, both primary and secondary circuit breakers are tripped. Compare this with Figure 9.7b, which provides independent arc flash detection for the downstream faults. The extension unit in Figure 9.7b will trip its associated feeder circuit breaker. Simultaneously, it communicates to the central unit that a downstream trip has occurred. If the fault is not cleared within the programmed time, the central unit will trip its associated circuit breakers, thereby providing coordinated arc flash backup protection and selective fault clearing.

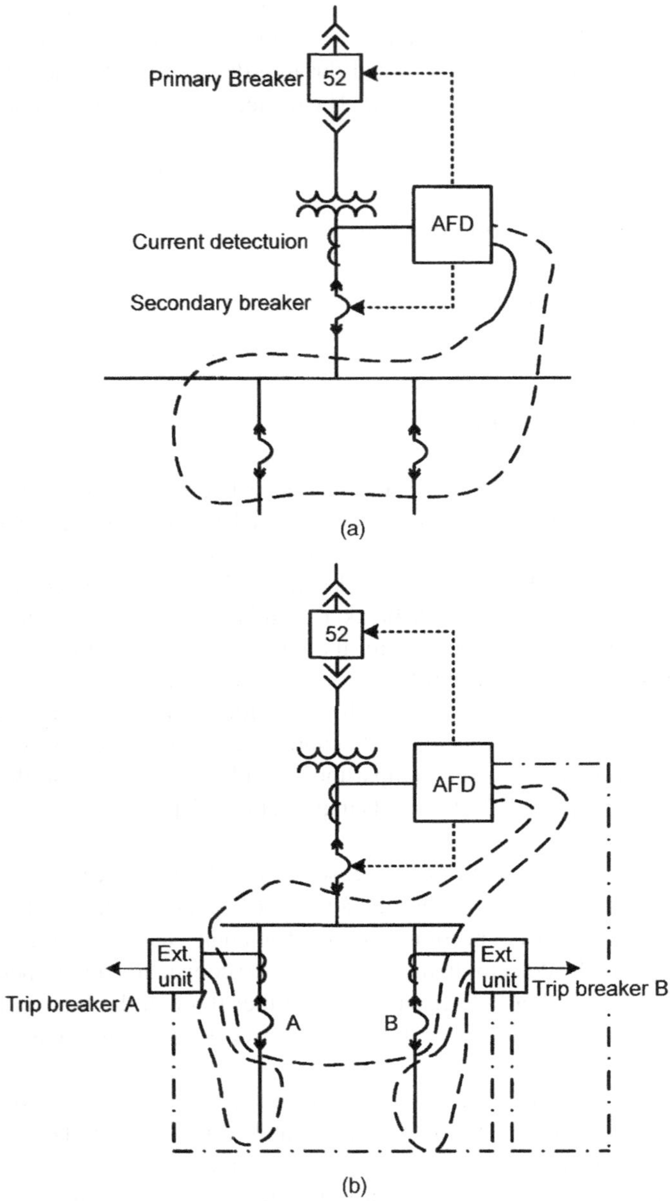


Figure 9.7. (a) Routing of fiber for nonselective AFD protection; (b) routing of fiber for selective AFD protection.

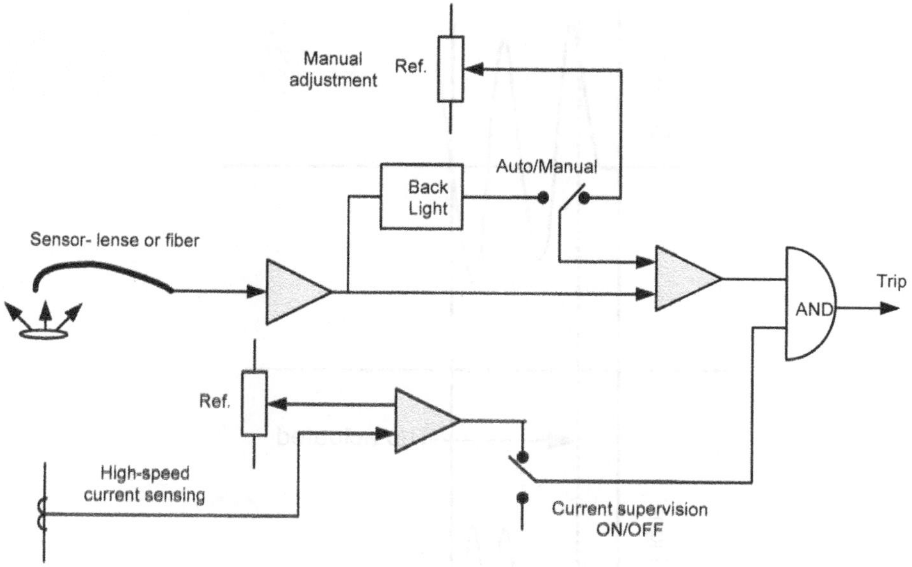


Figure 9.8. A logic circuit diagram of AFD system.

## 9.6 SUPERVISION WITH CURRENT ELEMENTS

Figure 9.8 shows a schematic diagram of the AFD system. This shows that the trip can be based upon the light sensing only, or it can be supervised through a high-speed current sensing. An obvious advantage is that low levels of phase and ground fault currents are sensed, limiting the equipment damage as well as arc flash hazard. In addition, the security of the system is enhanced. Figure 9.8 shows a selector switch to enable or disable the current supervision. The set current level can be even below the load current. Conventional current elements have a response time of the order of 6–20 ms. This delay is unacceptable, and for arc flash detection, high-speed overcurrent elements should act as fast as the arc detection. Figure 9.9 illustrates a detection time of approximately 1/4 of a cycle; based upon the current and light sensing. Some commercial products claim an even faster overall detection time, the output contact closing time is of the order of 5 μs.

## 9.7 APPLICATIONS

### 9.7.1 Medium Voltage Systems

AFD systems can be applied to:

- metal-clad and cubical type switchgear
- load interrupter switchgear
- medium voltage MCCs.

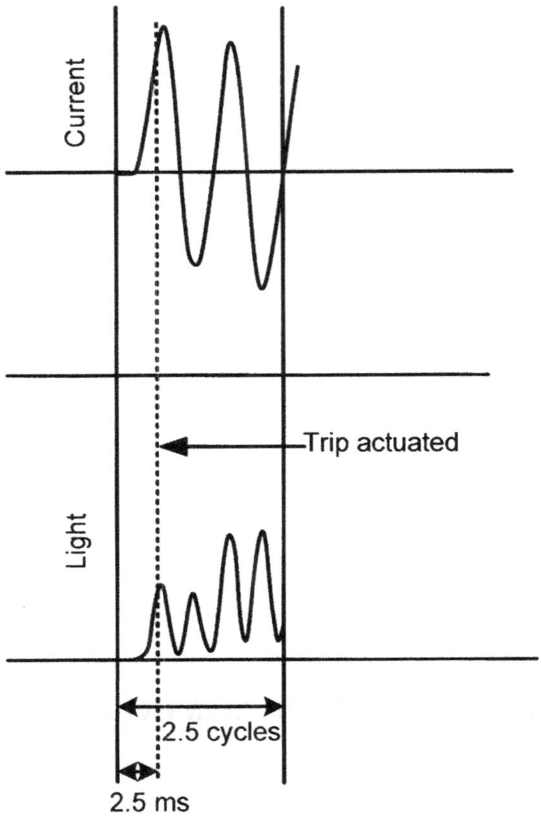


Figure 9.9. AFD with current supervision, operating time of output 2.5 ms or less, with digital output contact. With discrete output contact this time is of the order of 8 ms.

Vacuum or SF6 is the predominant current interruption technology for the medium voltage systems. The AC current interruption in these technologies takes place in sealed chambers, and the light flash detection is a reliable indicator of the arc flash event. In older air-magnetic circuit breakers, the arc is drawn in arc chutes and blown out by magnetic forces causing the arc to elongate and segmented. Most of the light is contained in the arc chutes, but the arc chutes are not sealed and leakage of light takes place. A question arises whether nuisance trips can occur due to escaped light?

High-current arc flash tests have been conducted on these circuit breaker types. The test results showed that for GE Magneblast™ circuit breakers, either the fiber or lens technology can be applied; the test currents were varied between 3 and 20 kA and with minimum to maximum light sensing. GE Circuit breaker arc chutes allow relatively little light to escape during normal fault clearing.

The Westinghouse 50DH™, 5-kV circuit breaker tests showed false operation with long fiber sensor technology, even with maximum light sensing; however, with lens-type sensors, satisfactory performance was obtained. The Westinghouse DHP circuit breakers arc chutes allow much light to escape [7]. When the fiber was located 2 ft away from the top of the arc chutes, the false operation could be prevented. Routing of the fiber in these types of circuit breakers should be carefully done and the installation tested when retrofits are planned.

### 9.7.2 Low Voltage Circuit Breakers

There has been some concern that low voltage circuit breakers produce enough ambient light while operating; yet AFD technology has been applied to low voltage circuit breakers and MCCs and commercial products are available. A light-sensing system should not be triggered by light emissions of properly interrupting low voltage circuit breakers. Filtering the light is an easily workable solution. Light filters can be designed that block the light from interrupting circuit breakers, as well ambient conditions.

### 9.7.3 Self-Testing of Sensors

Both point and fiber sensors can be tested by performing a loop-back test. A LED couples the test light to the loop or point sensor. For the loop sensor, the light travels through the fiber optic cable to the sensor in the relay. For a point sensor, the light travels through fiber optic cable and is emitted by the sensor dome. This emitted light by the sensor dome is picked up the adjacent fiber optic cable and returned to the sensor in the relay. For either type of sensors, the light sensed during the test is compared with the set limits, thus verifying the functionality of the light sensor relay and measuring the optic path attenuation. A test result outside the test limits can be alarmed. The fiber cuts, kinks, or disconnections can be detected. The self-test features are important to ensure the integrity of the AFD systems. Currently, there are no standards for testing the performance of AFD systems. The self-tests can be made so that the test signals and an arc flash event are distinguishable and no nuisance trip occurs, if an arc flash event occurs during the testing [8].

## 9.8 EXAMPLES OF CALCULATION

1. For a 13.8-kV resistance grounded system (ungrounded system for arc flash considerations) and applying IEEE 1584 parameters of working distance = 910 mm (36 in), gap length = 152 mm, Table 9.3 gives the arc flash calculation results with AFD relays. This can be compared with Table 8.1 for differential protection, under the same conditions. The arc flash incident energy is further reduced with AFD relays.
2. Table 9.4 provides similar calculation for 4.16- and 2.4-kV systems.
3. Table 9.5 shows calculations for low voltage systems.

## 9.9 ARC VAULT™ PROTECTION FOR LOW VOLTAGE SYSTEMS

The Arc Vault™ is GE's arc transfer system [9, 10] for the low voltage systems. The principle is to conduct the fault current away from an arcing fault. This is achieved by

TABLE 9.3. Arc Flash Calculation Results with AFD, Sensing Time = 0.25 Cycles, 13.8-kV Switchgear, Gap = 152 mm, Working Distance = 36″, Dimensions 26″W x 95″H, 96″D, EC = HCB

| Three-Phase Bolted Current (kA sym) | Arc Flash Current (kA sym) | Breaker Interrupt. Time Cycles | Arc Flash Time (Seconds) | Arc Flash Boundary (Inches) | Incident Energy (cal/cm$^2$) | PPE Category |
|---|---|---|---|---|---|---|
| 10 | 9.71 | 3 | 0.055 | 39 | 1.54 | 1 |
| | | 5 | 0.088 | 51 | 2.16 | 1 |
| 20 | 19.18 | 3 | 0.055 | 61 | 2.85 | 1 |
| | | 5 | 0.088 | 81 | 4.59 | 2 |
| 30 | 28.58 | 3 | 0.055 | 79 | 4.44 | 2 |
| | | 5 | 0.088 | 100 | 7.17 | 2 |
| 40 | 37.92 | 3 | 0.055 | 95 | 6.6 | 2 |
| | | 5 | 0.088 | 127 | 9.65 | 3 |
| 50 | 47.22 | 3 | 0.055 | 109 | 7.46 | 2 |
| | | 5 | 0.088 | 145 | 12 | 3 |

TABLE 9.4. Arc Flash Calculation Results with AFD, Sensing Time = 0.25 Cycles, 4.16- or 2.4-kV Switchgear, Gap = 104 mm, Working Distance = 36″, Dimensions 26″W x 95″H, 96″D, EC = HCB

| Three-Phase Bolted Current (kA sym) | Arc Flash Current (kA Sym) | Breaker Interrupt. Time Cycles | Arc Flash Time (Seconds) | Arc Flash Boundary (Inches) | Incident Energy (cal/cm$^2$) | PPE Category |
|---|---|---|---|---|---|---|
| 10 | 8.76 | 3 | 0.055 | 35 | 1.43 | 1 |
| | | 5 | 0.088 | 46 | 1.81 | 1 |
| 20 | 17.57 | 3 | 0.055 | 53 | 2.36 | 1 |
| | | 5 | 0.088 | 70 | 3.79 | 1 |
| 30 | 25.68 | 3 | 0.055 | 68 | 3.59 | 1 |
| | | 5 | 0.088 | 90 | 5.78 | 2 |
| 40 | 33.55 | 3 | 0.055 | 80 | 4.80 | 2 |
| | | 5 | 0.088 | 106 | 7.73 | 2 |
| 50 | 40.84 | 3 | 0.055 | 91 | 5.94 | 2 |
| | | 5 | 0.088 | 120 | 9.96 | 3 |

TABLE 9.5. Arc Flash Calculation Results with AFD, Sensing Time = 0.25 Cycles, 0.48-kV Switchgear, Gap = 32 mm, Working Distance = 24″, Dimensions 36″W x 90″H, 48″D, EC = HCB

| Three-Phase Bolted Current (kA sym) | Arc Flash Current (kA Sym) | Breaker Interrupt. Time Cycles | Arc Flash Time (Seconds) | Arc Flash Boundary (Inches) | Incident Energy (cal/cm$^2$) | PPE Category |
|---|---|---|---|---|---|---|
| 10 | 7.07 | 3 | 0.055 | 22 | 1.01 | 0 |
| 20 | 14.26 | 3 | 0.055 | 32 | 2.31 | 1 |
| 30 | 20.76 | 3 | 0.055 | 39 | 3.12 | 1 |
| 40 | 26.09 | 3 | 0.055 | 43 | 3.97 | 1 |
| 50 | 30.14 | 3 | 0.055 | 47 | 4.16 | 2 |

providing an alternate low impedance current path and to generate a second arc within an enclosed controlled environment. The system consists of light sensors, protective relay, current sensing, and a containment dome (arc vault), all connected and working together. The relay triggers the containment dome and sends a trip signal to the circuit breaker. A secondary arc is created within the arc vault, which can be located, a short-distance from the protected equipment and cable connected to it. The secondary arc in the arc vault continues until the main circuit breaker trips and deenergizes the system. The successful arc capture depends upon:

- A low impedance path is provided than the fault path. For the fault path, the source impedance acts in series with the arc fault resistance.
- The arc transfer must be fast. Transfer time will be impacted by the time it takes to sense and trigger the arcing fault capture and the commutation time. The minimum commutation time is achieved by limiting the inductance in the transfer circuit, and therefore the cable length connecting the equipment and arc vault must be small. The impedance between the arc vault and the source should not be significantly higher than the impedance between the fault arc and source.
- The transfer path should be more stable. In an enclosed volume, an arc is more stable than in the air, and the arc is easier to maintain. By transferring fault current to an enclosed volume with a specific electrode gap, a stable arc with predictable properties can be created.

Referring to Figure 9.10:

$$V_{\text{capture}} = V_{\text{sys}}\left(\frac{R_{\text{arc}}}{R_{\text{arc}} + Z_{\text{source}}}\right) \tag{9.1}$$

where $V_{\text{capture}}$ is the arc arc-capture device voltage, $V_{\text{sys}}$ is the system voltage, $Z_{\text{arc}}$ is the arc impedance, and $Z_{\text{source}}$ is the source impedance. The voltage divider network determines the voltage across the phase electrode gap. The plasma created by trigger arc must break down the dielectric strength of the air between the electrodes so that the available system voltage is sufficient to initiate arcing current between electrodes. It

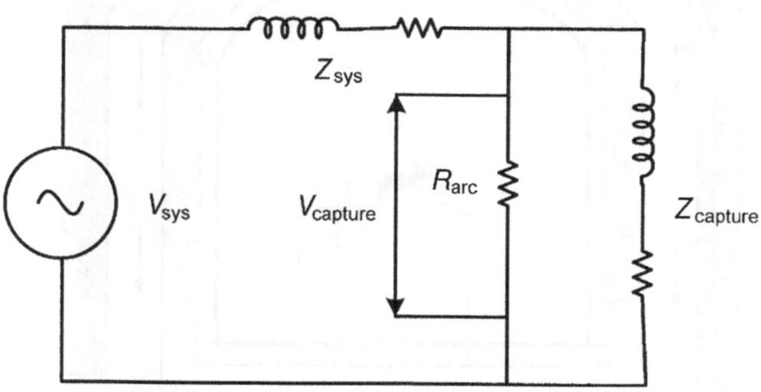


Figure 9.10. Transfer of arc from the equipment to a remotely located location-basic system schematic diagram.

will be difficult to trigger the arc when there is high source impedance (low short-circuit currents), which results in a lower arc voltage and lower arcing current. An analysis with 10-kA bolted fault current determined that the minimum available voltage at the electrodes was 250 V rms at any given point in one of the phase-to-phase voltages.

The current flowing through arc mitigation device after a successful transfer is given by:

$$I = \frac{V_{\text{sys}}}{Z_{\text{sys}} + Z_{\text{capture}}}\left(\sin(\omega t + \theta - \phi) - e^{-Rt/L}\sin(\theta - \phi)\right). \tag{9.2}$$

Where $\omega t$ is the sample time in radians, $\theta$ is the arcing current power factor, $\phi$ is the closing angle, $R$ is total system resistance, and $L$ is the total system reactance.

Figure 9.11 illustrates a cross-section through the arc vault; all the construction details are not shown. The electrode gap is the major component of controlled arc impedance and can be adjusted to fine tune the impedance and optimize performance in different circuits with varying short-circuit currents and source impedances. The electrode geometry is arranged to balance the magnetic forces on the arc, preventing it from lengthening the arc. The containment chamber must contain and cool the arc. The device must vent fast to prevent a pressure buildup that could destroy the vault. The electrode gap must maintain dielectric integrity under normal system operating conditions, and the containment device must not decrease the BIL of the equipment, though ANSI/IEEE standards do not specify the BIL testing of low voltage apparatuses.

The activation is done by breakdown of the air gap through a plasma gun, which is placed directly under the electrodes (not visible in Figure 9.11); the energy release by the plasma gun is approximately 60 J.

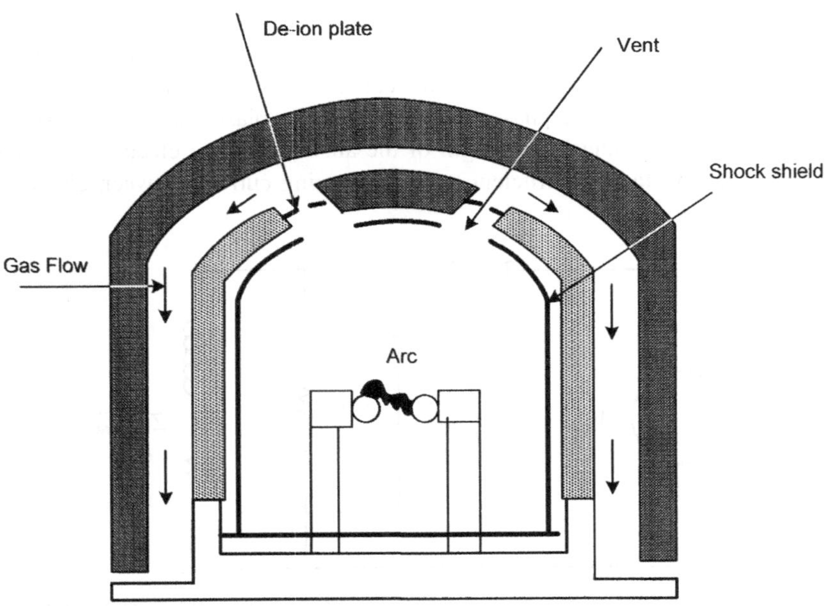


Figure 9.11. A cross-section through the arc vault—arc containment device.

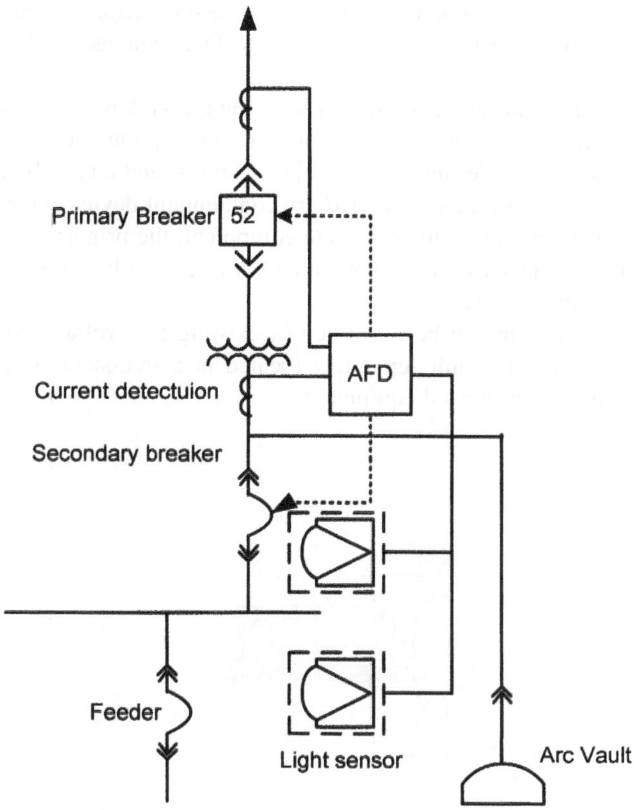


Figure 9.12. Arc vault, sensing and control system for system protected with transformer primary and secondary breakers.

## 9.9.1 Detection System

An arc flash detection system is required. As discussed previously, arc flash detection system are capable of differentiating between bolted and arc flash faults, as dead bolted currents do not produce arcing. Thus, light-based sensing is used. To protect the complete system, the light sensors must be placed throughout the system, which includes main bus compartment, cable run-ins, transition sections, and circuit breaker compartments. As discussed in Section 9.6, current supervision is provided, and the current sensor must be placed on the line-side (source side) of the switchgear. The sensing, triggering, and arc transfer require some time. The sensors can communicate enough light and current information to the relay in less than 2 ms after initiating fault current. The relay can issue a firing signal to the arc vault and a trip signal to the breaker in less than 6 ms and arc transfer time is less than 1 ms. Thus, the transfer time is of the order of 8 ms.

The arc vault system reduces the energy released by 63% or more compared with a bolted fault that will occur in a crowbar system. This reduces the stresses on the system components, such as transformers, circuit breakers, and bus bars. Due to transfer of arc energy to a remote device, the arc fault damage in the equipment is reduced.

Figure 9.12 shows an application to a low voltage system. To protect the entire equipment, the light sensors are placed throughout the equipment, which include the main bus compartment, cable runs-in, transition sections, and circuit breaker compartments. The location of current sensors and arc containment device determine the zone of protection. For protection of the complete equipment, the primary circuit breaker is required, and the current sensor and containment device must be on the line side of the main secondary circuit breaker.

The Arc Vault system can be retrofitted in existing low voltage circuit breakers. Figure 9.13 shows the arc vault separately located in a free-standing panel at some distance away from the protected equipment.

Figure 9.13. Arc vault located remotely in a separate cubicle. *Source*: Courtesy of GE.

## REVIEW QUESTIONS

1. Why is a loop fiber preferred? State one major precaution while using a radial long fiber with respect to its termination.
2. For an entirely new installation of a medium voltage vacuum circuit breakers lineup, compare bus differential relaying with AFDs.
3. In an AFD system, the faults in cable compartments can also be cleared fast, similar to bus compartment or breaker faults. Explain how it is achieved. Show in a sketch the routing of the fibers in the cable compartment and its connection to AFD relays.
4. The fault current in a 13.8-kV metal-clad switchgear is 40 kA rms symmetrical, while the maximum load current is 1200 A. At what current sensing level should the high speed overcurrent be set?
5. Arc flash incident energy is to be reduced in a 13.8-kV switchgear assembly consisting of vacuum circuit breakers. Which will be a better choice: bus differential protection or AFD relaying? Which is easier to implement? Describe relative merits and demerits.
6. In Question 5, the switchgear lineup consists of Westinghouse DHP switchgear. What will be a better choice of sensor technology?
7. Can AFD protection be obtained using one type of sensor technology? What will be the problems of locating fiber sensors in breaker compartments and lens sensors in bus compartments?
8. Describe six precautions when laying out sensing fibers in a switchgear assembly.
9. Referring to Figure 9.8, when will it be desirable to select manual mode of the background light biasing in an AFD system?

## REFERENCES


1. C. Inshaw and R. Wilson, "Arc flash hazard analysis and mitigation," A&M Relay Conference, Texas, 2006.
2. J. Buff and K. Zimmermann, "Application of existing technologies to reduce arc flash hazards," Proceedings of 33rd Annual Western Protective Relay Conference, Spokane, WA, 2006.
3. J.C. Das, "Protection planning and system design to reduce arc flash incident energy in a multi-voltage level distribution system to 8 cal/cm$^2$ or less—Part 1I Analysis," IEEE Trans. Ind. Appl., vol. 47, no. 1, pp. 408–420, Jan/Feb. 2011.
4. J.A. Kay, J. Arvola, and L. Kumpulanan, "Protecting at the speed of light: Combining arc flash sensing and arc-resistant technologies," in Conf. Record, IEEE Pulp and Paper Industry Technical Conference, San Antonio, pp. 112–118, 2010.
5. ABB, "REA Arc Protection System, Sensor Fiber Installation and Testing," Instruction Book 1VAD266601-MB, 2009.

6. ABB, "ABB Buyers Guide, REA-10-Arc Protection Relay," IMRS750929-MBG, May 1999.
7. R.A. Wilson, R. Harju, J. Keisala, and S. Ganeshan, "Tripping with speed of light: Arc flash protection," ABB publication: 1MRS756426, http://www.abb.com.
8. B. Hughes, V. Skendzic, D. Das, and J. Carver, "High-current qualification testing of an arc flash detection system," Schweitzer Engineering Laboratories, Inc., http:// www.selinc.com.
9. General Electric, "Out of the Box Thinking, Inside the Box Protection," GE Publication DEA-489A, January 2010.
10. G. Roscoe, T. Papallo, and M. Valdes, "Arc flash energy mitigation by fast energy capture," in Conf. Record, IEEE Pulp and Paper Industry Technical Conference, San Antonio, pp. 130–138, 2010.

# 10

# OVERCURRENT COORDINATION

The unit protection systems described in Chapters 8 and 9 are not applied throughout a distribution system, primarily because of cost considerations. Primary and secondary distribution buses, main utility interconnection transformers, generators, and large substation transformers and large motor are, generally, covered in differential zones, yet the distribution system downstream, like medium and low voltage motor control centers, panels, lighting circuits, and smaller distribution systems, have time–current-coordinated overcurrent protections. Extending the differential protection to low voltage switchgear buses for arc flash reduction is a recent innovation (see Chapter 14).

Even when unit protection systems are provided, the time–current coordinated systems are retained as backup systems. Though a primary protective system can fail, that is, an intended circuit breaker may fail to trip, this eventuality is not considered in the arc flash analysis.

The objectives of protective device coordination is to achieve maximum selectivity without much sacrificing the sensitivity, though, by the very nature of the coordination, the fault clearance times toward the source will increase. Properly applied instantaneous overcurrent settings can be of help for fast fault clearance and arc flash hazard reduction.

*Arc Flash Hazard Analysis and Mitigation*, Second Edition. J.C. Das.

Published 2021 by John Wiley & Sons, Inc.

An ideal time–current coordination is rarely achieved, as will be demonstrated with practical examples in this chapter. Though the setting ranges and characteristics of the overcurrent protective devices can be selected over a wide range (see Chapter 7), compromises may be required. In case these compromises are not acceptable, additional protective devices may be required.

Coordination of the protective devices is undertaken toward the commissioning stage of a power system; however, a proper selection of the protective devices in the design stage itself is important for a properly coordinated system.

## 10.1 STANDARDS AND REQUIREMENTS

The various protective elements in a distribution system are relayed circuit breakers, low voltage circuit breakers, low voltage and medium voltage fuses, electronic fuses, medium voltage and low voltage motor starting contactors, reduced voltage and soft starting schemes for the motors, and the like. These devices protect a variety of equipment, namely, transformers, cables, generators, and motors, with respect to continuous overloads and within their fault withstand capabilities and thermal damage curves.

The coordination must meet the specific requirements of the processes. While nuisance trips are not acceptable, an equipment damage due to lack of protection will result in prolonged shutdown. Coordination is not a substitute for lack of proper protection and system planning.

NEC [1] provides some guidelines for the overcurrent protection of conductors, transformers, and motors. OSHA follows NEC requirements. A coordination engineer must be aware of the requirements of overcurrent protection for individual pieces of electrical equipment before an attempt to coordinate the devices in an electrical system are made. Further, it is not only the overloads against which the protection is required, the protection must also address the available short-circuit currents and system isolations before the equipment short-circuit withstand capability is exceeded.

## 10.2 DATA FOR THE COORDINATION STUDY

Prior to starting the study, an extensive data collection effort is involved:

1. A single line diagram of the distribution system. As a minimum this should show all protective devices which are to be coordinated on a time–current basis. All the switching conditions and plant operations must be charted out, for example, loss of a source, alternate route of power, which breakers will be tripped, which breakers will be closed, how the system will be operated, which protective devices will go out of service, and which new protective devices will be brought into service should be known.
2. The equipment's rated kVA, short-circuit ratings, and load flow currents under normal and contingency conditions should be known. Sometimes, short-time overloads are allowed in the systems for continuity of operations. The

equipments do have a short-time overload capability that can be calculated according to guidelines established in various standards. For example, standard [2] details the guidelines for the loading of the liquid-immersed transformers, with and without derating in the life expectancy of the transformers.

3. Current transformer (CT) ratios, their burden, secondary resistances, and relaying accuracies data are required (see Chapter 12). In this chapter, it is assumed that CTs are selected properly for the overcurrent relaying applications in accordance with the requirements discussed in Chapter 12.
4. Specific time–current characteristics of all protective devices to be coordinated should be available.
5. Full load current, starting (locked rotor) current, thermal withstand curves of the motor (alternatively, at least hot and cold safe locked rotor times), motor protection fuse size for NEMA 2 [3] medium voltage starters should be available for plotting and coordination. For low voltage motors, the ratings and characteristics of fused starters, thermal magnetic breakers, or MCPs (motor circuit protectors, which have magnetic protection only) should be available.
6. For transformers, their percentage impedances, primary and secondary winding connections, and system grounding should be known. A computer program will generate a transformer frequent or infrequent fault withstand curve without manual calculations (see Chapter 11).
7. For conductors, short-circuit withstand curve should be available. The modern computer programs for coordination will plot a short-circuit curve without manual intervention based upon the input data of the conductors, their type, and installation methods.
8. The following short-circuit data should be available.
   - The first cycle asymmetrical current is required for instantaneous (also differential and distance) relays when the operation is fast and the asymmetry in the fault current should be accounted for. The asymmetry in the first cycle also impacts CT performance.
   - The maximum symmetrical current is required to establish CTI (coordination time interval); see section below. The minimum interrupting current is required to ascertain whether the circuit protection sensitivity is adequate.
   - The 30-cycle short-circuit current is required for application of time delay relays operating beyond 10–20 cycles. The induction and synchronous motor contribution to the fault currents decay to zero and generators are represented by transient or higher impedances related to their decaying short-circuit currents. See Chapter 6 for detailed discussions on the decaying short-circuit currents and arc flash hazard analysis.
   - For arc flash, we ignore the DC component. Also, the relay operating time is calculated with 85% reduced arc flash current. See discussions in Chapter 1.

It is evident that unlike other power system studies, lot of preparation and data collection is needed before starting a coordination study.

## 10.3 COMPUTER-BASED COORDINATION

Twenty years back, the coordination of overcurrent devices was carried out manually using a light-box; this had a translucent surface illuminated with lamps placed below in the box. The published curves by the manufacturer for the specific devices will be gathered, placed on the light box, and traced on a log-log paper. Appropriate coordination was ensured visually by adjusting the curves being plotted and analytically by calculating the time and current margins and tolerances from the characteristic plots. This was a rather tedious routine, yet academically instructive. *This methodology is no longer in use.*

Currently, the coordination is done using one or the other commercially available program. These programs will run the short-circuit currents in a given system configuration, input the results to the relay coordination program, and if there are system changes, the short-circuit currents are automatically updated in the coordination. The time–current plots will be correctly terminated at the short-circuit currents seen by the protective devices in the system.

The protective device data and characteristics (of relays, low voltage and medium voltage fuses, and low voltage circuit breaker trip programmers) are built in to the computer database. The software vendors have kept pace with the introduction of new devices and multifunction overcurrent relays with programmable characteristics and keep expanding these databases. Also, the computer software has the capability that a user can program the time–current characteristics of a protective device not included in the computer database and add it to the database library. Similarly, circuit breaker types, low voltage trip programmers, fuses, and other equipment data can be added to the library. Thus, it is much easier today to experiment with various characteristics and settings for coordination and study their effect on arc flash. When a time–current curve is plotted, it is possible to move it upwards, downwards, and sideways for the best coordination. The settings are automatically corrected and charted out with each movement and placement of the time–current curve.

Some programs are available that will also suggest the coordinating settings or provide a complete auto-coordination of the protective devices in a certain setup. Though these auto-set programs can be tried, a better coordination is often achieved with manual intervention. This is a field where experience of a coordination engineer comes to the forefront and cannot be substitutes with auto-coordination programs, especially in situations where compromises have to be made.

## 10.4 INITIAL ANALYSIS

An initial analysis of the protection should reveal that none of the system components are exposed to damaging overloads or short-circuit currents, which will go undetected under various operating conditions. All equipment should be applied within its assigned continuous and short-circuit ratings and thermal withstand capabilities, and meet the requirements of ANSI/IEEE, OSHA, NEC, NESC, and other national and international standards. While manipulating the settings for arc flash reduction, it is permissible to make an informed judgment with respect to lack of some selectivity, but not jeopardize the equipment protection.

## 10.5 COORDINATING TIME INTERVAL

The sequential operation of series connected protective devices depends upon maintaining a minimum coordinating time interval (CTI) throughout the operating range. A graphical representation of the time–current characteristics is an accepted method, though it is possible to determine selectivity by comparing at the most three critical values of the fault currents and ascertaining the relay operating times.

The CTI takes into account the following:

- Circuit breaker interrupting time
- Relay overtravel (also called impulse margin time), for electromechanical relays, explained next
- An arbitrary safety factor to account for CT saturation and setting errors.

### 10.5.1 Relay Overtravel

Consider the coordination between two relays R1 and R2 in series. Relay R1 is upstream of relay R2, with little intervening impedance and a fault occurs downstream of relay R2. Both the relays see approximately the same magnitude of fault current. The operation of relay R1 should be delayed by CTI, so that relay R2 clears the fault and relay R1 does not operate. The impulse margin time is defined as:

$$T_{IM} = T_{OP} - T_I, \tag{10.1}$$

where $T_{IM}$ is the impulse margin time, $T_{OP}$ is the time as read from the time–current curves of the relay for the specific current pickup and time dial setting, and $T_I$ is the minimum impulse time during which sufficient inertia is supplied to the disc to eventually cause the disc to close its contacts, even when the fault seen by the relay is removed, in this case by the operation of relay R2. This time is calculated by testing, and manufacturers can supply data for the exact coordination. This should be taken into consideration for the CTI. Table 10.1 provides the guidelines for selecting CTI.

## 10.6 FUNDAMENTAL CONSIDERATIONS FOR COORDINATION

- When protecting a delta-wye transformer, a 16% margin should be used between primary and secondary device characteristic curves; this is so because per unit primary current in one phase for a secondary phase-to-phase fault is 16% greater than the secondary current (see Chapter 11).
- In addition to short-circuit, a load flow study is required to determine normal and emergency load currents that the system should carry. The continuous current ratings of the equipment, for example, cables, circuit breakers, and transformers

TABLE 10.1. Coordinating Time Intervals

| Switching Device | Coordinating Time Interval |
|---|---|
| Relayed medium voltage circuit breakers | Very inverse and extremely inverse electromechanical relays<br>0.45 second: 8-cycle breakers (pre-1964 basis of rating)<br>0.40 second: 5-cycle breakers<br>0.36 second: 3-cycle breakers<br>Inverse time characteristics electromechanical relays<br>0.48 second: 8-cycle breakers (pre-1964 basis of rating)<br>0.43 second: 5-cycle breakers<br>0.39 second: 3-cycle breakers<br>(The relay impulse travel time is longer for inverse relays as compared with very inverse or extremely inverse relays)<br>Modern MMPR or solid-state relays<br>0.35 second: 8-cycle breakers (pre-1964 basis of rating)<br>0.30 second: 5-cycle breakers<br>0.26 second: 3-cycle breakers<br>Properly calibrated and field-tested relays<br>Reduce the above CTIs by another 0.05 second. Much less calibration and testing of the timing is required for modern MMPRs. |
| Relayed medium voltage circuit breakers and fuses | Electromechanical relays<br>0.20 second<br>MMPRs or solid state relays<br>0.1 second<br>Relay over-travel and breaker interrupting time is eliminated. |
| Low voltage relayed power circuit breakers | The interrupting time may safely be considered as three cycles (IEEE Guide 1584) Thus coordination with 0.1-second CTI possible. |
| Low voltage power circuit breakers with solid state trip programmers, see Figure 7.15 | The slight time margin provided by the manufacturer between short-time delay bands is adequate for coordination. Sometimes, these bands overlap when plotted graphically, yet coordination is achieved. Referring to Figure 7.15, which shows three ST bands, the coordination is achieved within the small gap shown between the four bands. |
| Fuses, medium voltage and low voltage | Manufacturers provide minimum melting and maximum operating time curves. There should be a margin of at least 10% between the maximum operating characteristics of the downstream fuse and the minimum melting characteristics of the upstream fuse. Coordination below 0.01 second on a time–current basis should not be attempted. |
| Instantaneous functions, electromechanical relays | The settings must recognize the asymmetrical nature of the fault, as the relays can operate equally well for AC and DC currents. Coordination without intervening impedance is not attempted. |
| Coordination instantaneous basis, current-limiting devices | See text. |

MMPRs, microprocessor-based multifunction relays.

should not be exceeded. It is understood that all the equipment are selected and designed to serve the continuous and emergency load currents, as a first protection requirement.

- All switching devices should be rated to meet the available short-circuit current requirements. The cables should be designed to withstand the required short-circuit currents in case of fault, for a duration dependent upon the protective device.
- Inrush currents of motors and transformers should not result in nuisance trips. The pickup settings should override these inrush currents.
- The individual load or branch circuit protection should meet the requirements of NEC. Sometimes to meet these requirements, higher cable sizes may be required.

**Example 10.1**

Consider a 2000-kVA, 13.8–0.48-kV transformer, of percentage impedance of 5.75% on 2000 kVA rating base, which is provided with primary overcurrent protection through a 150E fuse and has a secondary main circuit breaker of 3000 A provided with low voltage trip programmer. According to table 450.3(A) of NEC, the secondary circuit breaker should not have an overcurrent pickup setting of more than 125% of the transformer full load rated current of 2405.6 A, that is, a setting of 3007 A. Though the actual load current of the transformer may be lower than its rated current, say only 80%, yet the secondary cables between the transformer and 3000-A circuit breaker must be rated for a minimum ampacity of 3007 A for 125% settings.

### 10.6.1 Settings on Bends of Time–Current Coordination Curves

This situation is illustrated in Figure 10.1. The intended arcing times are A and B, but these can translate into A′ and B′ providing much larger time delays. Settings should ensure that calculated arcing currents are not on flexture of curves.

## 10.7 COORDINATION ON INSTANTANEOUS BASIS

This is a new frontier of much importance for arc flash hazard reduction. The selectivity is accessed from the time–current coordination curves (TCC) of devices plotted to the same scale and noting down the overlaps. The time–current coordination curves are not extended below 0.01 second, and when a current-limiting fuse or circuit breaker operates faster in subcycle range, the TCC plots cannot determine the selectivity with any accuracy. When there is device interaction, the TCC plots will not indicate how two fast acting devices will behave together when connected in series.

The current-limiting molded case circuit breaker are available, and even the normal molded case circuit breakers (MCCBs) not really marked as current limiting according to UL 489 may be current limiting to an extent. The circuit breaker instantaneous trip is sensitive to peak amperes, and the faults of same rms value but of different *X/R* ratio will be sensed differently by the circuit breaker trip systems, see Chapter 7.

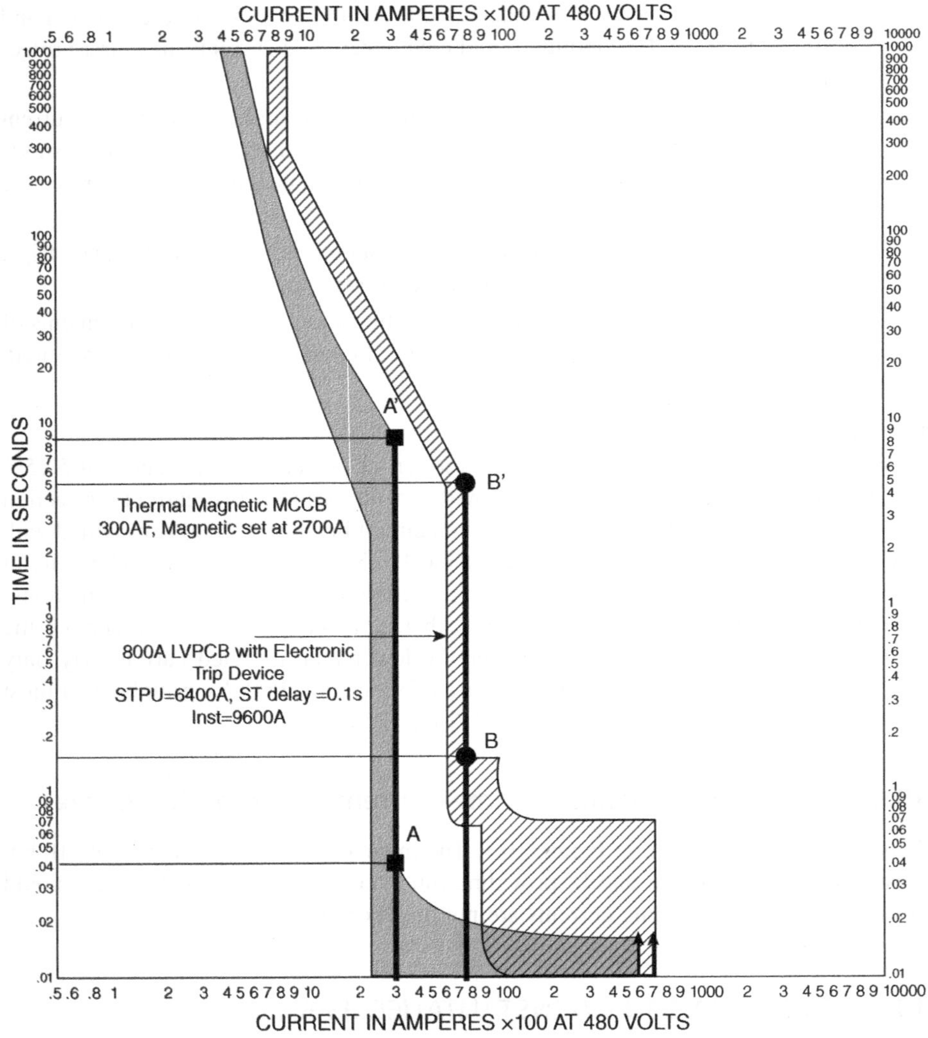


Figure 10.1. Settings on curve flexures, which can give increased fault clearance times.

The characteristics of current-limiting fuses are also discussed in Chapter 7. Above approximately one cycle, the selectivity between the two fuses is determined by maintaining a separation between the total clearing time of the downstream fuse and minimum melting curve of the upstream fuse.

The current-limiting behavior has the following advantages:

- Reduced let-through energy
- The downstream faults can be cleared faster.
- Selectivity can be obtained with proper application considerations.

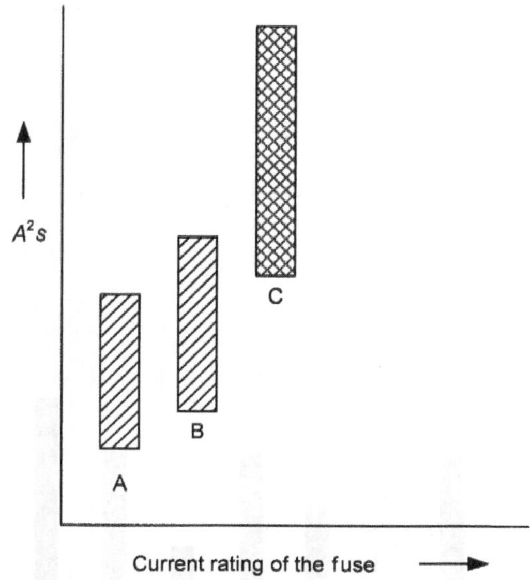


Figure 10.2. Selectivity criteria based upon $I^2t$ let-through of current-limiting devices.

### 10.7.1 Selectivity between Two Series-Connected Current-Limiting Fuses

Fuse melting and clearing $I^2t$ values can be compared to determine selectivity in their current-limiting range. The total $I^2t$ of the downstream fuse must be less than the melting $I^2t$ of the upstream fuse for events lasting less than 0.01 second. This is shown in Figure 10.2. The curves must be plotted for the same test power factor. The fuse of rating A will coordinate with the fuse of rating C, but not with the fuse of rating B. Another factor to be considered is the number of fuses in parallel clearing the fault. When two fuses in parallel clear the fault, these share line-to-line voltage and yield a lower arcing $I^2t$ than the single fuse. Figure 10.3 shows the let-through characteristics of medium voltage power fuses of 4–40 A of a manufacturer. The curves are based on the fuse being at an ambient temperature of 25°C and with no load; all variations should be negative. Figure 10.4 illustrates the let-through characteristics of RK5 time-delay fuses of a manufacturer.

### 10.7.2 Selectivity of a Current-Limiting Fuse Downstream of Noncurrent-Limiting Circuit Breaker

In this case, traditional TCC may be misleading to determine selectivity. If the fuse curve crosses the instantaneous foot of the circuit breaker curve, it is likely that the combination may not coordinate and a proper gap between the maximum operating time of the fuse and the instantaneous setting of the circuit breaker is required. Sometimes, it may be misleading to visually examine and place the fuse and circuit breaker curves with some separation gap at 0.01 second and conclude that the system is selective.

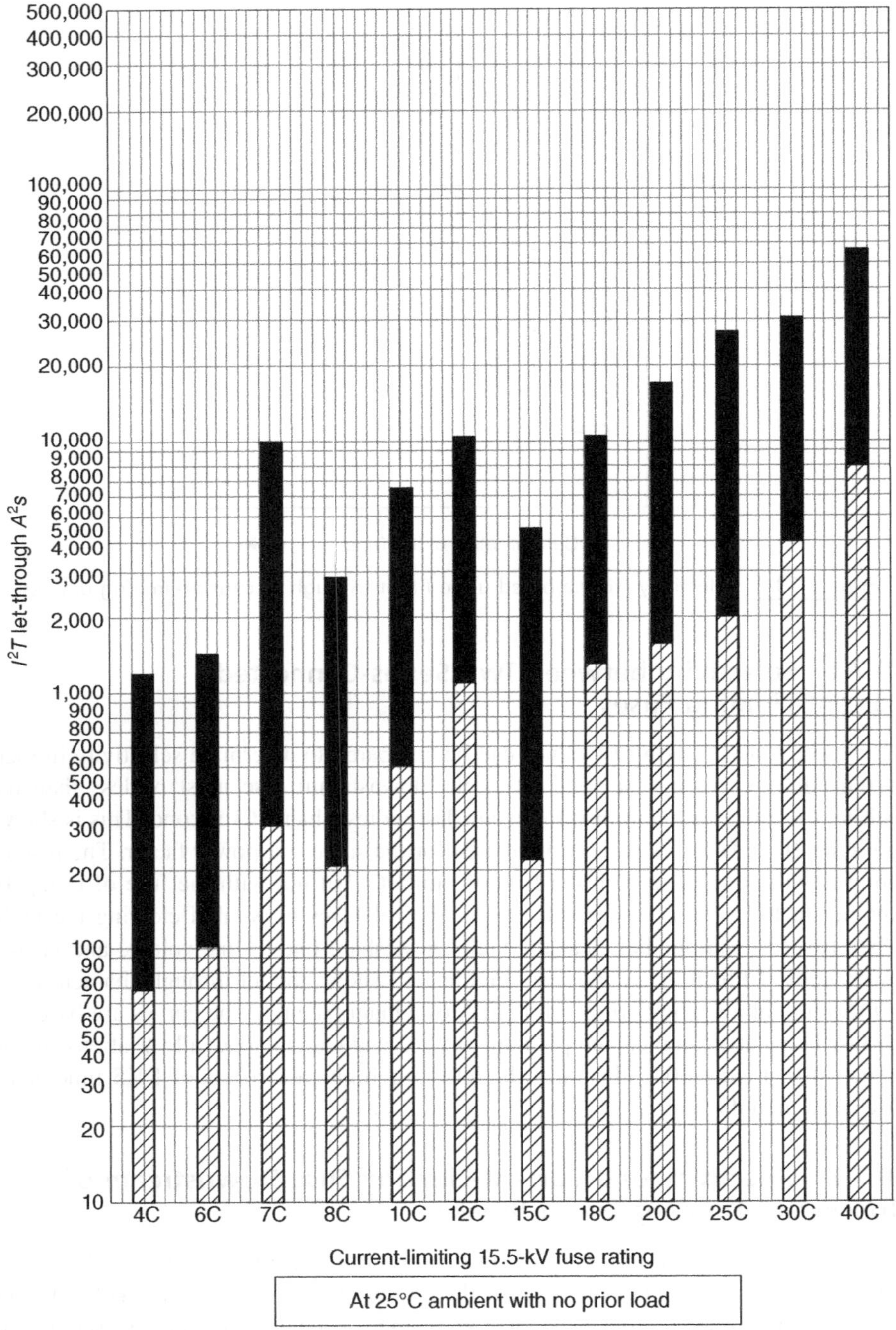


Figure 10.3. $I^2t$ let-through curves of medium voltage power fuses, 4–40 A, of a manufacturer.

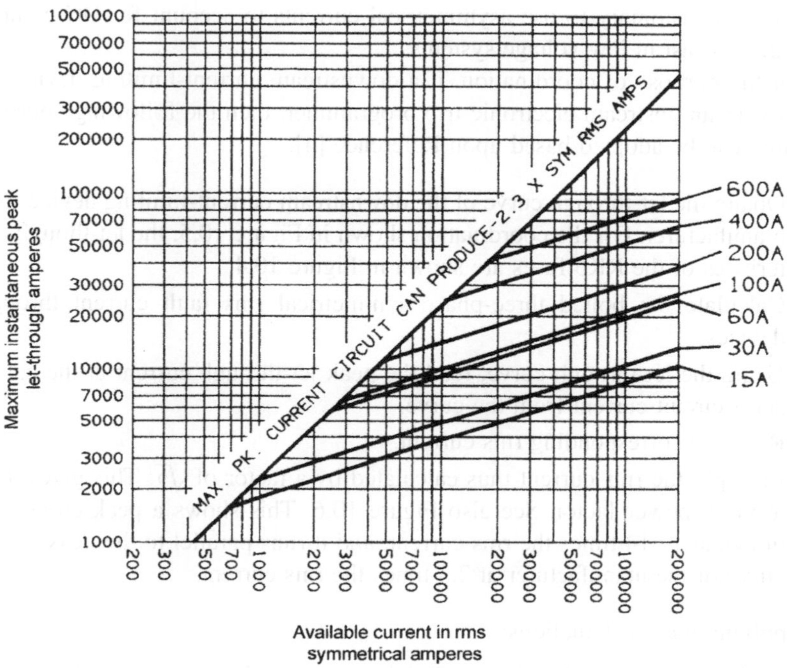


Figure 10.4. Peak let-through current of time-delay RK5 fuses of a manufacturer.

**Example 10.2**

Figure 10.5 depicts an upstream 800 AF LVPCB provided with a programmable trip device having LSI functions and a sensor and plug rating of 800 A and its coordination with a 100-A RK5 fuse on the downstream side. The upstream circuit breaker is set as follows:

Long-time pickup = 800 A (=1)
Long-time delay = 4 seconds
ST pickup = 3200 A (=4), $I^2t$ = In
ST pickup delay = 0.1 second
Instantaneous pickup = 9600 A (=12).

This figure shows that apparently, there is coordination between the fuse and the circuit breaker, which is true for time duration of 0.01 second; however, it will be demonstrated that on an instantaneous basis, these devices do not coordinate.

Circuit breaker instantaneous trip units are typically of magnetic or electronic types. Both these types' respond to instantaneous current values. Time is also influential as the current needs to last long enough to generate sufficient force in a magnetic trip to overcome friction forces and inertia. Similarly, electronic trip units may require several samples of currents, and may have filtering to minimize risk of nuisance trip. When the circuit breaker is set, a tripping threshold is defined, and if one or more samples of measured current exceed this set value, then the circuit breaker is tripped.

This may be somewhat confusing because the circuit breaker time–current characteristics are published by the manufacturers in terms of rms currents. For coordinating settings, it is customary to use asymmetrical currents to account for DC component, which decays fast in low voltage systems.

For the purpose of coordination, if a downstream current-limiting device is considered with an upstream electronic trip programmer, then the following coordination procedure can be adopted based upon Reference [4].

- Obtain the let-through curve of the downstream current-limiting device from the manufacturer. For the coordination shown in Figure 10.5, the let-though characteristics of the RK5 fuses are shown in Figure 10.4.
- Calculate the bolted three-phase symmetrical rms fault current through the devices.
- Using the let-through curve, read the peak let-through current at the calculated short-circuit current from the curve.
- Read the corresponding rms current.
- Multiply the rms current thus calculated by a factor of $\sqrt{2}$. This gives the peak at unity power factor. See also Figure 10.6. This shows a peak curve (dotted) drawn at 1.414 times the rms current, and it runs parallel to the peak let-though curve of the manufacturer at 2.3 times the rms current.

Applying these calculations:

Peak let-trough of 100-A fuse at 50 kA rms short-circuit current = 18.3 kA.

Peak at unity power factor $= 18.3/\sqrt{2} = 13.5$ kA.

Set the instantaneous of upstream circuit breaker at a minimum of 13.5 kA to coordinate with downstream current-limiting fuse. The coordination in Figure 10.5 shows instantaneous set at 9600 A. As this is <13,500 A, coordination will not be obtained in the instantaneous zone. The instantaneous setting of 800-A breaker should be raised.

The instantaneous settings in electronic trip programmers are generally specified in terms of sensor rating (sometimes, these can also be in terms of plug rating). Depending upon the circuit breaker type, the instantaneous settings range is up to 12 times the sensor rating and in some cases up to 15 times the sensor rating. To provide the required instantaneous setting of 13.5 kA, we must have a sensor rating of 1200 A. As a sensor rating of 1200 A cannot be accommodated in 800AF breaker, the breaker frame rating is changed to 1600 A, and it is provided with a 1200-A plug. For the low voltage systems, the long-time pickup must protect the conductors within their rated ampacities (Chapter 7). The new settings are:

Long-time pickup = 840 A (=0.7)
Long-time delay = 4 seconds
ST pickup = 3200 A (=4), $I^2t$ = In
ST pickup delay = 0.1 second
Instantaneous pickup = 14,400 A (=12).

This shows that the 800-A circuit breaker has to be replaced with a 1600-A breaker and required sensor and plug to accommodate new instantaneous settings. The coordination is shown in Figure 10.7: the coordination in the instantaneous zone will be achieved.

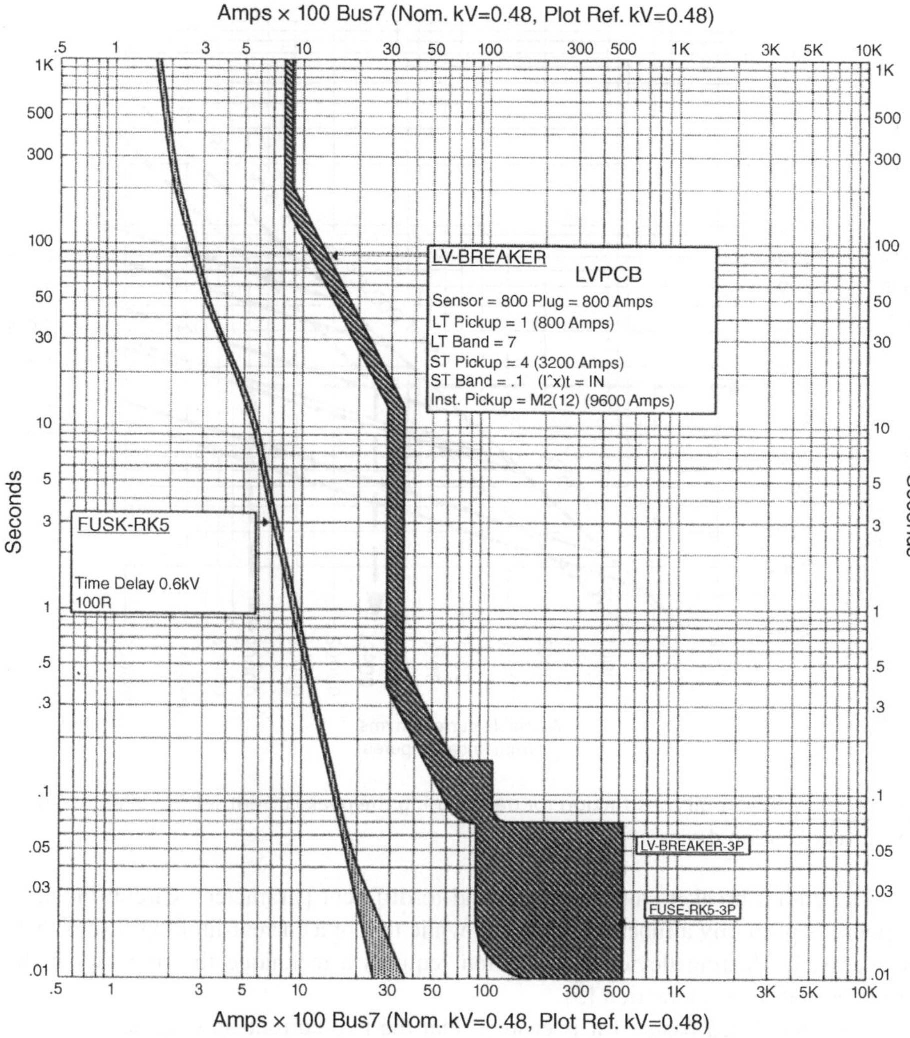


Figure 10.5. Coordination on instantaneous basis between an upstream LVPCB and downstream 100-A, RK5 fuse; the system is not selective on instantaneous basis (see text).

### 10.7.3 Selectivity of Current-Limiting Devices in Series

The following observations can be made:

1. In a series circuit, any current and energy limiting by either device will impact both the devices in series.
2. The device with the lowest current limiting threshold and fastest response will affect the current magnitude to operate the less sensitive and slower device.
3. The faster device limits the let-through energy, as well as let-through current.

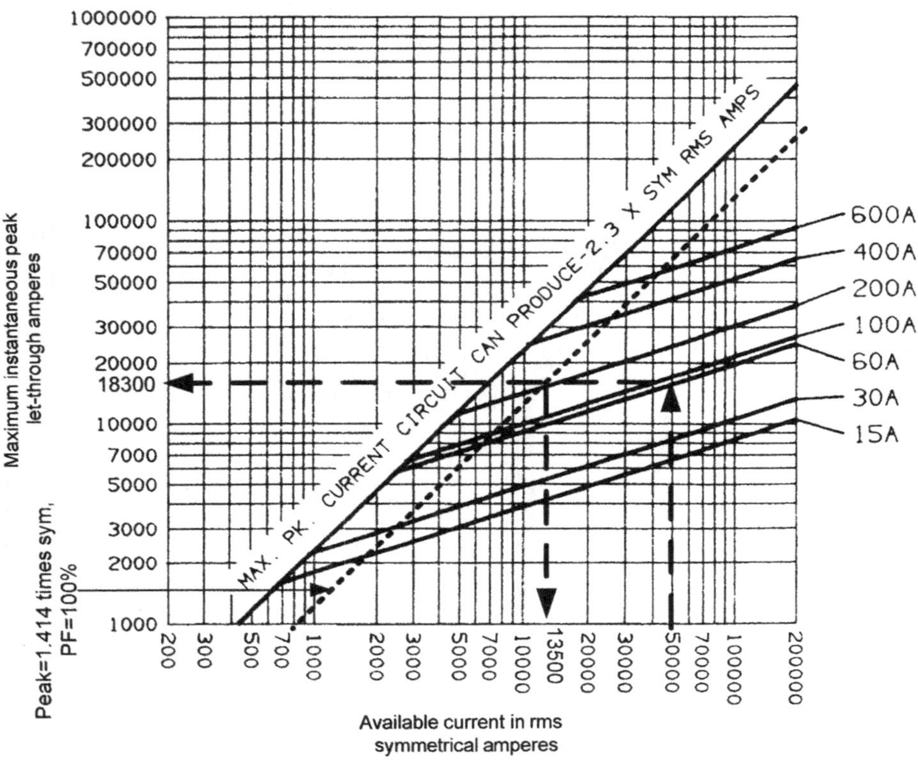


Figure 10.6. Graphical construction for determining the peak current setting of an upstream noncurrent-limiting device.

The circuit breakers and fuses respond to different parameters. Circuit breaker response is primarily a function of current while that of a current-limiting fuse is that of energy. Evaluating device coordination requires a technique that includes both variables and their interaction [5].

On a simplified basis, if it is ensured that the two current-limiting devices, say an upstream fuse and downstream circuit breaker are so sized, that for the maximum fault current at which the selectivity is desired, the upstream device does not see at all the peak let-though current or follow current of the downstream device; a coordination will be obtained. This method may give much larger rating of the upstream device, compared with the rigorous evaluation of the let-through current and energy.

An electronic trip implementing waveform recognition (WFR) can consider a combination of peak current and time to determine if the fault current shows a characteristic wave shape of an energy and current-limiting interruption. A trip device that is able to distinguish that waveform is energy limited can be set more sensitive than the one that considers peak current only, sometimes resulting in settings that are 50% of the peak current method alone. WFR is the specific to one manufacturer only [5].

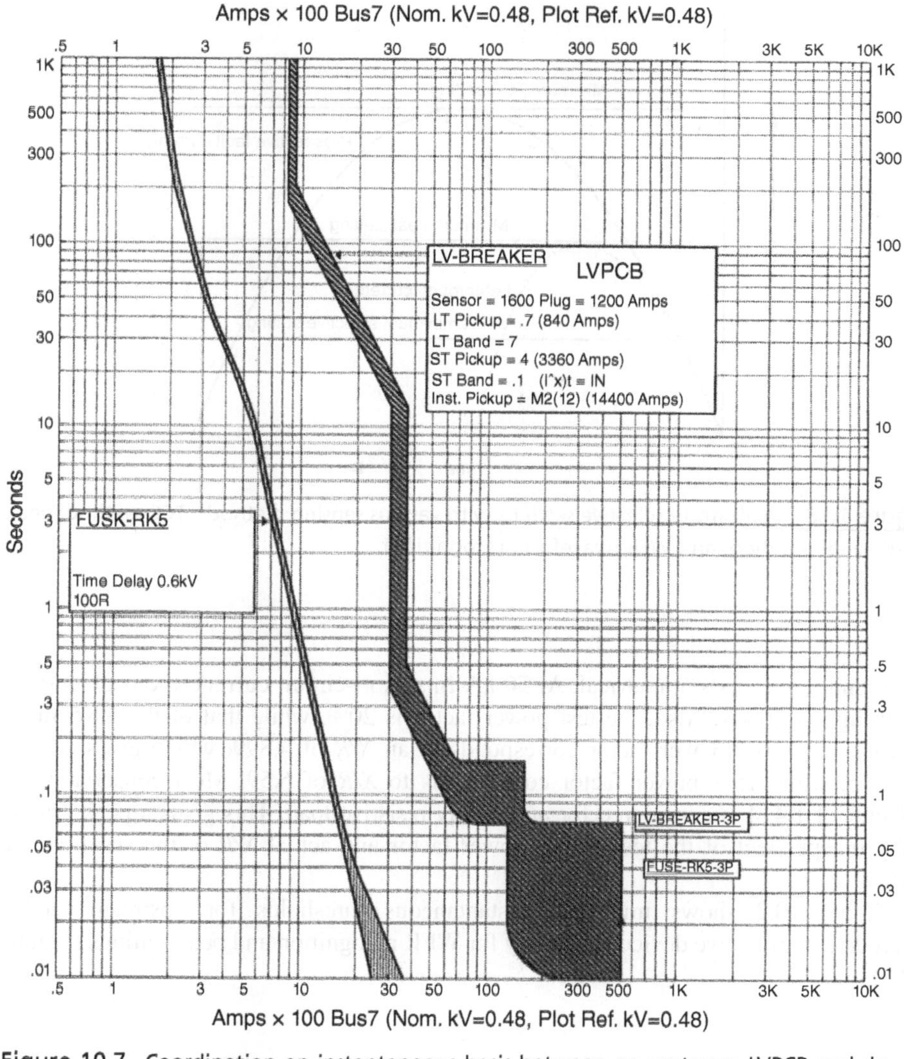


Figure 10.7. Coordination on instantaneous basis between an upstream LVPCB and downstream 100-A, RK5 fuse; the system is selective on instantaneous basis (see text).

Figure 10.8 shows a comparison of different selective circuit breaker settings. The minimum TCC setting can be identified as the setting by the prospective fault current peak, as determined from TCC curve considering DC offset. The minimum peak setting is based upon the downstream current-limiting overcurrent device. The WFR setting can be still lower, if the trip employs advanced waveform recognition or energy-sensing algorithm.

For current-limiting circuit breakers, the manufacturer's data should be applied. Figure 10.9 shows the curve for a current-limiting MCCB. The threshold current

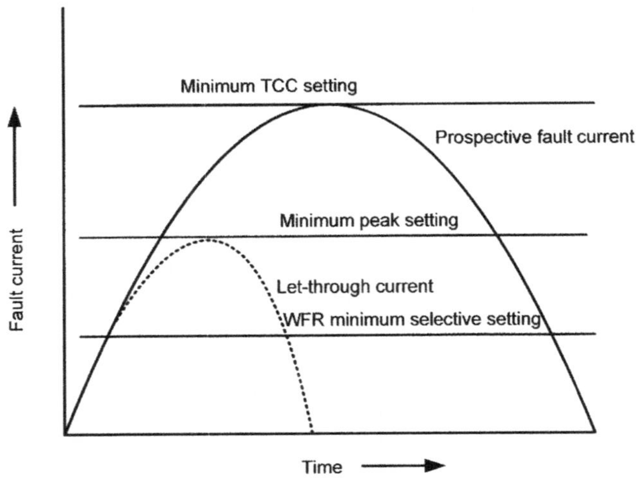


Figure 10.8. To illustrate selective settings with various sensing methods: TCC method, peak let-through method, and WFR (waveform recognition).

is around 9 kA rms symmetrical. At 50 kA rms short-circuit current, the peak current let-trough is 20 kA. Also, the test power factor is 20%, while that of the RK5 fuses is 15%. The 20% power factor corresponds to an *X/R* of 4.899, which gives a peak of 2.16, while 15% power factor corresponds to *X/R* of 6.59. Most conservatively, a factor of 1.414 is considered for an entirely resistive load. This gives a peak practically independent of the test circuit power factor and actual *X/R* of the short-circuit current.

Table 10.2 shows minimum instantaneous thresholds for upstream feeder overcurrent protective device selectivity for WFR recognition and peak settings (Figure 10.8).

## 10.8 NEC REQUIREMENTS OF SELECTIVITY

The following articles from NEC, 2011 relating to overcurrent selectivity are of interest:

*Article 240.12: Electrical System Coordination.* Where an orderly shutdown is required to minimize the hazards to personnel and equipment, a system of coordination based on the following conditions shall be permitted:

*Coordinated Short-Circuit Protection.* Here, the intention is that the faulty circuit is isolated by operation of the immediate overcurrent device, without impacting the overcurrent devices in the rest of the system.

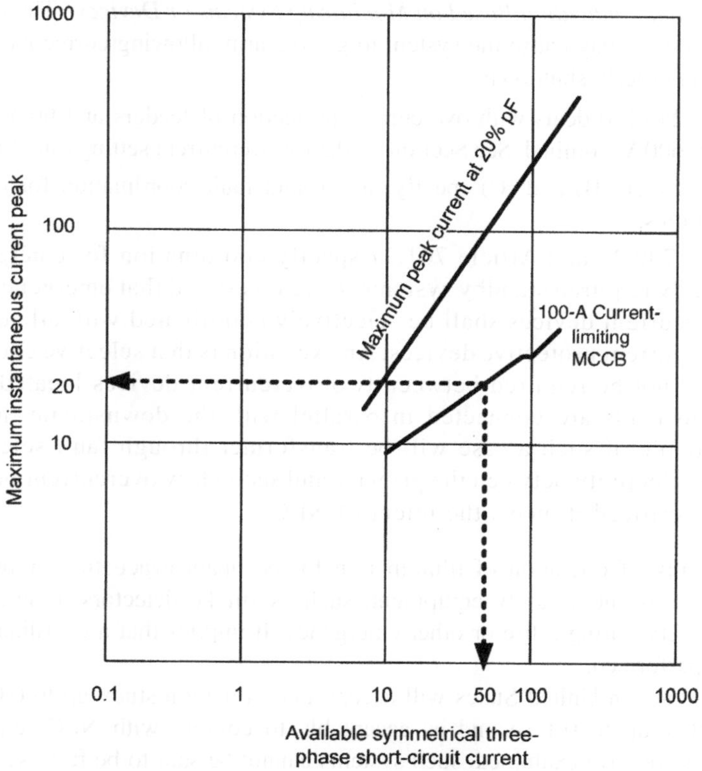


Figure 10.9. Peak let-through current of current-limiting 100 A MCCB.

TABLE 10.2. Typical Minimum Instantaneous Thresholds for Upstream Feeders Above Current-Limiting Overcurrent Protective Devices (Settings for Circuit Breakers Rated 480 V and Below, Settings for Fuses Rated 600 V and Below)

| Current-Limiting OCPD Type | Current-Limiting OCPD Rating | WFR TRIP[a, b] | Peak Sensing Trip for 55-kA Selectivity (Amperes) |
|---|---|---|---|
| MCP with current-limiters | 50 A | 3,800 | 9,740 |
| | 150 A | 10,750 | 19,200 |
| MCCB | 150 AF | 9,600 | 27,610 |
| | 600 AF | 20,350 | 33,810 |
| Class J time-delay fuse | 100 A | 3,810 | 7,730 |
| | 600 A | 26,730 | 28,330 |
| RK1 time-delay fuse | 100 A | 4,300 | 8,370 |
| | 600 A | 32,380 | 32,190 |
| RK5 time-delay fuse | 100 A | 8,820 | 13,520 |
| | 600 A | 57,120 | 47,000 |

[a] Selectivity will range up to the short-circuit rating of the lowest-rated device in pair, or withstand of the line side device, whichever is lower.

[b] Minimum setting assumes 10% tolerance.

*Source*: Adapted from Reference [4].

*Overload Indication Based on Monitoring Systems or Devices.* The monitoring system may cause the system to go to alarm, allowing corrective action for an orderly shutdown.

Article 240.100 deals with overcurrent protection of feeders and branch circuits, over 600 V, nominal. See Section 7.10.5 for overcurrent settings and fuse ratings.

Article 517.17(B) and (C) specify the ground fault coordination for health-care facilities.

Article 700.27 and Article 701.27 specify coordination for emergency and legally required standby systems. Here it is stated that emergency systems overcurrent devices shall be selectively coordinated with all supply side overcurrent protective devices. An exception is that selective coordination shall not be required between two overcurrent devices located in series if no loads are connected in parallel with the downstream device. An example of such a case will be transformer through fault selectivity. If the selectivity between the primary and secondary overcurrent protections is sacrificed it meets the intent of NEC.

Continuity of operation of illumination for occupant evacuation or maintaining continuity of essential safety equipment, such as smoke detectors, is necessary for occupant safety during a fire or other emergency. It implies that a coordination study should be performed.

Some states in United States will accept a coordination study up to 0.01 second, and the other up to 0.1 second as acceptable to comply with NEC requirements. However, as we have examined, such systems cannot be said to be fully selective.

### 10.8.1 Fully Selective Systems

Systems designed for higher level of selectivity may lead to higher arc flash energy. It becomes a challenge to have a fully selective system and also reduce arc flash energy simultaneously. This can only be achieved by selecting suitable devices that can coordinate on instantaneous basis.

The systems for complete selectivity may require:

- Higher sizes of cables and trip devices.
- Larger size (kVA rating) of equipment, more than what is required to serve the loads.
- The systems may be more expensive to implement.
- Selectivity between three low voltage circuit breakers in series is generally the maximum that can be attempted on instantaneous basis.
- If all the three circuit breakers see the same amount of short-circuit currents, that is, the interconnecting cables between these breakers are of short length, it will be all the more difficult to design for 100% *selectivity and simultaneously reduce the arc flash hazards.*

### 10.8.2 Selection of Equipment Ratings and Trip Devices

It is necessary to reiterate between the selection of breakers, their trip devices, and system design. Note that the 0.1 Ω current-limiting reactor is provided on the feeder circuit to the lighting transformer T2, in Figure 10.10. The calculated short-circuit level at the emergency panel is 14.9 kA, and at the MCC, it is 20.64 kA rms symmetrical with change over switch connected to normal power source. The 400-A

**Example 10.3**

Figure 10.10 illustrates an essential service distribution, where a selective coordination is required. A 400-A automatic changeover switch is provided between the emergency diesel generator and normal source of power. The essential loads are served from emergency panel, and only a 75-kVA lighting transformer is shown.

The normal power source is provided from a 1500-kVA, delta-wye, 13.8–0.48-kV transformer, 480-V system high resistance grounded. The transformer primary is protected with a current-limiting fuse of 150E, and there is a main secondary power circuit breaker of 2000 A, provided with electronic trip programmer with LSI functions. The feeder power circuit breakers of 800 A are also provided with electronic trip programmers. One such feeder circuit breaker serves a 480-V MCC, which in turn serves the emergency distribution panel through a breaker “BRK-PNL” and changeover switch of 400 A. The 75-kVA lighting transformer is provided with primary and secondary breakers and serves a lighting distribution board at 240 V.

The emergency generator is rated at 1.0 MVA, 0.85 PF, and has a main generator LVPCB, with an electronic trip programmer. A feeder circuit breaker from the generator panel connects to the 400-A change over switch as an alternate source of power.

When the system is served from a normal power source:

- Lighting board feeder circuit breaker of 30 A should fully coordinate with a 75-kVA lighting transformer secondary circuit breaker of 250 A.
- Coordination between the 75 kVA lighting transformer primary and secondary breakers is not necessary, according to NEC.
- 75-kVA lighting transformer primary circuit breaker should coordinate with MCC feeder circuit breaker, BRK-PNL.
- MCC feeder circuit breaker BRK-PNL should coordinate with low voltage feeder circuit breaker, BRK-SWITCH-FEEDER.
- The low voltage feeder circuit breaker BRK-SWITC-FEEDER should coordinate with the main 2000 A secondary circuit breaker of 1.5-MVA transformer.
- The main secondary circuit breaker should coordinate with 1.5-MVA transformer fuse.

In total, seven devices in series are required to be coordinated.

Similarly, when the system is served from the emergency generator, five devices in series must be coordinated.

Not only full coordination is required, but also arc flash hazard should remain low, preferably no more than HRC level 2.

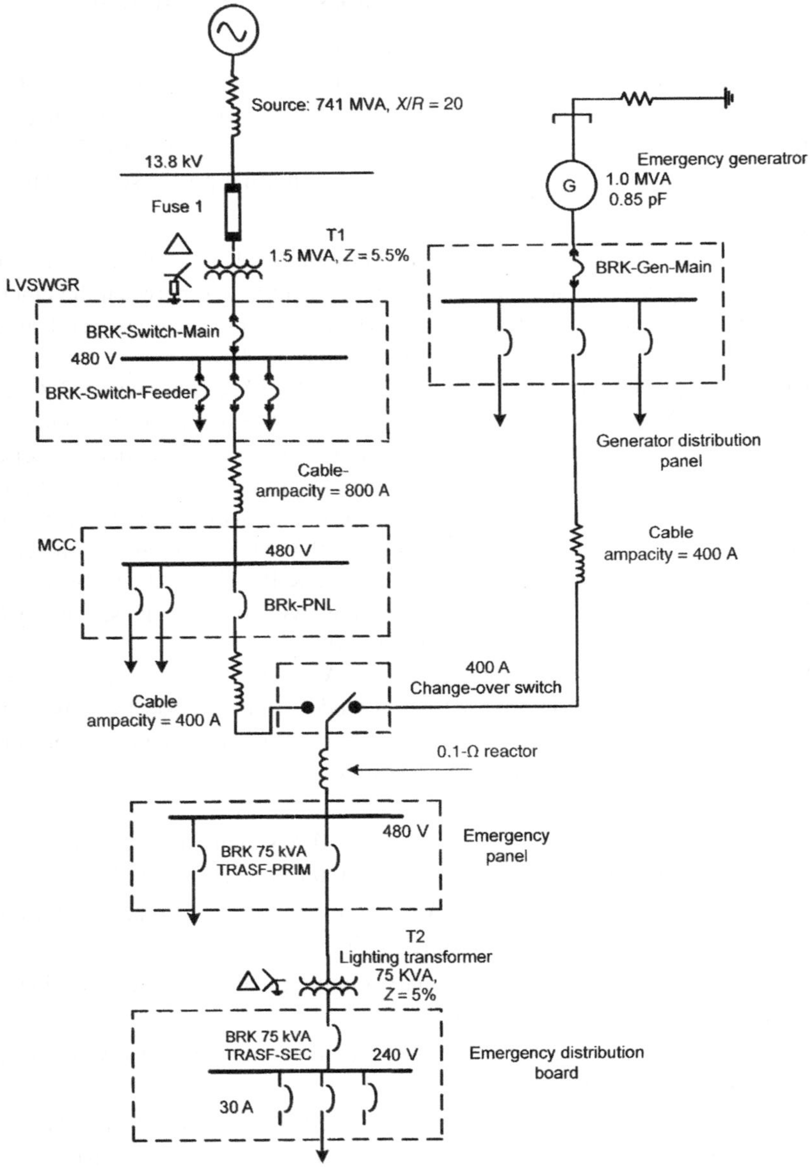


Figure 10.10. An essential service system configuration for full selectivity and arc flash hazard reduction.

It is obvious that current-limiting MCCBs are required for selectivity. A manufacturer's published data on his low voltage circuit breakers showing coordination on instantaneous basis is used to develop the settings. This manufacturer's data is not reproduced [6].

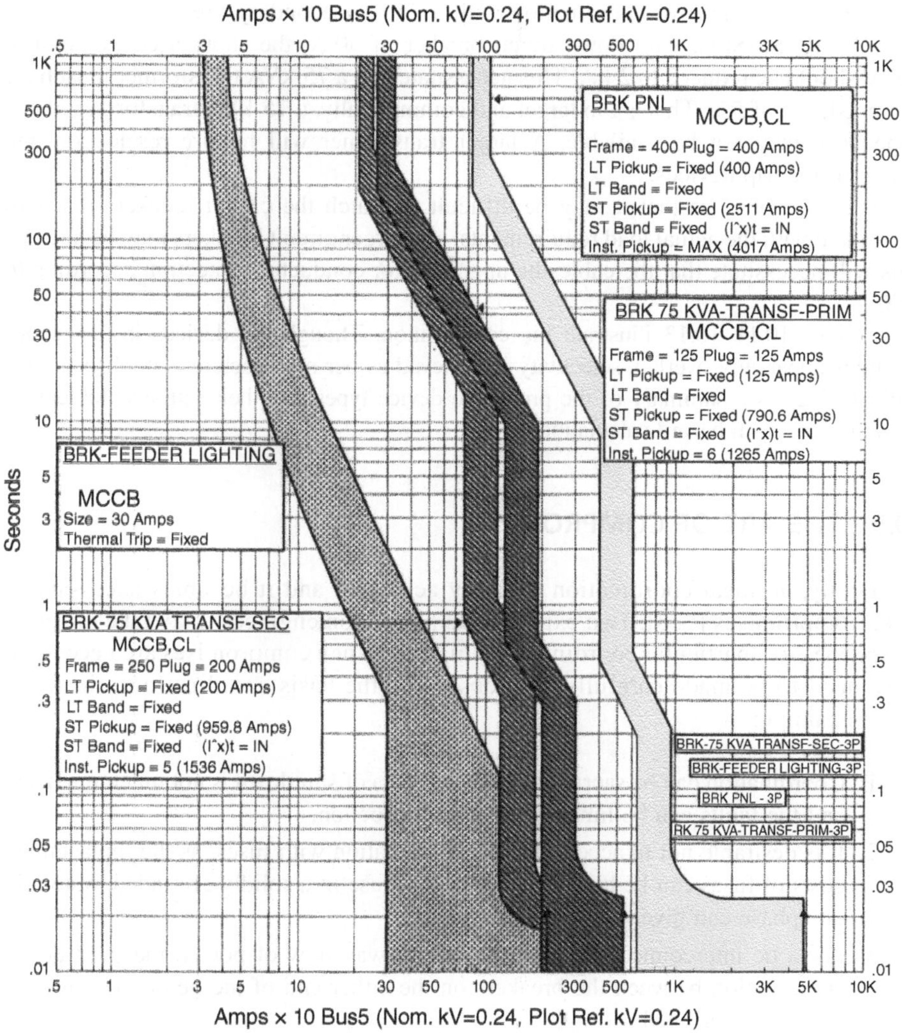


Figure 10.11. Time–current coordination of the OCPDs in Figure 10.10.

current-limiting MCCB, breaker BRK-PNL, to the emergency panel will coordinate with the feeder LVPCB on the low voltage switchgear, instantaneous set at maximum up to 65 kA maximum short-circuit current level, but the T2 transformer primary current-limiting breaker, BRK 75KVA-TRANSF-PRIM, of 150 A cannot be made to coordinate with MCC breaker for any currents above 3.212 kA on the instantaneous basis. The 0.1-Ω reactor reduces the short-circuit current on the source side of the T2 primary breaker to 2.37 kA. A perusal with the manufacturer's data shows that there is no combinations of tested breakers that can achieve coordination for up to 15 kA short-circuit currents and be of required size to protect the small T2 transformer of 75 kVA.

Similarly, to achieve coordination with a 250-A secondary breaker of T2, BRK 75KVA-TRASF-SEC, with downstream breaker of 30 A, the short-circuit current at 240 V should be reduced to 2 kA. This necessitate that a 75-kVA transformer must have an impedance of 5%. This is too large, and commercially, a 75-kVA transformer of 5% impedance may not be available. A larger transformer will require reselection and design of the trip devices.

It is also obvious that it may be difficult to match the circuit breakers that will coordinate with the actual load requirements, and larger circuit breakers may be needed. This will necessitate that the cables be upgraded for conductor protection according to NEC.

Figures 10.11–10.13 illustrate the coordination. Though the devices instantaneous functions apparently overlap, selectivity is achieved as per manufacturer's published data. Table 10.3 gives a summary of the protective device types and their settings. Table 10.4 shows the arc flash hazard calculations.

## 10.9 THE ART OF COMPROMISE

Practically, an ideal coordination is rarely achieved, and it becomes necessary to make compromises [7, 8]. Two experienced protection engineers may take different approaches to a particular coordination situation. When compromises are necessary, these need to be made carefully, still observing the basis coordination principles. As an example:

1. Coordination can be sacrificed when the area of shutdown is not much impacted and the faults can be limited to the faulty section.
2. It is desirable not to accept lack of coordination with fuses, as the replacement costs of fuses can be high. Operation of a fuse in a solidly grounded system in one phase can give rise to single phasing.
3. For a tie interconnecting circuit with two way flow of power, the overcurrent coordination between the breakers on the either end of the tie circuit can normally be sacrificed.
4. For the protection of a transformer, where both the secondary and primary protections are provided, the primary protection can cover, to an extent, the secondary faults, and is more important than the secondary protection. NEC allows only primary protection of the transformers, with no secondary protection, provided the requirements of NEC overcurrent settings are met.
5. For large motors, some lack of coordination at the motor starter level is acceptable rather than perpetuating it upstream.
6. Coordination should be obtained for the maximum fault currents in the system. However, bolted three-phase faults are rare. Some lack of coordination at the maximum fault currents can be acceptable.

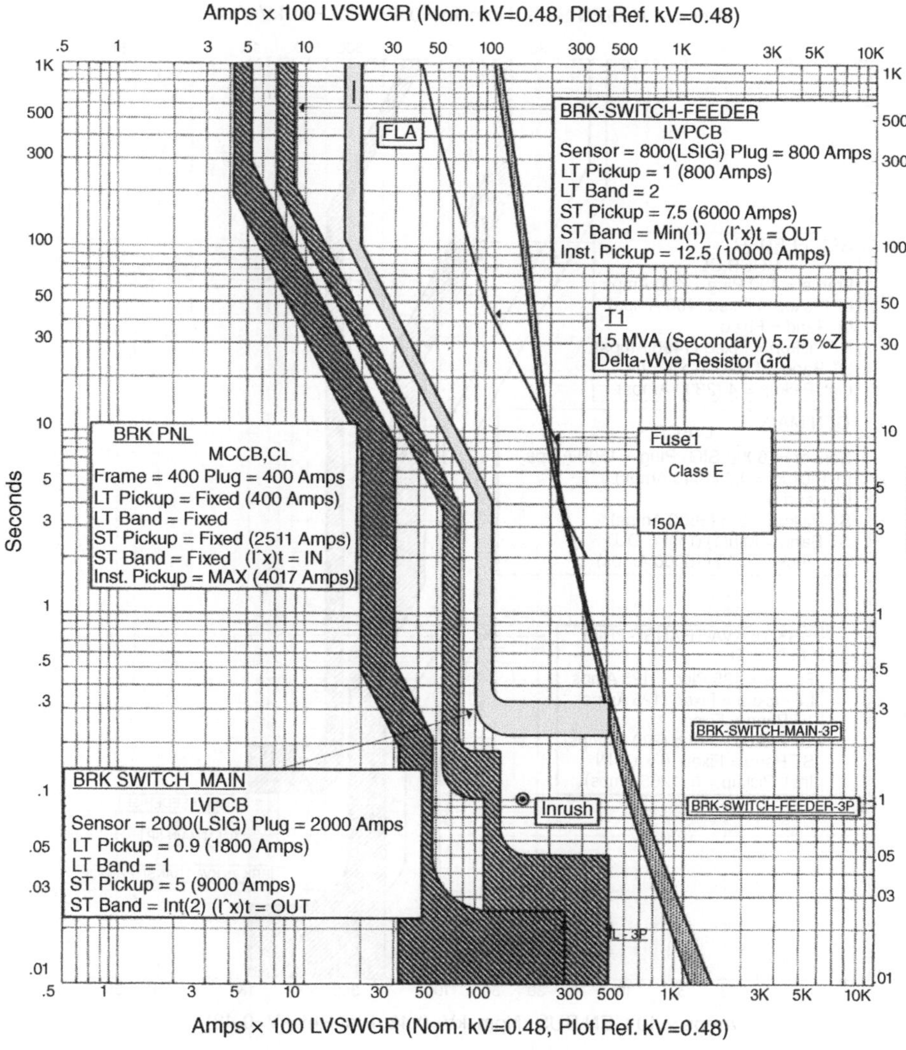


Figure 10.12. Time–current coordination of the OCPDs in Figure 10.10.

Figure 10.14 shows a medium voltage distribution system for coordination. This identifies the largest motor and the largest 4.16- and 13.8-kV feeder breakers to be coordinated. If the coordination is achieved with respect to the largest load, say a motor, served from a bus, it can be concluded that for smaller loads connected to the same bus, the coordination will be achieved; though some care is required in exercising this generality—for example, a smaller motor may have a much longer starting time.

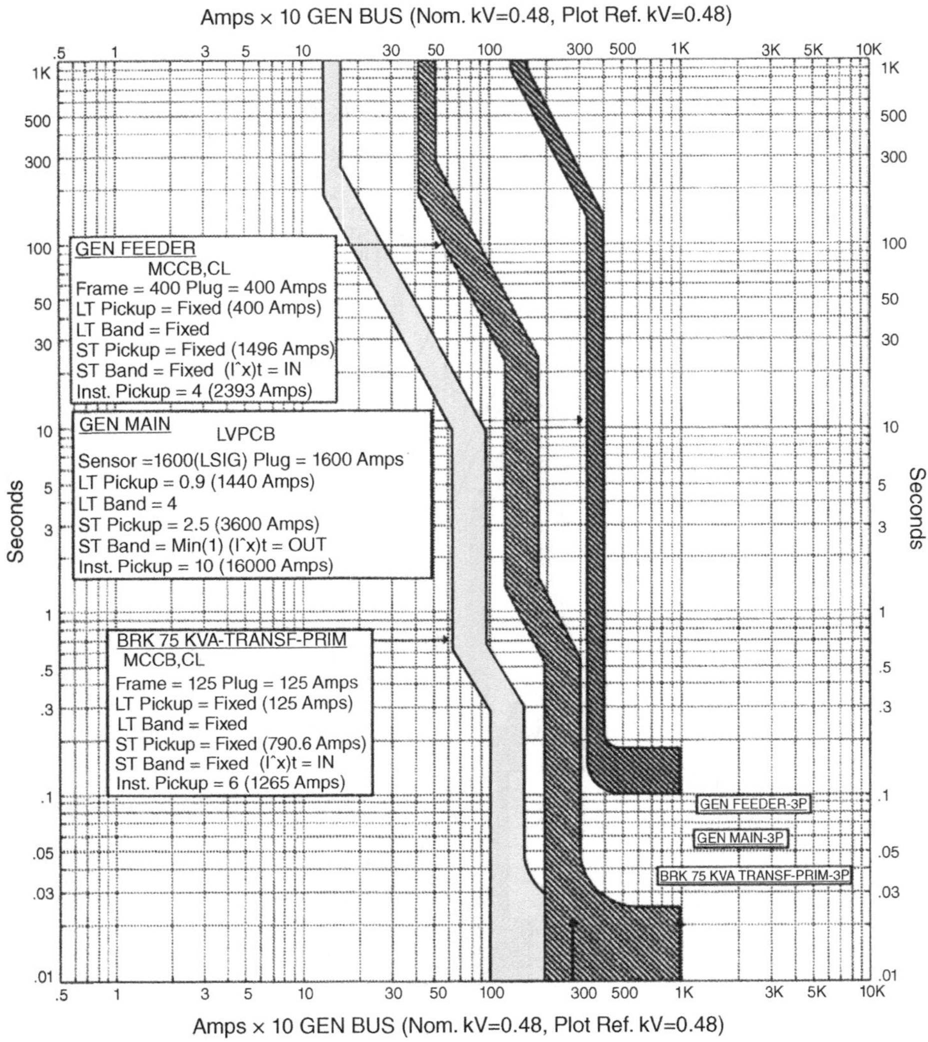


Figure 10.13. Time–current coordination of the OCPDs in Figure 10.10.

The 4.16-kV circuit breaker CB5 feeds a medium voltage MCC, bus rating of 1200 A, which has a large motor starter for 3000-hp motor. The transformer T2 is rated 5.0 MVA, and is protected by 300E current-limiting fuse.

Figure 10.15 shows the motor starting curve, the thermal withstand curves, and curves for motor fuse of 24R and transformer 300E primary fuse. As seen from this coordination plot, the motor fuse is properly selected for the application, and so also the 300E transformer primary fuse. However, the feeder breaker CB5 and main 4.16-kV secondary breaker CB4 overcurrent settings, howsoever manipulated, cannot fit between these two fuse curves.

TABLE 10.3. Circuit Breaker Selection and Trip Settings for Selective Coordination (Figure 10.10)

| Circuit Breaker Designation | Circuit Breaker Frame | Sensor/ Plug, Amps | Circuit Breaker Type | Trip Device | Coordinating Settings |
|---|---|---|---|---|---|
| BRK-switch-main | 2000AF | 2000/2000 | LVPCB | Digital LS | LT Pickup = 0.9, LT delay band = 1, ST. Pick up = 5, ST delay band = 2, $I^2t$ = Out, No instantaneous |
| BRK-switch-feeder | 800AF | 800/800 | LVPCB | Digital LSI | LT Pickup = 0.9, LT delay band = 2, ST. Pick up = 7.5, ST delay band = 1, $I^2t$ = Out, Instantaneous = 10,000 A |
| BRK-PNL | 400AF | 400/400 | CL MCCB | Digital LSI | LT Pickup = Fixed, LT delay band = Fixed, ST. Pick up = Fixed, ST delay band = Fixed, Instantaneous = 4,017 A |
| BRK 75KVA-transf-prim | 125A | 125/125 | CL MCCB | Digital LSI | LT Pickup = Fixed, LT delay band = Fixed, ST. Pick up = Fixed, ST delay band = Fixed, Instantaneous = 1,265 A |
| BRK 75KVA-transf-sec | 250A (240 V) | 250/200 | CL MCCB | Digital LSI | LT Pickup = Fixed, LT delay band = Fixed, ST. Pick up = Fixed, ST delay band = Fixed, Instantaneous = 1,536 A |
| BRK-feeder-lighting | 30A | – | MCCB | Thermal Magnetic | Fixed |
| BRK-gen main | 1600AF | 1600/1600 | LVPCB | Digital LS | LT Pickup = 0.9, LT delay band = 4, ST. Pick up = 2.5, ST delay band = 1, $I^2t$ = Out, No instantaneous |
| BRK-gen feeder | 400AF | 400/400 | CL MCCB | Digital LSI | LT pickup = fixed, LT delay band = Fixed, ST. Pick up = fixed, ST delay band = fixed, instantaneous = 2,393 A |

TABLE 10.4. Arc Flash Hazard Calculations, Example 10.3, ATS Connected to Normal Power Supply

| Equipment ID | Bolted Fault Current (kA) | Arcing Current kA (85% Ia) | Arcing Time (Seconds) | Working Distance (Inches) | Arc Flash Boundary (ft) | Incident Energy (cal/cm$^2$) | PPE |
|---|---|---|---|---|---|---|---|
| LV switchgear | 30.32 | 20.95 | 0.33 | 24 | 94 | 18.9 | 3 |
| MCC | 20.64 | 15.24 | 0.18 | 18 | 56 | 12.10 | 3 |
| Transfer switch | 14.88 | 9.11 | 0.025 | 18 | 11 | 0.51 | 0 |
| Emergency panel | 14.88 | 9.11 | 0.025 | 18 | 11 | 0.51 | 0 |
| Emergency distribution board, 240 V | 2.07 | 1.46 | 0.909 | 18 | 26 | 2.09 | 1 |

Dimensions: LV switchgear: 36″W × 90″H × 48″D, MCC 30″W × 90″H × 25″D″, EC = HCB, all others EC = VCB

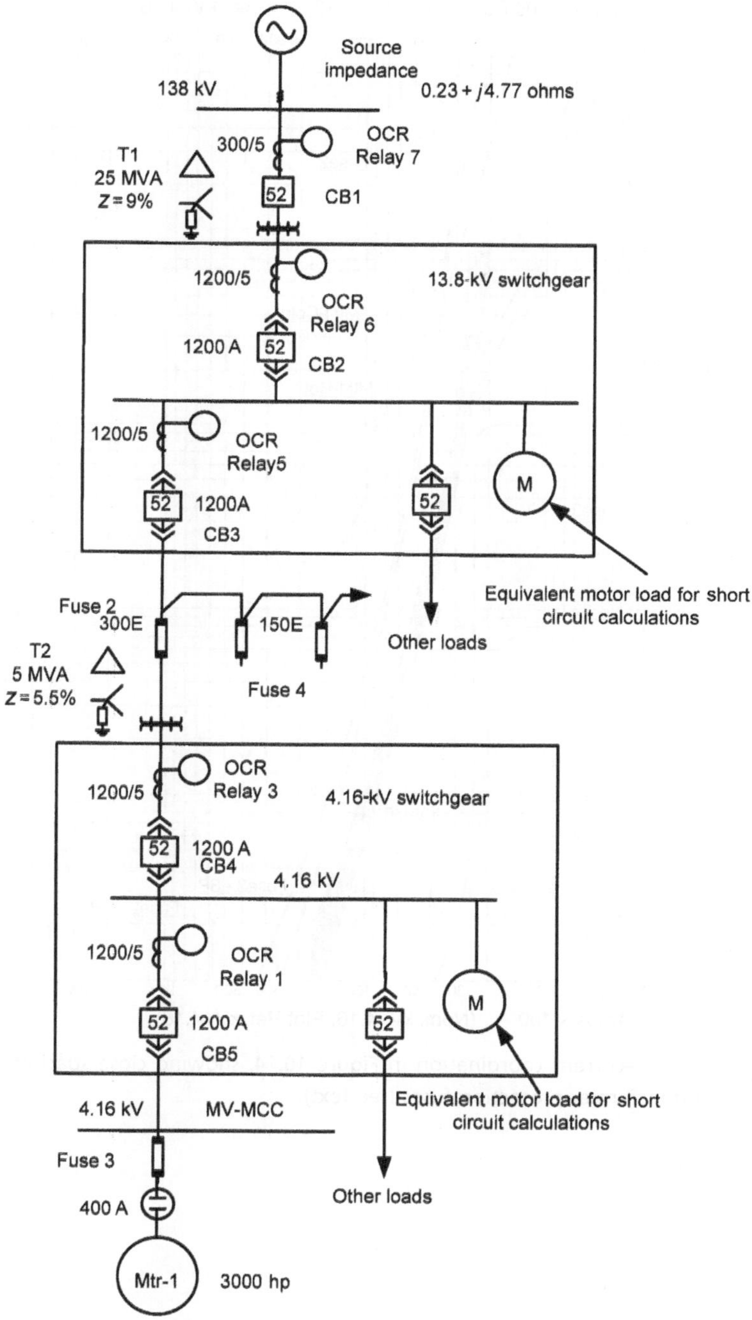


Figure 10.14. A medium voltage distribution system for selective overcurrent coordination.

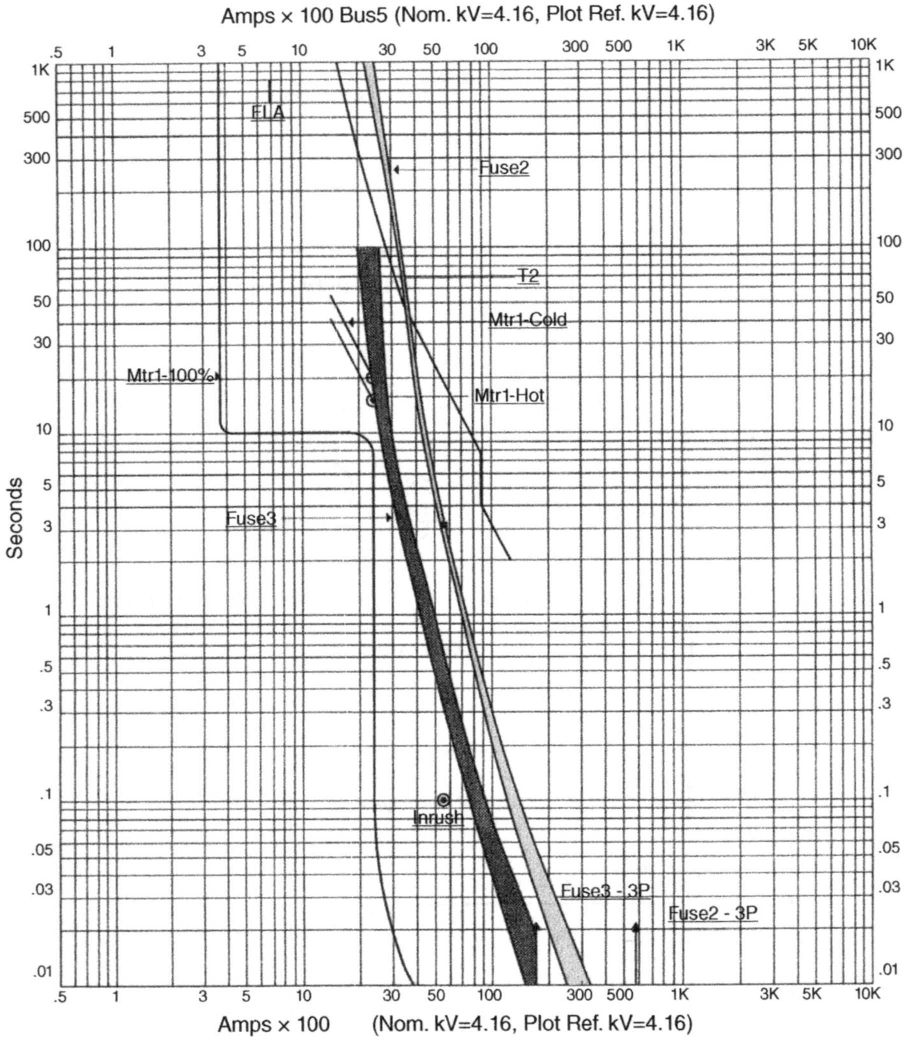


Figure 10.15. Time–current coordination in Figure 10.14, showing close spacing between the motor and transformer protection fuses (see text).

It will not be desirable to omit altogether feeder overcurrent, CB5, and main secondary circuit breaker CB4 overcurrent protections, and let the 300E fuse protect the 4.16-kV switchgear bus, medium voltage MCC bus, and feeder connections to the medium voltage MCC bus.

One possibility will be to replace the 300E fuse with an overcurrent relay for 5.0 MVA transformer primary protection, which can be set closer to the transformer T2 thermal damage curve. This can open up the space toward the right side of motor fuse and the transformer thermal withstand to accommodate the overcurrent settings on breakers CB4 and CB5.

It is also possible to eliminate the 300E fuse, as well as any overcurrent relay for the primary protection of 5-MVA transformer, and rely upon the protection provided by 13.8-kV breaker CB3 overcurrent relay settings. Let us apply this later approach.

The breaker CB3 serves other transformers too, daisy chained, of no more than 1750 kVA rating, which are protected with 150E current-limiting primary fuses. Consider that three such transformers, each protected with a 150E fuse, are in service.

Then, the settings on breaker CB3 should consider:

- That the primary protection settings are not exceeded more than six times the transformer full load current of 5-MVA transformer, as per NEC.
- The maximum overcurrent settings on the secondary breaker CB4 should not exceed three times the 5-MVA transformer full load current as per NEC.
- The instantaneous overcurrent should be set to account for all the feeder inrush currents, that is, all the transformers connected to beaker CB3 can be energized together. This inrush current is calculated as 6000 A.

Figure 10.16 depicts the revised coordination. Some coordination is sacrificed between the overcurrent relays 3 (breaker CB4) and 1 (breaker CB5) at the maximum short-circuit levels; but no coordination is sacrificed between the overcurrent relay 3 (breaker CB4) and 5 (breaker CB3). This is desirable, as breaker CB3 feeds an entire chain of transformers. The settings on breaker CB3 relay adequately protect the thermal withstand of 5 MVA transformer. The instantaneous is delayed by 0.05 second to coordinate with the motor circuit and other fuses. The instantaneous is set at 12 kA, which clears the relay 1 and relay 3 curves and allows for the feeder inrush current.

Proceeding further, the coordination with 25 MVA transformer T1 withstand and its primary and secondary relays is obtained adequately with appropriate CTI, as shown in Figure 10.17. A summary of the protective device settings are in Table 10.5.

This is illustrative of the compromises that can be made to the protective relaying and their settings. The strategy will vary from system to system, selected protected devices, loads to be served, and equipments to be protected.

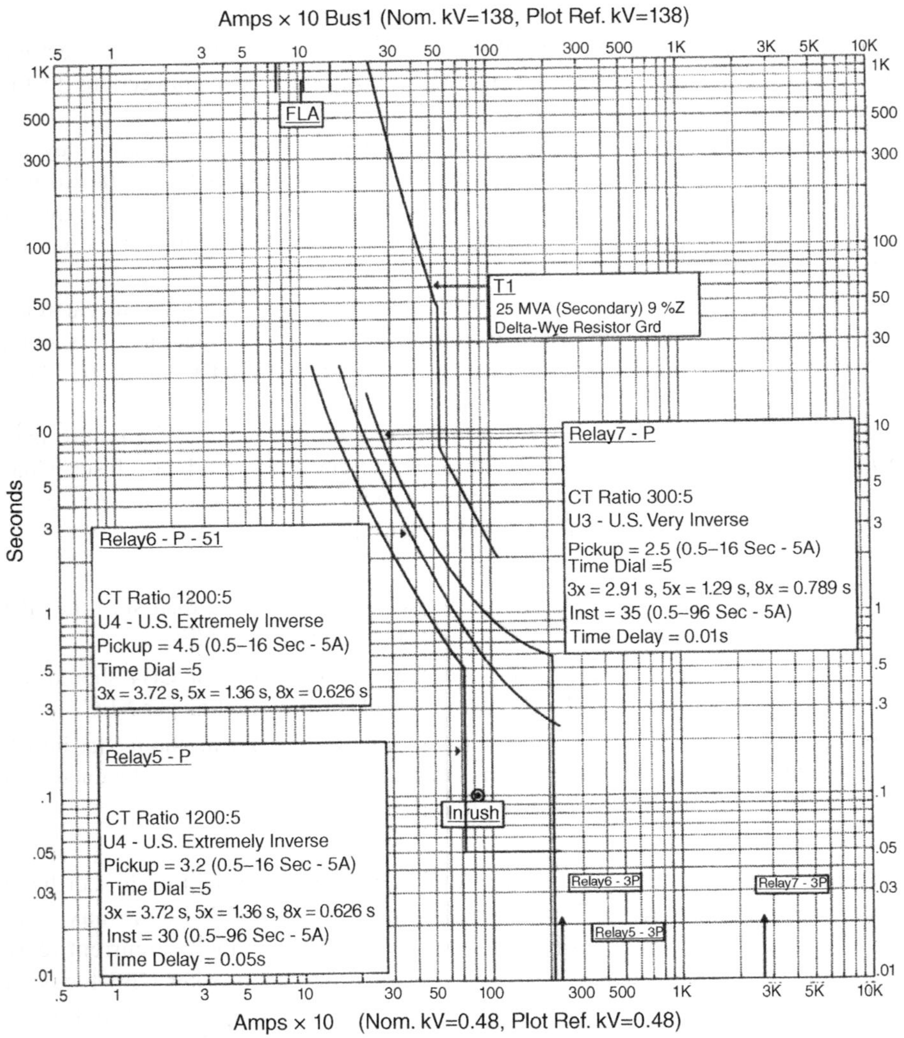


Figure 10.16. Time–current coordination achieved by omitting the 5 MVA transformer primary protection (see text).

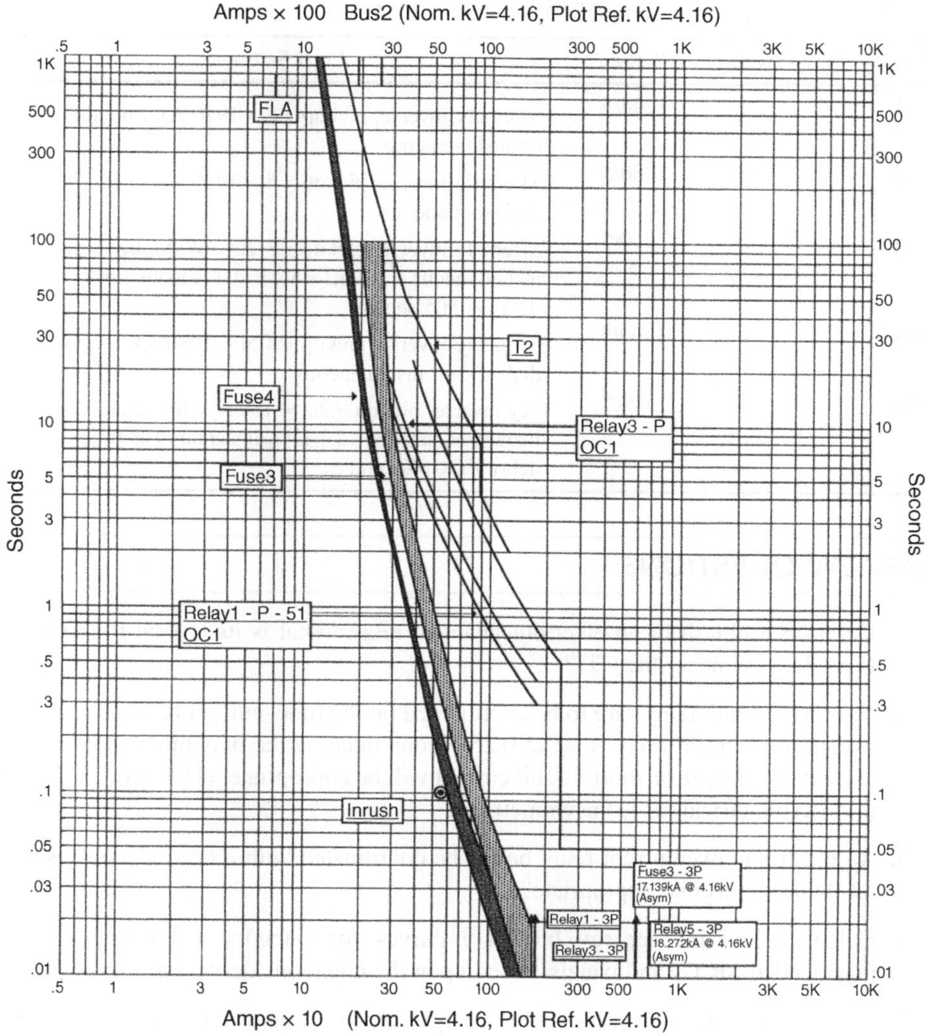


Figure 10.17. Time–current coordination for 25 MVA main transformer (Figure 10.14).

TABLE 10.5. Final Settings on OCPDs of Figure 10.14

| Breaker No. | Relay | CT Ratio | Final Settings |
|---|---|---|---|
| CB5 | 1 | 1200/5 | Extremely inverse, pickup = 8(= 1920 A), time dial = 3, no instantaneous |
| CB4 | 3 | 1200/5 | Extremely inverse, pickup = 8(= 1920 A), time dial = 4, no instantaneous |
| CB3 | 5 | 1200/5 | Extremely inverse, pickup = 3.2(= 768 A), time dial = 5, instantaneous = 30 (12 kA), instantaneous time delay = 0.05 second |
| CB2 | 6 | 1200/5 | Extremely inverse, pickup = 4.5(= 1080 A), time dial = 5, no instantaneous |
| CB1 | 7 | 300/5 | Very inverse, pickup = 2.5(= 150 A), time dial = 5, instantaneous = 35 (2.1 kA), instantaneous time delay = 0.01 second |

## REVIEW QUESTIONS

1. Describe backlash of an electromechanical relay. What is its typical time value? How does it impact the CTI?

2. Two overcurrent relays are to be coordinated on the time–current basis. The downstream overcurrent relay is set at 0.25 second delay at the maximum short-circuit current. What type of short-circuit current will be appropriate: (1) 30 cycle, (2) 3–4 cycle symmetrical, or (3) asymmetrical?

3. An electronic overcurrent relay has to be coordinated with a downstream fuse. Is a CTI of 0.1 second appropriate?

4. Calculate and plot energy boundary curves for category 4 for low voltage ungrounded MCCs and panels, based upon the equations derived in this chapter.

5. Describe how the arc flash incident energy can be lowered in low voltage switchgears by selecting an appropriate trip programmer with multiple ST bands?

6. What are the advantages of series connected and coordinated current-limiting devices for arc flash reduction?

7. What is meant by peak sensing as related to low voltage circuit breaker trip programmer's instantaneous settings?

8. A current-limiting fuse gives 40 kA peak at test power factor of 15%. What is the peak at unity power factor? Which peak value is used for coordination with series connected overcurrent protection devices?

9. Describe three areas where some lack of coordination will be acceptable with hand sketches of the device characteristics.

**10.** Why mostly ANSI/IEEE extremely inverse or very inverse characteristics are chosen for overcurrent coordination in power systems? Name an application for inverse and short-time characteristics.

**11.** When is the $I^2t$ function provided on low voltage trip programmers for short-time pickups most appropriate to apply?

## REFERENCES


1. NEC, NFPA 70. *National Electric Code*, 2011.
2. IEEE Standard, C.57.91 (Also Cor.1-2002). *IEEE Guide for Loading Mineral-Oil Immersed Transformers*, 1995.
3. NEMA, Industrial Control and Systems, Part ICS 2-324. *AC General Purpose High Voltage Contactors and Class E Controllers, 50 Hz and 60 Hz*, 1974.
4. V. Marcelo, H. Steve, and P. Sutherland, "Improving selectivity and arc-flash protection through optimized instantaneous protection settings," in Conf. Record, IEEE Pulp and Paper Industry Technical Conference, Nashville, TN, pp. 34–41, June 2011.
5. M. Valdes, T. Papallo, and A. Crabtree, "Method of determining selective capability of current-limiting overcurrent devices using peak let-through current," in Conf. Record, IEEE Pulp and Paper Industry Technical Conference, Birmingham, AL, pp. 145–153, June 2009.
6. GE, Publication DET 537. *GE Overcurrent Device Instantaneous Selectivity Tables*, 2007.
7. ANSI/IEEE Standard, 242. *IEEE Recommended Practice for Protection and Coordination of Industrial and Commercial Power Systems,* Chapter 14: Overcurrent Coordination, 1986.
8. J.C. Das, "Protective relay coordination—Ideal and practical," IEEE IAS Annual Meeting, San Diego, CA, pp. 1861–1874, 1989.

# 11

# TRANSFORMER PROTECTION

Transformer protection is of special significance for arc flash hazard reduction, as we will examine in this chapter. ANSI/IEEE standards publish the transformer frequent and infrequent fault withstand curves, and the provisions of NEC [1] for transformer protection are followed in the industry. NEC does not apply to utility installations, generating stations and substations distribution and metering designed, maintained, and operated by the utilities.

## 11.1 NEC REQUIREMENTS

NEC requirements for overcurrent protection of transformers and transformer vaults, including secondary ties, are contained in Article 450.2. It is important to note that the transformer secondary overcurrent protection is not required, provided certain conditions of installations in NEC are met. The recommendations for protection in supervised and "any location"; part of NEC table 450.3(A) are summarized in Tables 11.1 and 11.2.

*Arc Flash Hazard Analysis and Mitigation*, Second Edition. J.C. Das.

Published 2021 by John Wiley & Sons, Inc.

TABLE 11.1. Maximum Rating or Setting of Overcurrent Protection for Transformers Rated over 600 V

| Supervised Locations | | | | | |
|---|---|---|---|---|---|
| | Primary Protection over 600 V | | Secondary Protection over 600 V | | Secondary Protection 600 V or less |
| Transformer Rated Impedance | Circuit Breaker (%) | Fuse (%) | Circuit Breaker (%) | Fuse (%) | Circuit Breaker of Fuse (%) |
| Any | 300 | 250 | Not required | Not required | Not required |
| Not >6% | 600 | 300 | 300 | 250 | 250 |
| >6% but not >10% | 400 | 300 | 250 | 225 | 250 |

*Source*: Data from NEC [1].

TABLE 11.2. Maximum Rating or Setting of Overcurrent Protection for Transformers Rated over 600 V

| Any Location | | | | | |
|---|---|---|---|---|---|
| | Primary Protection over 600 V | | Secondary Protection over 600 V | | Secondary Protection 600 V or less |
| Transformer Rated Impedance | Circuit Breaker (%) | Fuse (%) | Circuit Breaker (%) | Fuse (%) | Circuit Breaker of Fuse (%) |
| Not >6% | 600 | 300 | 300 | 250 | 125 |
| >6% but not >10% | 400 | 300 | 250 | 225 | 125 |

*Source*: Data from NEC [1].

Also, for transformers of 600 V or less, secondary protection is not required, provided the primary protection is set at no more than 125% for currents of 9 A or more and 167% for currents less than 9 A.

Secondary protection of transformers over 600 V can be omitted if there are no more than six secondary circuit breakers, and the sum of the ratings of circuit breakers does not exceed 300% of the rated secondary current of the transformer. If there are secondary fuses and circuit breakers in combination, not exceeding six, the sum of their ratings should not exceed 250%.

The overcurrent protection of primary and secondary conductors of the transformers is covered in NEC Article 240. Again, it is permissible to omit the secondary protection of conductors provided certain specified conditions in NEC are met.

As a result of these provisions, many industrial systems do not have transformer secondary overcurrent protection. The secondary overcurrent devices and circuit breakers are eliminated as a cost-saving measure.

## 11.2 ARC FLASH CONSIDERATIONS

From the arc flash considerations, these provisions of NEC and industrial practice of omitting transformer secondary protection cannot be recommended. In Figure 11.1, consider a secondary fault anywhere within the zone shown in solid rectangular block, say faults at F1 and F2. With no secondary circuit breaker, faults at F1 and F2 are

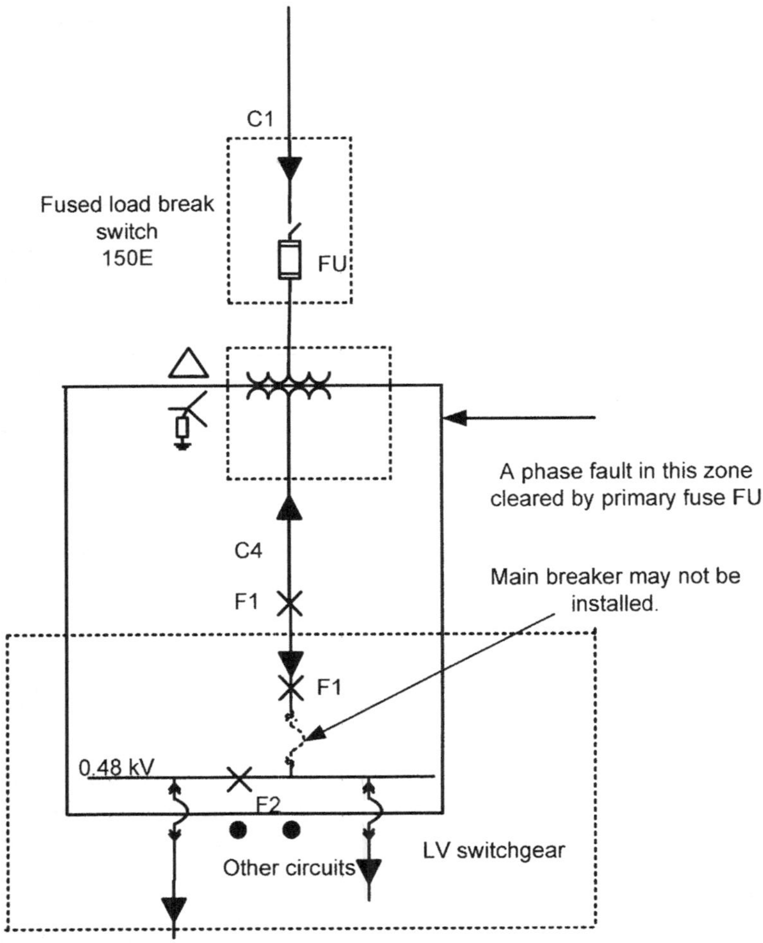


Figure 11.1. Faults in the secondary zone shown in solid block cleared by primary protection device.

cleared by the transformer primary fuse. Even if a main secondary circuit breaker is present, a fault in the incoming cable compartment or the main secondary breaker itself, fault at F1, must be cleared by the primary protection. A three-phase secondary fault will be reduced in the ratio of the transformation when reflected to the primary side of the transformer, that is, for a 13.8–0.48 kV transformation ratio; the fault current is reduced by a factor of 28.75. The fuse will operate with considerable time delay, releasing immense amount of incident energy, and arc flash hazard can be extremely high.

## 11.3 SYSTEM CONFIGURATIONS OF TRANSFORMER CONNECTIONS

Figure 11.2 shows some of the configurations for substation transformers in industrial environment.

**Radial System of Distribution (Figure 11.2a)**

A radial system of distribution is fairly common in the industry due to its cost saving advantages. As many as 10 or more unit substation transformers are daisy chained on to a single 13.8-kV feeder circuit breaker. From a protection and arc flash point of view, this is not a desirable system. For any fault in the 13.8-kV primary feeder cables, for a ground fault on the secondary sensed through a ground fault relay, operation of a sudden fault pressure relay in any of the transformers and a fault on any primary load interrupter switch must *all* trip the single 13.8-kV feeder breaker; resulting in a complete shutdown of all the transformers fed from the single 13.8-kV breaker.

**Primary Selective System (Figure 11.2b)**

In a primary selective system with redundant sources of power, a substation transformer can be connected to any of the two sources through interlocked selector switches that are of load-break type. Yet an entire shutdown of the load will occur, before a switchover can be made to the alternate source of power. From arc flash considerations, the situation is identical to that of Figure 11.2a, except that the primary switchover of power for a cable fault allows bringing the system online after a short interruption.

**Group Feed System (Figure 11.2c)**

In a group feed system, the transformer primary protection is grouped in one location through fused load interrupter switchgear. Again, for all secondary fault types, as discussed in Figure 11.2a, the primary circuit breaker has to be tripped, resulting in complete shutdown. The only difference is that for a number of transformers, primary fuse protection is grouped together.

**Dedicated Circuit Breakers (Figure 11.2d)**

Dedicated circuit breakers on the primary and secondary sides of each transformer are provided. All the protective devices shown in this figure, except the differential protection, is required to be provided to meet the requirements of FM (Factory Mutual) Global Property Loss Prevention Data Sheet [2] for transformers rated

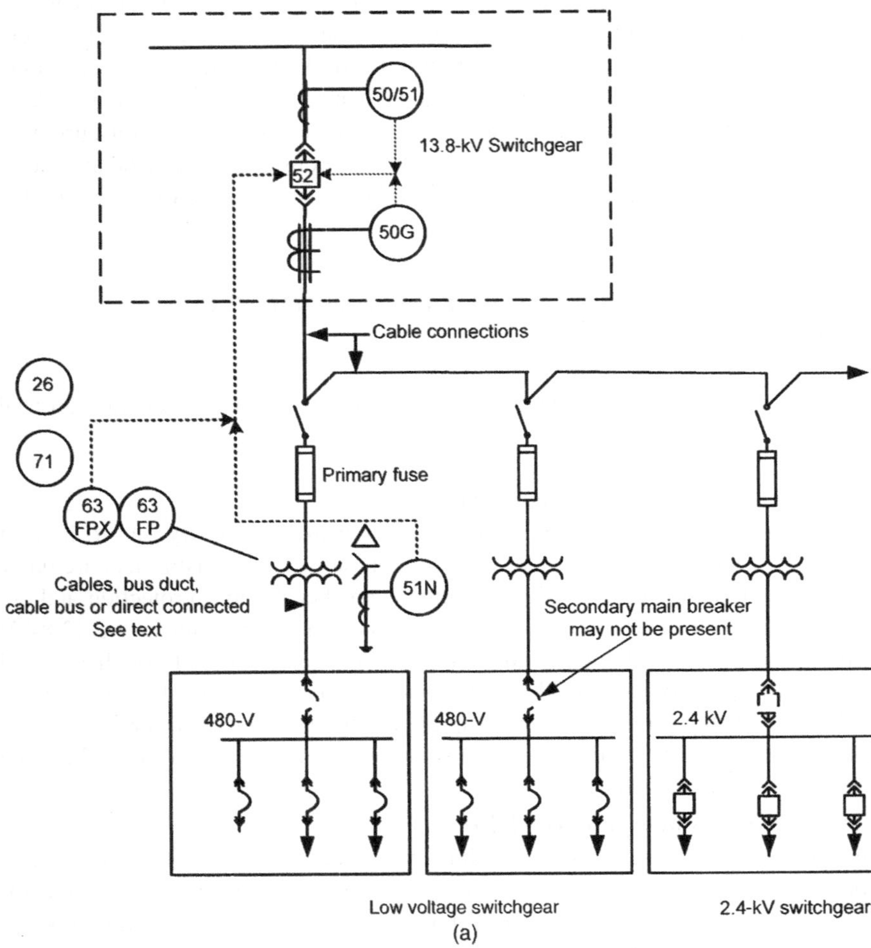


Figure 11.2. (a) A radial system of distribution, (b) primary selective system (protection and secondary distribution same as Figure 11.2a), (c) group feed system, (d) dedicated primary and secondary circuit breakers, (e) fixed mounted primary relayed breaker, and (f) secondary selective system.

1000–10,000 kVA located outdoors. Similar protection is required for transformers rated less than 1000 kVA when it creates fire hazard. Though expensive in terms of providing dedicated circuit breakers, it limits the area of shutdown, and transformer secondary and primary faults can trip out its respective circuit breaker. The addition of differential protection further enhances the protection and limits the arc flash damage.

**Fixed Mounted Primary Circuit Breaker (Figure 11.2e)**

The primary load-break switch-fuse protection is replaced with fixed mounted (non-draw out) or metal-clad draw-out circuit breaker and overcurrent relays. At

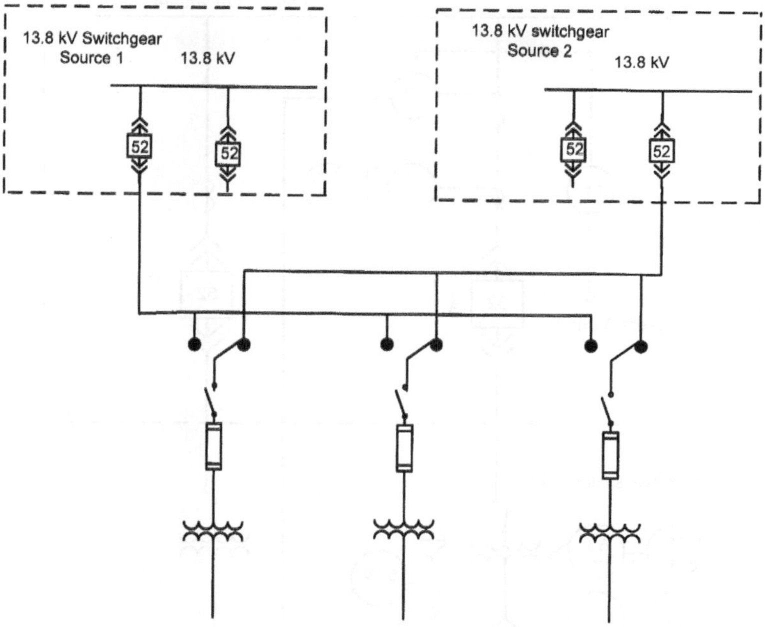


(b)

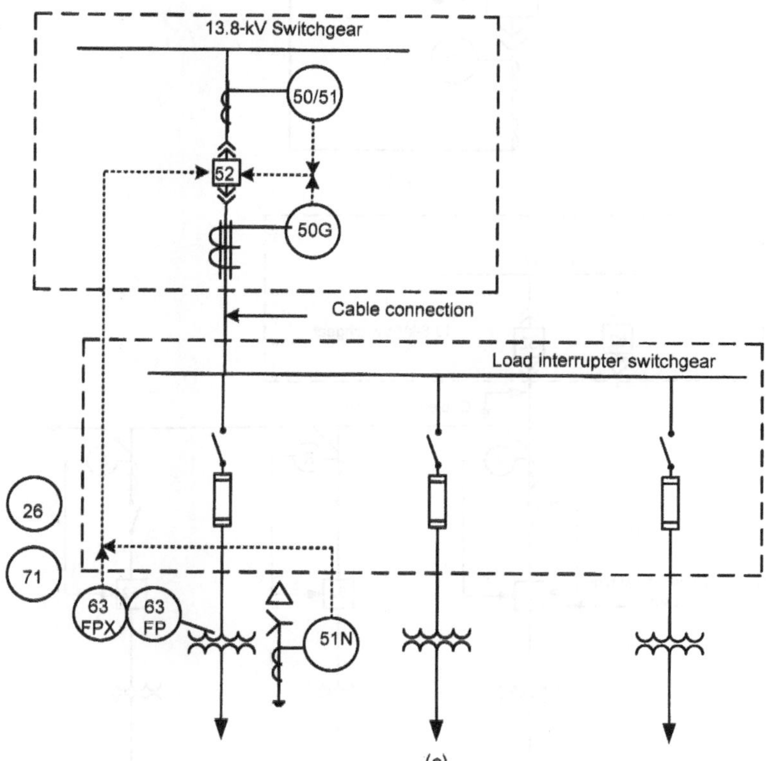


(c)

Figure 11.2. *(Continued)*

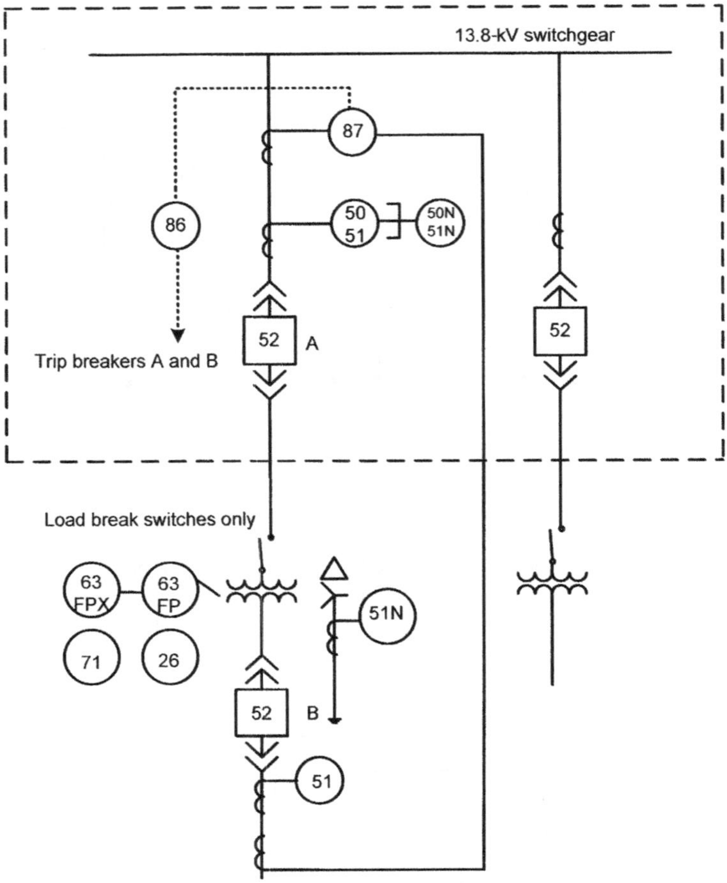


(d)

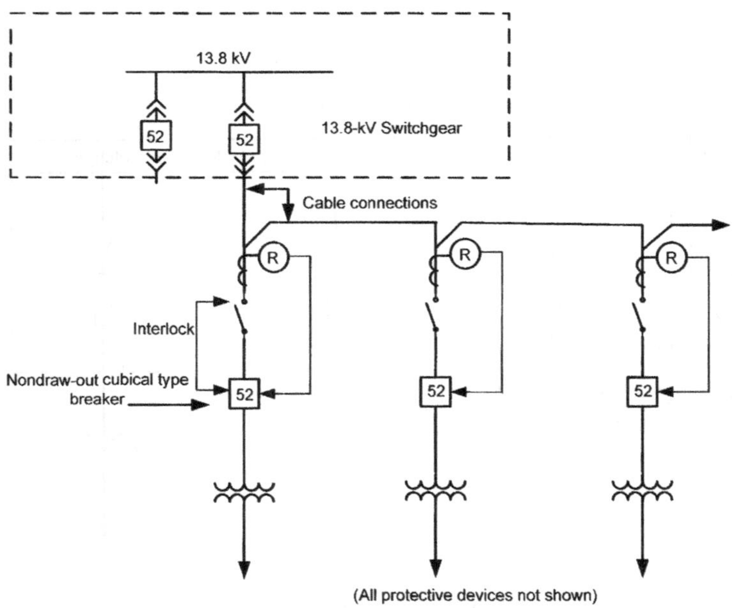


(e)

Figure 11.2. (*Continued*)

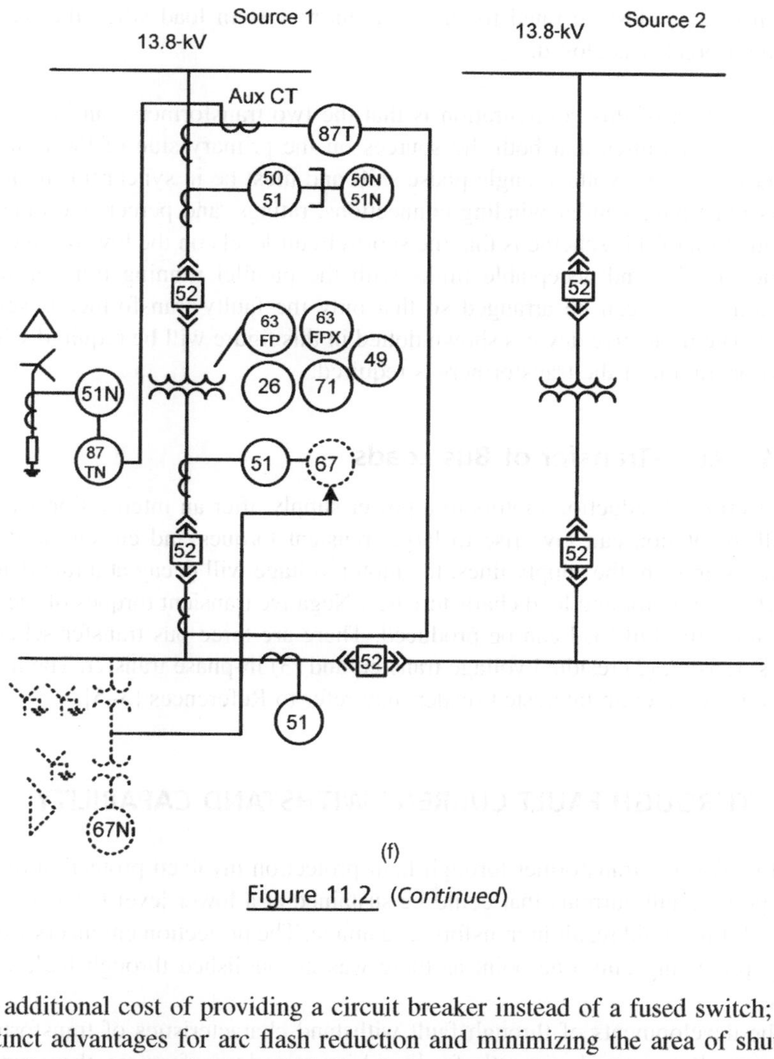


(f)

Figure 11.2. (*Continued*)

the additional cost of providing a circuit breaker instead of a fused switch; it has distinct advantages for arc flash reduction and minimizing the area of shutdown for a fault in a transformer. Only the faulty transformer will be isolated, leaving the rest of the daisy chained transformers in service.

**Secondary Selective System (Figure 11.2f)**

A further enhancement in protection and maintaining the continuity of power can be achieved through double-ended secondary selective systems. The system is operated with bus section circuit breaker normally open, each transformer supplying its bus load. All the protective devices shown in this figure are required to be provided for transformers rated above 10,000 kVA according to FM [2]. In case of failure of one of the transformers or primary source of power, the bus section switch can be closed, either manually, or an automatic bus transfer scheme can be arranged. In case of manual closing, the loads will be interrupted. With fast auto-transfer of power, it is possible to maintain the motor running loads. It implies that each transformer must be rated to carry the entire system load when the bus section circuit breaker is closed.

A variation of this configuration is that the two transformers can be operated in parallel. This requires that both the sources on the primary side of the transformers must have the same voltage angle phase shift and must be in synchronism, and transformers must have similar winding connections, ratings, and percentage impedances. The limitation of this scheme is that the short-circuit levels on the low voltage systems may increase beyond acceptable limits with the parallel running transformers. The protection system can be arranged so that only the faulty transformer is selectively isolated. The protective devices shown dotted in this figure will be required when such parallel operation of the transformers is required.

### 11.3.1 Auto-Transfer of Bus Loads

Reconnection of induction motors to a power supply after an interruption while these are still in rotation can give rise to large transient torques and current surges. After disconnection from the supply lines, the motor voltage will decay at a rate determined by inertia and motor and load characteristics. Negative transient torques of the order of seven times the full load can be produced. There are three bus transfer schemes: (1) fast bus transfer, (2) residual voltage transfer, and (3) in-phase transfer. The details are not discussed here; an interested reader may refer to References [3–5].

## 11.4 THROUGH FAULT CURRENT WITHSTAND CAPABILITY

Prior to 1978, the transformer through fault protection involved protection of a single ANSI point. Fault currents that could be sustained at a lower level for a considerable period of time could result in transformer damage. The protection engineers were satisfied by protecting only one point as there was no published through fault withstand curves.

The developments of through fault withstand characteristics of transformers has seen many changes, and, currently for liquid-immersed transformers, these are defined in IEEE standard [6] and are categorized as follows:

*Category I.* Single-phase 5–500 kVA, three-phase 15–500 kVA

*Category II.* Single-phase 501–1667 kVA, three-phase 501–5000 kVA

*Category III.* Single-phase 1668–10,000 kVA, three-phase 5001–30,000 kVA

*Category IV.* Single-phase >10,000 kVA, three-phase >30,000 kVA

ANSI standard [7] for Loading of Oil Immersed Distribution and Power Transformers provides the transformer short-time thermal overload capability as shown in Table 11.3.

TABLE 11.3. Transformer Short-Time Thermal Overload Capability

| Time | Times Rated Current |
|---|---|
| 2 seconds | 25 |
| 10 seconds | 11.3 |
| 30 seconds | 6.3 |
| 60 seconds | 4.75 |
| 5 minutes | 3.0 |
| 30 minutes | 2.0 |

*Source*: Based on ANSI/IEEE Standard [7].

Low values of currents of 3.5 or less times the normal base current of transformer may occur from overloads rather than faults, and for such cases, loading guides may indicate allowable time durations [7]. The per-unit short-circuits currents shown in through fault curves are balanced transformer winding currents. The line currents that relate to these winding currents depend upon transformer connections and the type of fault. The following explanations apply to each category.

### 11.4.1 Category I

The recommended duration limit is based on the curve in Figure 11.3. This curve reflects both thermal and mechanical damage considerations. The dot-dash portions of the curve cover transformer varying short-circuit withstand capabilities required by IEEE standard [6] up to a maximum of 40 times the normal current.

### 11.4.2 Category II

In Figure 11.4, fault frequency refers to the number of faults with magnitudes greater than 70% of the maximum possible. Figure 11.4a reflects both the thermal and mechanical damage considerations. It is applied for faults that will occur frequently, more than 10 times in the life of the transformer. Part of the curve is dependent upon transformer short-circuit impedance for faults above 70% of the maximum possible and is keyed to $I^2t$ of the worst case mechanical duty as shown by dashed curves for a few selected impedances. The remaining portion matches the thermal portion of curve for faults below the 70% level. Figure 11.4b which is the solid portion of the curve in Figure 11.4a reflects primarily thermal damage. This curve may be applied for less frequent faults typically no more than 10 in the life of the transformer.

### 11.4.3 Category III and IV

For these categories, it is required that the system short-circuit reactance based upon the available short-circuit current is added to the transformer short-circuit impedance. Consider a three-phase transformer of 10 MVA, 138–13.8 kV, which falls in Category III. If transformer impedance is 9%, and the 138 kV source has a short-circuit level of

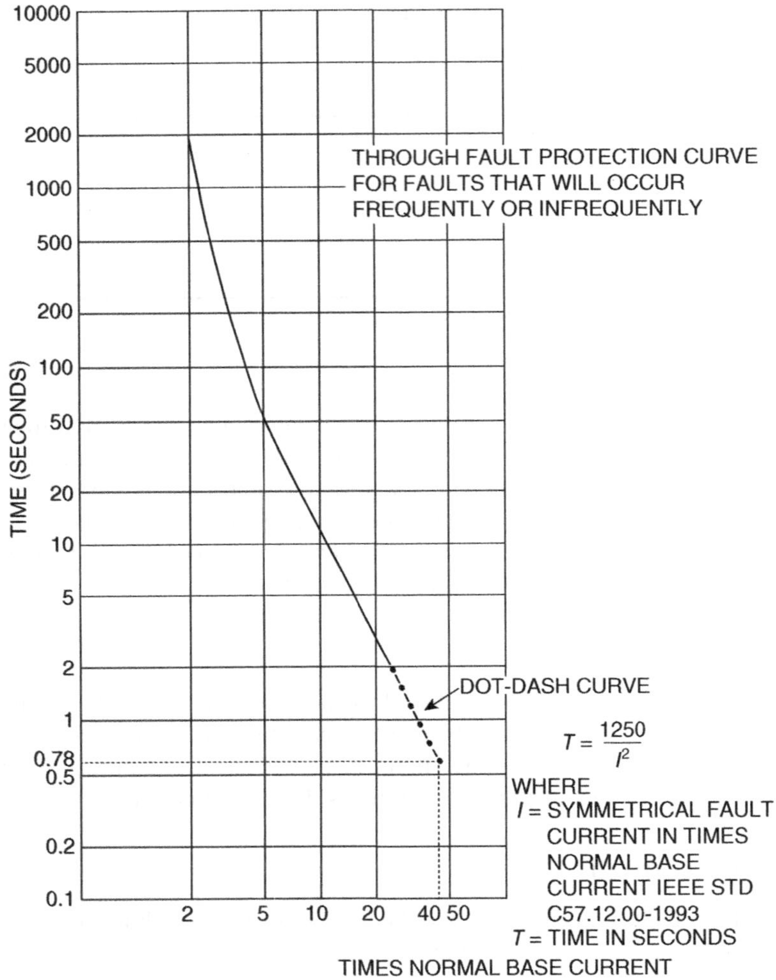


Figure 11.3. ANSI through fault withstand curve, category I transformers, liquid immersed.

5000 MVA, then the combined impedance on transformer 10 MVA base is 9.2% for the construction of the curves. In the absence of unavailable short-circuit data on the primary of the transformer, recommended values are provided in Reference [6].

***Category III.*** In Figure 11.5a, which reflects both the thermal and mechanical damage, part of the curve is dependent upon the transformer short-circuit impedance for faults above 50% of the maximum possible keyed to the [2] of the worst-case mechanical duty as shown by dashed curves for a few selected impedances. (For Category II, this transition point is 70%.) This curve is applied when typically more than 5 faults are expected in the life of a transformer. Again Figure 11.5b is applicable for faults less than 5 in the life of the transformer.

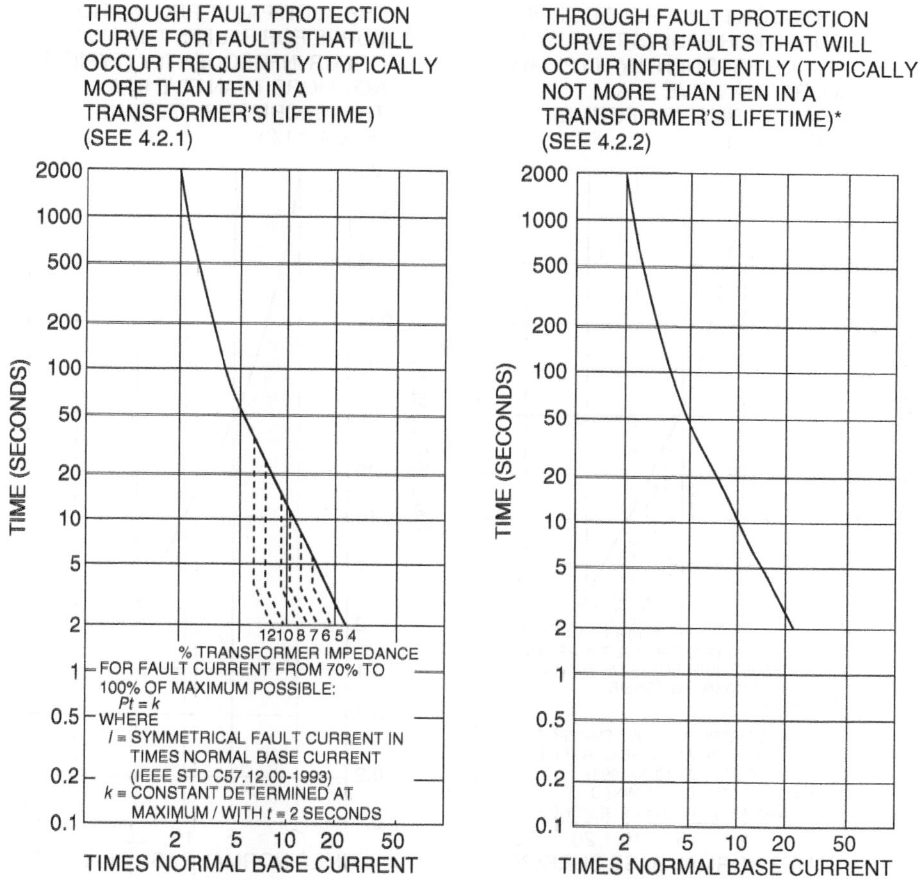


Figure 11.4. ANSI through fault withstand curve, category II transformers: (a) frequent fault curve and (b) infrequent fault curve, liquid immersed.

***Category IV.*** For Category IV transformers, there is only one set of curves, as shown in Figure 11.6. This represents both thermal and mechanical damage and is applied for all transformers in this category.

## 11.4.4 Observation on Faults during Life Expectancy of a Transformer

It is rather difficult to project how many faults will occur in the life expectancy of a transformer, as it depends upon a number of factors, for example, how carefully the transformer was chosen for the required load duty, its protection, and preventive maintenance. Generally, for transformers connected to overhead lines, frequent fault damage curve can be applied, while for transformers connected through cables in an industrial

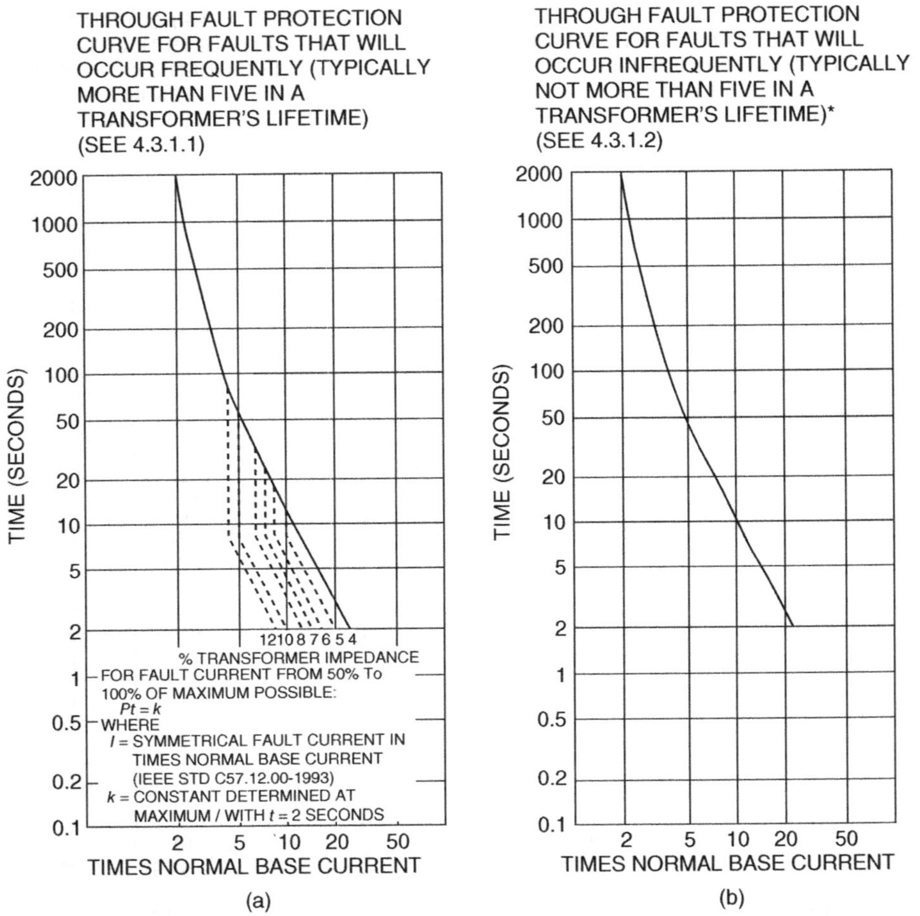


Figure 11.5. ANSI through fault withstand curve, category III transformers: (a) frequent fault curve and (b) infrequent fault curve, liquid immersed.

environment, less frequent fault damage curve can be applied. To be conservative, this book applies frequent fault damage curves of the transformers. It is also recognized that for some applications, the through fault withstand of the transformer provided in ANSI/IEEE standards may be exceeded. An example will be UAT (unit auxiliary transformer) to serve auxiliary generation loads for generating systems, with generator and transformer connected in a unit configuration directly to the utility source. These configurations are not discussed.

### 11.4.5 Dry-Type Transformers

It is recognized that dry type transformers differ considerably from liquid-immersed transformers, with respect to their withstand characteristics.

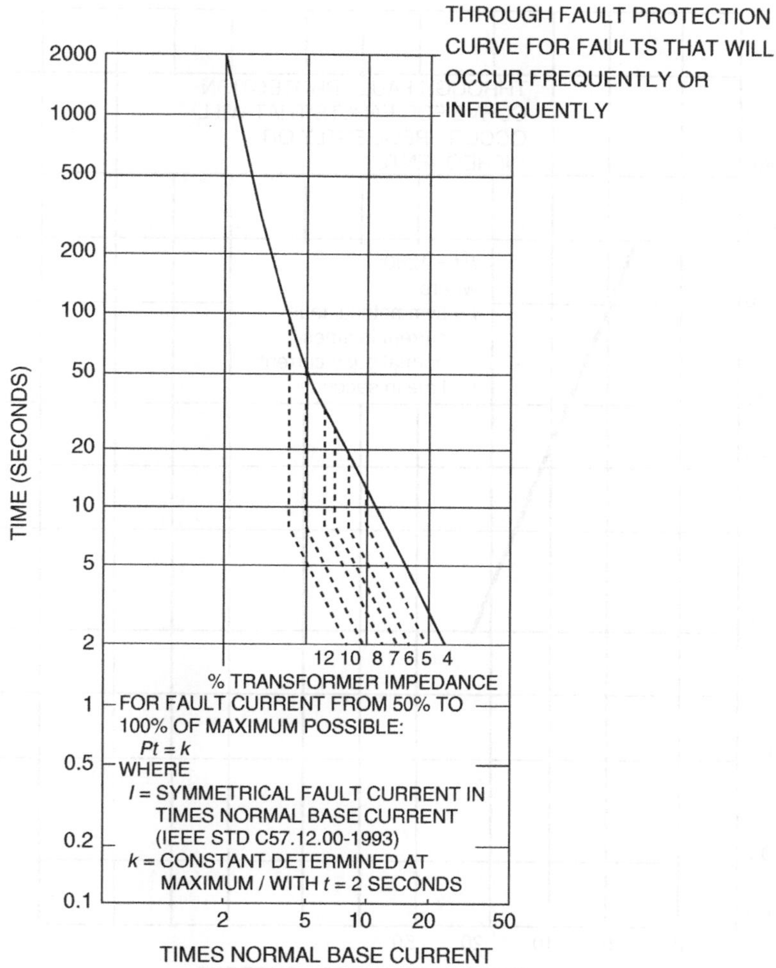


Figure 11.6. ANSI through fault withstand curve, category IV transformers, liquid immersed.

Dry type transformers can be designed for different temperature ratings, that is, 75, 90, 115, 130, and 150°C.

There is a significant variation in the construction of dry type transformers. The transient heating of liquid immersed transformers is considerably buffered by the insulating medium, providing a relatively long thermal time constant as compared to dry type transformers [8].

The following categories are applicable:

*Category I.* Single phase 15–500 kVA, three-phase 15–500 kVA (Figure 11.7).

*Category II.* Single phase 501–1667 kVA, three-phase 501–5000 kVA (Figure 11.8).

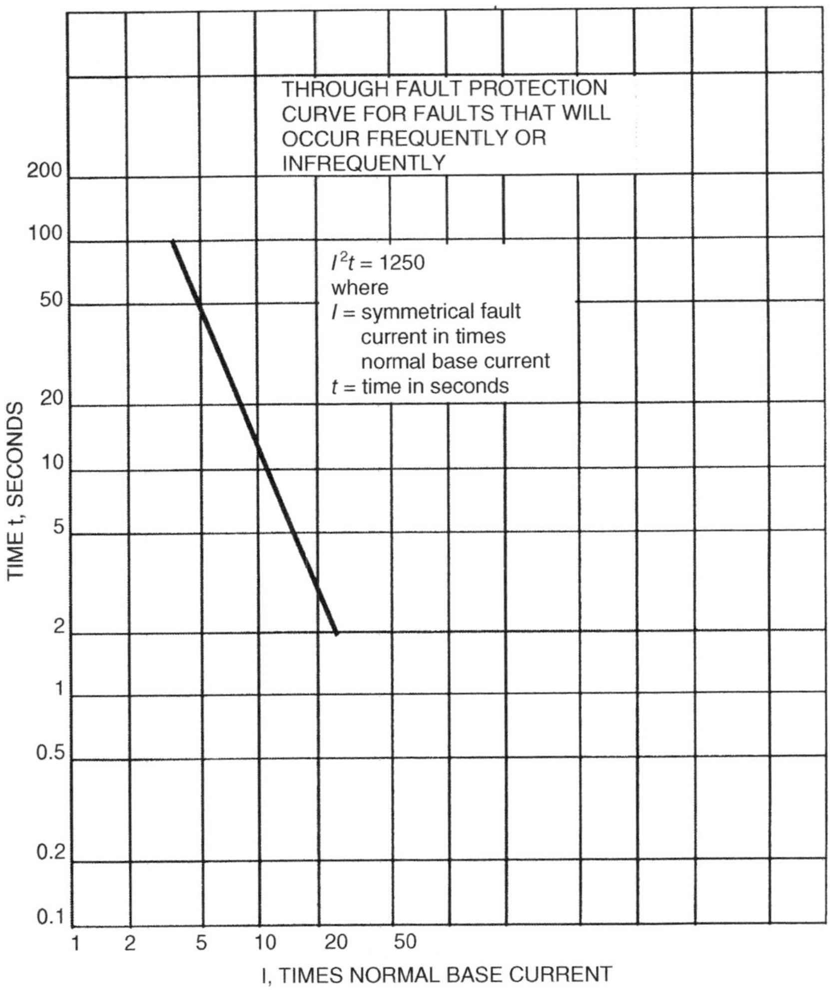


Figure 11.7. ANSI through fault withstand curve, category I, dry-type transformers.

Dry-type transformers of higher rating of 10 MVA and sometimes larger by special designs are commercially available and are in operation. In absence of any other data, Category II curve can be applied for these transformers too.

Thermal short-time overload capability for both category I and II is 2 seconds, at 25.0 times the rated current.

For category I transformers, the recommended duration limit is shown in Figure 11.7. This curve reflects both thermal and mechanical damage, and can be applied for faults that will occur frequently or infrequently.

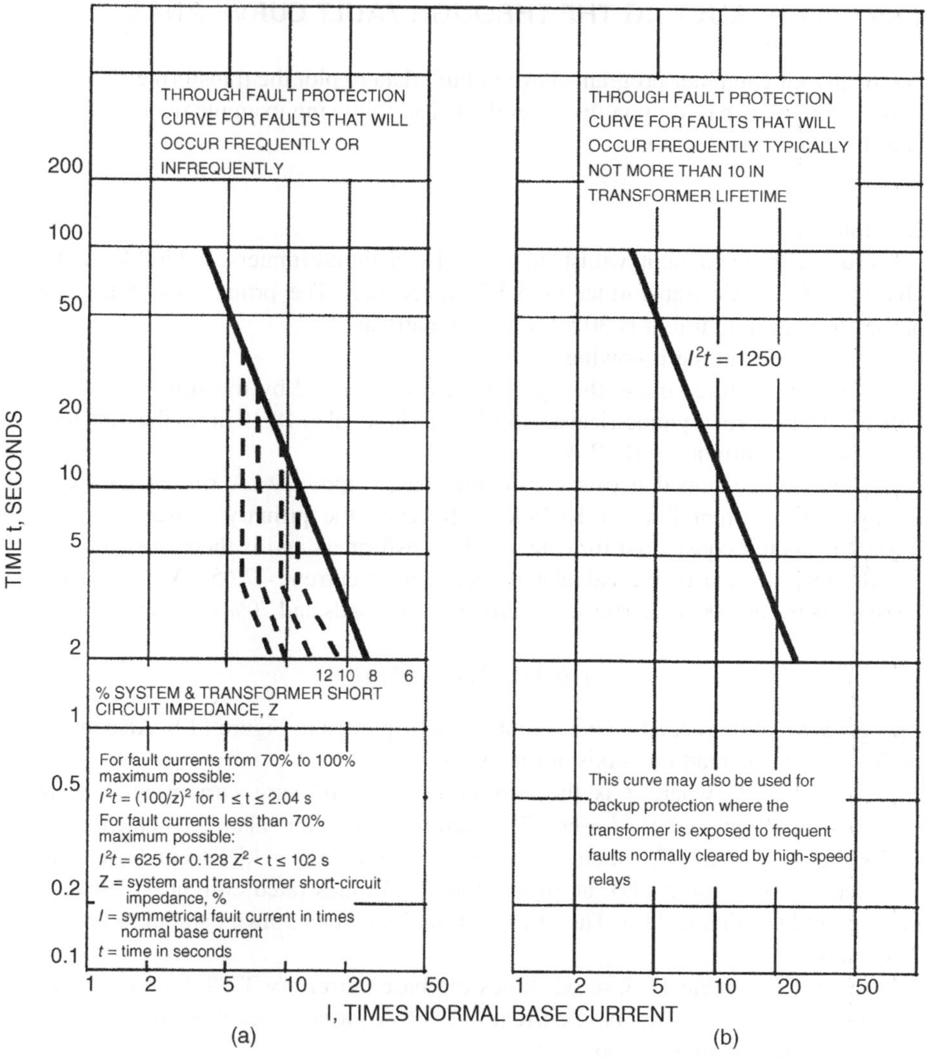


Figure 11.8. ANSI through fault withstand curve, category II, dry type transformers.

For category II transformers, the fault frequency is 70%; same as for liquid-immersed transformers, applicable for faults typically more than 10 in the life of the transformer. Figure 11.8 shows curves for category II dry-type transformers.

For dry-type transformers, the thermal curve with equation:

$$I^2t = 1250,$$

where $I$ is the symmetrical fault current in times the normal base current, and $t$ is the time in second applies to $t$ = 102 seconds in Figure 11.7. For dry-type transformers, the dash-dot portion of Figure 11.3 is absent.

## 11.5 CONSTRUCTING THE THROUGH FAULT CURVE ANALYTICALLY

Practically all computer programs have in-built data to plot the transformer frequent or infrequent curves based upon the input data. The following example details the procedure for hand construction.

**Example 11.1**

Construct a frequent fault withstand curve for a liquid-immersed 2500-kVA, three-phase, 13.8–2.4-kV transformer of 5.5% impedance. The primary (13.8-kV) three-phase short circuit current is 30 kA rms symmetrical.

This is done in the following steps.

Transformer three-phase through fault current limited by transformer impedance only is 10.94 kA rms symmetrical at 2.4 kV. This is equal to 18.19 times the transformer-rated full load current = 601.42 A.

If the source impedance is considered, based upon 30 kA rms at 13.8 kV, the through fault current falls to 10.28 kA. However, the primary source short-circuit impedance is not considered for category II transformers and is therefore ignored.

Seventy percent of the calculated short-circuit current = 7.65 kA, and $I^2t$ line is constant between 100% short-circuit current at 2 seconds and 70% short-circuit current:

$$(10.94)^2(2) = (7.65)^2(t_1)$$

This gives $t_1$ = 4.09 seconds. Thus, point C can be plotted in Figure 11.9. Also, a current of 7658 A can be read on $x$-axis at this point.

Using $I^2t = k$, where $I$ is the current in terms of base current at 2 seconds, that is = 18.19 pu, gives $k = 661.7$. Therefore, $t_1$ can also be calculated as: $(0.7 \times 18.19)^2(t_1) = 661.7$, which gives the same value of $t_1$ = 4.09 seconds, as before.

According to Table 11.3, at 10 seconds, 11.3 times rated current = 6796 A. This gives point D in Figure 11.9. The infrequent fault curve is drawn parallel to line AB to 2 seconds.

Again from Table 11.3, at 6.3 times the base current (= 3789 A), $t$ = 30 seconds (point E), at 4.75 times the base current (= 2856 A), it is 60 seconds, and at 3.0 times (= 1804 A), it is 5 minutes (point F).

Figure 11.9 shows the calculated curve.

### 11.5.1 Protection with Respect to Through Fault Curves

Standards do not establish guidelines that to what extent the through fault withstand curve of the transformers should be protected. The NEC primary protection settings as detailed above will not protect the through fault withstand curve over its entire range. See Figure 11.2. It is opined that fault currents in the range of 2–2.5 times are more of temporary overload nature. A high resistance fault will quickly breakdown, and the fault current will escalate. Consider, for example, NEC maximum permitted setting of six times the transformer full load current for primary protection (Table 11.1).

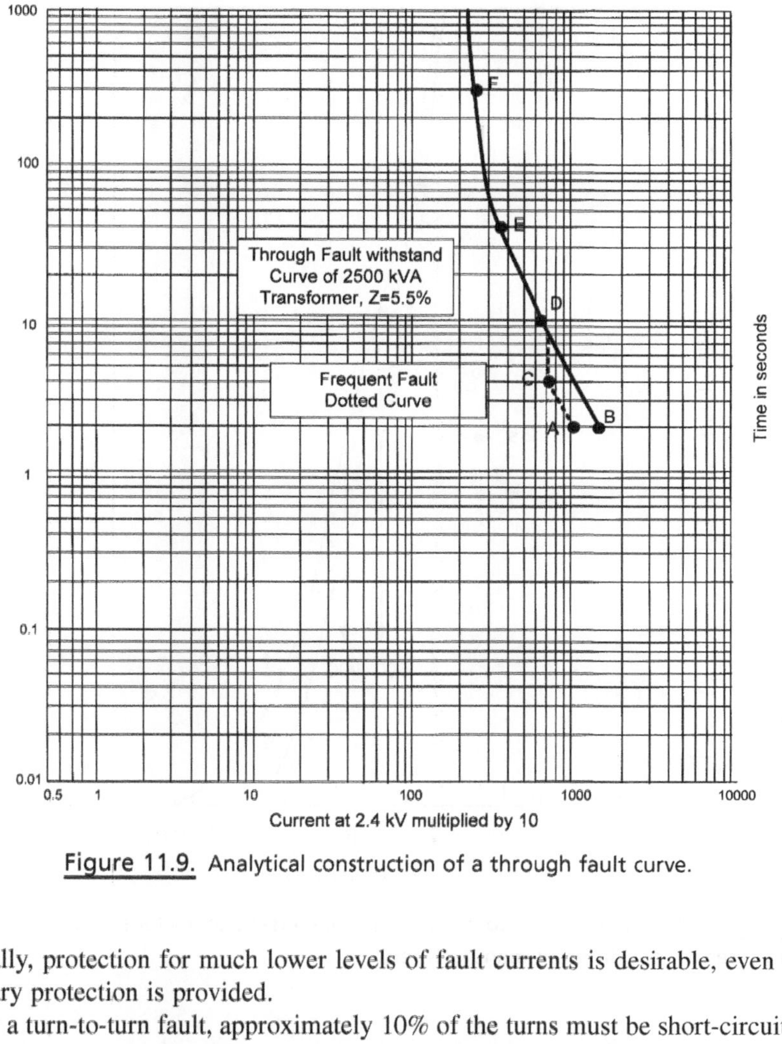


Figure 11.9. Analytical construction of a through fault curve.

Practically, protection for much lower levels of fault currents is desirable, even when secondary protection is provided.

For a turn-to-turn fault, approximately 10% of the turns must be short-circuited to circulate a fault current equal to the full load current of the transformer. Such a fault cannot be detected even with differential relays. Internal pressure rise relays (Fault pressure relays) are used. Again, it is not the scope of this book to explain all the aspects of equipment protection, other than phase over currents for the arc flash purposes.

## 11.6 TRANSFORMER PRIMARY FUSE PROTECTION

### 11.6.1 Variations in the Fuse Characteristics

Power fuses used for transformer primary protection can be current limiting or expulsion type. Class E power fuses rated 100E or less open in 300 seconds at a current level between 200% and 240% of their E rating. Fuses rated 100E or more open in 600

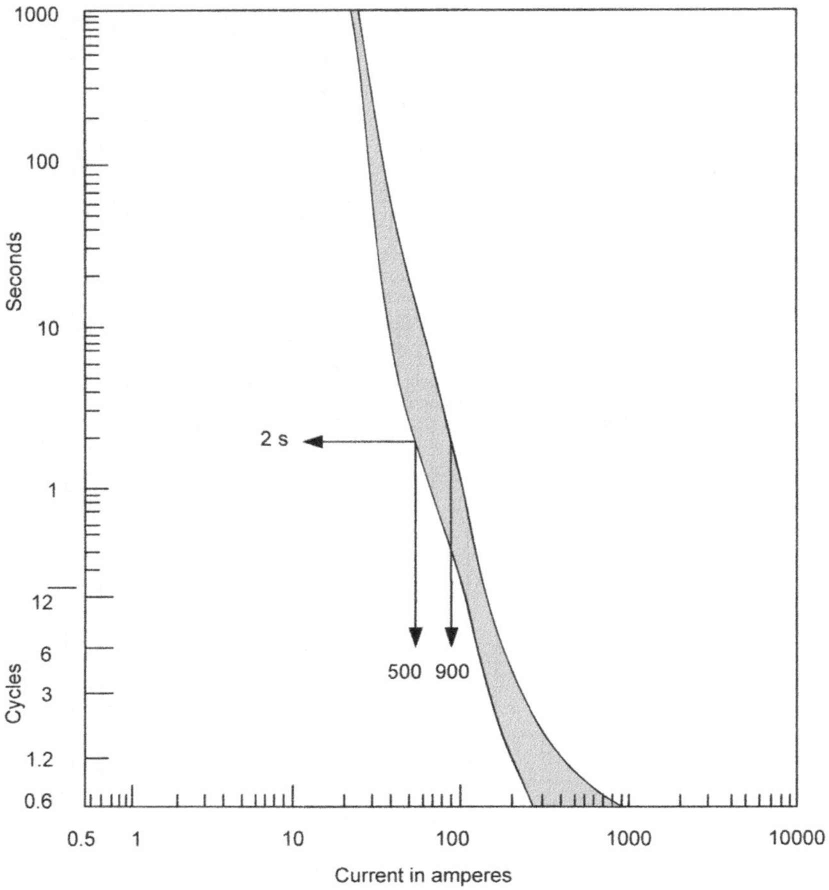


Figure 11.10. Variations in the total clearing time–current characteristics of 100E current-limiting fuses of four different manufacturers.

seconds at current level between 220% and 264% of their E rating [9]. The E rating also reflects 2 : 1 minimum melting current versus continuous current ratio, which is a design feature of the power fuse. The expulsion-type fuses are generally available in higher continuous ampere ratings, and current limiting fuses are available in higher short-circuit ratings (Chapter 7). *We limit our discussions to current limiting fuses.*

Only some typical operating parameters are fixed by the standards for the current limiting class E fuses. Figure 11.10 shows the variations in the total clearing time characteristics of 100E fuses, based upon the data of four different manufacturers. A spread of 500–900 A is seen at 2 seconds. (For expulsion type fuses, this spread is even more.) These variations in the time–current characteristics of fuses of the same specifications and ratings do not easily permit a generalization of the through fault protection of the transformers with fuses, though some guidelines are provided further.

### 11.6.2 Single Phasing and Ferroresonance

Operation of a fuse in one of the phases can give rise to single phasing. For example, for a line-to-line secondary fault in a delta-wye transformer secondary, the distribution of currents on the primary side are shown in Figure 11.11. The fuse in the phase carrying 1.0 per unit fault current will operate first, leaving two fuses in service. Thus, a phase of the transformer windings is energized through cable capacitance. This circuit can give rise to ferroresonance, as it involves excitation of one or more saturable reactors (transformer windings) through capacitance. When ferroresonance occurs, high peak voltages, irregular voltage, and current waveforms and loud noise in the transformer due to magnetostriction may be produced [10, 11].

### 11.6.3 Other Considerations of Fuse Protection

- Fuses should not operate on transient inrush switching current of transformers. The integrated heating effect of the switching current is represented by 8–12 times the transformer full load current for 0.1 seconds and 25 times the full load current for 0.01 second. For transformers of rating 10 MVA and above, it is safer to consider inrush current equal to 15 times the full load current to take account of residual trapped flux.
- The short-circuit rating of the transformer primary fused load-break switch should be selected compatible with the calculated short-circuit results at the point of application. The current limiting fuses generally have a short-circuit rating of 50 kA rms symmetrical and 80 kA rms asymmetrical. The switch short-circuits momentary (first cycle) and 10-cycle fault closing rating should be compatible.
- A current limiting fuse generates an arc voltage much higher than the system voltage (Chapter 7). This is of concern for coordination with surge arresters, which are provided for protection of transformer windings.
- The fuses will operate at their rated ampacity for 40°C ambient temperature. In an enclosure and in an outdoor location exposed to direct sunlight, the temperature may exceed 40°C, and derating will be required.
- Fuses tend to fatigue due to repeated inrush currents and transients below minimum melting operating limits. The time–current characteristics of a fatigued fuse may change, *and there is no way to ascertain these effects.* Proper margins are provided in the coordination.
- Table 11.4 shows the maximum available class E current limiting fuse sizes and the largest transformer for which these can be generally used, considering transformer inrush currents.

## 11.7 OVERCURRENT RELAYS FOR TRANSFORMER PRIMARY PROTECTION

Some of the problems inherent in the application of fuses are absent when relayed circuit breakers are used. This is a desirable trend in the industry, with respect to arc flash protection and limiting the area of shutdown. Some obvious advantages are:

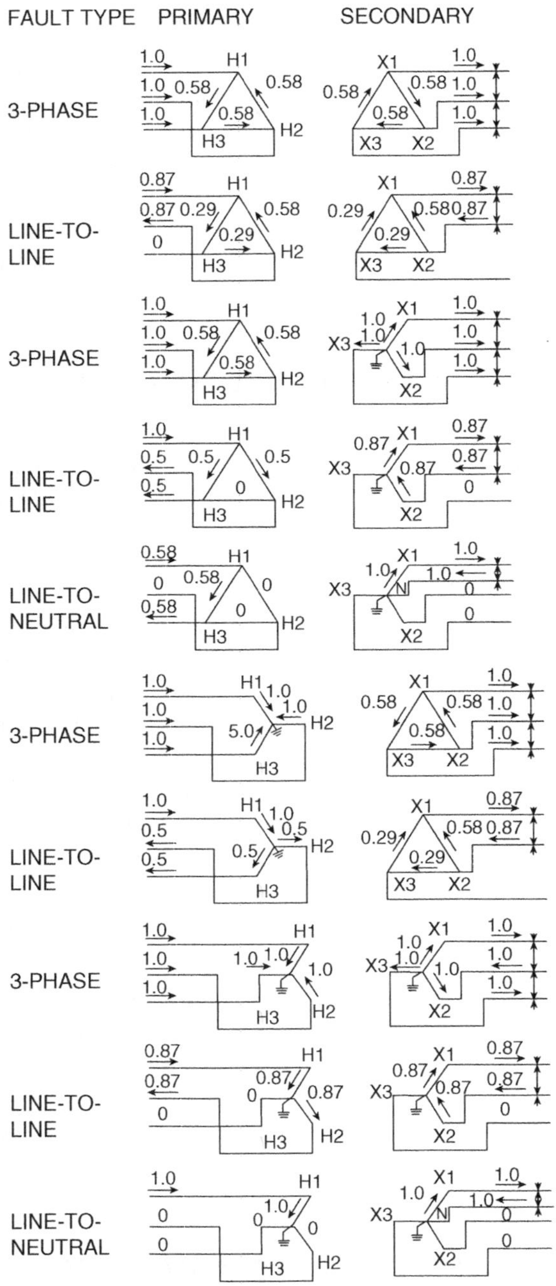


Figure 11.11. Line and winding currents in transformers for various secondary faults, different transformer winding connections.

TABLE 11.4. Maximum Ratings of Current Limiting Fuses and Protected Transformers, Primary Protection

| System Operating Voltage, kV | Maximum Available Fuse Size | Maximum Rating of Protected Transformers, $Z$ = 5.75% on base kVA |
|---|---|---|
| 2.4 | 750E, 5.5 kV | 2228/2600 |
| 4.16 | | 3862/4506 |
| 6.9 | 300E, 8.3 kV | 2560/2987 |
| 7.2 | | 2672/3117 |
| 12.47 | 300E, 15.5 kV | 4450/5190 |
| 13.2 | | 4898/5715 |
| 13.8 | | 5127/5975 |

- Single phasing is prevented.
- Possibility of ferroresonance is mitigated.
- Relay characteristics, properly calibrated and maintained, are not subject to change with time, and, therefore, a consistent performance is inherent.
- Selections of a wide range of time–current characteristics, which can be closely coordinated with transformer through fault, withstand curve, and downstream distribution.

A better through fault protection can be provided with relays as compared with fuses. The time–current coordination in Figures 11.12 and 11.13 provide a comparison: fuses versus relays for transformer primary protection. These figures show three-step coordination, a molded case circuit breaker at the MCC, a LVPCB feeder circuit breaker provided with programmable trip device, a secondary main LVPCB, and finally a delta-wye 2500-kVA, 13.8–0.48-kV transformer, which is solidly grounded on 480-V side. Its frequent fault and infrequent fault curves are shown in these figures.

Figure 11.12 shows a 175E class E fuse for the primary protection. There is a slight miscoordination with main LVPCB settings at the high level of short-circuit current; this is acceptable. The shifted through fault curve for single line-to- ground faults on the 480-V side is not protected at all. Compare this with Figure 11.13, where all other settings and coordination remains the same as in Figure 11.12, except that the primary fuse protection is replaced with an overcurrent relay device 50/51. A better coordination is achieved, lack of coordination with the main LVPCB is entirely eliminated, and the shifted 58% thermal damage curve of the transformer is fully protected.

## 11.8 LISTING REQUIREMENTS

Two listing agencies, FM Research Corporation and Underwriters Laboratories, Inc., list less flammable liquids for transformers. These liquids have a fire point of not less than 300°C. All the listing requirements are not enunciated here. The FM Research

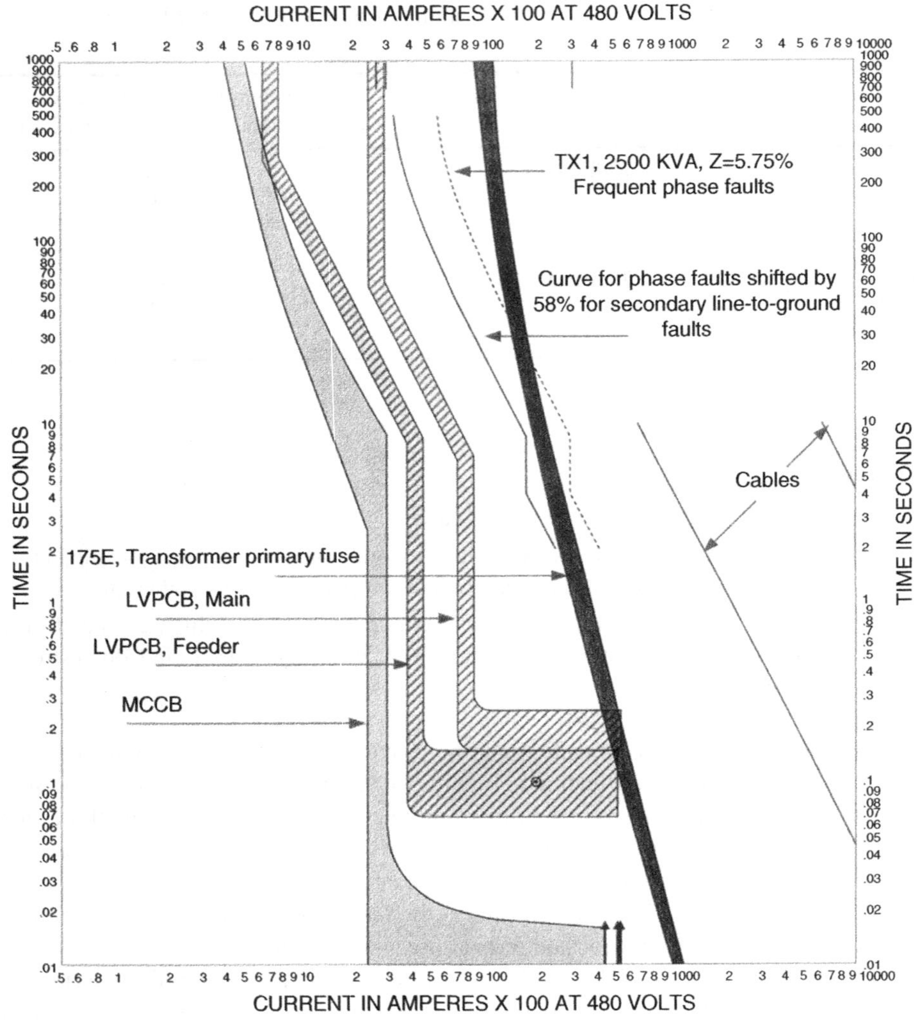


Figure 11.12. Through fault withstand curve protection of a 2500-kVA, Z = 5.75% transformer with a primary fuse of 175E.

listing is based on approved less flammable fluid in the tank that meets certain criteria (the Envirotemp FR3 Fluid is approved by FM). Pressure relief devices must be provided, enhanced electrical protection is required, spacing from adjacent combustibles must be provided in the event of a leak, and a liquid confinement area and drainage are required [2]. The UL listing is based on the requirements that no tank rupture or noticeable fluid leakage occur during low and high current arcing faults test.

The transformers with less flammable liquids may be installed in type I and type II buildings [1] without a transformer vault, provided transformers are not rated above 35,000 V primary and other conditions specified in NEC 450.23 (A) are met. FM Global

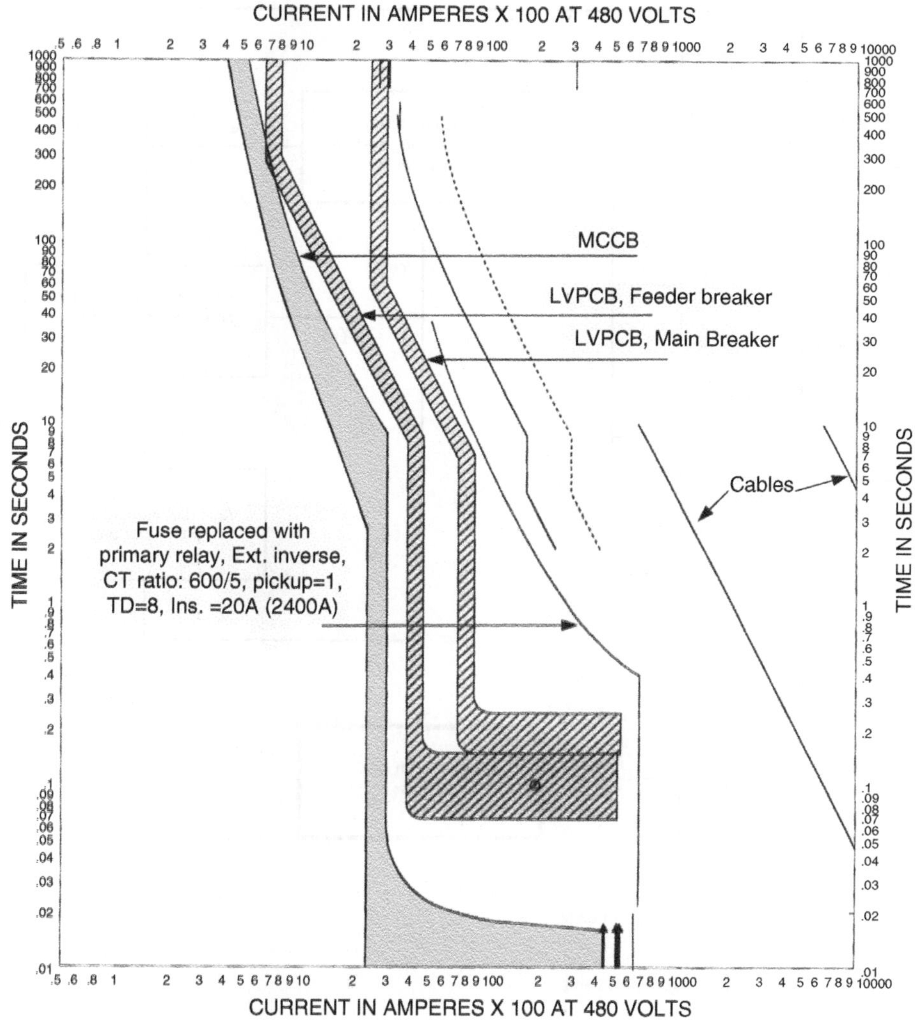


Figure 11.13. Through fault withstand curve protection of a 2500-kVA, Z = 5.75% transformer with primary relayed circuit breaker.

Property Loss Prevention Data Sheets [2] may also be seen. These specify protections for indoor and outdoor transformers. As a specimen, a flow chart of compliance with NEC 450.23 and FM Listing is shown in Figure 11.14, *for indoor installations* using less flammable liquid-immersed transformers; Figure 11.15 shows the UL listing requirements.

The UL listing is based upon the limitation of the fault energy due to primary current limiting fuses and transformer tanks capable of withstanding an internal pressure of 12 psig without rupture. Pressure relief devices must be provided. Table 11.5

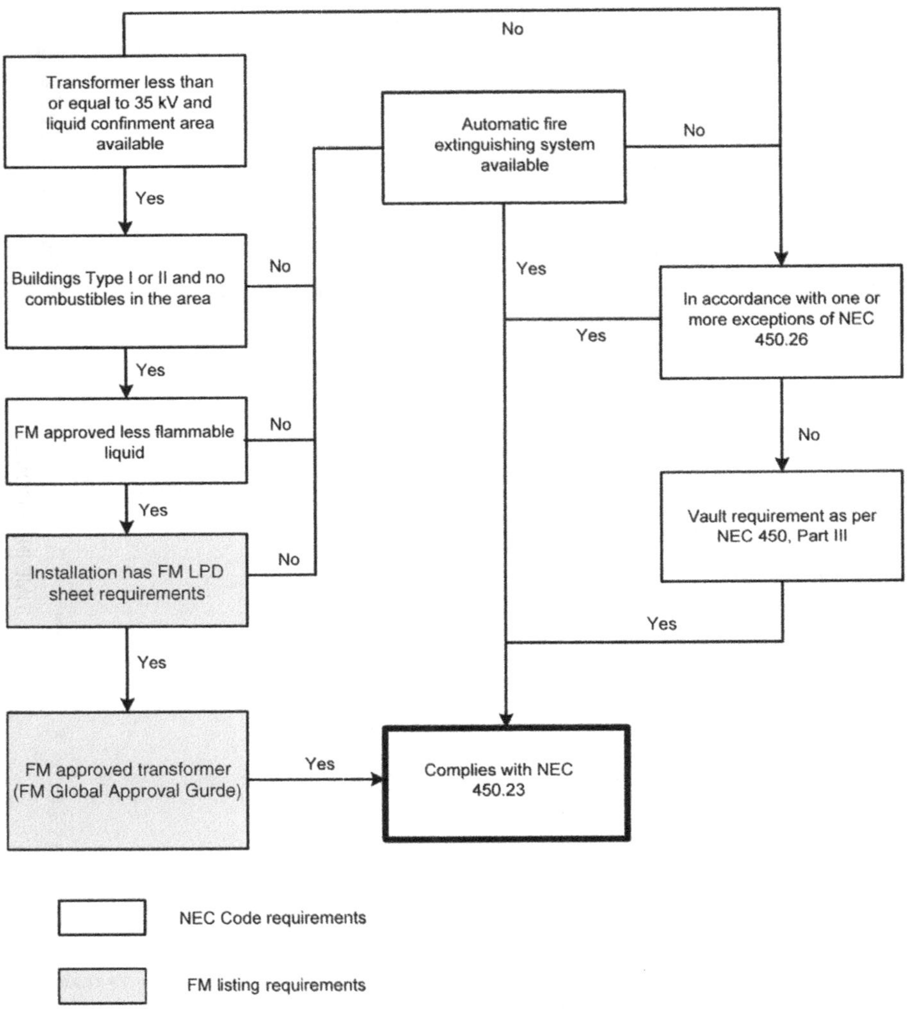


Figure 11.14. Less flammable liquid-immersed transformers compliance to NEC section 450.23 per FM listing, indoor installations.

shows the limiting fusing maximum $I^2t$ and minimum required pressure relief capability in SCFM (standard cubic feet per minute) at 15 psig. The opening pressure is 10 psig maximum. With respect to protection without current limiting fuses for the primary protection, it will be difficult to meet the requirements of maximum $I^2t$ ($A^2s$) in column 3 of the Table 11.5. As an example, consider a 2500-kVA transformer connected to a 13.8-kV primary system of short-circuit level 30 kA rms symmetrical on the transformer primary bushings. Considering an instantaneous relay with 1/2 cycle operating time and a 13.8-kV circuit breaker with interrupting time of three cycles, the $I^2t$ released is $90 \times 10^6$ $A^2s$. The permissible is $7.5 \times 10^6$ $A^2s$. This means that unless the available

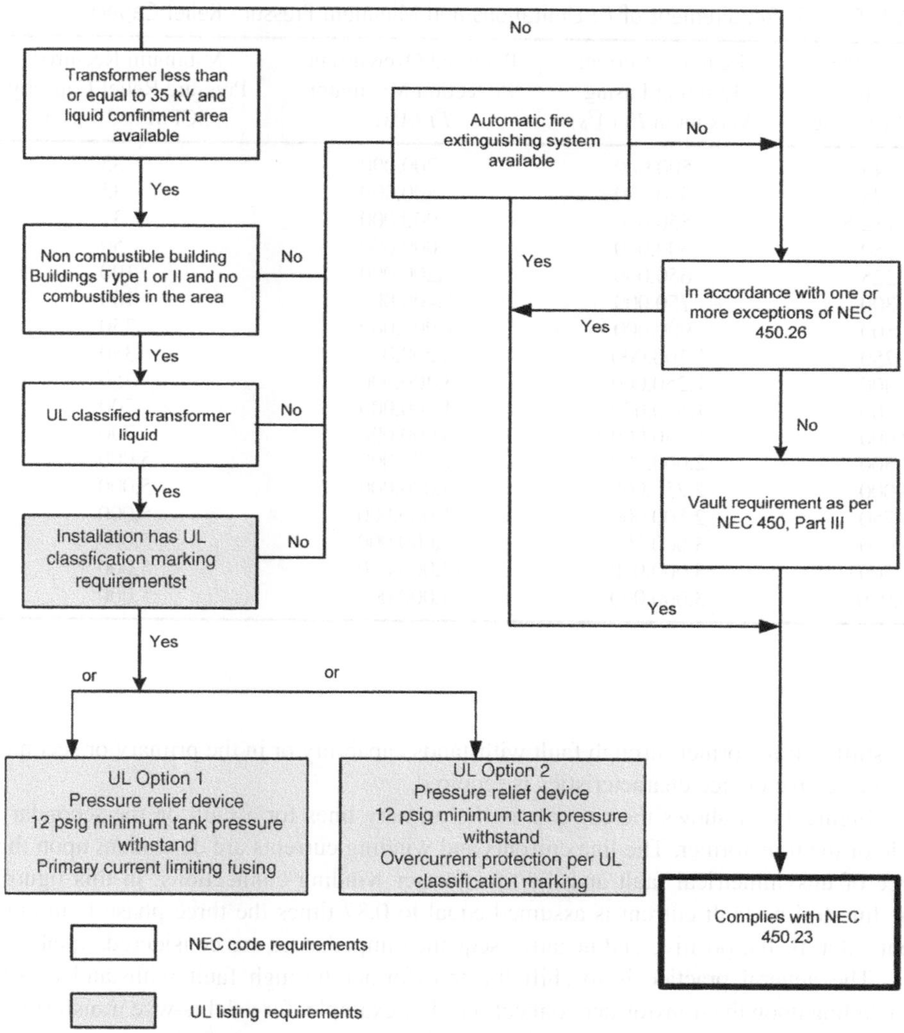


Figure 11.15. Less flammable liquid immersed transformers compliance to NEC section 450.23 per UL listing, indoor installations.

three-phase fault level is 11.3 kA maximum, the labeling requirements are not met with relayed circuit breakers.

## 11.9 EFFECT OF TRANSFORMER WINDING CONNECTIONS

A three-phase symmetrical fault on the secondary windings of a three-phase transformer gives rise to three-phase currents in the primary lines irrespective of the transformer connections. This primary fault current will be changed in the ratio of transformation.

TABLE 11.5. Requirement of $I^2t$ Limitations and Minimum Pressure Relief Capacity

| Three-Phase Transformer kVA Rating | Required Current Limiting Fusing Maximum $I^2t$ ($A^2s$) | Required Overcurrent Protection Maximum $I^2t$ ($A^2s$) | Minimum Required Pressure Relief Capacity SCFM at 15 psig |
|---|---|---|---|
| 45 | 500,000 | 700,000 | 35 |
| 75 | 500,000 | 800,000 | 35 |
| 112.5 | 550,000 | 900,000 | 35 |
| 150 | 600,000 | 1,000,000 | 50 |
| 225 | 650,000 | 1,200,000 | 100 |
| 300 | 750,000 | 1,400,000 | 100 |
| 500 | 900,000 | 1,900,000 | 350 |
| 750 | 1,100,000 | 2,200,000 | 350 |
| 1,000 | 1,250,000 | 3,400,000 | 350 |
| 1,500 | 1,500,000 | 4,500,000 | 700 |
| 2,000 | 1,750,000 | 6,000,000 | 700 |
| 2,500 | 2,000,000 | 7,500,000 | 5,000 |
| 3,000 | 2,250,000 | 9,000,000 | 5,000 |
| 3,750 | 2,500,000 | 11,000,000 | 5,000 |
| 5,000 | 3,000,000 | 14,000,000 | 5,000 |
| 7,500 | 3,000,000 | 14,000,000 | 5,000 |
| 10,000 | 3,000,000 | 14,000,000 | 5,000 |

No shift in transformer through fault withstands capability or in the primary or secondary protective device characteristics is required.

Figure 11.11 shows the currents in the primary lines for a fault on the secondary side of the transformer. The line currents and winding currents are dependent upon the type of unsymmetrical fault and the transformer winding connections. In this figure, the line-to-line fault current is assumed equal to 0.87 times the three-phase fault currents, that is, the positive and negative sequence impedances are considered equal.

The general practice is to shift the transformer through fault withstand curve, depending upon the transformer connections. For example, for a delta–wye transformer, wye windings solidly grounded, only two primary lines carry a fault current of 0.58 times the secondary line-to-neutral current, and therefore shift the transformer through fault withstand curve by 0.58 toward left in the time–current coordination plot (i.e., each current point on the curve is multiplied by a factor of 0.58).

Some authors are of the opinion that the transformer withstand curve is a fixed entity, and it should not be shifted. Rather than shifting the transformer withstand curve, shift the primary protective relay curve. As this protective device will operate slower, shift the curve by 1/058 (= 1.73) times toward the right in the time–current cooordination (TCC) plot.

Similar shifts for a line-to-line fault in a solidly grounded delta–wye transformer will be: (1) transformer withstand curve shift toward right by 16%, or the relay protective curve shift toward left, multiplied by a factor of 0.87 throughout.

Most computer programs for coordination shift the transformer withstand curves, rather than shifting the relay curves. Therefore, this method is adopted in this book.

## 11.10 REQUIREMENTS OF GROUND FAULT PROTECTION

For arc flash calculations, we consider three-phase faults. Yet many times the primary protection device is used to provide through fault protection for secondary unsymmetrical faults also, as there may not be a secondary ground fault protection. This omission of secondary ground fault protection is in contrivance of NEC requirements for services, Article 240.95, which states that ground fault protection will be provided for solidly grounded wye electrical services of more than 150 V to ground but not exceeding 600 V phase-to-phase for each service disconnect rated 1000 A or more; see also Section 4.12. This was first required in 1971, because of unusually high burnouts reported on these services. Further, it is specified that the ground fault protection will operate to open all ungrounded conductors (phase conductors) of the faulted circuit. The maximum setting is 1200 A, and the maximum time delay is 1 second for ground fault currents equal to or greater than 3000 A.

Listing Option B, FM Global *requires that all the following conditions must be complied with:*

- The transformer neutral ground fault relay as discussed above (device 51N in Figure 11.2d) should be provided.
- Indoor units >500 kVA and outdoor units >2500 kVA shall be equipped with alarm contacts on pressure relief device and a rapid pressure rise relay.
- Three-phase pad mounted and substation transformers shall be equipped with an oil level gauge. All transformers >750 kVA shall be equipped with liquid temperature indicator and pressure vacuum gauge.
- Transformer shall pass BIL (basic insulation level) test at a minimum tilt of 1.5° from vertical.

For high RG systems, no shifts in the thermal withstand curves for secondary ground faults are required.

## 11.11 THROUGH FAULT PROTECTION

### 11.11.1 Primary Fuse Protection

The extent of protection provided with respect to thermal damage curves of transformers using primary fuse protection is illustrated in Table 11.6. This table is constructed with the worst characteristics of the available class E fuses (Figure 11.10) and a number of coordination studies [12]. Delta primary and wye secondary transformers of various sizes connected to 13.8 kV primary voltage are considered. Transformers rated up to 2500-kVA have low voltage secondary windings, solidly grounded, and transformers of 3000 kVA and 5000 kVA have resistance-grounded (RG) medium voltage secondary windings. The symbol "N" means no protection is provided by the selected fuse. The numbers "5–15" signifies that transformer through fault withstand is protected

TABLE 11.6. Protection Provided by Primary Fuses: Frequent Fault and Infrequent Fault Damage Curves

| Transformer Three-Phase (kVA) | Maximum Current at Fan Cooled Rating | %Z on base KNAN | Fuse Rating Class E, Current Limiting | Frequent Fault Withstand | | | Infrequent Fault Withstand | | |
|---|---|---|---|---|---|---|---|---|---|
| | | | | 3-P | L-L | L-N | 3-P | L-L | L-N |
| 500/644 | 27 | 5.75 | 40 | 2-12 | 2-16 | N | 2-12 | 2-16 | N |
| 750/966 | 40 | 5.75 | 65 | 2 | 2-3&5-15 | N | 2-10 | 2-15 | N |
| | | 8 | | N | N | N | 2-10 | 2-15 | N |
| 1000/1288 | 54 | 5.75 | 80 | 2 | 2-3&5-15 | N | 2-10 | 2-15 | N |
| | | 8 | | N | N | N | 2-10 | 2-15 | N |
| 1500/1680 | 81 | 5.75 | 125 | 2 | 2-3&5-15 | N | 2-10 | 2-15 | N |
| | | 8 | | N | N | N | 2-10 | 2-15 | N |
| 2000/2576 | 108 | 5.75 | 150 | 2-20 | 2-28 | N | 2-15 | 2-25 | N |
| | | 8 | | N | 2-3&6-38 | N | 2-15 | 2-25 | N |
| 2500/3500 | 146 | 5.75 | 175 | 2-25 | 2-38 | N | 2-25 | 2-38 | N |
| | | 8 | | 2&9-25 | 2-3&6-38 | N | 2-25 | 2-38 | N |
| 3000/4200 | 176 | 5 | 200 | 2-50 | 2-60 | RG | 2-50 | 2-60 | RG |
| | | 8 | | 2-3& 7-50 | 2-60 | RG | 2-50 | 2-60 | RG |
| 5000/7000 | 293 | 5 | 300 | 2-40 | 2-60 | RG | 2-50 | 2-50 | RG |
| | | 8 | | 2-40 | 2-60 | RG | 2-50 | 2-50 | RG |

Note: All transformers are delta-wye connected at primary voltage of 13.8 kV. 3-P, three-phase fault; L-L, line-to-line fault; L-N, line to neutral fault; RG, resistance grounded; N, no protection is provided with respect to transformer withstand curve.

TABLE 11.7. Protection Provided by Fuses versus Relays for Transformer through Fault Withstand

| | Frequent Faults | | | Infrequent Faults | | |
|---|---|---|---|---|---|---|
| Primary Protection | 3-P | L-L | L-N | 3-P | L-L | L-N |
| Fuses | 2 | 2-35 | N | 2-15 | 2-35 | 2-7 |
| Relays | 2-100 | 2-200 | N | 2-100 | 2-200 | 2-40 |

between 5 and 15 seconds only. The calculations are made with transformer impedances as shown.

This table shows that no protection is provided by the transformer primary fuses for secondary line-to-neutral faults. Therefore, a transformer neutral connected ground fault relay is a must to protect the transformer (see Chapter 4). Also, the effect of transformer impedance is clearly visible in this table with respect to the protection; a higher transformer impedance lowers the extent to which the through fault curves can be protected.

### 11.11.2 Primary Relay Protection

The through fault protection calculated in Table 11.6 with fuses can be much improved with primary overcurrent relays. A comparison is shown in Table 11.7 [12]. Though the protection is improved, yet the need for a neutral connected relay for secondary L-N faults is apparent. Also, the listing requirements $I^2t$ need to be simultaneously considered.

## 11.12 OVERALL TRANSFORMER PROTECTION

The overcurrent protection discussed above does not provide thermal protection of the transformer. The conductor insulation in the transformers ages fast if the temperatures exceed the design temperatures. Winding temperature indicators with alarm, and sometimes for trip, are used, but these have limitations. With OLTC (on-load tap changing), the hotspot temperature may move from high to low voltage winding or vice-a-versa. Winding temperature indicator consists of a temperature-sensing bulb immersed in a well in the top layer of the insulating oil, and is heated by a replica of the load current of the transformer through a CT. It is calibrated for hotspot, but has limitations with respect to thermal protection of the transformer windings. Digital relays can be used for thermal winding protection, which can calculate the temperature on the primary, secondary, and tertiary windings. Transformers are provided with overexcitation, negative sequence protection, voltage controlled overcurrent relays, and mechanical detection of faults—like gas accumulator relay, gas detector relay, sudden gas/oil pressure relay, internal gas monitoring relay, oil temperature and oil level alarms, and the like. References [13–15] provide further reading on transformer protections.

## 11.13 A PRACTICAL STUDY FOR ARC FLASH REDUCTION

### 11.13.1 System Configuration

A radial distribution system is shown in Figure 11.16. A single 1200-A 13.8-kV circuit breaker serves four substation transformers, three transformers of 2000/2240 kVA 480 V and one of 7500/9375 kVA, 2.4 kV secondary. All transformers have 13.8-kV windings in delta connection and the secondary windings in wye connection. The 480-V transformer wye windings are high RG and 2.4-kV transformer wye windings are low RG through a 200-A resistor. Downstream distribution from one 2000-kVA low voltage transformerTX1 and 7500 2.4-kV transformer TX2 is shown. The 480-V transformer serves a lineup of low voltage switchgear provided with LVPCBs having electronic trip programmers with LSI functions. At first consider that *there are no main secondary circuit breakers either on 480 V or 2.4 kV transformers*, that is, circuit breakers BK5 and BK6 shown in Figure 11.19 below are not provided. The low voltage switchgear serves a number of low voltage motor control centers; one typical MCC is shown. For coordination purposes, we need to consider the largest motor/feeder on any MCC. In the industrial distribution system, generally, the 200-hp motor rating is the maximum, and higher rated motors are connected to medium voltage distributions. The 2.4-kV transformer serves a lineup of 2.4-kV metal-clad switchgear, which in turn serves some medium voltage MCCs. The major ratings of the equipment are shown in this figure. The interconnecting cable sizes are detailed in Table 11.8.

The 7500/9375-kVA transformer TX2 is protected by relay R1 on 13.8-kV circuit breaker BK1. The 2000/2240-kVA transformer TX1is protected with 150E current limiting fuse. The medium voltage MCC is served from a relayed feeder circuit breaker, BK4. The 2000-hp motor has a NEMA 2 fused starter, with a 700-A vacuum contactor and 36R fuse. The system ground fault protection is not shown. All switching devices are applied well within their short-circuit ratings. The three-phase bolted short-circuit level at 13.8-kV switchgear is 30 kA rms symmetrical.

This radial distribution system meets the requirements of NEC and is "well protected" with the protection devices and loads required to be served. Many industrial distribution systems implemented on this basis are in service. *However, in the analysis to follow, it is shown that this system is very deficient from the arc flash hazard mitigation consideration.*

### 11.13.2 Coordination Study and Observations

Let us consider fault locations shown from F1 through F11. For arc flash evaluations, faults in any location of the equipment assembly must be considered. Table 11.9 shows the fault locations and the protective devices that will clear the fault, when main circuit breakers are not provided.

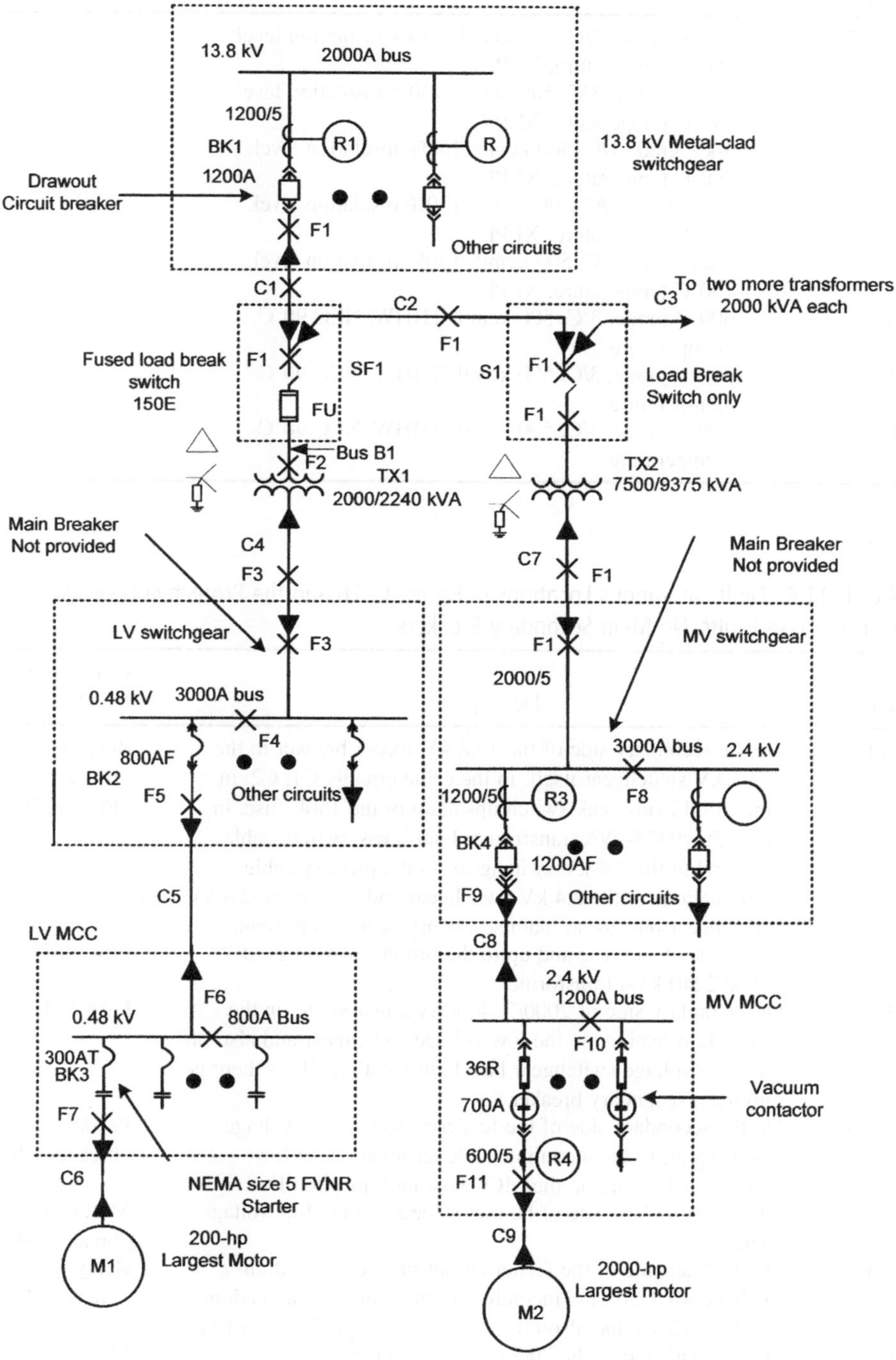


Figure 11.16. A low voltage and medium voltage distribution system for protective relay coordination and arc flash analysis.

TABLE 11.8. Cable Sizes and Lengths, Figure 11.16

| Cable Designation | Cable Description | Number in Parallel per Phase |
|---|---|---|
| C1, C2 | 15-kV grade, 3/C, 500 kcmil, 130% insulation level, 90°C temperature, XLPE | 2 |
| C3 | 15-kV grade, 3/C, 500 kcmil, 130% insulation level, 90°C temperature, XLPE | 1 |
| C7 | 5-kV grade 1/C, 500 kcmil, 130% insulation level, 90°C temperature, XLPE | 4 |
| C8 | 5-kV grade 1/C, 500 kcmil, 130% insulation level, 90°C temperature, XLPE | 2 |
| C9 | 5-kV grade 1/C, 500 kcmil, 130% insulation level, 90°C temperature, XLPE | 2 |
| C4 | 600-V grade, 3/C, 500 kcmil, THHW, NEC 90°C temperature | 7 |
| C5 | 600-V grade, 3/C, 500 kcmil, THHW, NEC 90°C temperature | 2 |
| C6 | 600-V grade, 3/C, 500 kcmil, THHW, NEC 90°C temperature | 1 |

TABLE 11.9. Faults at Various Locations in Figure 11.16 and the Protective Devices Clearing These Faults, No Main Secondary Breakers

| Fault at | Description | Fault Clearing Device |
|---|---|---|
| F1, F8 | On the secondary side of the 13.8-kV feeder breaker in the 13.8-kV switchgear itself, in the cable circuits C1, C2, in the fused load-break switch upstream of the 150E fuse, in the 7500/9375-kVA transformer load-break switch, cable C7, and to the 2.4-kV switchgear, in the primary cable terminations in the 2.4 kV switchgear and also in the 2.4 kV switchgear bus, as no main secondary beaker is present. | Relay R1, Feeder Breaker BK1 |
| F2 | Load side of the fuse and up to the primary windings of 2000/2240-kVA transformer. | Fuse SF1 |
| F3, F4 | On secondary side of 2000/2240-kVA transformer, on the cable terminations in the low voltage switchgear and also on the low voltage switchgear bus, fault location F4, as there is no main secondary breaker. | Fuse SF1 |
| F5, F6 | On the secondary side of the feeder beaker in low voltage switchgear, in the incoming cable terminations at low voltage MCC and on the MCC bus, fault location F6 | Feeder breaker BK2 |
| F7 | On the secondary side of he motor feeder in the low voltage MCC | Motor starter breaker BK3 |
| F9, F10 | On the load side of the feeder circuit breaker to medium voltage MCC, on the incoming terminations of the medium voltage MCC, and also on the MCC bus, fault location F11 | Relay R3, breaker BK4 |
| F11 | On the load side of the 2000-hp motor starter | Motor starter Fuse 36R |

Figure 11.17 shows the TCC plot of the 2000/2240-kVA transformer protective devices and Figure 11.18 shows the TCC plot for 7500/9500-kVA transformer. The following observations are of interest.

- The transformer ANSI frequent fault curves are protected.
- The system is well coordinated.
- The 150E fuse clears the transformer inrush point. There is a slight overlap with the feeder short-time delay setting at A, but this is acceptable. Transformer-rated

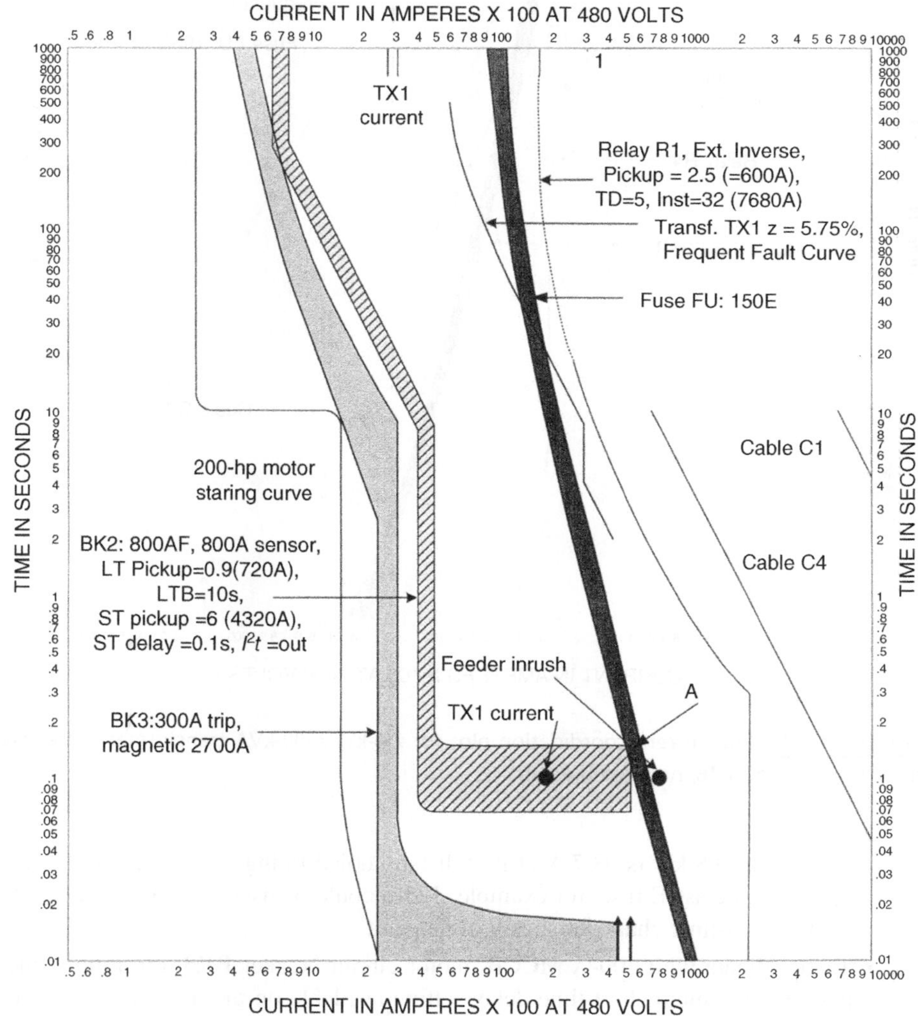


Figure 11.17. Time–current coordination plot for 480-V, 2000-kVA transformer protective devices in Figure 11.6, no main breaker.

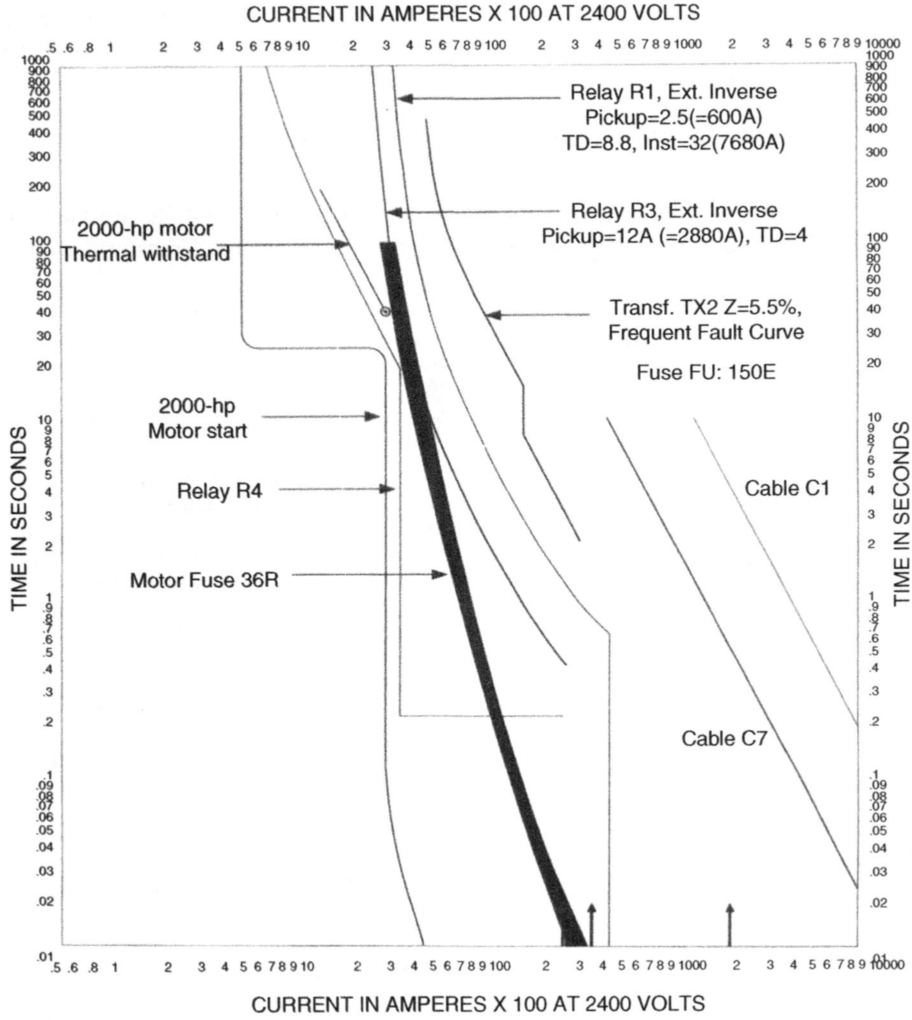


Figure 11.18. Time–current coordination plot for 2.4-kV, 7500-kVA transformer protective devices in Figure 11.16, no main breaker.

current at 13.8 kV is 93.7 A at its full fan cooled rating of 2240 kVA. Thus, a lower rated class E fuse, for example, 125E, could be used, but the overlap with feeder short-time characteristics will increase.

- To coordinate with 200-A MCCB, motor circuit breaker BK3, a setting higher than the minimum short-time delay setting available on the trip programmer of circuit breaker BK2 is not required.
- The motor starting curves are drawn for the actual motor starting time, driving its load inertia.

- The 2000-hp motor thermal curve is plotted based upon the manufacturer's data, and the multifunction relay R4 shown in Figure 11.18 admirably protect it.
- The settings on relay R1 protect the 7500/9375-kVA transformer as per NEC guidelines and allow the maximum system load equivalent to the total installed kVA rating of the transformer to be carried on continuous basis.
- The instantaneous settings on relay R1 take into account the total feeder inrush, that is, all the connected transformers will take their inrush currents on switching. This total inrush current is calculated as 6500 A at 13.8 kV.
- Curves for only a few of the cables are plotted as specimens.

### 11.13.3 Arc Flash Calculations: High Hazard Risk Category (HRC) Levels

Table 11.10 shows the results of arc flash hazard calculations according to IEEE Guide. The working distance—gap in millimeters according to equipment type—and the system grounding are shown in this table. The following observations are of interest:

- The currents shown are the ones that flow in the device that clears the fault. For example, a fault on low voltage switchgear is cleared by the transformer primary fuse, and the current through the fuse is only 1.37 kA. It is reduced in the transformation ratio, in this case by a factor of 28.75. Further, $I_a$ = 85% is used for calculation of arcing time.
- The operating times of the devices like fuses, low voltage trip programmers, and MCCBs are built into the curves published by the manufacturer (see Chapter 1). For relayed circuit breaker, the trip time and the circuit breaker opening time are shown separately. We have five cycle breakers, so the opening time is 0.083 second.
- The disadvantage of not having a main secondary circuit breaker is demonstrated by these calculations. As the secondary short circuit currents are reduced in the ratio of transformation, when reflected on the primary side, the primary protection device takes long time to clear the fault. In Table 11.10, the transformer primary fuse takes 35.76 seconds to clear a fault on the low voltage switchgear bus, releasing immense amount of incident energy (1729 cal/cm$^2$). This is not acceptable. Even if the arcing time is limited to 2 seconds, the incident energy release exceeds 40 cal/cm$^2$
- The personal protective equipment (PPE) required at medium voltage switchgear is 4, while at medium voltage MCC and low voltage MCC, it is 3.
- Multiple PPE levels can exist on the same equipment. For example, the medium voltage switchgear has a PPE of 4 for a bus fault, but for a fault in the load side terminals or outgoing cable connections, the PPE is 3.

Though from TCC plots in Figures 11.17 and 11.18, the protection seems to be adequate, but not so from arc flash reduction considerations. It becomes, therefore,

TABLE 11.10. Arc Flash Hazard Analysis (see Figure 11.16) and TCC Plots (see Figures 11.17 and 11.18), Dimensions: LV switchgear: 36"W x 90"H x48"D, MV switchgear: 26"W x95"H x96"D, EC = HCB

| Equipment Faulted | Voltage (kV) | Upstream Trip Device | Air Gap (mm) | Bolted Fault (kA) | Arc Fault (kA) | Trip Time (Second) | Opening Time (Second) | Arc Time (Second) | Arc Flash Boundary (Inches) | Working Distance (Inches) | Incd. Energy (cal/cm$^2$) | PPE |
|---|---|---|---|---|---|---|---|---|---|---|---|---|
| SF1 | 13.8 | R1 | 152 | 35.02 | 31.06 | 0.016 | 0.083 | 0.099 | 126 | 36 | 9.55 | 3 |
| S1 | 13.8 | R1 | 152 | 33.96 | 30.73 | 0.016 | 0.083 | 0.099 | 152 | 36 | 9.25 | 3 |
| Bus duct, B1 | 13.8 | FU | 152 | 35.16 | 31.78 | 0.01 | 0 | 0.01 | 56 | 36 | 2.52 | 1 |
| MV switchgear | 2.4 | R1 | 104 | 4.83 | 4.24 | 0.673 | 0.083 | 0.757 | 291 | 36 | 44.4 | Danger |
| MV MCC | 2.4 | R3 | 104 | 26.69 | 22.97 | 0.437 | 0.083 | 0.521 | 235 | 36 | 30.6 | 3 |
| Fault F11, MV MCC | 2.4 | 36R Fuse | 104 | 26.47 | 22.90 | 0.023 | 0 | 0.023 | 39 | 36 | 1.35 | 1 |
| LV switchgear | 0.48 | FU | 32 | 1.37 | 0.94 | 35.77 | 0 | 2* | 44 | 24 | 4.05 | 2 |
| LV MCC | 0.48 | BK2 | 25 | 36.06 | 24.96 | 0.15 | 0 | 0.15 | 66 | 18 | 16.9 | 3 |
| LV MCC Fault F7 | 0.48 | BK3 | 25 | 36.06 | 24.96 | 0.017 | 0 | 0.017 | 23 | 18 | 1.91 | 1 |

* Calculated for 2 seconds. The actual fault clearance time = 35.77, which releases immense incident energy.

necessary to reduce the incident energy by faster fault clearance times, which means additional protection devices have to be provided.

Next, we will demonstrate how the incident energy can be reduced so that a PPE of >2 is not required anywhere in the system.

### 11.13.4 Reducing HRC Levels with Main Secondary Circuit Breakers

In these calculations, main secondary circuit breakers, BK5 and BK6 are provided in Figure 11.16. This modification is shown in Figure 11.19 and the breakers properly coordinated. This is shown in Figures 11.20 and 11.21. The provision of a 2000-A circuit breaker BK6 requires the shifting of the circuit breaker BK1 relay R1 curve. The arc flash calculations are shown in Table 11.11. In this table, only the arc flash hazard for the faulted equipments that undergo a change due to provision of main circuit breakers is documented. The following are noteworthy:

- The incident energy at the medium voltage switchgear slightly increases, while at low voltage switchgear, it is much reduced from extreme hazard to PPE3.
- A fault at F3 in the incoming cable terminations of the low voltage circuit breaker BK5 will still be cleared by the primary fuse FU. Thus, though the incident energy is reduced on the rest of the switchgear, the main secondary circuit breaker is exposed to extreme hazard and cannot be maintained in the energized state.
- A similar situation is depicted for fault at F1 on the incoming side of the 2.4-kV circuit breaker BK6.

Thus, the installation of the main circuit breakers has much reduced the arc flash hazard on the main buses, but for a fault on the main circuit breaker itself, which is cleared by the primary protection, the incident energy release is not reduced, and these circuit breakers cannot be maintained in the energized state.

### 11.13.5 Maintenance Mode Switches on Low Voltage Trip Programmers

The trip programmers for low voltage circuit breakers can be provided with maintenance mode switch. This switch is being called with variety of names, and can be even remotely located. In the maintenance mode, entirely new settings can be activated, or only instantaneous setting can be activated to lower the arc flash hazard. For further discussions of the maintenance mode switch in the low voltage trip programmers and MMPR relays, see Chapter 14. This simple device allows the coordination to be maintained under normal operation, except that some coordination is sacrificed during maintenance mode by activating alternate settings. The premise is that under short-time maintenance, the likelihood of a fault is remote, and some coordination can be sacrificed. If a fault does occur, a larger area will be shut down, but the worker is protected.

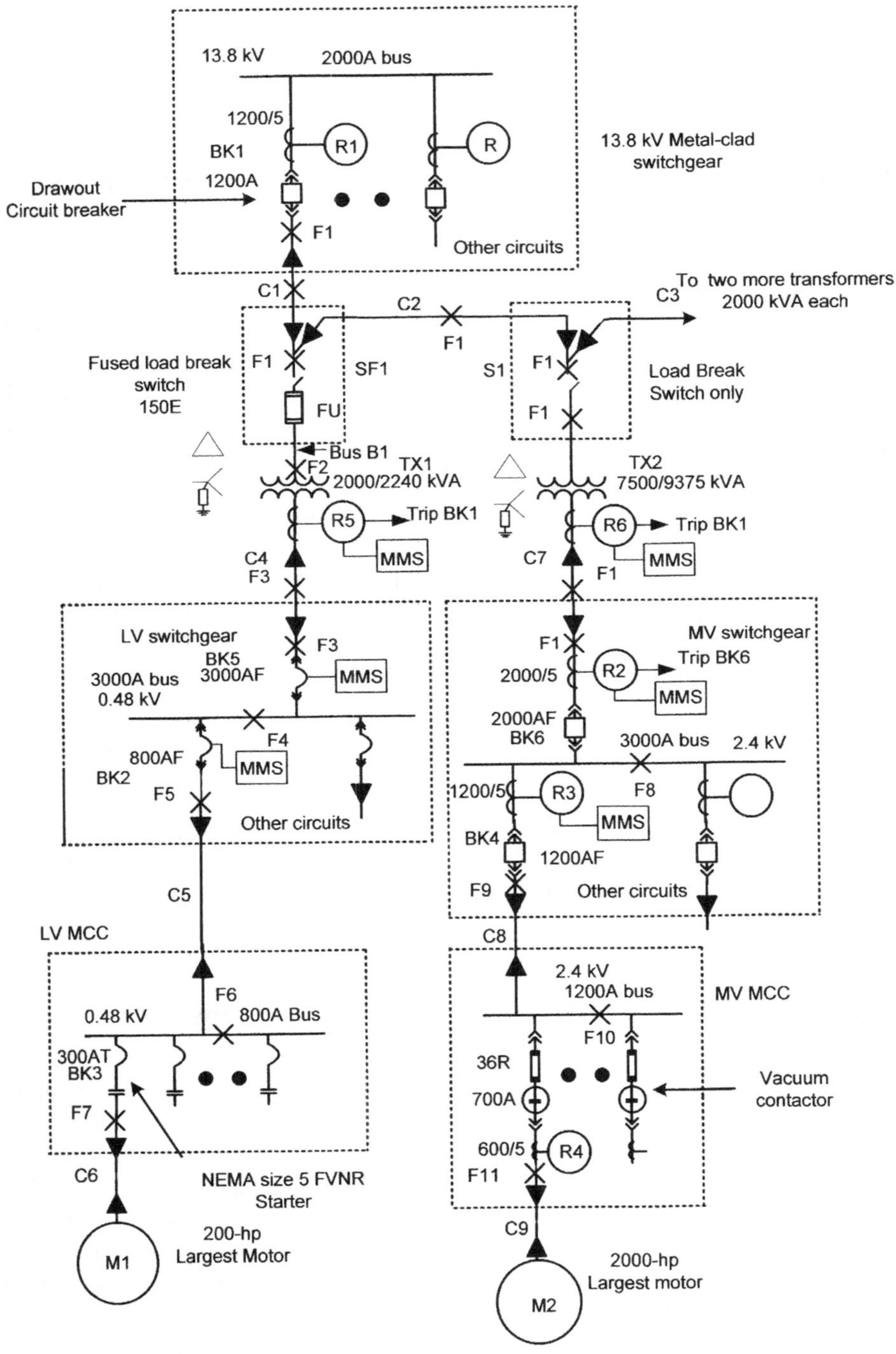


Figure 11.19. Modified distribution system of Figure 11.16, for arc flash reduction.

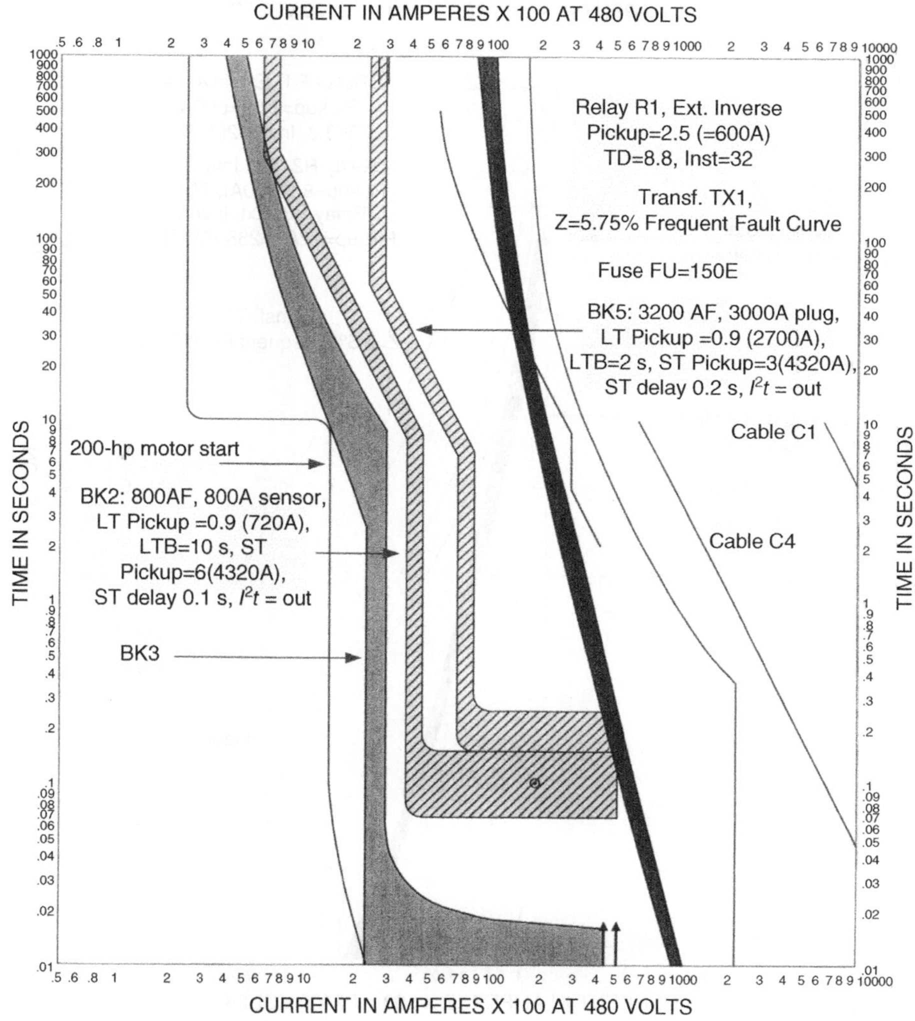


Figure 11.20. Time–current coordination plot for 480-V, 2000-kVA transformer protective devices in Figure 11.19 with main secondary breaker.

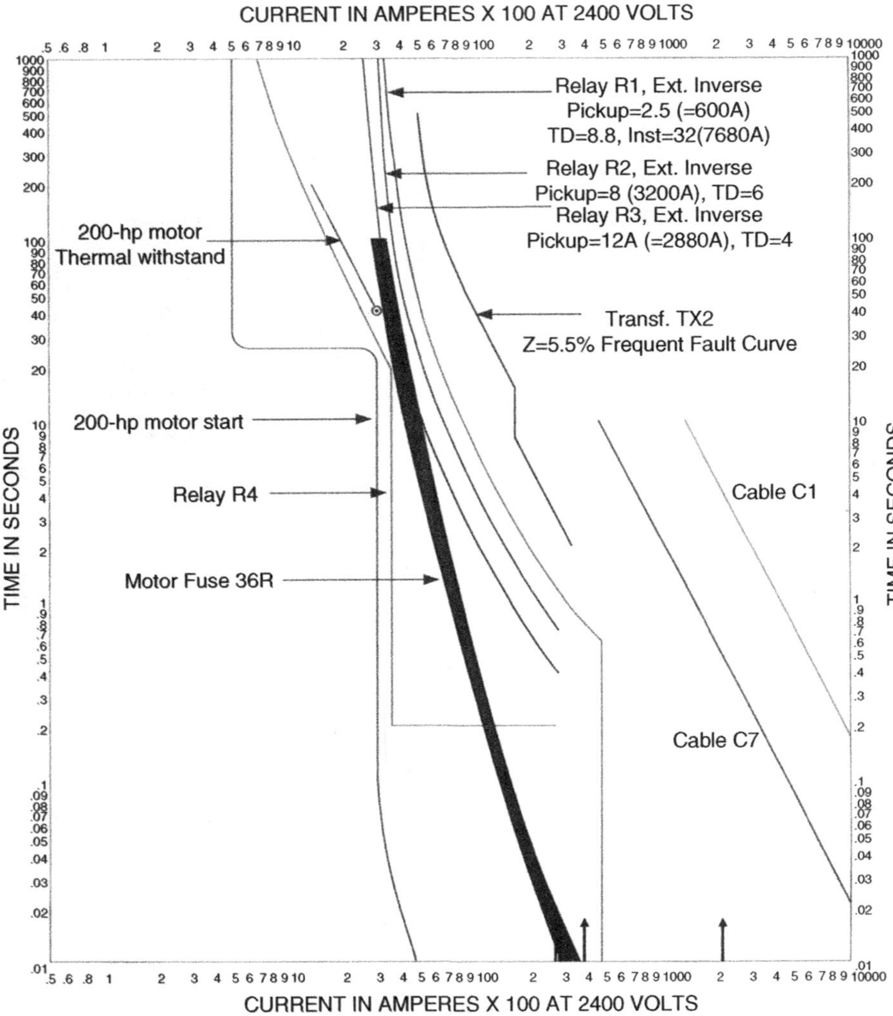


**Figure 11.21.** Time–current coordination plot for 2.4-kV, 7500-kVA transformer protective devices in Figure 11.19 with main secondary breaker.

TABLE 11.11. Arc Flash Hazard Analysis (see Figure 11.19) with Main Secondary Breakers and TCC Plots (see Figures 11.20 and 11.21, Dimensions: LV Switchgear: 36"W x 90"H x48"D, MV switchgear: 26"W x95"H x96"D, EC = HCB)

| Equipment Faulted | Voltage (kV) | Upstream Trip Device | Air Gap (mm) | Bolted Fault (kA) | Arc Fault (kA) | Trip Time (Second) | Opening Time (Second) | Arc Time (Second) | Arc Flash Boundary (Inches) | Working Distance (Inches) | Incd. Energy (cal/ $cm^2$) | PPE |
|---|---|---|---|---|---|---|---|---|---|---|---|---|
| MV switchgear | 2.4 | R2 | 104 | 27.81 | 23.89 | 0.722 | 0.083 | 0.805 | 310 | 36 | 49.4 | Danger |
| Breaker BK6 | 2.4 | R1 | 104 | 4.83 | 4.24 | 0.673 | 0.083 | 0.757 | 299 | 36 | 46.4 | Danger |
| LV switchgear | 0.48 | BK4 | 32 | 39.37 | 25.8 | 0.25 | 0 | 0.25 | 86 | 24 | 16 | 3 |
| Breaker BK5 | 0.48 | FU | 32 | 1.37 | 0.94 | 35.77 | 0 | 2* | 44 | 24 | 4.05 | 2 |

• Calculated for 2 seconds. The actual fault clearance time = 35.77, which releases immense incident energy.

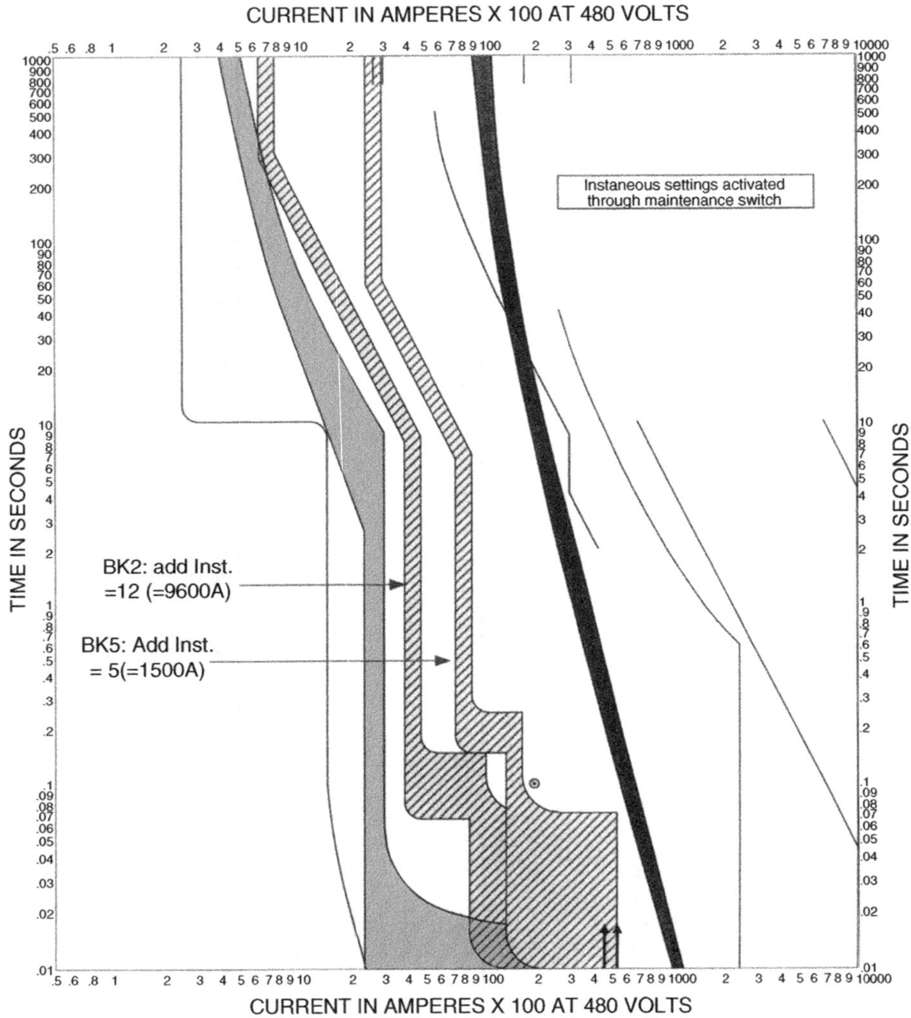


Figure 11.22. Time–current coordination plot for 480-V, 2000 kVA transformer protective devices in Figure 11.19 with main secondary breaker and instantaneous settings activated through maintenance mode switch; for arc flash reduction during maintenance.

The coordination with instantaneous settings on low voltage trip programmers is shown in Figure 11.22. Figure 11.23 shows the instantaneous settings activated through maintenance mode switch on relays R2 and R3. The switch can be wired into the inputs of microprocessor-based overcurrent relay (see Chapter 14).

Table 11.12 shows the HRC levels. These are reduced to HRC 2 for all fault locations, except for a fault on the main circuit breaker BK5 and BK6; hazard is not changed.

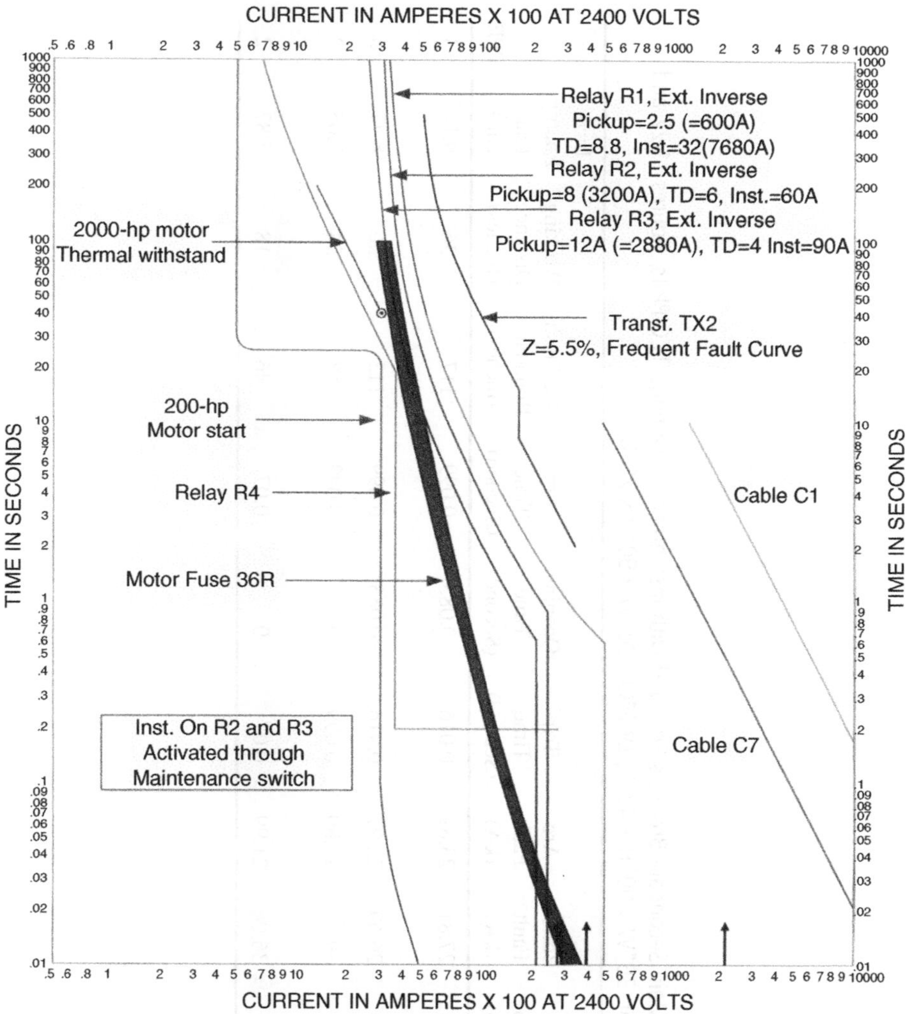


Figure 11.23. Time–current coordination plot for 2.4-kV, 7500-kVA transformer protective devices in Figure 11.19 with main secondary breaker and instantaneous settings activated through maintenance mode switch; for arc flash reduction during maintenance.

## 11.13.6 Addition of Secondary Relay

The system design can be modified, and the hazard on main circuit breakers are reduced with the modified configuration is shown in Figure 11.19. This figure shows that additional overcurrent relays R5 and R6 are provided on the secondary of transformers.

The current transformers actuating relays R5 and R6 are located in the transformer tank or in the transformer secondary air terminal compartment. The relays R5 and R6 are located in their respective switchgears. By moving the sensing current transformers

TABLE 11.12. Arc Flash Hazard Analysis, Main Secondary Breakers Provided with Instantaneous Settings, TCC Plots, Figures 11.22 and 11.23, EC=HCB, Dimensions: LV switchgear: 36″W x 90″H x 48″D, LV MCC: 30″W x 90″H x 25″D

| Equipment Faulted | Voltage (kV) | Upstream Trip Device | Air Gap (mm) | Bolted Fault (kA) | Arc Fault (kA) | Trip Time (Second) | Opening Time (Second) | Arc Time (Second) | Arc Flash Boundary (Inches) | Working Distance (Inches) | Incd. Energy (cal/ $cm^2$) | PPE |
|---|---|---|---|---|---|---|---|---|---|---|---|---|
| MV switchgear | 2.4 | R4 | 104 | 27.81 | 23.89 | 0.016 | 0.083 | 0.099 | 117 | 36 | 9.14 | 3 |
| MV MCC | 2.4 | R3 | 104 | 26.69 | 22.97 | 0.916 | 0.083 | 0.099 | 112 | 36 | 8.47 | 3 |
| LV switchgear | 0.48 | BK2 | 32 | 39.37 | 25.80 | 0.099 | 0 | 0.099 | 58 | 24 | 7.05 | 2 |
| LV MCC | 0.48 | BK2 | 25 | 36.06 | 21.99 | 0.07 | 0 | 0.07 | 46 | 18 | 7.87 | 2 |

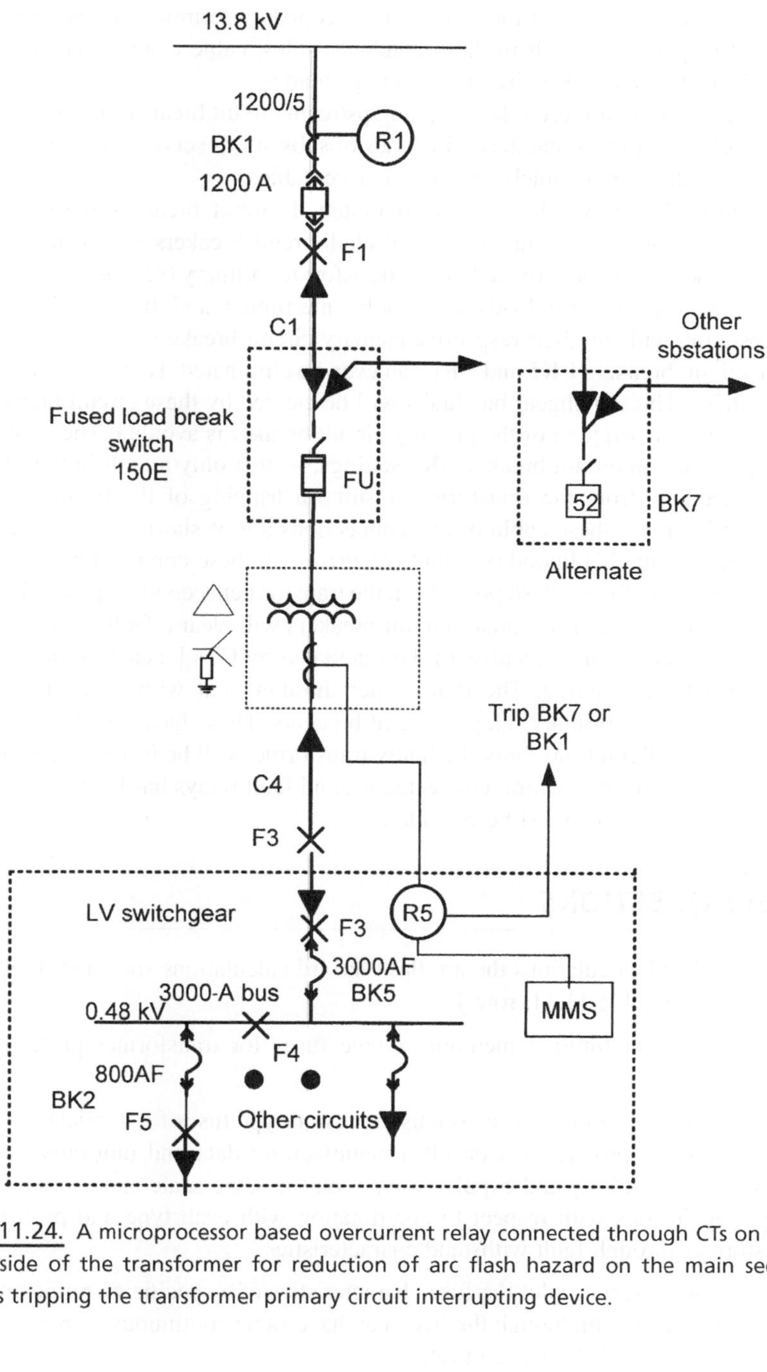


Figure 11.24. A microprocessor based overcurrent relay connected through CTs on the secondary side of the transformer for reduction of arc flash hazard on the main secondary breakers tripping the transformer primary circuit interrupting device.

to the transformer tank or in the transformer secondary terminal compartment, these relays will respond to a fault in the secondary cables, cable connections, and reduce HRC to 2 on BK5 and BK6 circuit breakers primaries.

It is necessary that these relays trip an upstream circuit breaker, in this case circuit breaker BK1. This means that the entire loads of substations served form circuit breaker BK1 will be interrupted, which may not be acceptable.

Instead of fused switches, (1) fixed mounted circuit breakers with interlocked disconnect switches or (2) draw-out metal clad circuit breakers or (3) fused 13.8 kV vacuum contactors can be provided on the transformer primary (see Figure 11.24). With this configuration, the entire loads need not be interrupted, and the secondary relays in each substation will trip their respective primary circuit breakers.

The circuit breakers BK5 and BK6 can even be eliminated. However, it is desirable to retain these. The switchgear bus faults will be cleared by these circuit breakers, and for such faults, the tripping of the primary circuit breaker is avoided. The relay R4 can be coordinated with circuit breaker BK5 settings, so that only a fault in the secondary cable connections from the transformer results in tripping of the upstream primary device. Furthermore, these cable or bus connections are of short length and mechanically protected—the likelihood of a fault occurring on these connections is remote.

To summarize, for arc flash protection, the transformer secondary protection should not be omitted. The main secondary circuit breakers will clear a fault on the secondary switchgears. In addition, overcurrent relays actuated by CTs located on the secondary of transformer are required. The transformer through fault withstand curves can be better protected with primary relayed circuit breakers. These have an advantage that in radial system configurations, only the faulty transformer will be isolated. The necessity of providing transformer neutral connected ground fault relays has been demonstrated. The listing requirements must be considered.

## REVIEW QUESTIONS

1. Verify by hand calculations the arc flash hazard calculations shown in Table 11.10, row 2, and in Table 11.11, row 1.

2. Compare current limiting medium voltage fuses for transformer protection with overcurrent relays.

3. Compare current limiting fuses versus expulsion-type fuses for application of transformer primary protection. Consult a manufacturer data and plot curve of 200-A fuses, current limiting and expulsion type, on the same scale and same plot. Draw some conclusions with respect to coordination with each type and protections of transformer through fault withstand characteristics.

4. Why it is necessary to plot the fuse characteristics of a specific manufacturer being applied in the system, though the fuse may have same continuous current rating and voltage ratings and short-circuit ratings.

5. A three-phase, 1500-kVA delta–wye transformer is solidly grounded on the wye side 480-V windings. Describe NEC protection requirements for ground fault. In case

this protection is not provided, will a primary fuse protection or overcurrent relay protection will be better choice?

**6.** Draw a phasor diagram and connections of a wye–delta transformer, with a three-phase, line-to-line phase, and single-phase to-line faults on the delta side. Draw magnitudes of currents in transformer windings and also the primary and secondary lines.

**7.** What is preferable of the two from the arc flash reduction considerations: (1) a main secondary circuit breaker, (2) a current transformer operated relay, current transformer located in the transformer terminal compartment?

**8.** A three-phase, 13.8 kV–0.48-kV, 1500-kVA transformer serves a lineup of 480 V switchgear. A main secondary circuit breaker is provided. Can this circuit breaker be maintained in the energized condition?

**9.** Describe the compromise that is inherent when maintenance mode switches are used to lower arc flash hazard levels by activating instantaneous protections.

## REFERENCES

1. NEC, *National Electric Code, NFPA 70*, 2011.
2. F.M. Global, *Property Loss Prevention Data Sheets*, 5-4, Jan. 1997.
3. C.C. Young and J. Dunki-Jacobs, "The concept of in-phase transfer applied to industrial systems serving essential service motors," AIEE Trans., vol. 79, pp. 508–516, Jan. 1961.
4. J.C. Das, *Transients in Electrical Systems*, Chapter 17, McGraw-Hill, New York, 2010.
5. D.L. Hornakand and D.W. Zipse, "Automated bus transfer control for critical process industries," IEEE Trans. Industry Appl., vol. 27, no. 5, pp. 862–871, Sept./Oct. 1991.
6. IEEE, C57.109. *IEEE Guide for Liquid Immersed Transformer through-Fault Current Duration*, 1993 (R2008).
7. IEEE Standard, C.57.91 (Also Cor.1-2002). *IEEE Guide for Loading Mineral-Oil Immersed Transformers*, 1995.
8. IEEE, C57.12.59. *Guide for Dry Type Transformer through Fault Current Duration*, 1989.
9. ANSI Standard, C37.46. *American National Standard for High-Voltage Expulsion and Current Limiting Type Power Class Fuses and Fuse Disconnect Switches*, 2000.
10. R.H. Hopkinson, "Ferroresonance during single phase switching of three-phase distribution transformer banks," IEEE Trans., vol. PAS-84, pp. 289–293, 1965.
11. D.R. Smith, S.R. Swanson, and J.D. Borst, "Overvoltages with remotely switched cable fed grounded wye-wye transformers," IEEE Trans. PAS, vol. PAS-94, pp. 1843–1853, 1975.
12. J.C. Das, "Overcurrent relays versus current limiting power fuses for transformer primary protection," in Conf. Record, 48th Annual Georgia Tech Protective Relaying Conference, 1994.
13. IEEE, C57.1200. *General Requirements for Liquid Immersed Distribution, Power and Regulating Transformers*, 2010.
14. IEEE, C37.91. *IEEE Guide for Protective Relay Applications to Power Transformers*, 2008.
15. ANSI/IEEE, ANSI/IEEE Std. 242, *IEEE Recommended Practice for Protection and Coordination of Industrial and Commercial Power Systems*, 1986.

# 12

# CURRENT TRANSFORMERS

For the relaying applications, the high system voltages and currents are reduced through the potential and current transformers (CTs) (called instrument transformers). These also protect and insulate the personnel and relays from high voltages and currents. The performance of instrument transformers is critical in protective relaying, as the reduced secondary currents and voltages should be an exact replica of the primary quantities both under steady-state and transient conditions; say under fault conditions when large magnitudes of currents flow. These primary currents and voltages should be applied to the relays reduced in exact proportion and without any waveform distortion. We will see that this ideal situation is not always practically possible, and saturation of the CTs has to be allowed for. Generally, the reduced voltages are 120 or 67 V, and the CT secondary current ratings are 5 or 1 A. We alluded to CT saturation and their accuracies in earlier chapters. This chapter provides conceptual base for proper applications of CTs for relaying quantities, considering steady-state and transient behavior. The voltage transformers are not discussed, as for relaying purposes, it is the accuracy and saturation of CTs that is of paramount operation. A reader may pursue Reference [1] for connections, ratings, and general nomenclature of PTs (potential transformers) for Group 1 through 5 connections and ratings.

*Arc Flash Hazard Analysis and Mitigation*, Second Edition. J.C. Das.

Published 2021 by John Wiley & Sons, Inc.

TABLE 12.1. Standard Accuracy Class for Metering Service and Corresponding Limits of Transformer Correction Factor (0.6 to 1.0 Lagging Power Factor of Metered Load)

| | Voltage Transformers (at 90% to 110% of Rated Voltage) | | Current Transformers | | | |
|---|---|---|---|---|---|---|
| | | | At 100% Rated Current | | At 10% Rated Current | |
| Metering Accuracy Class | Minimum | Maximum | Minimum | Maximum | Minimum | Maximum |
| 0.3 | 0.997 | 1.003 | 0.997 | 1.003 | 0.994 | 1.006 |
| 0.6 | 0.994 | 1.006 | 0.994 | 1.006 | 0.998 | 1.012 |
| 1.2 | 0.998 | 1.012 | 0.998 | 1.012 | 0.976 | 1.024 |

*Source*: Reference [1].

## 12.1 ACCURACY CLASSIFICATION OF CTs

### 12.1.1 Metering Accuracies

The instrument transformers have two distinct accuracy classifications: Metering and Relaying. Table 12.1 from Reference [1] shows metering class accuracies for the voltage and CTs. A metering CT must reproduce faithfully the primary quantities, and it is immaterial if it saturates on heavy fault currents. In fact, this saturation means that there will be little output from the CT, and the meter connected to the secondary of the CT will not be subjected to high magnitudes of secondary currents. On the contrary, a relaying class CT must reproduce the high primary fault currents accurately without saturation for proper relay operation. Referring to Table 12.1, for revenue metering, sometimes, even better accuracies than 0.3 metering class are demanded, where large power supplies to a consumer are to be metered. Generally, such revenue metering instrument transformers do not have any other secondary burden, except the meter to which these are connected.

A single set of CTs are often used both for relaying and metering in industrial relaying applications. *While relaying class accuracy is acceptable for metering, the metering class accuracy is not acceptable for relaying.*

### 12.1.2 Relaying Accuracies

Relaying CTs have C, X, or T classifications [1, 2]. C classification applies to transformers in which the leakage flux in the core does not have an appreciable effect on the ratio, and the ratio correction can be calculated. The excitation curve is plotted on a log-log paper between secondary exciting current and voltage. A typical excitation curve for class C transformers with nongapped cores is shown in Figure 12.1. The data represents secondary exciting rms currents by applying rms voltage to the CT secondary

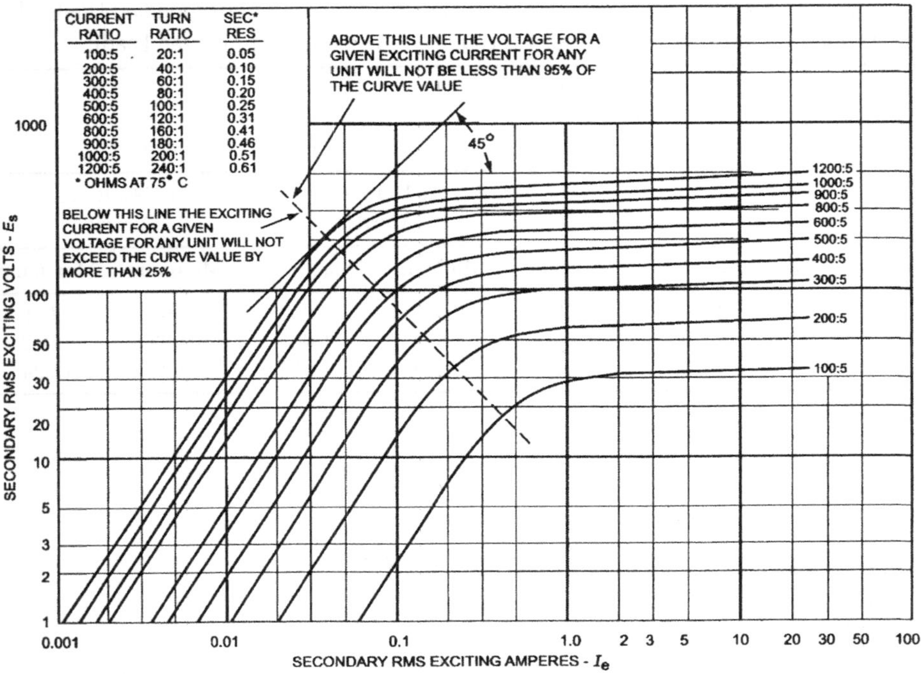


Figure 12.1. Excitation characteristics of C-type current transformers, showing knee-point voltage.

windings, with primary open circuited. This gives the approximate exciting current requirements for a secondary voltage. The *knee point* is defined as the point where tangent is 45° to the abscissa. For CTs with gapped cores, this angle is 30°. The maximum excitation values above knee are shown in Figure 12.1.

***The Knee-Point Voltage, $V_k$.*** This is defined as the sinusoidal voltage applied to the secondary terminals, all other windings open circuited, which, when increased by 10%, causes excitation current to increase by 50%.

## 12.1.3 Relaying Accuracy Classification X

The accuracy classification *X* is user defined for a specific condition, where the requirements are given as follows:

$V_k$, the minimum knee-point voltage;

$I_k$ = maximum exciting current at $V_k$; and

$R_{ct}$ = maximum allowed secondary winding resistance, measured with DC current, corrected to 75°C.

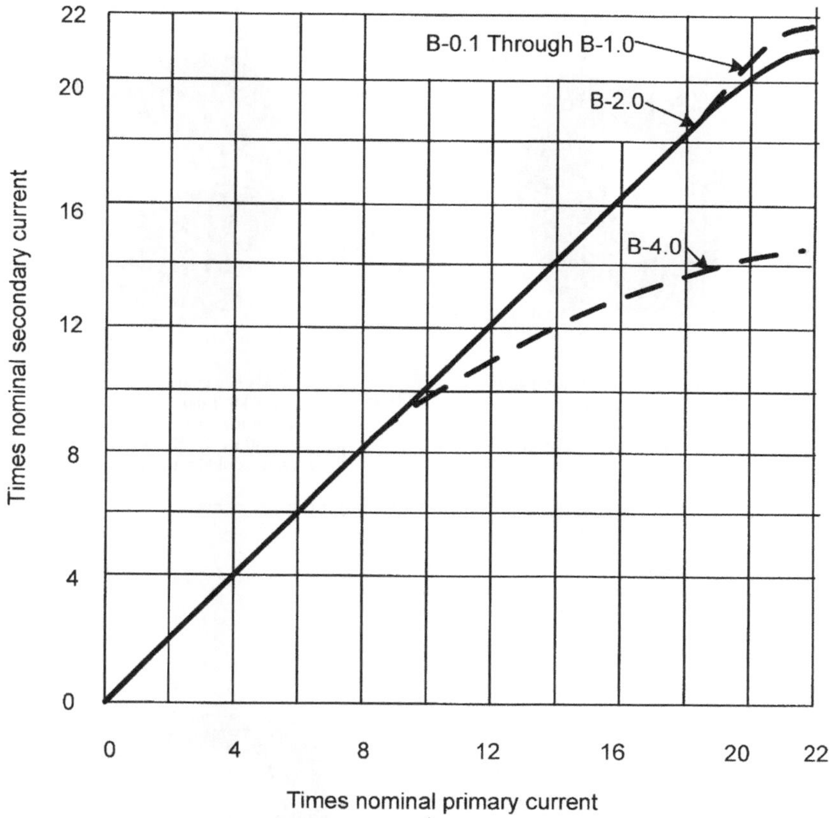


Figure 12.2. Excitation characteristics of T-type current transformers.

### 12.1.4 Accuracy Classification T

Class T transformers have appreciable leakage flux, and the ratio correction must be determined by test. Typical ratio curves for class T transformers are plotted over the range from 1 to 22 times normal primary current for all standard burdens that cause a ratio correction of 50% [1]. Wound-type transformers are sometimes used as auxiliary CTs for ratio matching, or these can be separate CTs in themselves, available for low voltage to high voltage applications. Figure 12.2 shows the excitation curves of a class T CT.

## 12.2 CONSTRUCTIONAL FEATURES OF CTs

Figure 12.3 shows constructional features of some CTs. In the window-type CT, the primary conductor consists of just one single conductor passing through the window (Figure 12.3a). The CTs located in indoor metal-clad switchgear, and those located on

(a) A window type CT, window diameter up to 8", ratios up to 5000: 5, accuracy up to C800 Single ratio, double ratio, or multiratio

(b) A fuly insulated CT up to 34.5 kV, ratios up to 4000 : 5, accuracy up to C800 (Say BIL =110 kV for 13.8 kV) Single ratio, double ratio, or multiratio

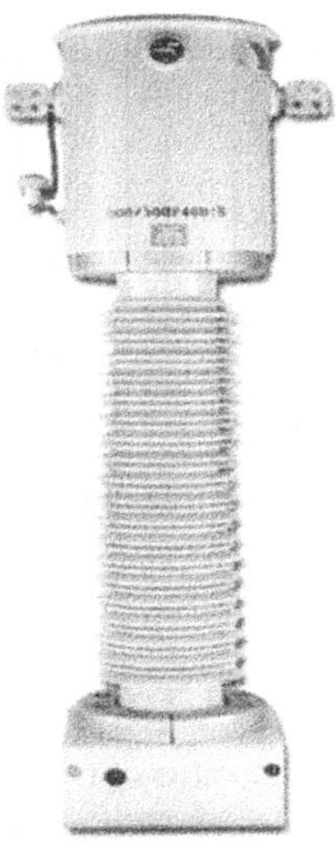

(c) Oil-filled head type outdoor CT, 25–161 kV, accuracy up to C800. Single ratio, double ratio, or multiratio

(d) An encapsulated generator CT, window size up to 35", ratio 40,000 : 5, accuracy up to C800 Single ratio, double ratio, or multiratio

Figure 12.3. Construction of various CT types: (a) window-type CT, (b) a fully insulated window type CT, (c) oil-filled outdoor CT, and (d) an encapsulated generator CT for mounting on isophase bus.

transformer and outdoor high voltage circuit breaker bushings are window type. These do not have high BILs. A fully insulated window-type CT is shown in Figure 12.3b. For high voltage outdoor substation applications, a window-type CT mounted in a oil-filled tank and installed on a separate steel framework with in and out high voltage connections can be applied (Figure 12.3c). Figure 12.3d shows a CT of large window diameter and ratio that can be mounted on generator isophase bus ducts. The window type CTs, generally, meet the above definition of knee-point voltage for class C.

## 12.3 SECONDARY TERMINAL VOLTAGE RATING

A class C or T transformer will deliver to the standard burden a secondary terminal voltage at 20 times the secondary current, without exceeding 10% ratio correction. Furthermore, the ratio correction should be limited: 3–10% from rated secondary current to 20 times rated secondary current at the specified rated burden [1]. For example, for a C200 CT, the ratio correction should not exceed 10% at 20 times the rated secondary current at a standard 2 Ω burden, at 0.5 power factor. (2.0 Ω multiplied by 100 A equals 200 V). If the relaying class accuracy is C200, and the CT secondary current is 1 A, then the burden to develop secondary terminal voltage will be 200 V/(1 A × 20) = 10 Ω.

When multiratio CTs are applied, care has to be exercised that the accuracy quoted by the manufacturer applies to full secondary winding.

Table 12.2 shows the limits of ratio errors for relaying class CTs, and Table 12.3 shows the standard burdens and secondary terminal voltages.

TABLE 12.2. Ratio Errors for Relaying Accuracy CTs

| Relaying Accuracy Class | At Rated Current | At 20 times Rated Current |
|---|---|---|
| C and T Classifications | 3% | 10% |
| X classification | 1% | User defined |

*Source*: Reference [1].

TABLE 12.3. Class C or T Relaying Accuracy: Secondary Voltage and Relaying Burden

| Secondary Terminal Voltage | Burden Designation | Resistance (Ω) | Inductance (mH) | Impedance (Ω) | Total Power (VA at 5 A) | Power Factor |
|---|---|---|---|---|---|---|
| 100 | B-1.0 | 0.5 | 2.30 | 1.0 | 25 | 0.5 |
| 200 | B-2.0 | 1.0 | 4.60 | 2.0 | 50 | 0.5 |
| 400 | B-4.0 | 2.0 | 9.20 | 4.0 | 100 | 0.5 |
| 800 | B-8.0 | 4.0 | 18.40 | 8.0 | 200 | 0.5 |

*Source*: Adapted from Reference [1].

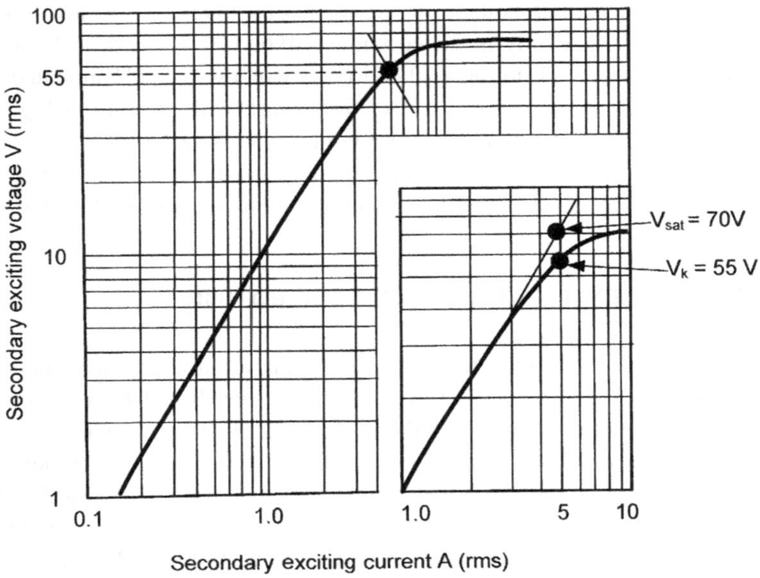


Figure 12.4. Excitation characteristics of a 100/5 core balance CT. $V_k$ is the knee-point voltage, $V_{sat}$ is the saturation voltage.

### 12.3.1 Saturation Voltage

Saturation voltage is defined as the voltage across the secondary winding at which peak induction just exceeds the saturation flux density. It is not the same as the knee-point voltage. This difference can be illustrated graphically. Consider the point on the excitation curve where the straight line relation just starts to deviate. It is found graphically by locating the intersection of the straight portions of the excitation curve on log-log axes. This is clearly shown in Figure 12.4. $V_k$ is the knee-point voltage, and $V_{sat}$ is the saturation voltage.

### 12.3.2 Saturation Factor

Saturation factor is the ratio of the saturation voltage to the excitation voltage. It is an index of how close to saturation a CT is applied in a given application.

These definitions and concepts are important for the analysis of the saturation of CT discussed further.

## 12.4 CT RATIO AND PHASE ANGLE ERRORS

The phasor diagram of a CT can be constructed much alike a power transformer; Figure 12.5 is applicable for class C CTs. The primary current in a CT is not determined by the secondary load, and there is no appreciable voltage across the primary terminals of the CT. To magnetize the core, a small current $I_0$ flows, which can be resolved into two components; $I_m$, the magnetizing current, which is in phase with the magnetic flux, and $I_e$, the eddy current, which is in quadrature. $I_e$ is required to meet the hysteresis and eddy current loss, given by the following equations.

$$\begin{aligned} P_h &= K_h f B^s \\ P_e &= K_e f^2 B^2, \end{aligned} \tag{12.1}$$

where $P_h$ and $P_e$ are the hysteresis and eddy current loss, respectively; $f$ is the frequency, $B$ is the flux density, $K_h$ and $K_e$ are constants, and $s$ is Steinmetz exponent, which varies from 1.5 to 2.5 depending upon the core material; generally, it is equal to 1.6

The magnetizing impedance of the CT, obtained by dividing the excitation voltage and excitation current (Figure 12.1), is not constant. It is highly nonlinear, changing from a high value at low excitations to low value at high excitations.

This gives rise to ratio error. The ratio correction factor, RCF is defined as:

$$RCF = \frac{I_p}{nI_s}. \tag{12.2}$$

Where $I_s$ is the secondary current, and $n$ = secondary turns. The flux $\phi$ produces a voltage $E_2$ in the secondary. The voltage $V_s$ at the secondary terminals of the CT is $V_s$, obtained by subtracting the vectors of voltage drops $I_sR_s$, and $I_sX_s$, where $R_s$ and $X_s$ are the resistance and reactance of the secondary burden plus CT resistance and reactance. The secondary winding leakage reactance of the CT is small. It is shown exaggerated in Figure 12.5 for clarity. This gives rise to a phase angle error $\beta$, given by,

$$\beta = \tan^{-1}\frac{I_m}{nI_s}. \tag{12.3}$$

The relaying standard burdens are at a power factor of 0.5 (Table 12.2). Figure 12.5 shows that if the actual burden is resistive, the CT error will be much less.

IEC [3] defines a composite error, given by the equation:

$$\frac{100}{I_p}\sqrt{\frac{1}{\tau}\int_0^{\tau}(nI_s - I_p)^2\,d\tau}. \tag{12.4}$$

This may be considered to take into account ratio and phase angle errors as well as waveform distortion, Figure 12.6. The IEC designations of the CTs are based upon the composite error. For example, a 5P30 CT means a protection class CT with composite error no more than 5% and VA burden of 30 VA.

ANSI definition of TCF (total correction factor) is:

$$\text{TCF} = \text{RCF} - \beta/2600, \tag{12.5}$$

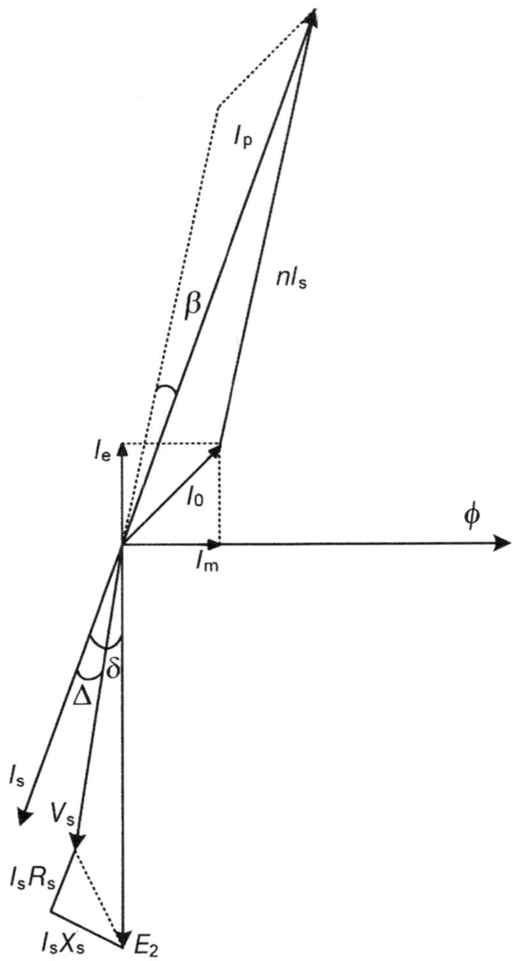


Figure 12.5. Phase diagram of a class C current transformer.

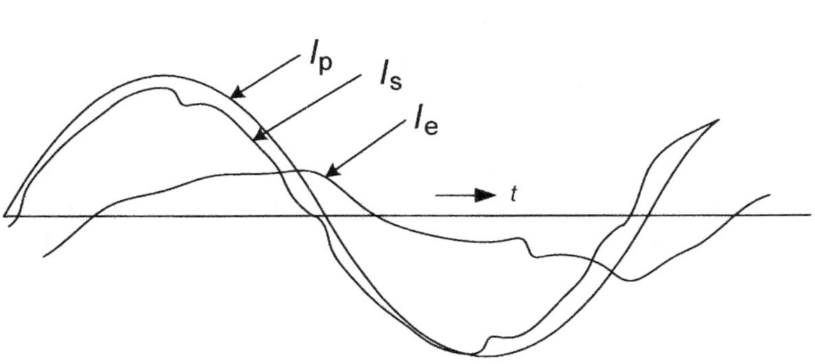


Figure 12.6. Illustrates the composite error according to IEC standards. *Source*: Reference [3].

where RCF is the ratio correction factor and β is the phase angle in minutes. This is of importance for metering class CT's. Similarly, for voltage transformers:

$$\mathrm{TCF} = \mathrm{RCF} - \gamma / 2600, \tag{12.6}$$

where γ is the phase angle in minutes for the voltage transformers.

Reference [1] gives the following expression for RCF:

$$\mathrm{RCF} = \frac{I_{\mathrm{st}}}{I_{\mathrm{s}}}, \tag{12.7}$$

where $I_{\mathrm{st}} = I_{\mathrm{s}} + I_{\mathrm{e}}$.

## 12.5 INTERRELATION OF CT RATIO AND C CLASS ACCURACY

The following considerations can be postulated.

1. When the CT ratio is low, the ampere-turns (denoted as $\widehat{A}$) are low. In the window-type CT, the primary consists of a single turn, the ampere-turns = primary amperes. (The use of the term $\widehat{A}$ is no longer in use, instead MMF is used. However, for illustrative purposes, the use of ampere-turns is retained.)
2. The low $\widehat{A}$ makes the design of high accuracy CT difficult, as the flux produced is low.
3. The larger the dimension of the window, the lesser is the flux that will be produced by the same primary current.
4. A larger core cross-section is required as flux is equal to flux density multiplied by the area of the cross-section of the core.
5. A larger core length means that the magnetization and core loss components of the current will increase, which in turn increases ratio and phase angle errors.
6. The $\widehat{A}$ required to establish a certain amount of flux in an air gap are higher by a factor of hundreds. Thus, the flux must be increased to account for even a small air gap in the CT of a split-core construction.
7. The introduction of an air gap in the core of the CT leads to fringing and leakage of flux, which needs to be controlled. The winding reactance is a function of the leakage flux.
8. The higher the accuracy, the greater is the voltage across the secondary windings that should be produced at a minimum error, that is, the saturation of the magnetic material should not be reached.

Items 6 and 7 apply to split core CTs. The split core construction allows ease of installation without dismantling, say around a bus bar. Thus, there are limitations in obtaining high accuracy with low ratio CTs.

Analytically, the MMF (magnetomotive force) is written as:

$$\text{MMF} = \oint H dl, \tag{12.8}$$

where $H$ is the magnetic intensity and $\oint$ means that integration is taken all the way around the closed circuit. For a solenoid, MMF = $NI$, where $N$ is the number of turns and $I$ is the current. The flux $\phi$ is given by:

$$\phi = \frac{\text{MMF}}{S} = \frac{\text{MMF}}{l / \mu\mu_0 a}, \tag{12.9}$$

where $S$ is the reluctance of the magnetic circuit, $l$ = length of the circuit in meters, $a$ is the area of cross section in m$^2$, $\mu$ = relative permeability of the CT core magnetic material, and $\mu_0$ = permeability of air:

$$\mu_0 = 4\pi 10^{-7}. \tag{12.10}$$

This gives flux in webers. Also:

$$\phi = Ba, \tag{12.11}$$

where $B$ is the flux density in Wb/m$^2$.

The secondary voltage induced in the CT core is given by,

$$E_2 = -N\frac{d\phi}{dt}. \tag{12.12}$$

Rewriting and adjusting for the units, the steady state secondary rms voltage is,

$$E_2 = 4.44 N\phi f 10^{-8} = 4.44 NBaf \times 10^{-8}\ V. \tag{12.13}$$

The magnetic field produced in air due to an infinitely long straight conductor carrying a current $I_p$ consists of concentric circles, which lie in a plane perpendicular to the axis of the current carrying conductor, and have its center on its axis. The magnetic filed intensity, $H$, at a distance $r$ from the conductor, is given by

$$H \propto \frac{I_p}{r}. \tag{12.14}$$

Also

$$\phi = Ba = \frac{Ha}{\mu\mu_0}. \tag{12.15}$$

When a CT of the *lower current ratio is* to be designed, the primary ampere turns are small. Thus, the magnetic flux produced in the core by the primary single turn conductor is small. A linearity of the excitation curve (Figure 12.1) should be obtained for certain accuracy. The secondary turns must be increased, or the cross-sectional area of the core must be increased. The turns are fixed by the transformation ratio, that is, for a 100:5 CT, $n = 20$. This limits the secondary voltage and the C class accuracy that can be obtained. From Equation (12.11), a core of higher cross-section area and lower reluctance is required.

Thus, the maximum secondary VA burden and accuracy is controlled mainly by the primary ampere turns.

For split-core CTs, introducing an air gap in the core makes the design all the more difficult. The permeability of iron is approximately 600 times that of the air and more ampere turns (MMF) are required for the flux to cross even a small air gap. The fringing and leakage effect must be considered, so that leakage reactance is small. Thus, the design is an optimization of number of conflicting parameters.

Table 12.4 shows the normal C class accuracies of window-type transformers that are normally provided and that can be provided for medium voltage switchgear. In metal-clad switchgear, the CTs are located on the circuit breaker spouts (Figure 12.7). Two sets of CTs of normal accuracy can be provided on the source and load side of the circuit breaker. With high accuracy CTs, the width of the CT increases, and it may not be possible to accommodate two CTs on source and load side.

## 12.6 POLARITY OF INSTRUMENT TRANSFORMERS

We have shown the polarity of CTs in many figures in the previous chapters by a solid square (■); it is also denoted by an "X" sometimes. The polarity indications shown in Figure 12.8 is applicable to both current and voltage transformers. Figure 12.8a

TABLE 12.4. C Class Accuracy Window Type CTs Normally Provided on Medium Voltage Switchgear and Also Higher Accuracy CTs That Can Be Provided

| Ratio | Accuracy Class Normally Provided | Accuracy Class That Can Be Provided |
|---|---|---|
| 50:5 | C10 | C20 |
| 75:5 | C10 | C20 |
| 100:5 | C10 | C20 |
| 150:5 | C20 | C50 |
| 200:5 | C20 | C50 |
| 300:5 | C50 | C100 |
| 400:5 | C50 | C100 |
| 600:5 | C100 | C200 |
| 800:5 | C100 | C200 |
| 1200:5 | C200 | C400 |
| 1500:5 | C200 | C400 |
| 2000:5 | C200 | C400 |
| 3000:5 | C200 | C400 |
| 4000:5 | C200 | C400 |

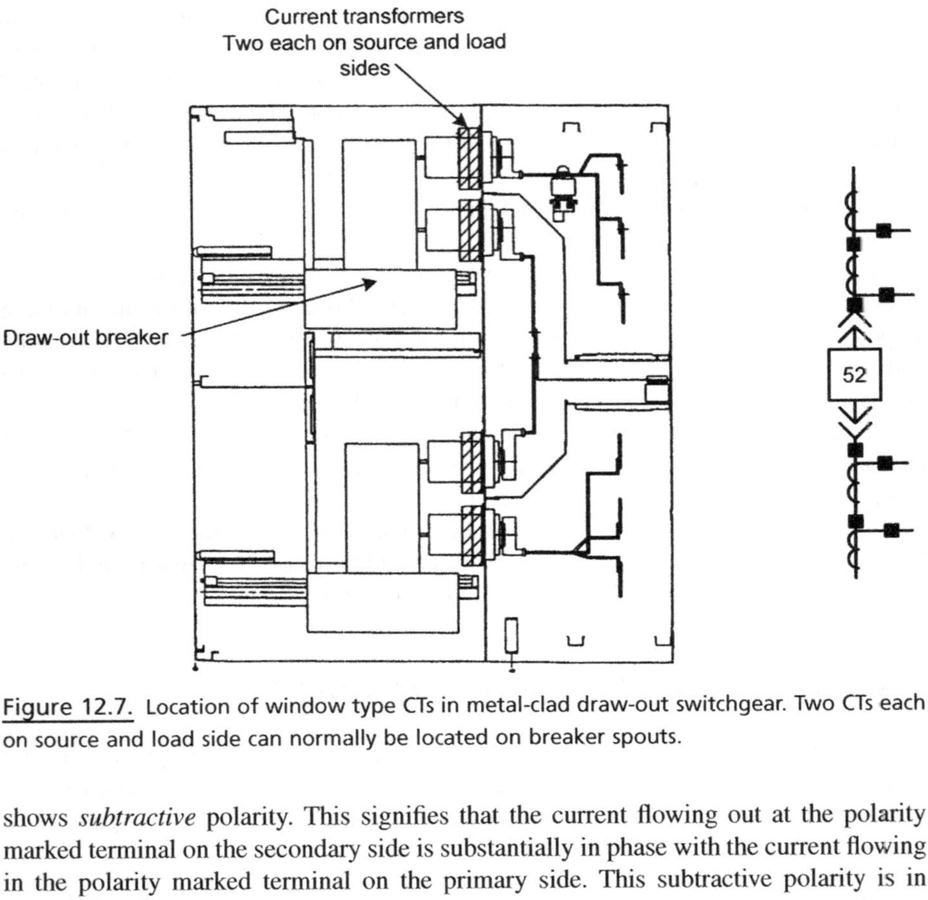


Figure 12.7. Location of window type CTs in metal-clad draw-out switchgear. Two CTs each on source and load side can normally be located on breaker spouts.

shows *subtractive* polarity. This signifies that the current flowing out at the polarity marked terminal on the secondary side is substantially in phase with the current flowing in the polarity marked terminal on the primary side. This subtractive polarity is in common use.

Figure 12.8b shows the additive polarity; the voltage drop from the polarity marked terminal to the nonpolarity marked terminal on the primary side is substantially in phase with the voltage drop from the polarity marked to nonpolarity marked terminals on the secondary side.

*According to ANSI standards, all power and distribution transformers and also dry type power transformers have subtractive polarity.*

## 12.7 APPLICATION CONSIDERATIONS

The following generalities apply to selection of Class C accuracy CTs.

### 12.7.1 Select CT Ratio

This is dependent upon the desired sensitivity of relaying and also ratio-balancing that is required in some relaying applications, like ground fault differential relaying. For

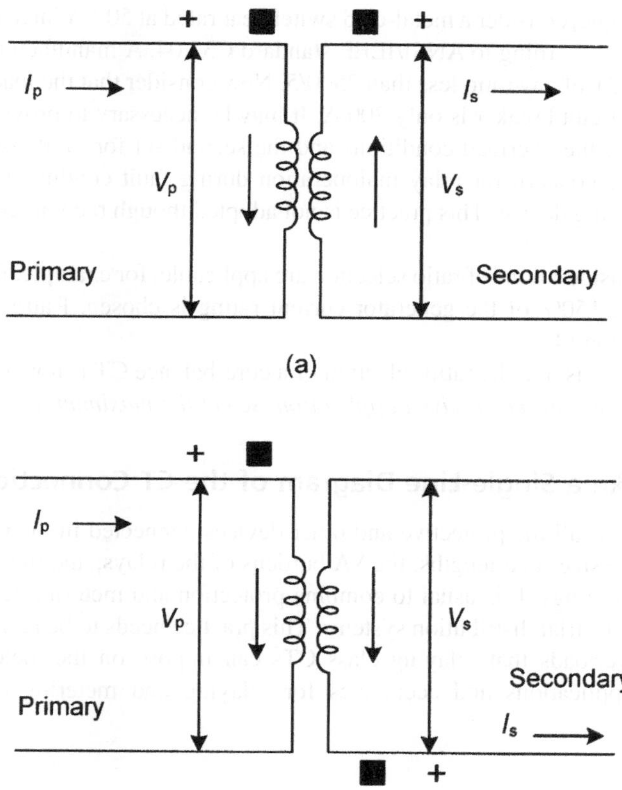


Figure 12.8. (a) Subtractive polarity and (b) additive polarity, shown by darkened square blocks.

example, many MMPRs (microprocessor-based multifunction relays) will correct the CT mismatch in differential scheme up to a certain level only.

Another important consideration for selection of ratio is the accuracy of the CT. By definition, Class C accuracy holds for only 20 times its primary current, at rated burden (thereafter nonlinearity can set in and the errors are unpredictable). Now consider a switchgear phase CT of 600/5, and the primary symmetrical three-phase short-circuit current of 40 kA. The maximum short-circuit current to which 600/5 CT should be applied is 12 kA, and the minimum ratio that should be selected for 40 kA primary current application should be 2000/5. A calculation can be made when the actual burden is other than the standard burden. It is not unusual to see this misapplication in almost every switchgear installation. A switchgear manufacturer will provide a CT ratio as demanded by the customer, and this qualification of 20 times maximum current is ignored.

As an example, consider a metal-clad switchgear rated at 50 kA interrupting, K = 1 (see Chapter 5), according to ANSI/IEEE standard C37.04. A manufacturer should not install a phase CT of any ratio less than 2500/5. Now consider that the load to be served from a feeder circuit breaker is only 200 A. It may be necessary to provide two sets of CTs: one set for the overload conditions and the second set for fault conditions. This may resolve the concern for relay maloperation during fault conditions, but requires additional metering device. This practice is not adopted though recommended in Reference [2].

Further considerations of ratio selection are applicable, for example, for generators, a CT a ratio of 150% of the generator current rating is chosen. Ratio selection also impacts saturation [4].

An exception is that the ratio selection of a core-balance CT is not dependent upon the phase load current. *For a given application, select the maximum permissible ratio.*

### 12.7.2 Make a Single-Line Diagram of the CT Connections

This should show all the protective and other devices connected in the CT circuit, the secondary cable sizes and lengths, the VA burdens of the relays, and the instruments at the calculated settings. It is usual to combine protection and metering functions in the same CTs in industrial distribution systems. This practice needs to be reviewed in terms of transient overloads that relaying class CTs can impose on the metering circuits. Further, the applications and accuracies for relaying and metering class CTs are different.

### 12.7.3 CT Burden

Accurately calculate the CT secondary burden as reflected on the CT terminals, including the CT secondary windings. Though the power factor of the burden is often neglected, for accuracy, it is desirable to include it. The burden is usually expressed in VA at a certain power factor, or it can be expressed as impedance in R + jX format.

### 12.7.4 Short-Circuit Currents and Asymmetry

Accurately calculate the short-circuit currents on the primary side and their asymmetry and the fault point *X/R* ratio. For time-delayed devices, it is necessary to calculate 30-cycle currents.

### 12.7.5 Calculate Steady-State Performance

The excitation curve of the CT, similar to Figure 12.1, is required from the manufacturer. Consider a 600/5 CT in Figure 12.1. Its knee-point voltage can be read approximately = 90 V. Corresponding to 10-A excitation current, the voltage is 200 V. The knee-point voltage is approximately 46% of the excitation voltage corresponding to 10 A excitation, and the excitation voltage at 10 A current is also the C-rating. Thus, for the steady-state performance, *as a rule of thumb,* the calculated secondary impedance, including CT winding resistance, when multiplied by the CT secondary current under maximum fault condition, should not exceed the CT C rating. Vectorial calculation is required.

### 12.7.6 Calculate Steady-State Errors

This is illustrated in Example 12.1 below and also in examples in References [5, 6]. Once the secondary voltage is known, the excitation current can be read from the CT excitation curve and the error calculated.

**Example 12.1**

A 600/5 ratio CT has the following devices connected in the secondary circuit:

- A very inverse electromechanical relay, set at a tap of 4-A time delay overcurrent, no instantaneous. According to the manufacturer, the relay burden at the 4-A tap is 2.38 VA, power factor = 68%.
- CT secondary loop impedance corrected to 75°C = 0.0820 + j0.0062 Ω.
- An ammeter VA burden = 0.05 VA at unity power factor.
- A wattmeter, VA burden = 077 W at 4 A and power factor = 0.54.
- CT secondary winding resistance, (reactance ignored) = 0.31 Ω (see Figure 12.1).

The total secondary impedance is: 0.5663 + j0.1262 Ω = |0.5802|Ω. See Table 12.5 for the calculations.

The primary short-circuit current is 18 kA rms. Thus, the maximum secondary current is 150 A.

Therefore, the secondary voltage developed across CT windings is 84.9 V, ignoring reactance, and 87.02 V if the reactance is considered. Table 12.5 also calculates the

TABLE 12.5. Calculation of CT Secondary Resistance and VA Burden (Example 12.1)

| Device | Specified Burden Data | Burden, VA at 150 A | Impedance, $R + jX$ (Ω) |
|---|---|---|---|
| Overcurrent relay | 2.38 VA at 4.0-A setting, 68% PF | 3347 | $0.101 + j0.1090$ |
| CT secondary leads | $0.0820 + j0.0062$ Ω | 1850 | $0.0820 + j0.0062$ |
| CT secondary resistance | 0.31 Ω | 6975 | $0.31 + j0$ |
| Ammeter | 1.05 VA,5 A at unity PF | 945 | $0.042 + j0$ |
| Wattmeter | 0.82 VA at 5 A and 0.94 PF | 740 | $0.031 + j0.011$ |

burden at the maximum current of 150 A. The secondary impedance based on the calculated burden is:

$$\frac{13857}{(150)^2} = 0.616\Omega,$$

which gives an excitation voltage of 92.38 V.

The excitation current read from Figure 12.1 for 600/5 CT at this excitation voltage is approximately 0.08 A.

Therefore, the percentage ratio error is:

$$\left(\frac{0.08}{150}\right) \times 100 = 0.053\%.$$

This is acceptable. The excitation current will increase steeply, if the secondary voltage increases.

**Example 12.2**

In Example 12.1, the overcurrent relay is provided with an instantaneous function, which is set at 50 A. Then, in the above calculation of CT secondary connected resistance, we add the additional burden of the instantaneous element, say 50 VA at 50-A pickup and at 80°. The impedance of 0.003 + j0.197 is added to the calculations of impedance in Example 12.1, and the total secondary impedance becomes = 0.5693 + j0.3232 Ω = 0.6546 Ω.

However, to calculate the secondary voltage, we do not consider the maximum fault current of 150 A. We consider the maximum setting of the instantaneous overcurrent element = 50 A. Then, the secondary excitation volts = 32.73 V. The ratio error will be even smaller as compared with Example 12.1.

**Example 12.3**

This example addresses a major problem of CT saturation, when an auxiliary CT is used to step up the current for ratio matching. The configuration is shown in Figure 12.9. Figure 12.9a illustrates a residual connection, and an auxiliary CT of ratio 1/30 in interposed to increase the sensitivity of ground relay pickup, shown as GR in this figure. The effective CT ratio as seen by the relay GR becomes 100/5. Compare this with Figure 12.9b, which shows a zero sequence CT of ratio 100/5, connected to GR, as an alternative connection. Thus, the relay GR should see the same effective secondary current in either of the two connections, but this is not true.

First, consider the connection in Figure 12.9b. The ground fault current is 400 A, and the excitation characteristics of 100/5 core balance CT is in Figure 12.4. Then, the CT secondary current $I_s = 20$ A. Assuming a CT secondary winding resistance $R_s = 0.5\ \Omega$ and secondary burden $R_b = 0.5\ \Omega$, the total burden is 1.0 Ω. The burden will be at a certain power factor, which we ignore in this example. The CT secondary voltage $V_s$ is given by:

$$V_s = I_s(R_s + R_b + jX_b). \tag{12.16}$$

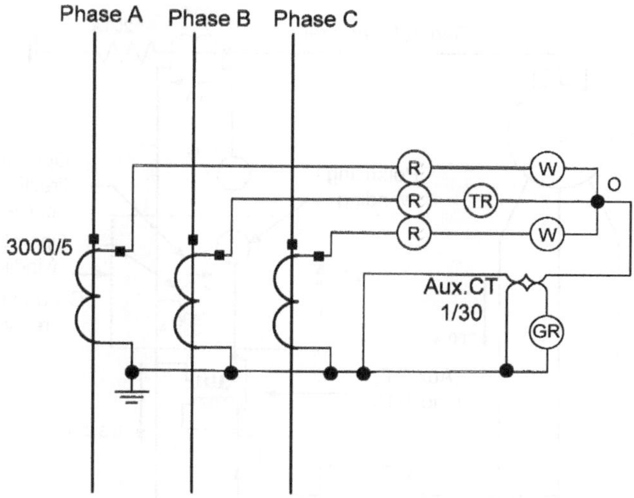


(a)

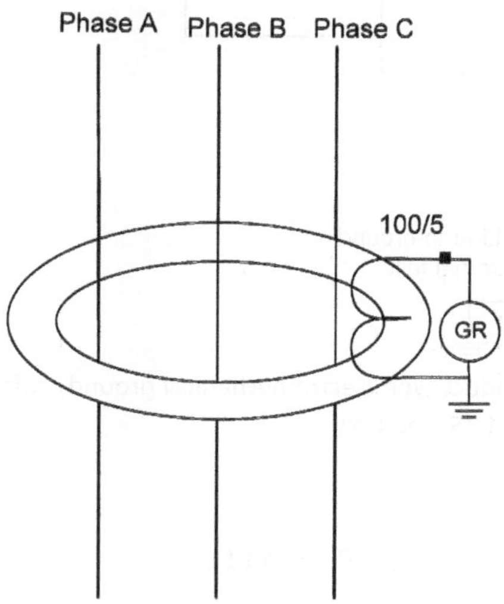


(b)

Figure 12.9. (a) Residual connection of a GR fault relay through auxiliary wound-type step-up CT of ratio 1/30; (b) equivalent connection through a 100/5 core balance CT.

Where $R_s$ is the CT secondary resistance, $R_b$ is the CT secondary burden. Ignoring the power factor, $V_s = 20$ V. Thus, even a C20 rating is adequate. From Figure 12.4, the excitation current = 1.0 A, which gives a ratio error of 5%.

Again, considering 400 A ground fault current and residual connection of the CTs in Figure 12.9a, the secondary current in auxiliary CT windings is 20 A. This being a wound-type CT, its secondary resistance will be higher. Considering the same burden and same secondary resistance for comparison, a total of 1 Ω, when reflected on the primary side becomes,

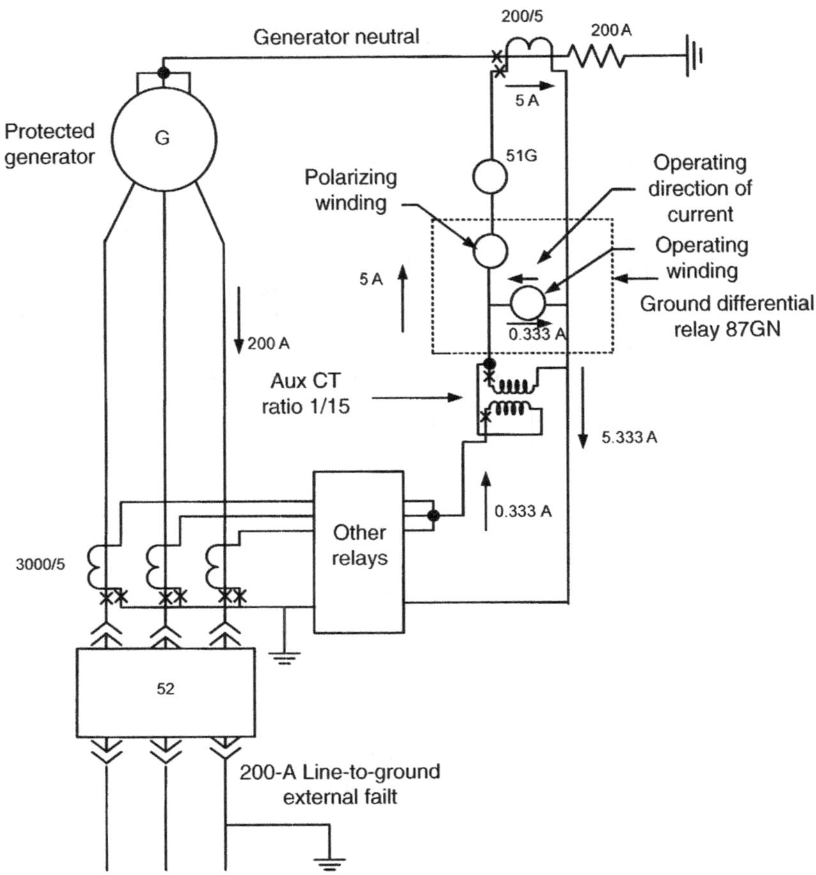


Figure 12.10. Connections of a product-type electromechanical ground fault relay, showing auxiliary ratio matching CT of ratio 1/15. See text.

$$R_p = n^2 \times (R_s + R_b) = 900\Omega. \tag{12.17}$$

The phase CTs should not saturate for a voltage of 900 × 0.67 = 600 V, ignoring all other burdens. This is rather a large secondary voltage, and the phase CT's must be designed for C800 or better accuracy. Reference [7] recommends that the auxiliary CTs should be avoided in a step-up configuration, though it has been an industry practice. An example of step up connection for differential ground fault protection is shown in Figure 12.10. Note the connections of auxiliary ratio balancing CT and the flow of the currents in the differential product type relay for a 200-A external fault. The MMPR relays for this application internally balance the CT mismatch and provide more sensitive and stable differential protection.

It is not unusual to see metering class or T class CTs applied as step up CTs, and the resultant nuisance trips.

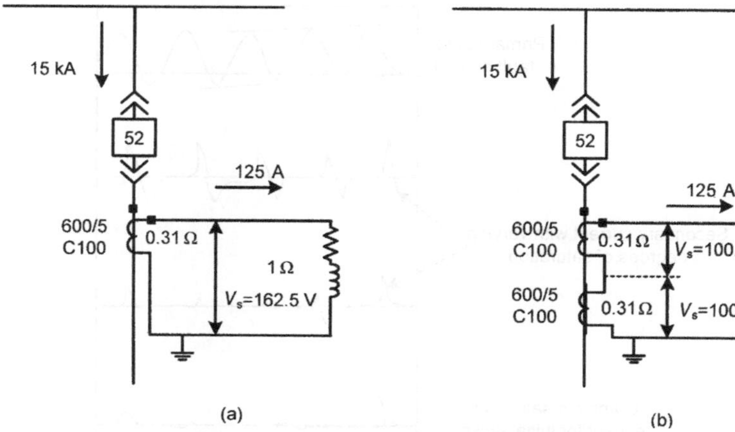


Figure 12.11. (a) Connection of a single 600/5 ratio CT, secondary voltage 162.5 V; (b) connection of two identical 600/5 CTs in series, the secondary voltage across each CT reduced to 100.6 V.

## 12.8 SERIES AND PARALLEL CONNECTIONS OF CTs

The CTs can be connected in series and parallel. Figure 12.11a shows a single CT of the ratio 600/5, secondary resistance 0.31 Ω, and the CT leads and devices connected to CT secondary have a burden of 1.0 Ω. The CT secondary voltage is 162.5 V, which gives high excitation currents. If two similar CTs are connected in series with the same secondary burden, Figure 12.11b, the secondary voltage across each CT winding reduces to 100.6 V.

The CT burdens can also be reduced by using:

- Multifunction microprocessor relays (MMPR). As an example, the relay burden is only 0.1–0.2 VA.
- A larger-size CT secondary leads can be used to reduce their burden.
- The electronic meters have a small requirement of CT burden, typically 0.1–0.2 VA.

A CT of higher ratio, considering the sensitivity of the pickup desired and the available relay setting range, may be possible to be selected. The CT of higher ratio will have higher winding resistance, but the CT secondary current will be reduced, which has a more pronounced effect in reducing the secondary voltage. Also, a CT of higher ratio can have a higher C class accuracy.

The other CT specifications are continuous thermal rating factor, short-time rating, and BIL, which are not discussed here.

## 12.9 TRANSIENT PERFORMANCE OF THE CTs

The above analysis of steady state analysis of calculation is not adequate. The transient performance and saturation on short-circuit asymmetry should be considered. This may

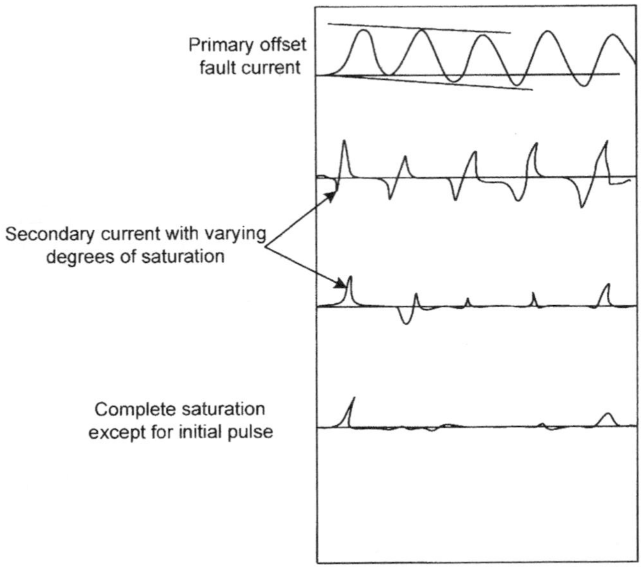


Figure 12.12. Oscillogram showing progressive saturation of a CT. *Source*: Reference [8]. Reproduced with permission from John Wiley & Sons.

even result in nonoperation of the instantaneous devices [4, 8]. This is of special importance for differential and instantaneous relaying.

The CT saturation is addressed in References [4–14]. The integrity of protection can be seriously jeopardized and nuisance trips can occur, if the CTs are not properly selected for the application, and saturation characteristics are not accounted for. Recommendations of the manufacturers for a certain application must be followed (Chapter 8).

Figure 12.12 [8] shows the saturation of a CT. The top curve shows the asymmetrical fault current and the bottom curves show progressive saturation.

A completely saturated CT does not produce an output except during the first pulse, as there is a finite time to saturate and desaturate. The transient performance must consider DC component, as it has more pronounced effect in producing severe saturation of the CT as compared with AC component.

### 12.9.1 CT Saturation Calculations

In the calculation in above examples, in order to avoid saturation, the CT secondary saturation voltage $V_x$, was calculated based upon the following equation:

$$V_x \geq I_s Z_s, \tag{12.18}$$

where $Z_s$ is the total secondary burden including CT resistance. When the offset waveform concept was introduced, the following equation was introduced:

$$V_x \geq K I_s Z_s. \tag{12.19}$$

The following values of $K$ have been used:

$$K = 1.6$$

$$K = 2$$

$$K = 2\sqrt{2}.$$

In the mid-1980s, Zocholl and Kotheimer published their papers, References [9–11]. Later, the IEEE Power Engineering Society Relay Committee addressed this topic and in 1996 IEEE Standard 110–1996 and formalized the work of these publications [7]. This standard recommends the (1 + *X/R*) method. To avoid saturation with DC component of the fault current, the following equation holds:

$$V_x > I_s Z_s \left( 1 + \frac{X}{R} \times \frac{R_s + R_b}{\sqrt{(R_s + R_b)^2 + (X_b)^2}} \right), \tag{12.20}$$

where *X/R* is calculated at the fault point. Equation (12.20) takes into account the inductive component of the CT burden. If this is ignored, the simplified equation is:

$$V_x > I_s Z_s \left( 1 + \frac{X}{R} \right). \tag{12.21}$$

As the CT saturation increases, so does the secondary harmonics, before the CT goes in to completely saturated mode. Harmonics of the order of 50% 3rd, 30% 5th, 18% 7th, 15% 9th, and higher order harmonics may be produced. These can cause improper operation of the protective devices. Thus, the two interrelated issues that need to be addressed are:

- Evaluation of the saturation of the CT
- Effect of distorted waves and harmonics on the operation of the protective devices.

## 12.9.2 Effect of Remanence

When a magnetic material is subjected to an alternating exciting current, the hysteresis loop is traced (see Figure 12.13). The relation is different with increasing and decreasing values of magnetic intensity. This is due to the irreversible process that results in energy dissipation, produced as heat. The first time the magnetic core is excited, neutral or virgin curve OA is produced, but it can not be reproduced in the reverse direction. There is some magnetism left as the MMF drops to zero, given by *oe*, the residual magnetism. To bring it to zero, a reverse MMF = *of* or *og* must be applied, called the coercive force. The area under the curve, and *oe, of,* and *og* depend upon the magnetic material.

This is very akin to power transformers, where the residual magnetism can give higher inrush currents and saturation on switching.

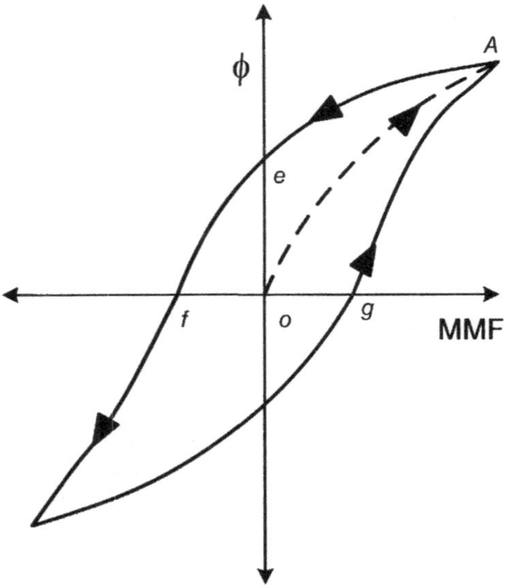


Figure 12.13. Hysteresis loop on magnetization of a magnetic material.

In a CT, to avoid saturation due to remanence:

$$V_x > \frac{I_s Z_s \left(1 + \frac{X}{R} \times \frac{R_s + R_b}{\sqrt{(R_s + R_b)^2 + (X_b)^2}}\right)}{1 - \text{per unit remenance}}. \tag{12.22}$$

## 12.10 PRACTICALITY OF APPLICATION

Consider that in a 13.8-kV system, the available three-phase fault current is 35 kA and the *X/R* ratio is 30. In fact, the fault point *X/R* at the primary voltage of distribution in an industrial power system can be as high as 80, when the distribution system contains large generators. The *X/R* ratio of a generator of 80 MVA can be close to 100. Also, short-circuit current limiting reactors have high *X/R* ratios to limit the fundamental frequency copper loss.

For 35 kA of fault current, $X/R = 30$, to meet the criterion of 20 times the current to limit ratio error to 10%, let us consider a CT of ratio 2000:5, secondary resistance = 1.15 Ω, and a secondary burden of 0.550 Ω, total secondary burden = 1.7 Ω. Then the secondary voltage calculated from various equations cited above is:

- $V_s = 102$ V, Equation (12.18)
- $V_s = 163.2$ V, $K = 1.6$, Equation (12.19)

TABLE 12.6. Calculated CT Secondary Voltage (Volts) Using Factor (1 + *X/R*), and Ignoring Remanence and No CT Secondary Burden Except the CT Secondary Winding Resistance

| System Short-Circuit Current in kA rms sym. | Minimum Required CT Ratio | CT Winding Resistance, Ω | Secondary Voltage in Volts | | | |
|---|---|---|---|---|---|---|
| | | | *X/R* = 15 | *X/R* = 20 | *X/R* = 30 | *X/R* = 50 |
| 20 | 1000/5 | 0.51 | 816 | 1071 | 1581 | 2601 |
| 30 | 1500/5 | 0.84 | 1344 | 1764 | 2604 | 4284 |
| 40 | 2000/5 | 1.15 | 1840 | 2415 | 3565 | 5865 |
| 50 | 2500/5 | 1.50 | 2400 | 3150 | 4650 | 7650 |

TABLE 12.7. Calculated CT Secondary Voltage (Volts) Using Factor (1 + *X/R*), Ignoring Remanence and Standard CT Secondary Burden

| System Short-Circuit Current in kA rms sym | Minimum Required CT Ratio | Standard CT Burden, Ω All CTs of C200 Accuracy | CT Winding Resistance, Ω | Secondary Voltage in Volts | | | |
|---|---|---|---|---|---|---|---|
| | | | | *X/R* = 15 | *X/R* = 20 | *X/R* = 30 | *X/R* = 50 |
| 20 | 1000/5 | 2 | 0.51 | 4,016 | 5,271 | 7,781 | 12,801 |
| 30 | 1500/5 | 2 | 0.84 | 4,544 | 5,964 | 8,804 | 14,484 |
| 40 | 2000/5 | 2 | 1.15 | 5,040 | 6,615 | 9,765 | 16,065 |
| 50 | 2500/5 | 2 | 1.50 | 5,600 | 7,350 | 10,850 | 17,850 |

- $V_s = 204$ V, $K = 2$, Equation (12.19)
- $V_s = 288$ V, $K = 2\sqrt{2}$, Equation (12.19)
- $V_s = 4611$ V, Equation (12.20)
- $V_s = 7685$ V, with 0.4 per unit remenance, Equation (12.22).

The (1 + *X/R*) factor gives high CT secondary voltages, impractical to meet in real-world applications. The maximum C class accuracy described in the standards is 800. *This clearly demonstrates that calculation of saturation using 1 + X/R method is not a real-world situation* [14]. Further, Table 12.6 shows the calculated secondary voltages for various primary fault currents and *X/R* ratios, ignoring all secondary burden except the CT winding resistance itself. The secondary voltages are too high for practical selection of an accuracy class, even for the *X/R* = 15. Table 12.7 shows similar calculations, with standard CT burden of C200 accuracy CTs. The secondary calculated voltages are even higher, though no remenance is considered.

The maximum ANSI rating, that is, C800, can be selected for all CTs. Reference [7] suggests to use identical CTs and match the knee-point voltages in differential relaying application, so that these have the same saturation characteristics. References [9–14] further investigate the impact of CT saturation on protective relaying.

A MMPR has internal matching auxiliary CTs, filters, and A/D converters with scaling of output. Apart from rigorous CT saturation simulations using EMTP (Electromagnetic Transient Program), efforts are directed toward development of software and other programs to calculate the impact of CT saturation on protective relaying.

## 12.11 CTs FOR LOW RESISTANCE-GROUNDED MEDIUM VOLTAGE SYSTEMS

In industrial medium voltage low resistance grounded systems, the fault point *X/R* ratio for a ground fault is low, that is, the zero sequence reactance is small and the resistance predominates. The ground fault current can be calculated using symmetrical component method. As a specimen calculation, the following data is presented for an industrial system, maximum ground fault current = 400 A, using rigorous calculations:

$$Z_0 = 2.5116 + j0.0290$$

$$Z_1 = 0.0005 + j0.0233$$

$$Z_2 = 0.0008 + j0.0221.$$

Based upon these sequence impedances, the *X/R* ratio of a single line-to-ground fault is 0.029, that is, power factor is 99.95%, and the fault current is practically in phase with the voltage. For three-phase fault, the *X/R* from above data is 46.6. For the short-circuit calculations, according to ANSI/IEEE standards, the *X/R* ratio is calculated from separate *R* and *X* networks. For protective relaying, it is appropriate to calculate it from complex impedance. Note the low *X/R* ratio and low magnitude of line-to-ground fault current. Thus, the requirements of CT accuracy are minimal. For example, a 400/5 CT, the maximum fault current limited to 400 A, and with 1.0 Ω total burden, including CT resistance and lead length resistance, need to have an accuracy of C10 only.

## 12.12 FUTURE DIRECTIONS

In Chapter 8, we discussed the CT saturation and its impact upon low impedance bus differential relaying. It was shown that even with heavy CT saturation a proper operation is obtained with the algorithms built into the relay to account for CT saturation.

This is the direction in which the relaying technology is progressing. CT saturation cannot be avoided in all applications. The CT saturation reduces the apparent current seen by the relay. This can delay operation of time overcurrent elements, and the instantaneous elements may not operate at all.

MMPRs employ digital filtering [13] to obtain phasors that eliminate DC component and harmonics. It is important to employ instantaneous elements that operate on the fundamental in absence of saturation, but respond to peak currents during saturation.

Figure 12.14 shows that the fundamental is severely reduced in a severely saturated waveform, a 100/5 ratio C50 CT, with 40-kA fault current. The magnitude of

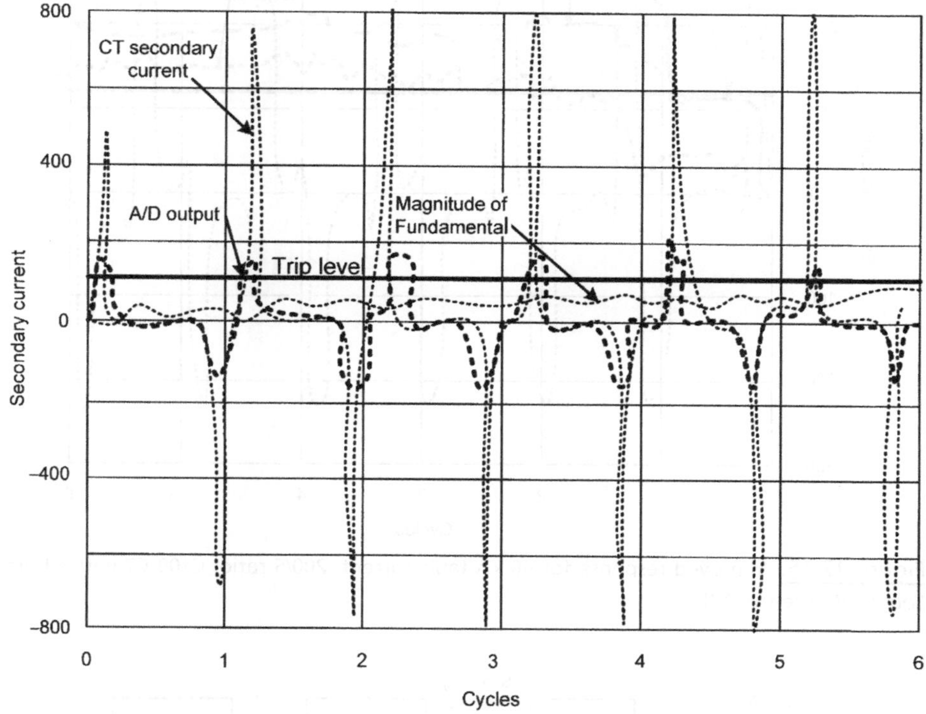


Figure 12.14. Output of a 100/5 ratio C50 CT for a 40-kA fault current, severely saturated CT. *Source*: Reference [13].

the fundamental frequency current in severely saturated CT waveform is poor representation of the actual fault current. The digital filters cannot make accurate measurements once the saturation sets in. The improved response of the RMS, peak, and cosine filters, with same fault current but CT of ratio 200/5 and C100, is shown in Figure 12.15. Both peak and rms filter respond quickly to a fast-rising signal and exhibit a high transient response as these respond to DC component in the asymmetrically offset waveform. The cosine filter responds to the fundamental frequency component of the signal and is slower, but has admirable characteristics with respect to DC offset and removal of harmonics. Combining a bipolar peak detector with a cosine filter provides an efficient solution to instantaneous element. Figure 12.16 shows instantaneous function logic of a modern MPPR using cosine peak adaptive filter. The cosine filter supplies magnitude of normal sine wave operation and bipolar peak detector for the saturated waveforms. By incorporating a method to determine when the CT is saturated, the adaptive element can switch the elements such that they operate on the input of cosine filter or on the output of bipolar detector as appropriate. The adaptive overcurrent element determines which filter to use by means of saturation detector, which operates when the harmonic content of A/D converter output exceeds a threshold called the harmonic distortion index. Also, the bipolar detector is enabled only when phase pickup setting is greater than 40-A secondary, 5-A relay. This ensures that bipolar peak detector is active in conditions where CT saturation is likely to affect overcurrent operation [15].

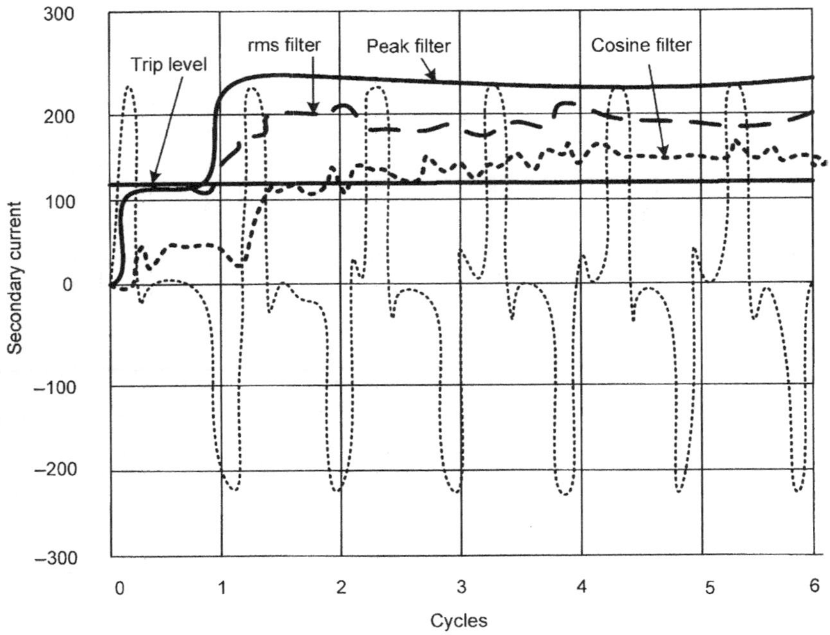


Figure 12.15. Improved response for 40 kA fault current, 200/5 ratio, C100 CT with filters. *Source*: Reference [13].

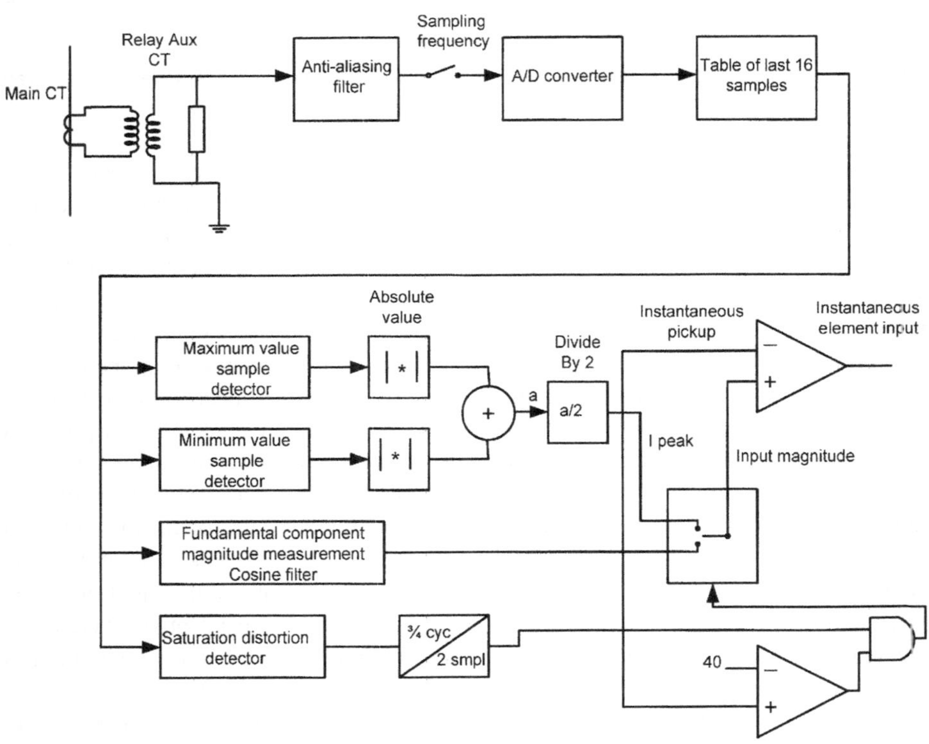


Figure 12.16. Adaptive overcurrent element block circuit diagram showing bipolar and cosine filters with CT saturation distortion detector. *Source*: Reference [15].

It should not, however, be implied that a CT can be randomly chosen for the MMPRs. The relays are tested for required performance, and the recommendations of the manufacturer for a particular relay type and application should be followed.

## REVIEW QUESTIONS

1. Specify the minimum CT ratios for primary fault currents of 10, 20, 30, and 40 kA to limit ratio error to less than 10%.
2. How can remenance impact the CT saturation?
3. Comment on the impracticality of selecting CT accuracy class based upon $(1 + X/R)$ factor.
4. A CT is required for a 13.8 kV system, and short-circuit at the point of application is 30 kA, $X/R = 10$. Considering a CT winding resistance of 0.7 Ω, relay burden of 0.1 Ω, and lead burden of 0.2 Ω, calculate the required CT ratio for C400 CT based upon Equations (12.18) and (12.21)
5. Repeat problem 4 for $X/R = 20$ and 30.
6. What is the expected fault point $X/R$ ratio near an industrial bus connected 80 MVA, 13.8 kV, 0.85 generator?
7. Describe without mathematical expressions why a low-ratio CT cannot be constructed in higher C class accuracies.
8. Describe five important steps for selection of CTs for an application.
9. What are the additional considerations for selection of a CT for differential protections?
10. A CT is connected to serve only nondirectional overcurrent relays. The relays are connected to nonpolarity marked terminal, and the polarity marked terminal is grounded. Is this connection acceptable?
11. Explain why the secondary windings of a CT should not be open circuited when the primary windings are carrying current.
12. How is the saturation of CTs accounted for in the modern MMPRs?

## REFERENCES


1. ANSI/IEEE Standard, C57.13. *Requirements for Instrument Transformers*, 1993 (R 2008).
2. ANSI/IEEE Standard, C57.13.2. *Conformance Test Procedures for Instrument Transformers*, 2005.
3. IEC, 60044-1. *Instrument Transformers, Part 1 (Current Transformers)*, 1997.

4. J.R. Linders, "Relay performance considerations with low ratio CT's and high fault currents," IEEE Trans. Industry Appl., vol. 31, no. 2, pp. 392–405, March/April 1995.
5. Westinghouse, *Applied Protective Relaying*, Westinghouse Electric Corporation, Newark, NJ, May 1982.
6. ANSI/IEEE Standard, 242. *IEEE Recommended Practice for Protection and Coordination of Industrial and Commercial Power Systems*, 1986.
7. IEEE Standard, C37.110. *IEEE Guide for Application of Current Transformers Used for Protective Relaying Purposes*, 1996.
8. J.C. Das and J.R. Linders, "Power system relaying," in *Wiley's Encyclopedia of Electronic and Electrical Engineering*, vol. 17, pp. 71–84, Wiley, New York, 1999.
9. S.E. Zocholl, W.C. Kotheimer, and F.Y. Tajaddodi, "An analytical approach to the application of current transformers for protective relaying," in 43rd Annual Georgia Tech. Protective Relaying Conference, pp. 1–21, May 3–5, 1989.
10. S.E. Zocholl and W.C. Kotheimer, "CT performance in critical relay application," in 44th Annual Georgia Tech. Protective Relaying Conference, pp. 1–14, 1991.
11. S.E. Zocholl and D.W. Smaha, "Current transformer concepts," in 45th Annual Georgia Tech. Protective Relaying Conference, pp. 1–16, 1992.
12. J.C. Das and R. Mullikin, "Design and application of low ratio high accuracy split-core, core-balance current transformer," IEEE Trans. Industry Appli., vol. 46, pp. 1856–1865, Sept./Oct. 2010.
13. J. Hill and K. Behrendt, "Upgrading power system protection to improve safety, monitoring, protection and control," in Conf. Record, IEEE Pulp and Paper Industry Technical Conference, Seattle, Washington, pp. 77–87, 2008.
14. R.E. Cossé, D.G. Dunn, and M. Spiewak, "CT saturation calculations: Are they applicable in the modern world?—Part I: The question," IEEE Trans. Industry Appl., vol. 43, no. 2, pp. 444–452, March/April 2007.
15. SEL, *Current Transformer Selection Criteria for Relays with Adaptive Overcurrent Elements. Application Guide*, Vol. 3, Publication AG2005-04.

# 13

# ARC-RESISTANT EQUIPMENT

Arc-resistant switchgear may be defined as the equipment designed to withstand the effect of internal arcing fault by meeting the testing requirements of IEEE Guide C37.20.7-2007 [1]. The 2001 issue of this guide did not cover low voltage power circuit breakers, which are added in 2007. Also, Corrigendum 1 [2] corrects the technical errors found in the test guide during use in laboratories concerning arc initiation in LV testing and supply frequency for all equipment.

In the 1970s, an interest arose in Europe in evaluating electrical equipment under internal arcing, which culminated in IEC standard [3]. This knowledge spread to North America and was used as a basis for EEMAC G14-1, 1987, Procedure for Testing the Resistance of Metal-Enclosed Switchgear under Conditions of Arcing Due to Internal Fault [4]. The development of IEEE Guide [1] heavily rests on Annex AA of IEC [3] and incorporates many of the refinements originated in EEMAC G14-1. IEC 298 has been renumbered as IEC 62271-200 [5] in 2003. It specifies the requirements for factory assembled metal-enclosed switchgear and control gear for alternating currents at rated voltage above 1 kV and up to and including 52 kV. It covers both indoor and outdoor installations and frequencies up to and including 60 Hz. IEC Standard Reference [6]

*Arc Flash Hazard Analysis and Mitigation*, Second Edition. J.C. Das.

Published 2021 by John Wiley & Sons, Inc.

pertains to arc resistance low voltage MCCs. The arc-resistant switchgear has been more popular in Europe than in North America.

The arc-resistant construction can be applied to:

- medium voltage switchgear
- medium voltage MCCs
- low voltage switchgears
- low voltage MCCs.

## 13.1 CALCULATIONS OF ARC FLASH HAZARD IN ARC-RESISTANT EQUIPMENT

It is recognized that even if arc-resistant equipment is specified, supplemental power system protection, like AFD and differential protection systems, should be provided to reduce the arcing time and incident energy release. *The calculation of incident energy, hazard risk category and PPE category for arc-resistant equipment and their reductions follows the same procedure as for conventional equipment.*

### 13.1.1 Probability of Arcing Fault

Reference [1] states that there is little likelihood of an internal arc in equipment meeting the requirements of IEEE Standard C37.20.1-2002, C37.20.2-1999, and C37.20.3-2001, which pertain to metal-clad and metal-enclosed switchgear [7–9] though the possibility cannot be entirely eliminated. There is even less likelihood of an internal arc in equipment that has insulted bus, compartmentalization, barriers, and interlocks, for example, metal-clad switchgear [8].

If any panels or doors are opened for maintenance, which are not intended to be opened, the equipment is no longer arc resistant. Depending upon the testing and construction, it is possible to open, say the low voltage compartment of arc-resistant switchgear, without impacting the arc-resistant nature of the equipment. Nevertheless, all maintenance tasks cannot be carried out simply because the equipment is arc resistant. NFPA 70E [10] specifies an hazard risk category of 4 for the arc-resistant switchgear, type 1 or 2, with clearing time of <0.5 second for insertion or removal of circuit breaker from cubicles with door open. (This should be analyzed case by case depending upon the protection and system fault levels.) Rarely, the circuit breakers will be locally racked out in modern work environment. Industrial establishments are resorting to remote racking, and the presence of a worker near the equipment is not required except to engage the remote racking mechanisms.

Thus, arc-resistant equipment is no longer arc resistant if any doors or panels are opened, which are not intended to be opened. So long as such panels or doors are not opened, the arc flash hazard outside the equipment can be considered zero. Some arc flash hazard does exist around conventional equipment with the door closed. Chapter 1 states the new interpretation provided by NFPA committee members that: "Normal operation of the enclosed electrical equipment, operating at 600 V or less, *that has been properly installed and maintained by qualified persons* is not likely to expose the employee to an electrical hazard."

Therefore, a question arises weather it is necessary to invest 20–40% more capital outlay for arc-resistant equipment. In Europe, the emphasis has been on arc containment and protection. In North America, the impetus has been to lower the arc flash hazard by system innovation and protection systems rather than reinforcing the equipment to withstand the internal arcing faults.

## 13.2 QUALIFICATIONS IN IEEE GUIDE

IEEE guide [1] states that a guide cannot mandate or define equipment ratings. A standard specifies mandatory requirements and performance levels that must be met, verified, and tested for. A guide provides recommendations and procedures, leaving performance levels to the customer and manufacturer. IEEE guide qualifies that such equipment cannot ensure *total personal protection under all circumstances that may exist at the time of internal arcing fault.* It goes on to clarify that it is not intended to provide additional degree of protection to operating personnel, who in the normal performance of their duties would be required to alter the equipment from its normal operating conditions. Examples of personnel activities or installations conditions *not* covered include:

- Personnel on top of switchgear
- Any activity that requires personnel to be elevated above base level of switchgear, that is, use of a ladder, lift, or a catwalk
- Switchgear installed on an open grating
- Installations over a cable vault large enough for personnel to enter the vault
- The selection of equipment according to the guide does not imply protection from equipment damage or ensure continued operation without disruption of electrical service.

As the conventional equipment is not arc resistant, a closed door or panel can shatter and release the arc flash products, causing injury to a worker who happens to be in the vicinity of the energized equipment. Though this eventuality may be small, it cannot be entirely ruled out. The fact that with all the specified constraints for arc-resistant equipment, the incident energy outside an arc-resistant enclosure is zero, so long it is operated as per the test conditions, even though a worker may be present in the close proximity of it has benefits to the operating personnel. No special PPE is required for the presence of a worker in the vicinity of the arc-resistant equipment. In other words, if an electrical room houses all arc-resistant equipment, an operator can carry out visual inspection, go around the equipment, take meter readings, or even open a low voltage compartment, depending upon the arc-resistant design and testing, as discussed further.

## 13.3 ACCESSIBILITY TYPES

The IEEE Guide [1] accessibility types are the following.

### 13.3.1 Type 1

Switchgear with arc-resistant designs and features at the freely *accessible front of the* equipment only.

### 13.3.2 Type 2

Switchgear with arc-resistant designs and features at the freely *accessible exterior (front back and sides)* of the equipment only.

The accessibility is designated by selecting the type (either type 1 or type 2) and by adding the appropriate suffix to designate the specific performance tested. For the basic tests described, a suffix A is used, that is, type 1A or 2A, to describe the basic rating.

The other suffixes are as follows below.

### 13.3.3 Suffix B

This designates equipment where normal operation involves opening the doors or cover of the compartments, especially identified as low voltage, control, or instrumentation compartments. Testing procedures, not reproduced here, are described for this suffix. The suffixes will be 1B or 2B. Suffix 1B lays down that the arcing does not cause holes in the freely accessibly front of the enclosure or in the walls isolating the low voltage control or instrument compartments. Suffix 2B lays down that the arcing does not cause holes in the freely accessibly front, sides, and rear of the enclosure or in the walls isolating the low voltage control or instrument compartments. Smoke ingress in the compartment under evaluation is acceptable. Distortion of internal compartment walls, without signs of burning from a direct arc strike, that is, a hole, or opening in the walls, that is, torn seams are not the cause of equipment failure.

Meeting the requirements of this evaluation additionally qualifies the equipment for installation of instrument and control devices on the door or cover of the compartment tested without retesting the switchgear for the condition where the door or cover is present with devices mounted on it.

### 13.3.4 Suffix C

This suffix is designated for equipment where isolation from effects of an internal fault is desired between all adjacent compartments within a switchgear assembly. This does not imply that equipment may be operated with doors, covers, or panels opened. Suffix C designation is not applicable to equipment using open bus or open frame construction. The suffix 1C implies that the arcing does not cause holes in freely accessibly front of the enclosure or in the walls separating the compartments in which arc is initiated from all adjacent compartments. The suffix 2C implies that the arcing does not cause holes in the freely accessibly front, sides, and rear of the enclosure or in the walls separating the compartments in which arc is initiated from all adjacent compartments.

The exception is that in a metal-clad design, a fault in the main bus bar compartment of the vertical section is allowed to propagate into the main bus bar compartment of the adjacent section, if the bus bars are in the same circuit, but not if the bus bars are in the different circuits.

This requires some explanation.

If the switchgear is of metal-enclosed design or nonsegregated construction (see IEEE standards for the different constructional features of metal-enclosed and metal-clad switchgear), an arc flash occurrence can propagate within the enclosure, and the only protection is pressure relief vents.

In metal-clad design, the protection is maintained because all compartments are block segregated and can confine generated arc energy to that zone. The four zones are:

- main bus compartment
- fused contactor compartment
- power cable compartment
- low voltage compartment.

This, however, does not imply that there will be no pressure vents.

### 13.3.5 Suffix D

This suffix is designated for equipment designed for installations where some external surfaces of equipment are inaccessible and there is no need to use a type 2 design. Thus, this suffix applies to accessibility type 1 only. The arcing does not cause holes in the freely accessible front and any other surface under evaluation of the enclosure. The evaluated surfaces must be listed on nameplate:

SR, right side

SL, left side

R, rear.

As an example, the designation can be Type 1D-SR. Additional specific tests are applicable for the suffixes.

The manufacturer should establish a minimum spacing between the room walls and the equipment. A minimum ceiling distance shall also be established, even if an external exhaust system is used.

## 13.4 IEC ACCESSIBILITY TYPES

IEC standard defines Internal ARC Classification (IAC), which makes allowance for overpressure acting on covers, doors, inspection windows, and the like.

*Type A.* Restricted to authorized persons only.

*Type B.* Unrestricted accessibility, including that of general public [5].

The IEC accessibility types are for the equipment manufactured according to IEC specifications, for example, IEC [6] applies to the MCCs manufactured according to IEC 61439 [11]. These are not interchangeable with the equipment manufactured according to ANSI/IEEE and NEMA standards.

## 13.5 ARC-RESISTANT RATINGS

Arc-resistant ratings can be obtained by two methodologies, discussed as follows.

### 13.5.1 Duration Ratings

The ratings are based on rating the equipment to withstand an internal arcing fault for the specified duration. The maximum system voltage and maximum available prospective three-phase fault current are part of the duration rating.

This method provides the user more flexibility in selecting the protective relaying systems and ensuring that the faults are cleared within the arc duration rating. The selection of the upstream circuit breaker is still critical, and an arc flash analysis and relay coordination study is required to determine the arcing current levels and the arcing time. A fault on the main breaker itself, when located in the arc-resistant equipment, will be cleared by an upstream breaker abundantly discussed in the previous chapters. If this breaker is included in the differential relaying scheme or AFD protection zone, including the incoming cable connections and circuits, then the arc flash time duration can be reduced consistent with the protection. But if the time–current coordination is relied upon to clear the fault in the arc-resistant rated equipment, the extended arcing time withstands may be required.

IEEE guide [1] specifies a preferred rating of 0.5 second. With AFD or differential systems covering the entire arc-resistant equipment, the time duration can be reduced. It has much impact on the cost of the equipment. The following considerations apply:

- It may be more economical and safe to add high-speed protection and reduce the arc withstand time, than to specify equipment with higher arc withstand times.
- The higher is the arc fault current and the longer is its duration, the more robust the equipment has to be. The internal damage to the equipment will be higher with higher arc withstand times.
- With higher arcing times, it will be necessary to replace or refurbish a larger portion of the equipment after an arc fault event.
- The process downtime is another consideration.

Such equipment should carry a name plate showing:

- accessibility type
- tested arcing short-circuit current, kA rms
- arc duration, ms.

### 13.5.2 Device-Limited Ratings

The ratings are based upon the internal arcing fault being limited by a specific device. The location of the device is important, for example, a current limiting fuse will only limit the energy downstream of it. The protection depends upon the function of a protective device. Thus, the protection device should be reliable and comply with industry standards.

The application will meet the requirements, so long as:

- The system voltage does not exceed the tested voltage.
- The short-circuit current does not exceed the tested current.
- The specific type of device and the settings are used.

The device types may be:

1. Current limiting fuses
2. Fast acting circuit breakers
3. Conventional circuit breakers with trip programmers
4. Specific overcurrent relays with recommended settings
5. Additional testing is required for current limiting devices as the specific performance may vary under different circuit conditions.

**Example 13.1**

This example illustrates the principle behind device-limited ratings and may not be construed to apply to any specific situation. Consider four devices for 480-V three-phase 60-Hz applications:

a 600-A class J fuse

a 2000-A class L fuse

a 600-A MCCB, thermal magnetic trip, magnetic pickup set at 3000 A.

A 2000-A LVPCB, with electronic trip programmer, set as: Long time pickup = 1 (2000 A), long time delay band = 2, short-time pickup = 4 (8000 A), short-time delay band set at minimum and instantaneous pickup = 10 (20,000 A). The characteristics of these four devices are plotted in Figure 13.1.

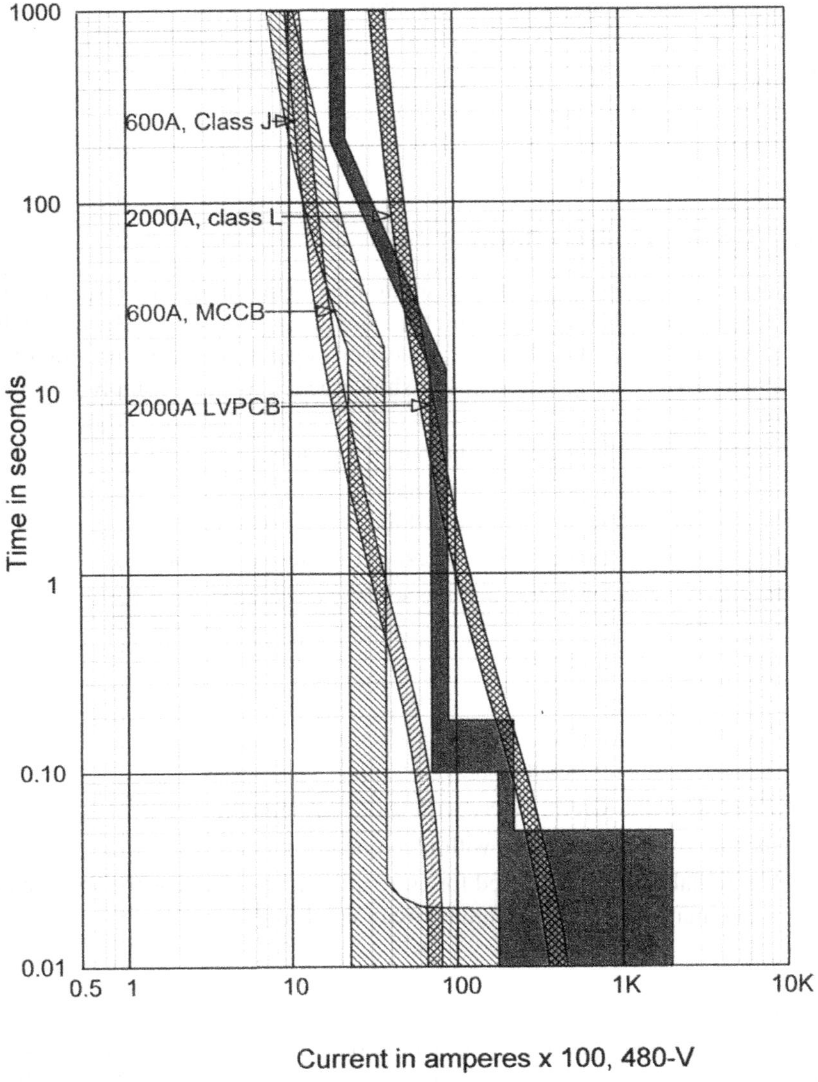


Figure 13.1. Time–current characteristics of four low voltage devices for analysis of device limited ratings of arc-resistance equipment.

Figure 13.2 shows the incident energy release of these four devices as the bolted three-phase short-circuit current is varied from low values, say, 5–100 kA. The calculations are according to IEEE Guide 1584, with a gap of 25 mm and working distance of 455 mm. Detailed step-by step results are not presented, and the plots in Figure 13.2 represent the ultimate incident energy calculations in cal/cm$^2$. The following observations are of interest:

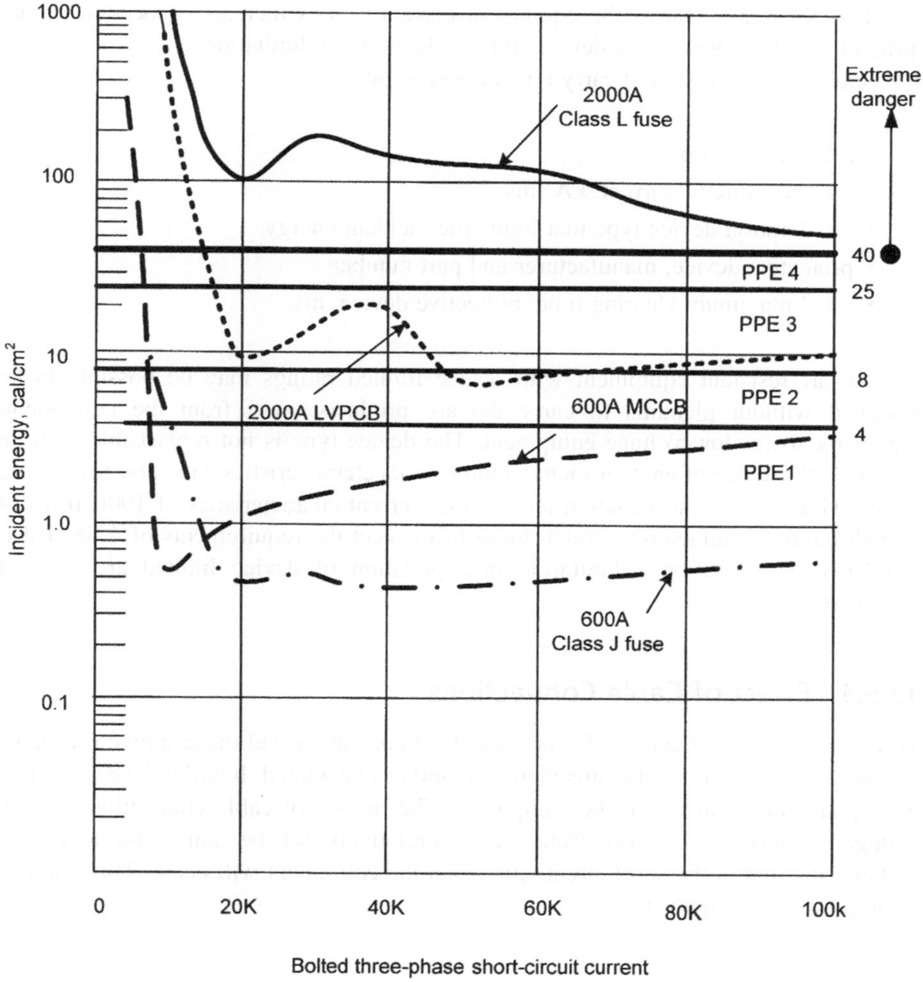


Figure 13.2. Calculated incident energy release versus three-phase bolted short-circuit current for the four devices in Figure 13.1.

- The incident energy release of a 2000-A class L current limiting fuse remains above 40 cal/cm$^2$ (extreme hazard). At low levels of short-circuit currents, say at 10 kA, the incident energy release is >1000 cal/cm$^2$. This fuse rating and type is, therefore, unsuitable for any low voltage system energy limitation.
- The 2000-A LVPCB incident energy release goes to PPE level 2 for short-circuit currents, approximately >46 kA. By choosing a much lower instantaneous setting, the situation can be improved, though coordination with other overcurrent devices must be considered simultaneously.
- The 600-A MCCB and 600 A class J fuse give much reduced incident energy release.

This illustrates that as the equipment current ratings increase, it becomes more difficult to select and choose devices for incident energy limitation.

Such equipment should carry a nameplate showing:

- accessibility type
- arc short-circuit current, kA rms
- arc duration device type that limits the incident energy
- protective device, manufacturer and part number
- rated maximum clearing time, protective device, ms.

An arc-resistant equipment with device limited ratings may be possible to be designed without plenums to carry the arc products away from the equipment, especially so for low voltage equipment. The device type is not replaceable with an "equivalent" device of another manufacturer, as its characteristics can vary. See Figure 11.10, which shows the variation in the time–current characteristics of 100E fuses of four different manufacturer, though these fuses meet the requirements of ANSI/IEEE standards. This can be a limitation in application of device limited arc-resistant equipment.

### 13.5.3 Effect of Cable Connections

When the protective device is located some distance away and the equipment is cable connected, the effect of cable impedance should be considered. It will reduce the bolted three-phase fault current at the equipment. The impact of cable connections in low voltage systems will be more pronounced as relatively, for the same cable lengths, a higher reduction in the short circuit current at the equipment will occur. This requires a careful arc flash analysis.

## 13.6 TESTING ACCORDING TO IEEE GUIDE

The IEEE guide [1] provides details of test arrangements for all equipment and specifics for equipment covered in IEEE standards—arc initiation, current limiting devices, placement of test indicators, and test conditions—voltage, current, frequency, and the like. These are not reproduced here. An interested reader may peruse the referenced material. The assessment of test results is briefly described.

### 13.6.1 Criterion 1

The properly latched or secured doors, covers, and so on do not open. Some bowing or buckling and distortion are permitted provided no part comes as far as the position of the indicator mounting racks or walls on any accessed surface. For an installation mounted closer to the wall than tested, the additional criteria are:

1. The permissible distortion is less than the intended distance to the wall.
2. Exhausting gasses are not directed to the wall.

### 13.6.2 Criterion 2

No fragmentation of enclosure occurs within the time period specified for the test, though ejection of small parts, up to an individual mass of 60 g, from any assessed external surface above a height of 2 m and from any surface not under assessment, is acceptable. No restriction is placed on the number of parts ejected.

### 13.6.3 Criterion 3

***Assessment of Burn Through.*** It is assumed that any opening in the switchgear caused by direct contact with the arc will also ignite an indicator mounted outside of switchgear at that same point. As all the area cannot be covered with indicators, any opening in the area under assessment that results in direct contact with the arc is considered cause of failure. Opening above the indicator mounting rack height (2 m) that do not cause ignition of horizontally mounted indicators are ignored.

*Accessibility Type 1.* That arcing does not cause holes in the freely accessible front of the enclosure.

*Accessibility Type 2.* That arcing does not cause holes in the freely accessible front, sides, and rear of the enclosure.

### 13.6.4 Criterion 4

That no indicators ignite as a result of escaping gases. Indicators ignited as a result of burning of paint or labels, glowing particles, and so on, are excluded from this assessment. Holes in horizontally mounted indicators caused by particles that do not ignite the indicator are ignored.

Surface discoloration or charring that does not result in the glowing or flaming of the indicator cloth is allowed.

### 13.6.5 Criterion 5

This criterion requires that all grounding connections remain effective.

The guide qualifies that all possible conditions and circumstances that can affect performance of equipment are not addressed. Some by-products, like toxic gases and sound pressure—which must be considered by the user when equipment is assessed—are not addressed.

The indicators mentioned in Criterion 4 require some explanation. During testing, all energy release external to the equipment is monitored using special rack mounted indicators strategically placed around outside surface of switchgear. Figure 13.3 shows a typical indicator frame arrangement. This supports a special material untreated for

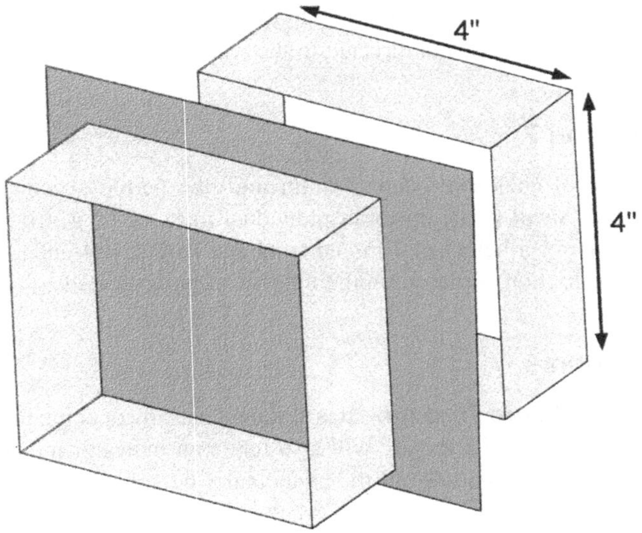


Figure 13.3. A typical indicator frame used in arc-resistance testing.

fire retarding and 100% black cotton with a density of 150 g/m$^2$. The weave is defined as Cretonne, which is a strong fabric with a cotton wrap and woven weft.

It is, therefore, evident that after an internal arc flash event, the arc-resistant equipment may require considerable maintenance before putting back in service. The integrity of restoring the equipment back to pre-fault even remains uncertain. It is prudent that the manufacturer should be contacted, and the equipment receives a thorough examination with replacement of parts as required. *Note that retesting in the field is not possible.*

### 13.6.6 Maintenance

The manufacturer should identify special characteristics of the equipment and details of maintenance procedures that are required. The equipment should be maintained according to manufacturer's recommendations.

## 13.7 PRESSURE RELIEF

The pressure inside the enclosure rises at a rate of 18 lbf/in$^2$/ms. There is an initial compression stage after the arc is initiated. This is due to the oxygen and other materials being consumed by the arc. The vaporization of these materials quickly pressurizes the enclosure. As the pressure builds, the pressure relief devices arrests its continuous rise (Figure 13.4). At this point, a large amount of thermal energy has been created. See

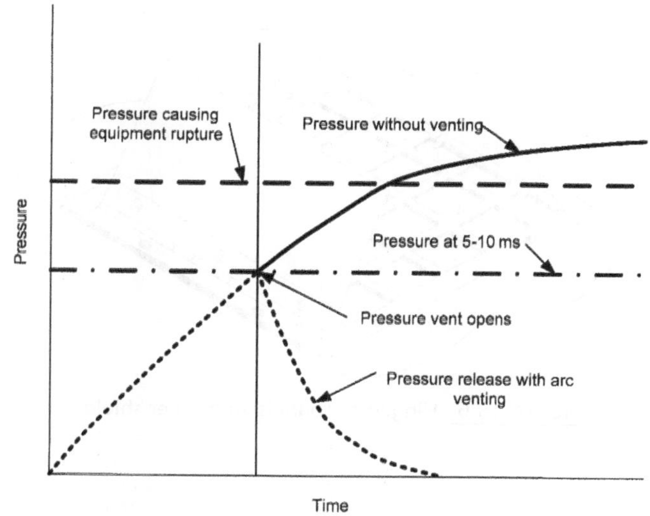


Figure 13.4. Pressure buildup after an arc flash event, with and without venting.

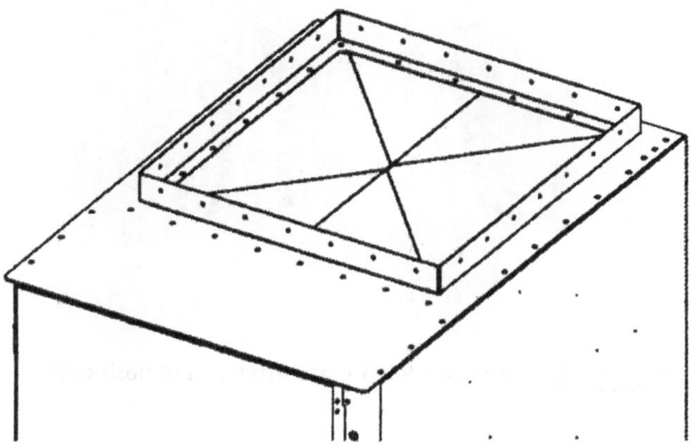

Figure 13.5. Flap plates on an arc-resistant equipment, normally closed.

also Figure 1.2. The maximum pressure inside the equipment is a function of the pressure relief system used.

The manufacturers design a unique system for pressure relief. These pressure relief devices open fast to limit the damage resulting from internal pressure during compression stage of the fault.

Figure 13.5 shows a rupture relief plate, and Figure 13.6 shows hinged arc relief panels with inner shields. Figure 13.7 is the picture of a rupture plate after an arc fault event.

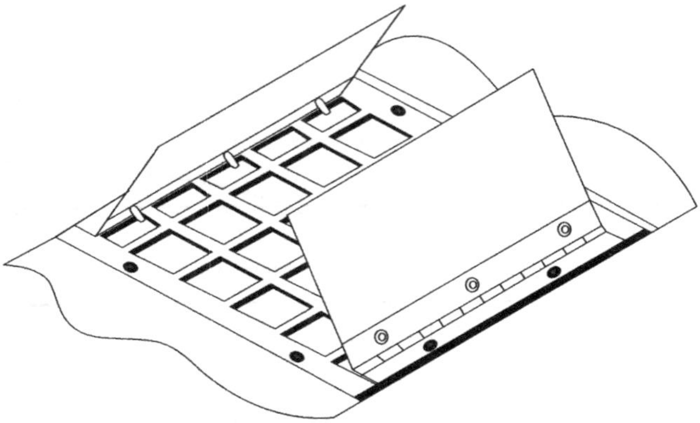

Figure 13.6. Hinged arc panels with inner shield.

Figure 13.7. Pressure relief plate after an arc flash event.

## 13.8 VENTING AND PLENUMS

The two options are:

1. Vent into the surrounding area.
2. Conduct the arc products outside the electrical room to a safe area.

### 13.8.1 Venting into Surrounding Area

This is a common technique to vent the gases into the area above the switchgear. No object, such as building structure, pipes, ducts, conduits, wire ways, and lighting fixtures, should enter the minimum clear space specified by the manufacturer. Minimum ceiling and wall clearances are required. The clear space requirements are typically much greater than the space requirement with plenum installations. The specific problems could be:

- The arc products are at a temperature of 16,000–35,000°F. The objects in the direct path may be instantaneously vaporized unless the arc gasses have sufficiently cooled before a contact is made.
- The hot gasses could be redirected and splash down on a worker or anyone else present around the equipment.
- Considerations should be given to the materials located above the enclosure. Cables and other flammable materials could be ignited.

Then, further considerations of building construction apply:

- Adequate volume to absorb pressure wave
- Structural capability to withstand pressure wave
- The ability of doors, windows, and other openings to withstand pressure wave
- The ability of the roof, ceiling, and fittings, located in the path of gasses released from switchgear pressure vents, to remain intact.

Figure 13.8 shows arc-resistant MV enclosure with arc chimneys [12].

For the low voltage MCCs, plenums, ductwork, and chimneys are not normally required to channel the arc flash products.

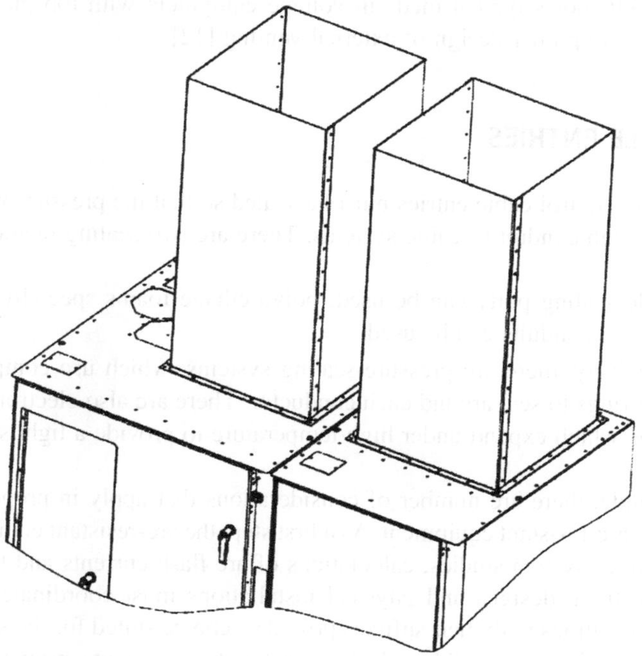

Figure 13.8. Typical arc-resistant medium voltage enclosure with arc chimneys. *Source*: Reference [12].

### 13.8.2 Plenums

The plenums will facilitate the channeling of arc products to a safe and controlled location. The plenum run outside the electrical rooms, and the containment area must be according to manufacturer's recommendations. Some considerations are:

- The release area must be free of inflammable materials and common traffic.
- The number of turns, bends, and lengths of plenums will be dictated by the layout of the electrical rooms and must be considered in the planning stage.
- The plenums normally exhaust from the top of the equipment, and there should be attachment structures to support the weight of the plenums.
- A large number of bends in the duct may impede the venting, and the manufacturer should be consulted for the acceptable layout.
- The space on the top of the room to run the plenum should be examined.
- The exhaust duct should not allow external air to enter the switchgear.

If the duct passes through a wall, it may require fire rating equal to the rating of the wall. The duct should prevent external fires from penetrating the wall at the point where the duct passes through the wall while providing an exhaust path for the arc gases to exit.

Considerations must be applied to safeguard the nearby equipment from exhaust gases and pressure waves.

Figure 13.9 shows typical medium voltage equipment with top plenums. Figure 13.10 shows an improper design of external venting [12].

## 13.9 CABLE ENTRIES

The power and control cable entries must be sealed so that the pressure wave does not propagate through conduit or cable systems. There are two sealing methods:

- Suitable sealing putty can be used; polyurethane foams specially designed for cables and conduits can be used.
- Alternatively, there are pressure-sealing systems, which use compression style sealing rings to seal around each conductor. There are also electrometric sealing systems which expand under high temperature to provide a tight seal.

To conclude, there are number of considerations that apply in proper design and application of arc-resistant equipment. As a first step, the arc-resistant equipment should be specified after system studies, calculations of arc flash currents and their duration. Then, the electrical designs and physical installations must coordinate. The various types of accessibilities with their suffixes provide a choice suited for the specific needs. Technoeconomical factors, building designs, installations, layouts, proper maintenance, training, and operation—all enter into a well-informed decision. Also it is pertinent that the qualifications attached to the arc-resistance equipment in IEEE guide [1] are

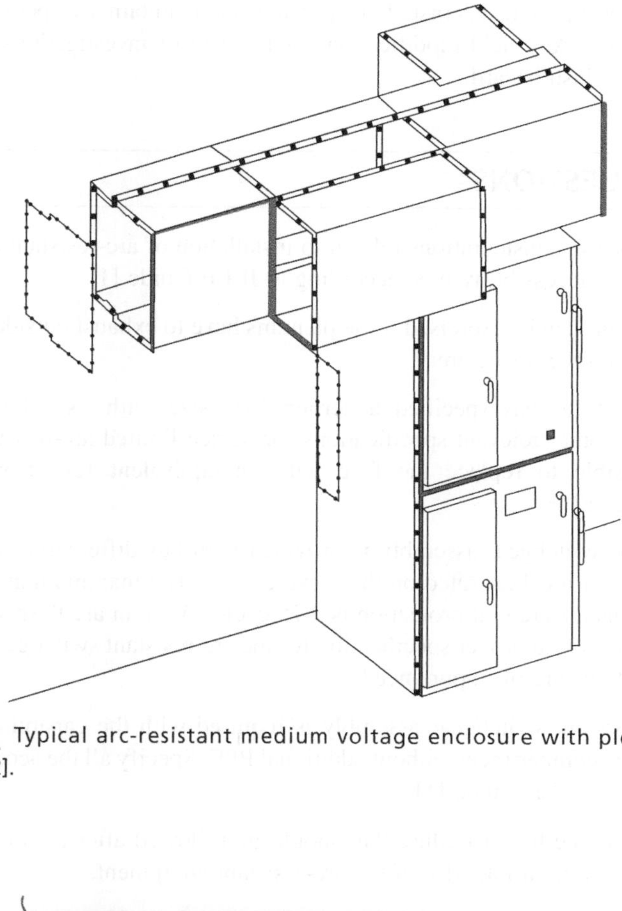

Figure 13.9. Typical arc-resistant medium voltage enclosure with plenum. *Source*: Reference [12].

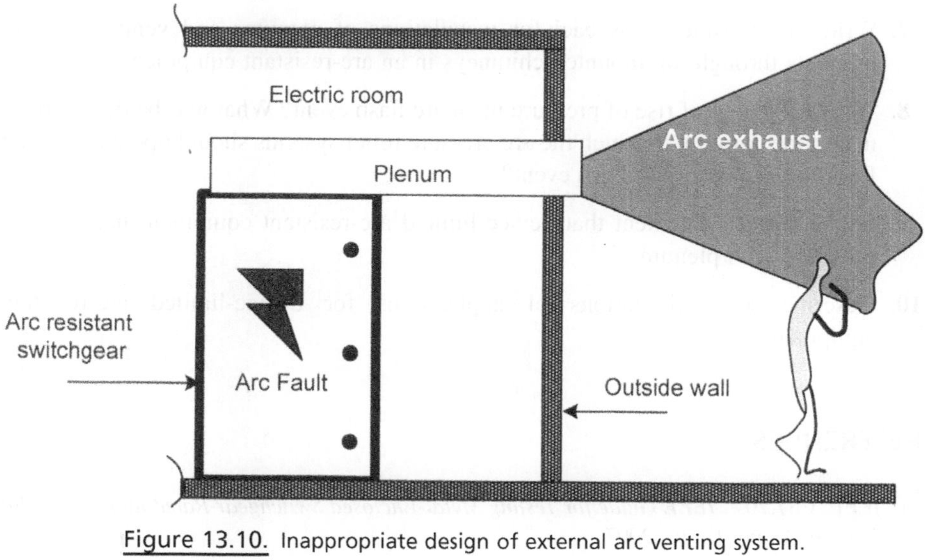


Figure 13.10. Inappropriate design of external arc venting system.

observed and the equipment is installed, operated, and maintained as per manufacturer's recommendations. Any field modifications, without prior investigations, can result in un-intended arc flash hazard.

## REVIEW QUESTIONS

1. Describe some considerations related to installation of arc-resistant equipment for selecting an accessibility type according to IEEE Guide [1].
2. What options can be exercised if the plenums have to exhaust outside the electrical room in a public traffic area?
3. A manufacturer has specified a certain fuse size with its catalogue number, rating, and other relevant specifications for device-limited arc-resistant ratings. Is it permissible to replace the fuse with an equivalent fuse from a different manufacturer?
4. A 13.8-kV switchgear assembly is provided with bus differential protection, and 15-kV circuit breakers rated on three-cycle basis. The maximum arcing time considering bus differential protection is 3.75 cycles. Will an arc flash withstand time of four cycles be a proper specification for the arc-resistant switchgear? What other considerations are of importance?
5. An arc-resistant switchgear assembly is required with the capability to open the low voltage compartment without additional PPE. Specify all the accessibility types according to IEEE Guide [1].
6. Briefly describe the procedure that should be followed after an arc flash event to restore the original integrity of the arc-resistant equipment.
7. Write six considerations each,for installations of plenums and venting the arc products through top-mounted chimneys in an arc-resistant equipment.
8. What is the rate of rise of pressure in an arc flash event? What will be the approximate time duration in which the arc pressure relief systems should operate, starting from the instant of arc flash event?
9. Is it a correct statement that device-limited arc-resistant equipment need not be provided with plenums?
10. Describe some limitations of applications for device-limited arc-resistant equipment.

## REFERENCES

1. IEEE, C37.20.7. *IEEE Guide for Testing Metal-Enclosed Switchgear Rated up to 38 kV for Internal Arcing Faults*. 2007.

2. IEEE, PC37.20.7 Cor 1/D4. *Draft Guide for Testing Metal-Enclosed Switchgear Rated up to 38 kV for Internal Arcing Faults—Corrigendum 1*, 2007.
3. IEC Standard, 298-1994-11. *High Voltage Switchgear and Control Gear—Part 200: AC Metal-Enclosed Switchgear and Control Gear for Rated Voltages above 1 kV and up to and Including 52 kV*, 1981.
4. Electrical Equipment Manufacturer's Association of Canada (EEMAC), G14-1-1987. *Procedure for Testing the Resistance of Metal Clad Switchgear under Conditions of Arcing Due to an Internal Fault*, 1987.
5. IEC Standard, 62271-200-2003-1. *High Voltage Switchgear and Control Gear—Part 200: AC Metal-Enclosed Switchgear and Control Gear for Rated Voltages above 1 kV and up to and Including 52 kV*, 2003.
6. IEC/TR, 61641. *Enclosed Low-Voltage Switchgear and Control Gear Assemblies—Guide for Testing under Conditions of Arcing Due to Internal Fault*, 2008.
7. IEEE Standard, C37.20.1. *IEEE Standard for Metal-Enclosed Low-Voltage Power Circuit Breaker Switchgear*, 2002.
8. IEEE Standard, C37.20.2. *IEEE Standard for Metal-Clad Switchgear*, 2012.
9. IEEE Standard, C37.20.3. *IEEE Standard for Metal-Enclosed Interrupter Switchgear*, 2001.
10. NFPA, NFPA 70E. *Electrical Safety in Workplace*, 2009.
11. IEC, 61439. *Low-Voltage Switchgear and Control Gear Assemblies—Part 1: General Rules*, 2009.
12. J. Kay, "Considerations for installing and applying arc resistant low-voltage and medium-voltage control equipment in forest product industries," in Conf. Record, IEEE Pulp and Paper Industry Conference, Birmingham, AL, pp. 114–120, June 2009.

# 14

# RECENT TRENDS AND INNOVATIONS

New products and innovations are marking advancements in arc flash containment and providing a safer work environment for the workers engaged in maintenance of the electrical equipment in the energized state. For example, Arc Vault™ (GE), described in Chapter 9, is a recent innovation. Also some old established concepts are being revisited. Coordination of instantaneous devices in series, discussed in Chapter 10, is a new frontier, and so also the developments of current limiting low voltage molded case circuit breakers. Then there are innovations in the construction of low-voltage MCCs, new breed of low-voltage trip programmers, remote racking of MCC buckets and the like.

## 14.1 STATISTICAL DATA OF ARC FLASH HAZARDS

Some statistical data of the arc flash hazard collected form 91 sites is shown Figure 14.1 from Reference [1]. This depicts the number of buses at different voltage levels in the electrical distribution systems. Large numbers of buses, 84.7%, are at low voltage levels. Figure 14.2 shows the arc flash energy of buses, for example, it shows that 5% of the total buses have hazardous incident energy level of 40–100 cal/cm$^2$ and on 1%

*Arc Flash Hazard Analysis and Mitigation*, Second Edition. J.C. Das.

Published 2021 by John Wiley & Sons, Inc.

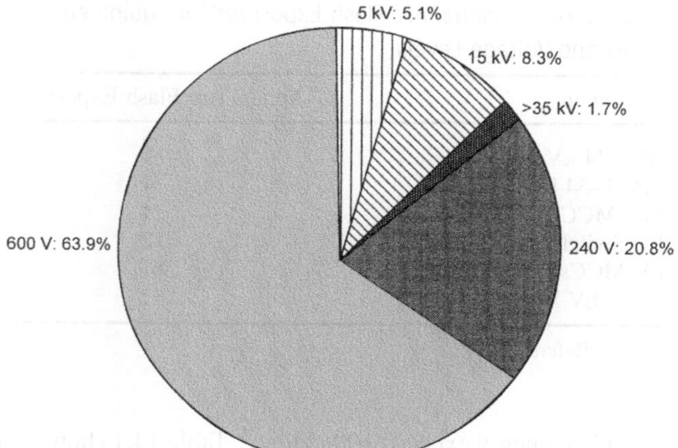


**Figure 14.1.** Statistical data of arc flash hazard with respect to voltage levels in industrial systems. *Source*: From Reference [1].

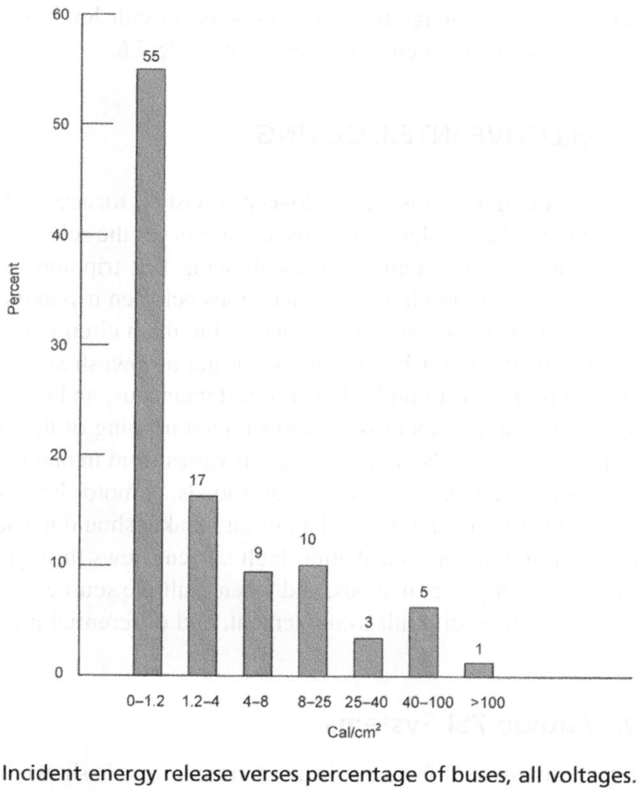


**Figure 14.2.** Incident energy release verses percentage of buses, all voltages. *Source*: Reference [1].

TABLE 14.1. Annual Arc Flash Exposures for Equipment Types and Voltage Levels

| Equipment | Annual Arc Flash Exposures |
|---|---|
| HV(>34 kV) | 2 |
| MV(1–34 kV) | 4 |
| MV MCCs | 24 |
| LV Switchgear | 12 |
| LV MCCs | 365 |
| Other LV Equipment | 52 |

*Source*: Reference [1].

of the buses the incident energy exceeds 100 cal/cm$^2$. Table 14.1 shows annual exposure for each type of equipment. This shows that the annual exposure on low voltage MCCs is 365, on other low voltage equipments 52, and on low voltage switchgear 12. The total yearly exposures are 459, out of which 429 are on the low voltage equipments, that is, 93.4%. The exposure on low voltage MCCs alone is 79.5%.

This correlates well with the product innovations and advancements in industry, which are concentrated on low voltage equipments. In the design stage, by simply reducing the ratings of low voltage transformers, short-circuit levels can be reduced, which in turn impact the incident energy levels. See Table 7.6.

## 14.2 ZONE-SELECTIVE INTERLOCKING

Zone-selective interlocking (ZSI) is an old concept revisited for arc flash reduction. It can also be applied to medium voltage systems and preserves the selective coordination between main, tie and feeder circuit breakers allowing fast tripping between device desired zones. This is done through wired connections between trip units and relays. If a feeder detects a fault, it sends a restraint signal to the main circuit breaker, but for a fault on the bus, the main circuit breaker does not get a downstream restraint signal and trips without delay. The restraint logic is not instantaneous, and there is some time delay associated with it, so that there is no unrestrained tripping of the main. For conservatism, a delay of 20 ms can be added, though it varies from manufacturer to manufacturer. Also, care has to be exercised with motor loads. A motor load will contribute to the bus short-circuit current, and the feeder circuit breaker should not send a restraint signal upstream when the motor contribution fault current flows through it. There can be more than one source of power to a bus, and when multiple sources feed into a fault, the zone interlocking will be difficult to implement, and differential protection can be adopted.

### 14.2.1 Low Voltage ZSI Systems

Typical low voltage distribution from a substation transformer is shown in Figure 14.3. The feeder circuit breakers and main circuit breaker have ZSI. For a fault at F1,

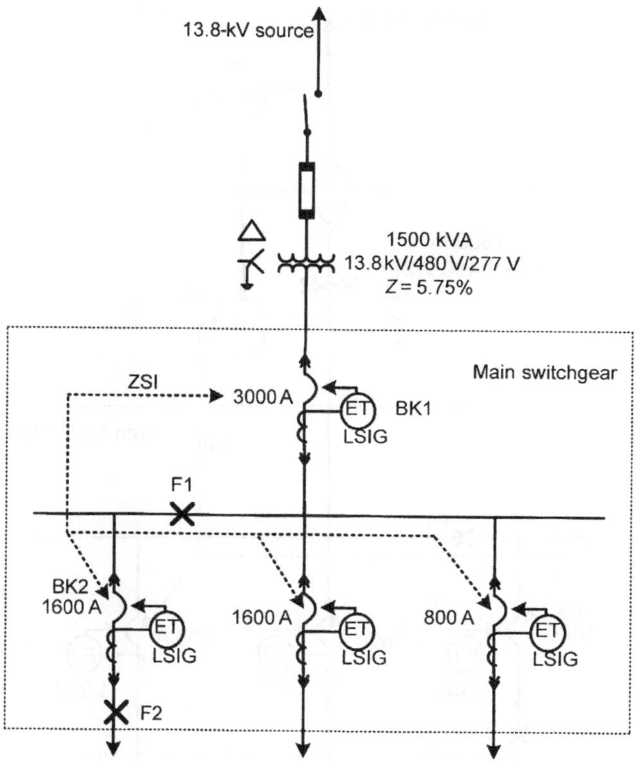


Figure 14.3. Zone interlocking between feeder and main secondary circuit breaker of a low voltage transformer.

the main circuit breaker does not receive a restraint signal, and it trips out with no intentional delay. For a fault at F2, the main circuit breaker receives a blocking signal and trips on short-time delay, allowing the feeder circuit breaker to first clear the fault.

Figure 14.4 shows yet another application. Consider that the current sensors (CTs) connected to 50/51 relay R1 are mounted in the transformer secondary terminal compartment, and the relay itself is located on the low-voltage switchgear. For a fault on the load side of the feeder circuit breaker F2, it sends a signal to relay R1, and the standard settings are applicable for coordination. For a bus fault at F1, there is no ZSI signal, and the R1 uses a definite time delay, which is many times faster than the delay necessary for coordination.

In summary, a coordination step between two devices in series is much reduced, lowering the arc flash hazard.

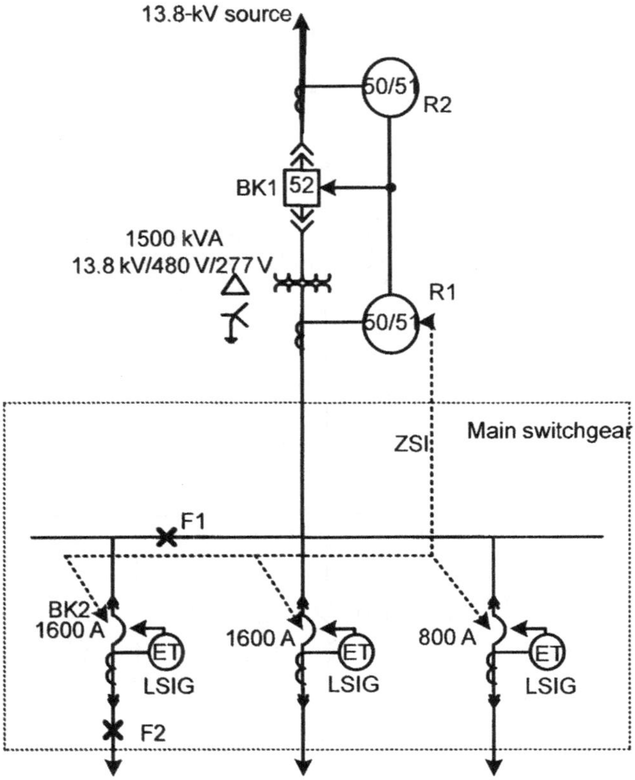


Figure 14.4. Zone interlocking between feeder and main secondary circuit breaker and also primary breaker of a low voltage transformer (see text).

**Example 14.1**

Figure 14.5 shows a section of a distribution from a 1500-kVA substation transformer. The main circuit breaker BK1 and feeder circuit breakers BK2 are zone interlocked. Figure 14.6 shows three-step coordination. The 400-A circuit breaker BK3 feeding the panel is a current limiting circuit breaker. Circuit breakers BK1 and BK2 are LVPCBs with electronic trip programmers.

The coordination shown in Figure 14.6 illustrates the zone interlocking features. For a bus fault, no signal to block the trip is received from feeder circuit breakers BK2, and circuit breaker BK1 clears the fault with its short-time delay band moved down as shown in dotted lines, which shows a delay of 20 ms. In coordination, a conservative approach is taken, and the maximum clearing time of the short-time band is considered, in this case 0.07 ms. This means that instead of clearing the arcing current in time A in 0.34 second, it is cleared in time B, 0.07 second.

For a fault on the downstream of circuit breakers BK2, the main circuit breaker BK1 does not receive any restraint signal and the normal coordination applies. That

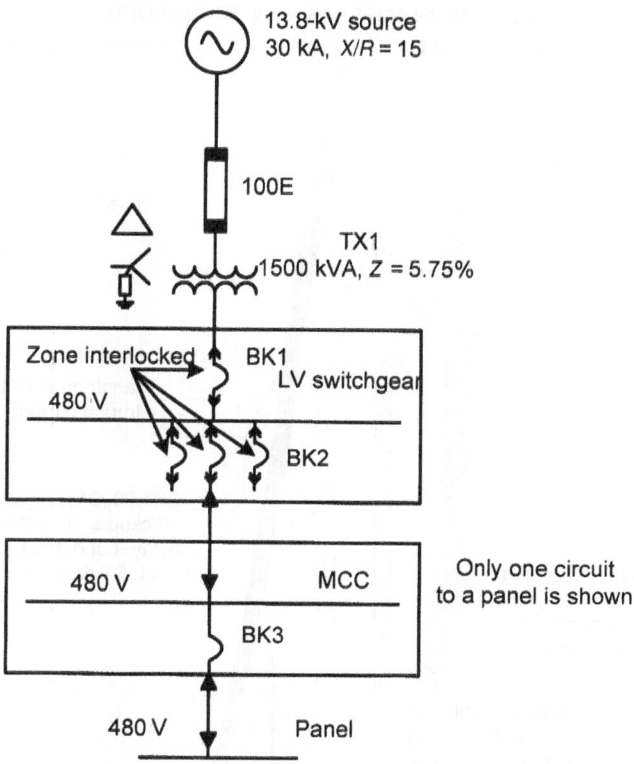


Figure 14.5. A low voltage distribution system for study of zone interlocking.

means the characteristic stays where it is, as shown in the solid lines. Thus, the feeder circuit breaker clears the fault selectively in 0.05 second, point C.

Table 14.2 shows the impact of the zone interlocking on the arc flash reduction. Without zone interlocking, the incident energy release on the low-voltage switchgear is 15.3 cal/cm$^2$. With zone interlocking, it is reduced to 3.1 cal/cm$^2$.

In Figure 14.6, it seems that there is no coordination between the instantaneous settings on circuit breaker BK2 and BK3. However, circuit breaker BK3 is a current-limiting circuit breaker, and the coordination on instantaneous basis is discussed in Chapter 10. Though these two circuit breakers do not seem to coordinate in the TCC plot, but on a current let-through basis, these do.

**Example 14.2**

This example illustrates the problem that can occur with zone interlocking when large motor loads are present. Consider the system configuration shown in Figure 14.7. The coordination is shown in Figure 14.8. The 200-hp motor starting curve is plotted. Also, a curve illustrating the starting load plus the running load of 7, 100-hp motor is plotted. The short-circuit current profile of the motor loads crosses the feeder short-time setting

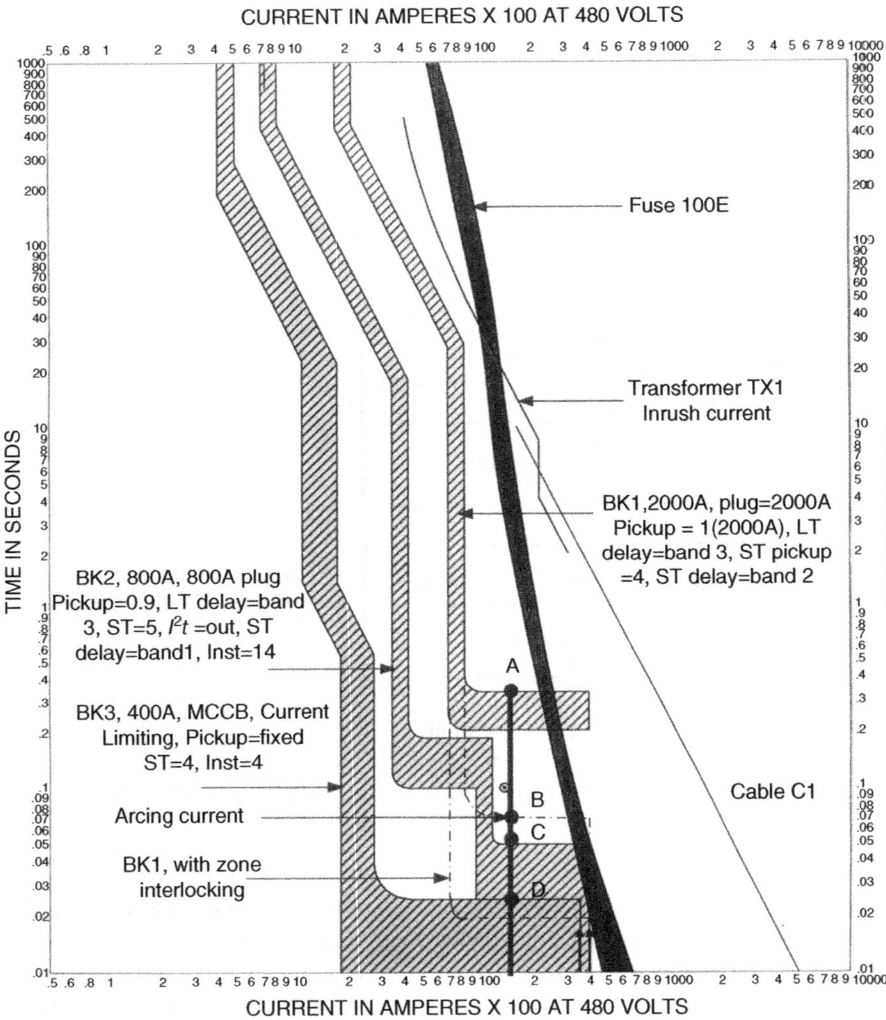


Figure 14.6. Time–current coordination of devices in Figure 14.5, with and without zone interlocking.

band 1. This could result in tripping of the feeder circuit breaker BK2. There are two solutions to this problem:

1. The short-time band of the feeder circuit breaker relay can be raised so that it clears the motor fault decrement curve. This implies that the incident energy release and arc flash hazard will increase.
2. The second method utilizes the microprocessor-based technology to sense the direction of the fault current, as discussed in Section 14.3.

TABLE 14.2. Arc Flash Hazard Analysis with TCC Plots Shown in Figure 14.6, Example 14.1, Dimensions: LV switchgear: 36″W × 90″H × 48″D, LV MCC: 30″W × 90″H × 25″D, EC = HCB, EC = VCB for Panel

| Equipment Faulted | Voltage (kV) | Upstream Trip Device | Air Gap (mm) | Bolted Fault (kA) | Arc Fault (kA) | Trip Time (Second) | Opening Time (Second) | Arc Time (Second) | Arc Flash Boundary (Inches) | Working Distance (Inches) | Incident Energy (cal/cm$^2$) | PPE |
|---|---|---|---|---|---|---|---|---|---|---|---|---|
| LV switchgear without zone interlocking | 0.48 | BK1 | 32 | 30.28 | 20.93 | 0.34 | 0 | 0.34 | 95 | 24 | 19.5 | 3 |
| LV switchgear with zone interlocking | 0.48 | BK1 | 32 | 30.28 | 20.93 | 0.07 | 0 | 0.07 | 44 | 24 | 4 | 1 |
| LV MCC | 0.48 | BK2 | 25 | 28.49 | 20.35 | 0.05 | 0 | 0.099 | 49 | 18 | 4.1 | 2 |
| Panel | 0.48 | BK3 | 25 | 28.49 | 20.35 | 0.025 | 0 | 0.07 | 36 | 18 | 3.6 | 1 |

*Notes*: Calculations remain unchanged for LVMCC and Panel, with or without zone interlocks.

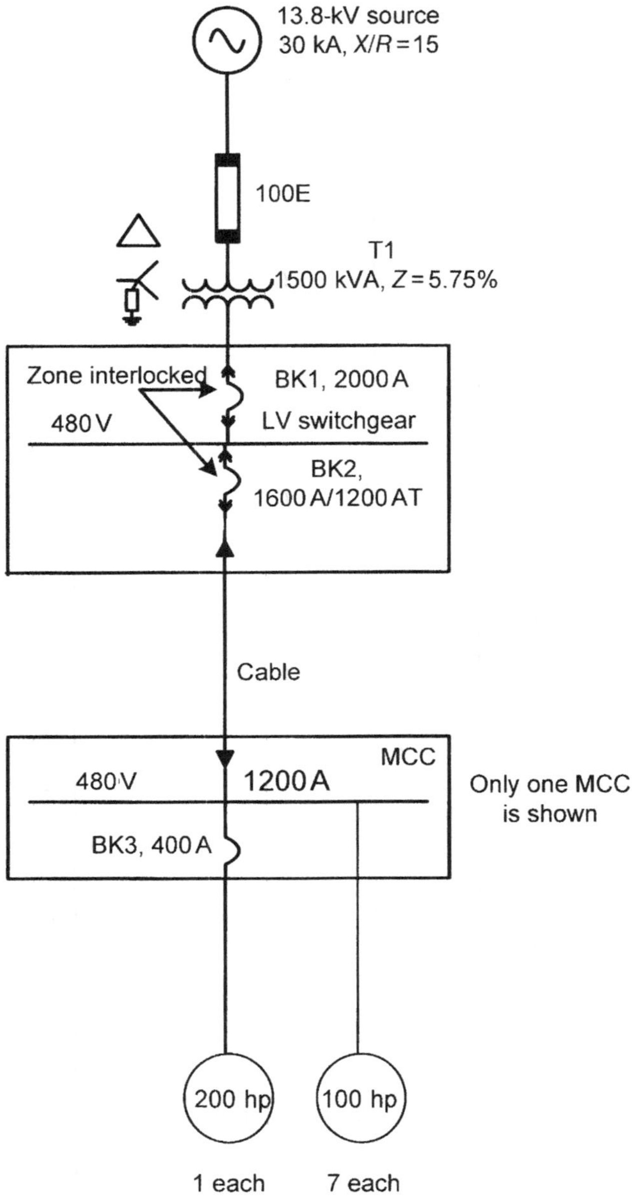


Figure 14.7. A low voltage distribution system, with considerable rotating motor loads downstream for study of zone interlocking.

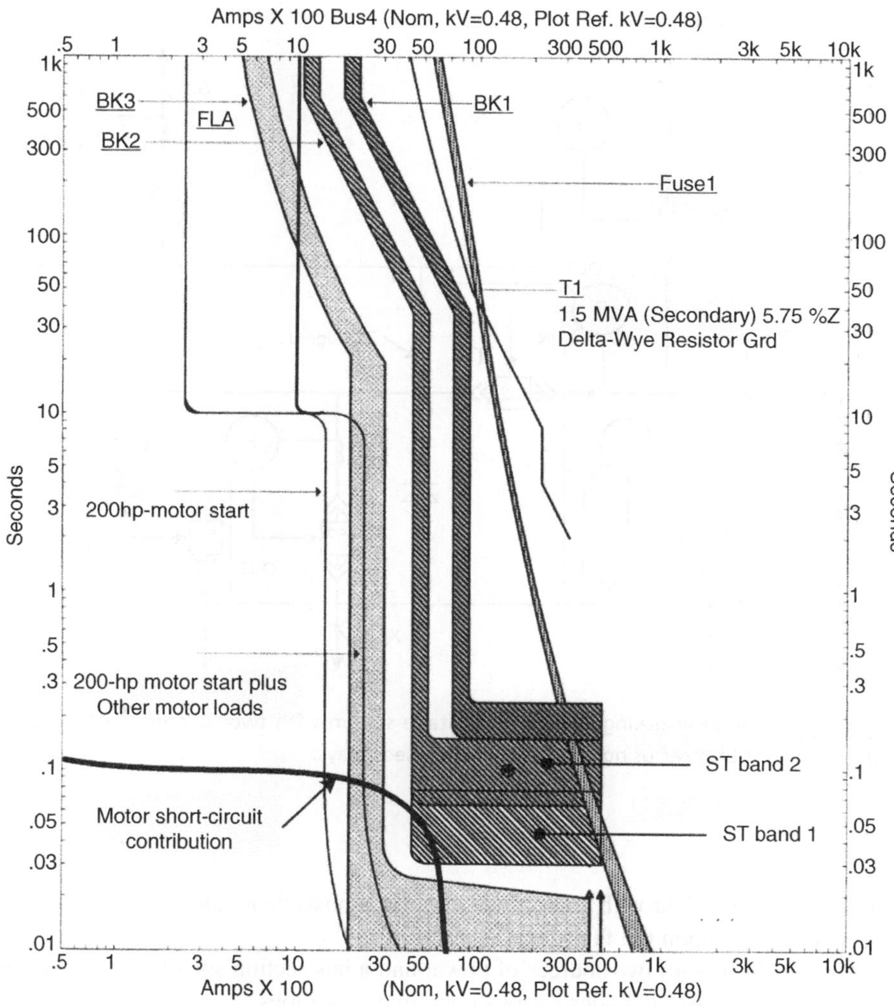


Figure 14.8. Time–current coordination of devices in Figure 14.7, showing motor contribution that impacts the settings.

## 14.2.2 Zone Interlocking in Medium Voltage Systems

Similar concepts can also be applied to zone interlocking in medium voltage systems. Figure 14.9 shows one such configuration. Here, the MMPR relays shown as *R* have the adaptability to receive a signal from the downstream relay, when the fault is downstream. Hard-wired or fiber optic communication channels can be used. The upstream relay can bring altogether new settings or block a particular setting for a downstream fault. This can simply be achieved through the programming capability of these relays

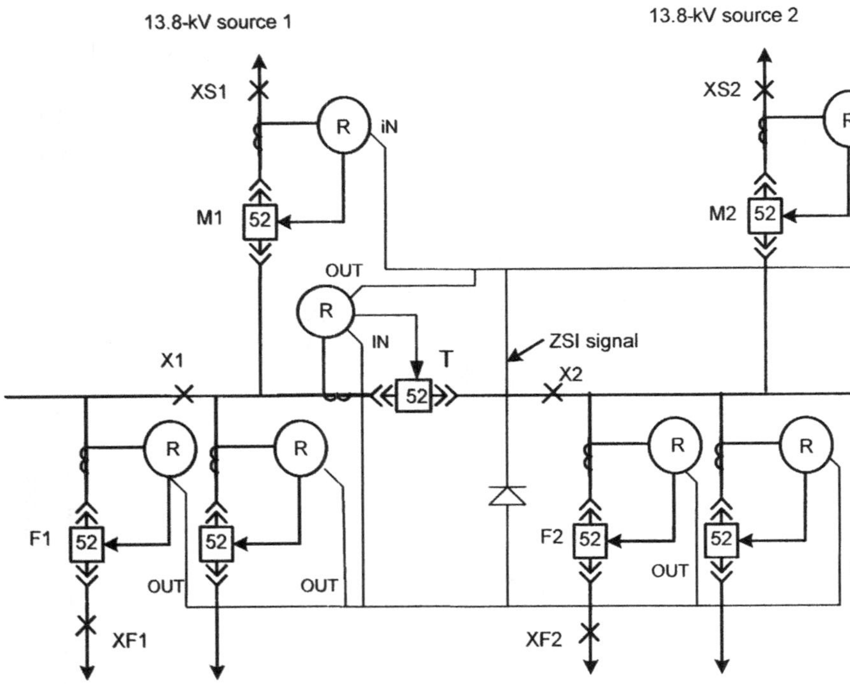


Figure 14.9. Zone interlocking in medium voltage system with two sources of power and bus section circuit breaker using microprocessor-based relays.

and Boolean algebra. Through appropriate logic, it is possible to alter settings or block settings depending upon the fault locations.

In Figure 14.9, with two sources of power and a bus section switch, the following operation must be ensured under various operating conditions.

### Bus Section Circuit Breaker T Open, Circuit Breakers M1 and M2 Closed

- Relay on feeder F1 detects the faults XF1. It sends a zone interlocking signal to relay on M1, and this will block tripping of M1 unless coordination protection times are exceeded. This does not impact the tie circuit breaker, as it is open.
- Similar operation occurs for a fault at XF2.
- For the fault at X1, the fault is ignored by feeder relay, and no zone interlocking signal is transmitted to M1. The relay at M1 will trip instantaneously, limiting the fault damage.
- Similar operation is achieved for a fault at X2.
- Faults at XS1 and XS2 must be cleared by upstream relays.

**Bus Section Circuit Breaker T Closed, Circuit Breakers M1 M2 Closed**

- A fault at XF1 is fed from both the sources. Relay at F1 provides restrain signal to relay at M1, bus section circuit breaker T, and also to M2. In addition, relay at bus tie breaker provides restraint signal to M1 and M2. The fault at F1 is therefore cleared by the settings on relay at feeder circuit breaker F1. Circuit breakers M1, M2, and T are blocked with the restraint signals.
- Similar operation results for a fault at XF2.
- Fault at X1 is fed from both the sources through M1, M2, and T. Now the bus section circuit breaker relay will provide restraint signals to M1 and M2. The bus section circuit breaker T relay will trip instantaneously, as no restraint signal will be sent from the feeder circuit breaker relays. After the bus section circuit breaker is opened instantaneously by the operation of its relay, the relay at M1 trips instantaneously, as there is no restraint signal from the tie circuit breaker.
- Similar operation occurs for a fault at X2.
- The fault at source XS1 is fed from the two sources, as the bus tie circuit breaker T is initially closed. Once the bus tie circuit breaker opens as described above, the fault will be ultimately cleared by the source 1 upstream relays. Thus, the power to bus 2 is not interrupted.
- Similar operation occurs for a fault at XS2.
- This illustrates selective tripping and only the faulty bus is isolated, maintaining operation on the healthy section of the bus.

**Bus Tie Circuit Breaker T Closed, M1 Closed and M2 Open**

- For a fault at XF1, there is only one source through M1 feeding the fault. The feeder relay still restrains relays at M1, M2, and T. The restraining signal does not have an effect on M2 as it is already open. The feeder relay F1 will clear the fault instantaneously.
- For a fault at XF2, feeder relay restraints the relays at M1, M2, and T. The bus tie relay also restrains M1 and M2. The restraint signals do not impact M2 as it is already open. Thus, feeder relay clears the fault XF2 instantaneously.
- For a fault at X1, no restraint signal is transmitted to M1, and it clears the fault instantaneously.
- For a fault at X2, relay at bus section circuit breaker provides restraint to M1 and M2. Again M2 is already open. The bus tie circuit breaker T will trip instantaneously.
- Faults at XS1 and XS2 must be cleared by the upstream source relays.

Similar operation occurs when tie circuit breaker is closed, M2 is closed, and M1 is open.

This description shows the complexity of applying ZSI to double-ended substations. The ZSI can be applied to phase as well as ground faults. Where the number of sources is more than two and there is more than one bus section or a switched tie circuit, differential protection is appropriate.

## 14.3 MICROPROCESSOR-BASED LOW VOLTAGE SWITCHGEAR

Low voltage power circuit breakers with redundant CPU (central processing unit)-based protection, monitoring, control, and diagnostic functions, which replace discrete devices and hard wiring, are available. These provide enhanced reliability and arc flash reduction. Synchronization between two CPUs is maintained through a hardwired sync clock, and the CPUs critical controls are powered through in-built redundant UPS systems. For arc fault reduction, bus differential algorithm with zone-based overcurrent protection is used. For work near the equipment, reduced energy let-through mode can be selected through a remotely located switch. This enables more sensitive protection settings to lower the arc flash incident energy.

### 14.3.1 Microprocessor-Based Switchgear Concept

Traditionally, the trip system of a circuit breaker react to the protective relays designed to actuate that circuit breaker, say an overcurrent relay will trip the circuit breaker when the current exceeds the set points. The microprocessor-based concept can control every circuit breaker in the assembly in the system and simulate exact conditions at that moment. For example, if it can be ascertained that a fault is downstream of a feeder circuit breaker, the source side circuit breaker can be blocked from tripping. The concept is to control each circuit breaker with full information about every type of current flow in the system.

Thus, the system knows the magnitude and location of a fault at every instant. The selectivity need not be sacrificed.

A microprocessor takes all pertinent system information and is able to process it simultaneously for the complete line up of switchgear. All the current and voltage information is synchronized and is available in one place; calculations can be made using simultaneous data samples or rms values calculated over time.

Consider that the differential current at a bus is required to be calculated in a simple radial system. Data samples of the currents can be taken flowing in all the circuit breakers at a certain time. Proper polarity assignment allows a differential calculation to be made with every data sample and current magnitude only. As an example, Figure 14.10 depicts data samples at an instant for the currents flowing in three circuit breakers, not to scale. The calculations can be corrected for expected processing and signal error. Using standard CTs sufficient resolution to detect fault currents lower than the rating of the bus being protected can be achieved. This calculation need be performed for a short duration of 1–1.5 cycles and confirms with high degree of certainty the magnitude and location of the fault current.

As a result, the same hardware providing normal overcurrent, ground fault, and ZSI can provide differential protection.

The differential protection range is limited to 10 times the rated current of any CT in the system to avoid saturation of the CTs (see Chapter 12). Say, for a 4000-A bus, a fault current >40 kA is required before the differential protection is made inactive, and for 800-A circuit breaker, the differential protection is inactive at 8.0 kA. These currents are high enough to be picked up in the short-time pick up range of the

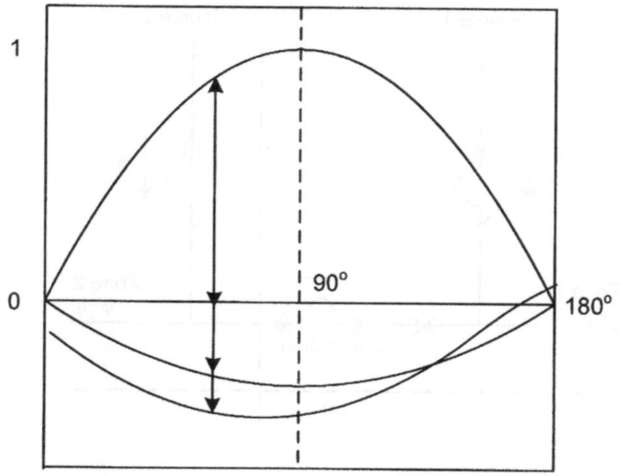


Figure 14.10. Instantaneous values of short-circuit wave forms from three source contributions ascertained through instantaneous sampling.

low-voltage trip programmers. The differential protection is complimented by short-time zone interlocking, providing a complete range of protection from low level fault currents to the maximum available fault currents.

### 14.3.2 Accounting for Motor Contributions

If the ZSI is made to recognize the direction of flow of the fault current, a satisfactory operation with respect to motor contributions and multiple sources of fault currents can be ensured. The traditional zone interlocking systems do not detect the fault current direction, but with the microprocessor-based technology, it can be achieved.

This is illustrated in Figure 14.11a, which shows two source circuit breakers and a bus tie circuit breaker. An algorithm is created similar to the partial differential relaying (see Chapter 8). To reiterate, partial differential calculations are based upon the main and tie circuits, ignoring the feeders. Define a current direction as positive for the currents flowing into the bus and negative for the currents flowing out of the bus. Figure 14.11a shows that the current flows for a fault on the left side, ignoring the feeder fault currents. Two zones left and right are shown with dotted lines. For the left zone, the current flows from the two mains and tie are considered positive. The right zone, the source 2 fault current is positive and that flowing out of the tie is negative. Therefore, the fault is identified to be on the left side of the bus.

Now consider the fault on the feeder. Applying the same convention, if the fault current flows toward the bus, say from a motor contribution, it is positive, and if it flows out from the bus to the feeder, it is negative. This identifies a feeder fault.

The rotating motor loads on one feeder will feed into the bus, and consider that a fault is located on the second feeder, Figure 14.11b. In this case, the flow of current into the bus is positive from one feeder, and flowing out of the bus is negative from the other feeder, which identifies the fault location.

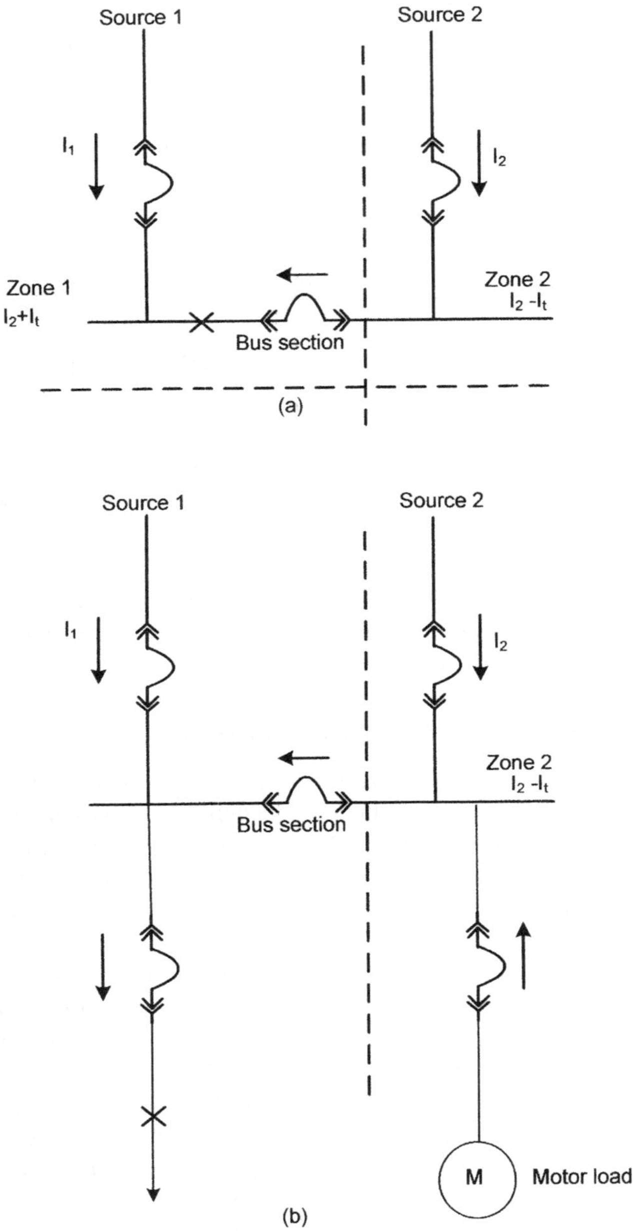


Figure 14.11. (a) Location of a fault through sensing the fault current direction; (b) with motor loads.

The following equation can be written:

$$I_r = \sum_{1-p} I_m - \sum_{1-q} (I_t D_t), \tag{14.1}$$

where $I_r$ is the residual current for the partial differential zone, $I_m$ is the current from the mains, $I_t$ is the current from the tie, $D_t$ is the reference direction unit vector for the tie, and $p$ and $q$ are the numbers of mains and ties feeding the partial differential zone. $I_r$ is compared with the largest current in the system multiplied by a factor of 1.5.

It may happen that one or more current readings exceed 10 times, then the short-time pickup has to selectively clear the fault. Here, it becomes necessary to find a large phasor, and all currents in the short-time pickup range are compared with this reference phasor to determine the relative direction.

For each phase, a reference current and angle is calculated. Then, to ascertain whether the other fault currents in the partial differential zone are in phase is computed by the equation:

$$\Delta\theta = < I_\alpha - < I_\beta, \tag{14.2}$$

where $<I_\beta$ is the reference and $<I_\alpha$ is the phase angle of the current being considered. For the opposing current, the expected angular difference is of the order of 180°. Feeder currents that are only in the short-time pickup range are considered. A feeder current of 10 times will be above the short-circuit contribution of the motor, and is indicative of a fault downstream of the feeder. The feeder current is also compared to a reference phasor to determine its location. This enables are all circuit breakers in an assembly to be fast tripped intelligently, reducing the arc flash hazard.

### 14.3.3 Faults on the Source Side

There remains one more consideration to complete this picture. A fault on the source itself will remain indeterminate, as neither of the sectionalized buses will show the fault location on them. A further algorithm to sense the directions of the fault current through the main circuit breakers is required. This is achieved by assessing the past data prior to the fault. Then a comparison can be made with prefault voltage to pre- and postfault current. Looking at the phase angle of the fault current with the phase angle of the prefault current can detect the reversal of current:

$$150° \le \alpha_I - \beta \le 250°, \tag{14.3}$$

where $\alpha_I$ is the angle of positive sequence current phasor, and $\beta$ is the prefault voltage positive sequence phasor. The angle for a forward current will be 0–90° for a purely resistive to purely reactive current. A reversal of current will add 180° and would range from 180 to 270°. The prefault positive-frequency voltage phasor will require compensation for any mismatch between sampling rate and frequency [2–4].

The CTs can be located on the source or load side of the circuit breakers. For differential protection, the placement of current sensors is critical. For incoming mains

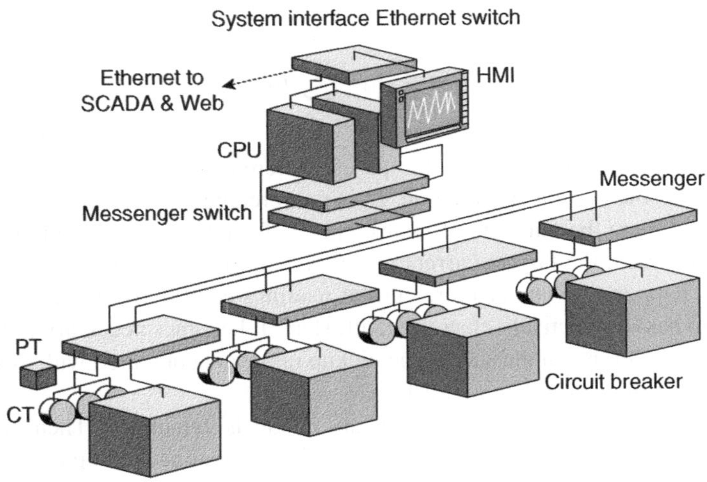


Figure 14.12. A block circuit diagram of microprocessor-based low voltage switchgear. *Source*: Courtesy of General Electric.

from the transformers, the sensors can be located on the source side of the circuit breaker cubical.

### 14.3.4 Arc Flash Hazard Reduction

With the algorithms as described above, the detection time is of the order of 25 ms while maintaining selectivity. The energy let-through from a fault is greatly reduced. Faster diagnostics and trouble shooting and redundancy of two synchronized microprocessor provide further reliability. The system architecture is shown in Figure 14.12.

## 14.4 LOW VOLTAGE MOTOR CONTROL CENTERS

Table 14.1 shows that 79.5% arc flash exposure occurs on the low voltage motor control centers (MCCs). The MCCs in the United States are UL listed to meet UL standard 485 [5]. The work of IEEE Working Group 1683—Standard for "MCCs Rated up to 1000 Volts with Requirements Intended to Reduce Injuries" is in progress. Reference [6] summarizes this work.

Process industries are predominantly dependent on low-voltage MCCs. Often a shutdown of an MCC can result in shutdown of a large section of the processes. Thus, the maintenance persons are faced with the task of troubleshooting or testing when the MCC is energized.

The doors in MCC compartments are interlocked to prevent access to energized parts with the disconnect device in the "ON" position. An interlock bypass mechanism is provided for the qualified personnel to access the live parts. Thus, maintenance work is carried out with the door open in the energized state. The MCC bucket assemblies with plug-in-stab contacts can be withdrawn with the door open and MCC energized.

Removal of the MCC bucket is usually followed with a bench test. Unless care is exercised in handling the drawn-out bucket, and while work is being performed, an undesirable force on the stab fingers can distort them. This can result in misalignment on insertion and diminishing the reliability.

### 14.4.1 Desirable MCC Design Features

Based upon the market research conducted by one manufacturer, the following are the desirable design features of the MCCs.

- The improper components should be possible to be rejected in the assembly or during maintenance.
- Testing of all voltage sources present in the unit should be provided on the dead-front cover.
- Ensure that all joints have been properly torqued at the factory assembly.
- Provide finger safe terminals throughout, preventing exposure to live parts.
- Provide voltage indication on the outside of the unit.
- All components must be properly rated.
- Provide insulation and isolation of bus with shutters.
- Provide facilities for remote racking.
- A manufacturer should provide correct installation and maintenance manuals.
- Fuses and high current-limiting circuit breakers shall be used.

### 14.4.2 Recent Design Improvements

Manufacturers are constantly modifying the designs for reliability and safety. Some of these innovations are:

***Heavily Latched Doors.*** Arc flash hazard is more hazardous to a worker with the MCC door open. Even in MCCs not specifically designed to meet arc resistant categories according to ANSI/IEEE standard [7] (see Chapter 13), much improvements in securing and latching the MCC doors are implemented.

***Main Current-Limiting Fuses or Circuit Breakers.*** The MCCs are, generally, provided only main incoming cable lugs and no main circuit breakers or fuses. For a fault on the horizontal or vertical buses, and pull-apart MCC stab-on terminals, the protection provided by an upstream circuit breaker in the low-voltage switchgear line-up is relied upon. Provision of a main current-limiting device can assure faster clearing times and reduction of HRC; however, it may be a problem to coordinate it with the downstream MCCBs and upstream circuit breaker.

Figure 14.13 illustrates a low voltage MCC, which serves a motor of 200-hp protected with a 400 A MCCB. First, consider that the MCC does not have any main fuses and is protected with an upstream 800-A LVPCB, BK2, provided with a trip

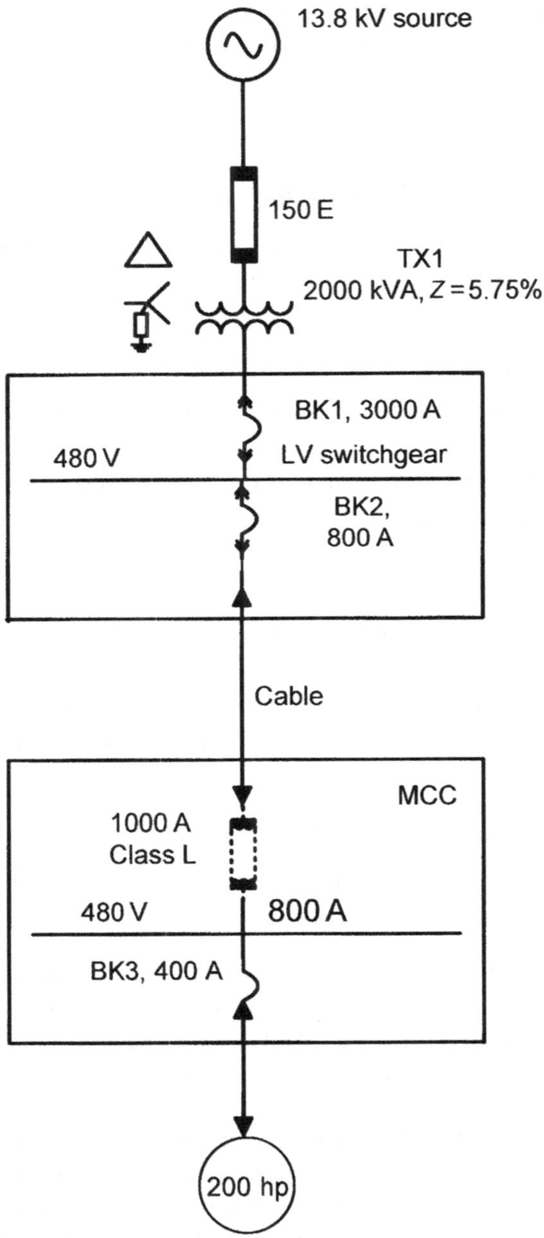


Figure 14.13. A low voltage distribution system with provision of 1000-A class L fuses on the incoming of a low voltage MCC.

programmer having LSI functions. The 2000 kVA transformer main secondary circuit breaker BK1 is rated 3000 A. The coordination is shown in Figure 14.14.

Now consider that a 1000-A class L current-limiting fuse is provided on the incoming of the MCC. The coordination is shown in Figure 14.15. If an 800-A fuse is used, there is a clash even with 400-A MCCB in the lower current range of around 1 kA. The fuse size is raised to 1000 A to avoid this lack of coordination, yet it does not coordinate in the instantaneous zone with 400 A MCCB. Also, it does not fully coordinate with circuit breaker BK2, and any manipulation of coordination settings cannot alter this situation. The long time delay and ST pickup of main circuit breaker BK1 are raised as compared in Figure 14.4, to coordinate with 1000-A fuse. Though this lack of coordination is acceptable, as in any case the power to the MCC will be interrupted whether the fuses operate or the upstream circuit breaker BK2 operates, but replacement of fuses and inventory requirements must be considered.

For a solidly grounded system, the line-to-ground current will be high, and the operation of a fuse in one faulted phase can give rise to single phasing and potential motor damage as discussed before.

Table 14.3 provides a comparison of the arc flash energy reduction on MCC with the provision of 1000-A main circuit fuses. The incident energy is reduced from 11 cal/cm$^2$ to 3.3 cal/cm$^2$.

***Door-Mounted Voltage Test Block.*** MCCs can be provided with an externally mounted voltage test block, accessible with door closed, to ascertain the presence of voltage before any tag-out procedure for maintenance.

Figure 14.16 shows a fixed voltage indicator mounted on dead front MCC door. One indicator for the main bus and another on load side of disconnects can be provided.

***Remote Racking Capability and Safety Shutters.*** Much like medium voltage switchgear, the MCCs' buckets can be drawn out remotely. It has unit stab assembly that allows this operation. Also, safety shutters will close the live spouts on draw-out. On the top of the unit is a designated space for mechanical racking mechanism through a motor operator. If a flash over did occur, the operator will be outside the arc flash boundary. Figures 14.17 and 14.18 show the remote racking devices. Figure 14.19 illustrates the bucket position indicators and internal shutter position.

***Improved Low Voltage Motor Protection.*** A new breed of protective devices is replacing the motor thermal overload elements. As an example, a motor protection device of a manufacturer can provide the following functions.

- enhanced thermal modeling
- RTD overtemperature
- current unbalance
- under power, phase reversal, PT fuse failure
- capability to accept zero-sequence CTs for sensitive ground fault protection

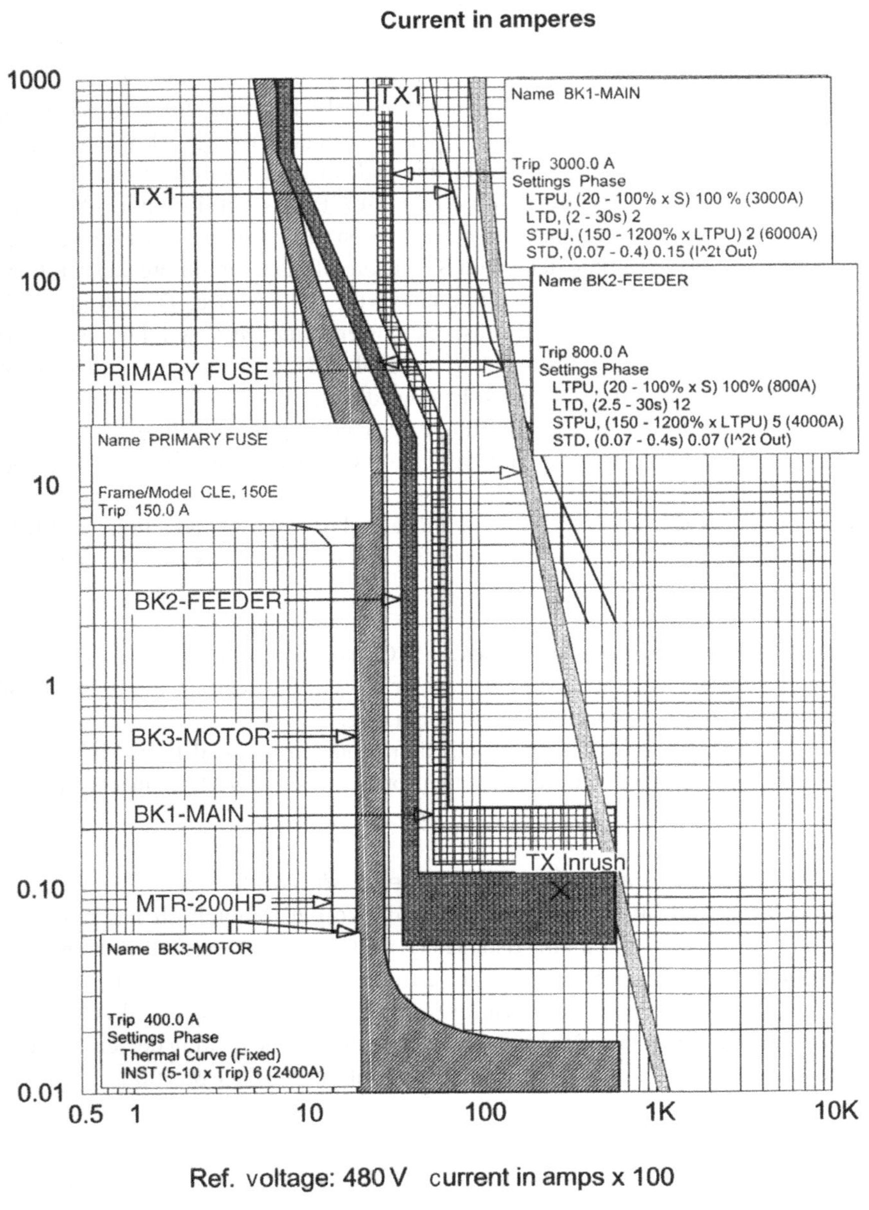


Figure 14.14. Time–current coordination of the devices in Figure 14.13, without 1000-A MCC fuses.

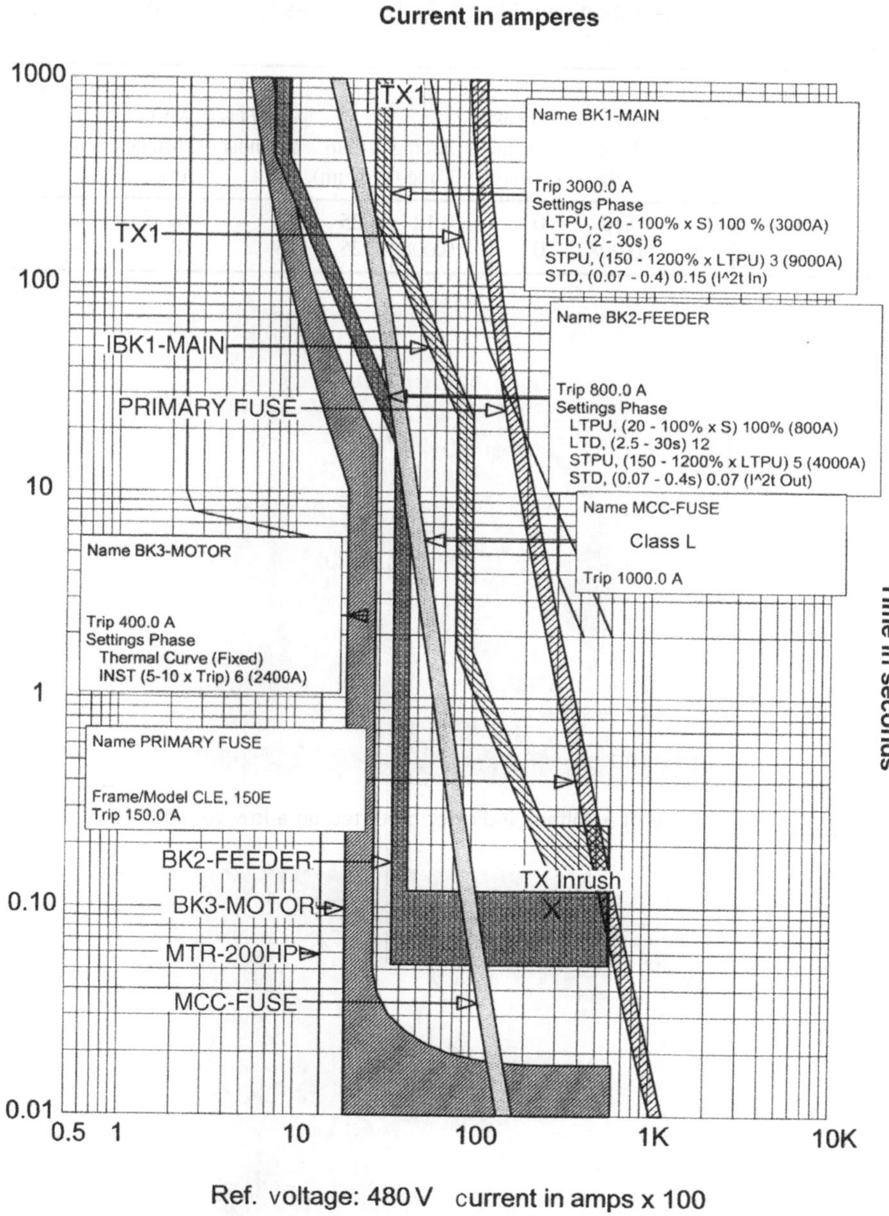


Figure 14.15. Time–current coordination of the devices in Figure 14.13, with 1000-A MCC fuses.

TABLE 14.3. Arc Flash Reduction with Main MCC Fuses, Class L, 1000 A, 480-V, LV MCC: 30"W × 90"H × 25"D, EC = HCB

| Protective Device | Bolted Fault current (kA) | Arcing Current (kA) | Tripping Delay (s) | Breaker Opening Time (s) | Arcing Time (s) | Gap (mm) | Arc Flash Boundary (in) | Working Distance (in) | Incident Energy (cal/cm$^2$) | PPE |
|---|---|---|---|---|---|---|---|---|---|---|
| BK1 | 35.02 | 24.90 | 0.12 | 0 | 0.12 | 25 | 59 | 18 | 13.2 | 2 |
| Fuse | 35.02 | 24.90 | 0.008 | 0 | 0.008 | 25 | 15 | 18 | 0.88 | 0 |

Figure 14.16. Picture of a voltage indicator mounted on a low voltage MCC door.

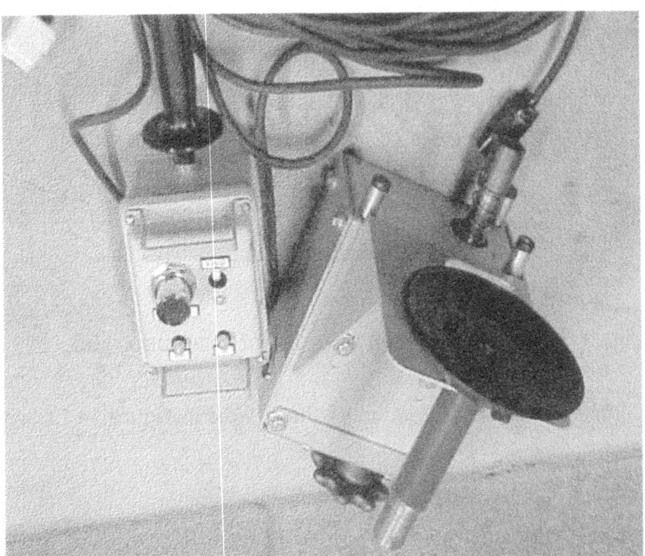

Figure 14.17. Portable remote racking system for low voltage MCC buckets.

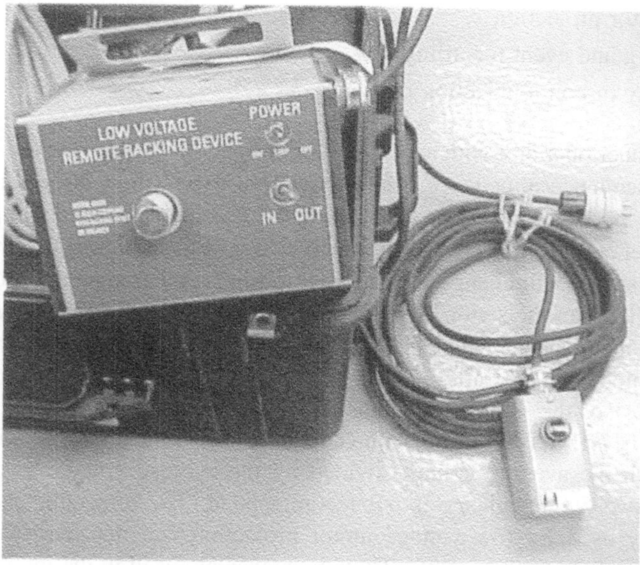


Figure 14.18. Another version of a portable remote racking system for low voltage MCC buckets.

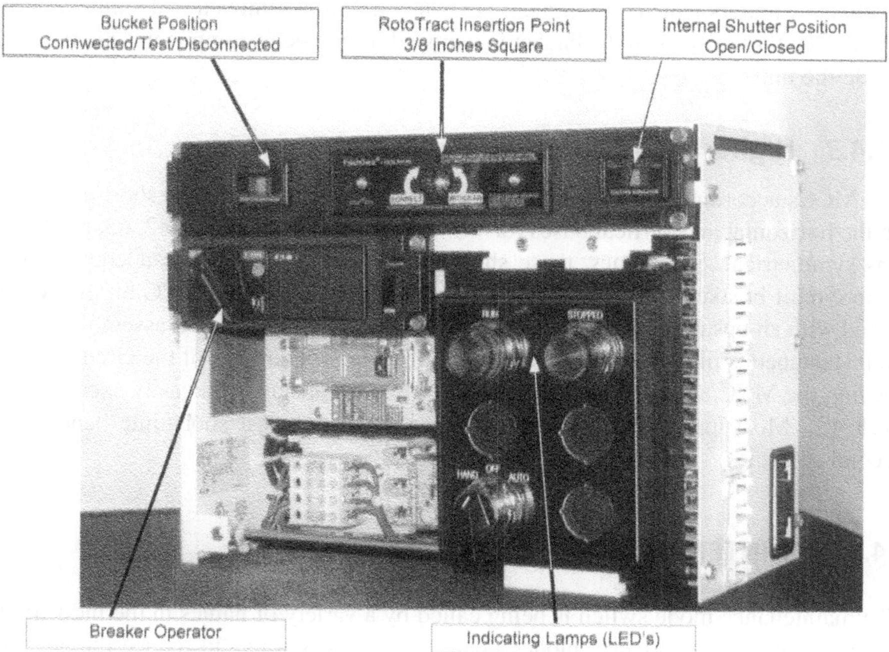


Figure 14.19. Picture of a low voltage MCC bucket, showing safety shutter position, bucket position, and remote racking insertion point. *Source*: Courtesy of Eaton Electric Corporation, reproduced with permission.

- thermistor protection
- metering and event recording
- a variety of communication protocols.

For an induction motor with a locked rotor current of six times the full-load current, the negative sequence impedance is approximately one-sixth of the positive sequence impedance. A 5% negative sequence component in the supply system will produce 30% negative sequence current in the motor, which gives rise to additional heating and losses. The rotor resistance will change with respect to high rotor frequency, and rotor losses are much higher than the stator losses. A 5% voltage unbalance may give rise to 38% negative sequence current with 50% increase in losses and 40°C higher temperature rise as compared with operation on a balanced voltage with zero negative sequence components. Also, the voltage unbalance is not equivalent to the negative sequence component. The NEMA definition of percentage voltage unbalance is maximum voltage deviation from the average voltage divided by the average voltage as a percentage. Operation above 5% unbalance is not recommended. The bimetallic thermal relays, conventionally used for low voltage motor protection, are not sensitive to negative sequence and ground fault currents. Though extensively used in the industry, these are a possible cause of motor winding failures.

***Insulation and Isolation.*** The horizontal bus and the labyrinth vertical bus are insulated. Shutters isolate the vertical bus when a unit is removed and shutters on the unit isolate the stabs when the bucket is removed. Finger-safe components are provided inside the units.

### 14.4.3 Higher Short-Circuit Withstand MCCs

The MCCs according to UL 485 [5] require a three-cycle three-phase short-circuit test on the horizontal and vertical buses for which these are rated, that is, 42, 65, or 100 kA rms symmetrical. Sometimes, these short-circuit ratings may not be adequate, as no main circuit breaker is normally provided on the MCC, and the MCC buses are protected with an upstream circuit breaker in the low-voltage switchgear assembly. In case an instantaneous function is not provided on the trip programmer of the circuit breaker feeding the MCC, the time duration of short-circuit withstand can easily exceed three-cycle [8]. Most low-voltage trip programmers have a lowest short-time band of 0.1 second (6 cycles).

## 14.5 MAINTENANCE MODE SWITCH

The maintenance mode switch is being called by a variety of names in the industry. Its operation in the maintenance mode activates alternate settings to reduce the arc flash hazard. These are being provided on low voltage trip programmers and can be remotely located. For example, if a maintenance mode switch is intended to lower the arc flash hazard on the MCC, it can be located on the MCC itself, though it will actuate to bring

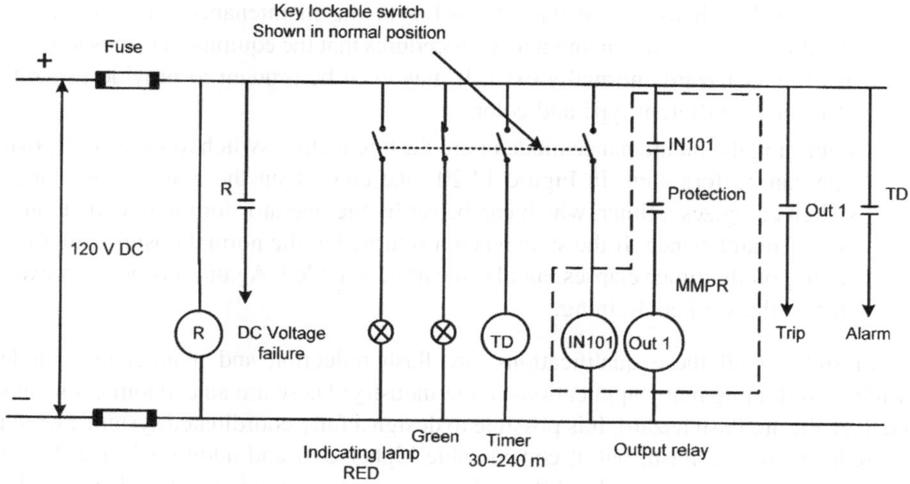


Figure 14.20. A control circuit diagram for connections of maintenance mode switch.

the new settings on low voltage trip programmer mounted on upstream circuit circuit breaker, in the low voltage switchgear.

A simple circuit of connecting a maintenance mode switch wired in a MMPR is shown in Figure 14.20. The operation of the switch actuates the input relay IN101, and an AND gate can be formed with the desired protection setting, with output directed to "Out 1." Also, a timer circuit is activated, and the desired time for the maintenance activity can be set on it. To prevent unauthorized operation, the switch can be key-interlocked. Also, it is possible to control a number of such switches simultaneously through a programmable controller (see Chapter 16).

In spite of all these safeguards, the following observations are of interest:

- The selectivity is compromised in the maintenance mode. If a fault does occur, though the operator is protected (which is more important), a larger process area will be shutdown. This may take increased time and manpower to restore the processes resulting in loss of revenue.
- Even when some coordination is sacrificed, it should be possible to carry out the normal running loads or start a motor. This requires that a careful coordination study is made and it brings about the issue of reliability—that the alternate settings are not grossly lower or inadequate to sustain even the load operations.
- A worker attending to energized equipment must be prepared for a trip and its resulting consequences, though the probability is small.
- Care has to be exercised that the proper maintenance mode switch is activated. This requires clear labeling and operating instructions. (A single Ethernet switch can be programmed to actuate a number of devices; see Chapter 16.)
- It is evident that two sets of arc flash calculations are required to be run, one with the maintenance mode switch and the other without it. A dilemma occurs on how the equipment should be labeled for arc flash hazard. If the equipment

is labeled with lower arc flash hazard with the maintenance mode switch, it should be a part of the maintenance procedures that the equipment is thus labeled, and higher hazards normally exist. It may even be prudent to provide two sets of labels of different type and color.

- Returning the maintenance mode or arc flash reduction switch to its normal position can be forgotten. In Figure 14.20, one contact on the maintenance mode switch energizes a timer, which can be set by the operator for the intended duration of maintenance. If the switch is not returned to the normal position after the setting on the timer elapses, an alarm can be sounded. Again, this adds an extra step in the workers training.

In spite of all these qualifications, arc flash reduction and maintenance mode switches are finding many applications in the industry. These are an economical means to reduce the arc flash hazard. It is possible to design a fully coordinated systems, which also reduces the arc flash, but at considerable higher cost and additional or different protective devices, for example, differential and unit protection systems. This will be especially difficult to implement in an existing distribution system. Many examples in this book have demonstrated the efficacy of reducing the arc flash hazard through maintenance mode switch.

## 14.6 INFRARED WINDOWS AND SIGHT GLASSES

Infrared scans are powerful diagnostic tools and are performed with the equipment energized. It can locate high resistance connections or "hot spots." The tests are performed with the equipment energized and when carrying normal load currents. It is a comparative type of test in which the thermographer performs a scan to look for areas that appear brighter (hotter) as compared with similar areas being scanned. Infrared camera cannot see through panel covers, and installations of infrared sight glasses are used. The construction of these windows and sight glasses does not put the thermographer or the equipment at risk. The two distinct products are:

- Viewing windows: These are designed to provide accidental impact and only provide visual inspection. These are not designed for withstanding the internal pressure due to arcing, and provide no arc flash protection.
- IR ports may have a mesh or grill and open hole that allow IR radiation to pass through; these provide no protection in the event of an arc flash.
- The infrared inspection window include locking security covers and require a trained authorized person to remove the covers before an IR scan. This is required to protect the optic material from day-to-day impacts. The cover provides additional protection during arc flash.

The mandatory testing requirements of IEEE C37.20.2 [9] are that the transparent covering of an observation opening must be secured in place so that it cannot be easily displaced in service and should meet the requirements of the following:

- The viewing panes should not shatter or crack or become dislodged when both sides of the viewing pane in turn are subjected to tests below:
- A force of 445 Newtons (100 lb-ft) should be exerted in a plane perpendicular to the surface in which the viewing pane is mounted. This force should be evenly distributed over an area of 16 $in^2$, and if the viewing pane is less than 16 $in^2$, the force should be evenly distributed over the viewing area and held for 1 minute.
- The viewing pane is subjected to an impact of 3.4 J (2.5 lb-ft) using a steel ball weighing approximately 0.54 kg and measuring approximately 50 mm in diameter.
- If viewing pane is intended to be exposed to insulating oil in a tank, it should be made of material that is resistance to insulating oil.
- Category A equipment should have lockable covers over viewing panes, if viewing panes are furnished. Category A enclosures are intended to provide protection against contact with enclosed equipment in ground level installations subject to deliberate unauthorized acts by members of unsupervised general public.

IEEE Standard [9] is not written for arc-resistant switchgear.

Figure 14.21 shows an IR sight glass, tested to be arc-resistant up to 40 kA arc flash current, with field of view equal to six times the cabinet depth (see Figure 14.22). The other tested particulars are the following:

Figure 14.21. Arc fault resistance infrared sight glass. *Source*: Courtesy of Fluke IR Windows.

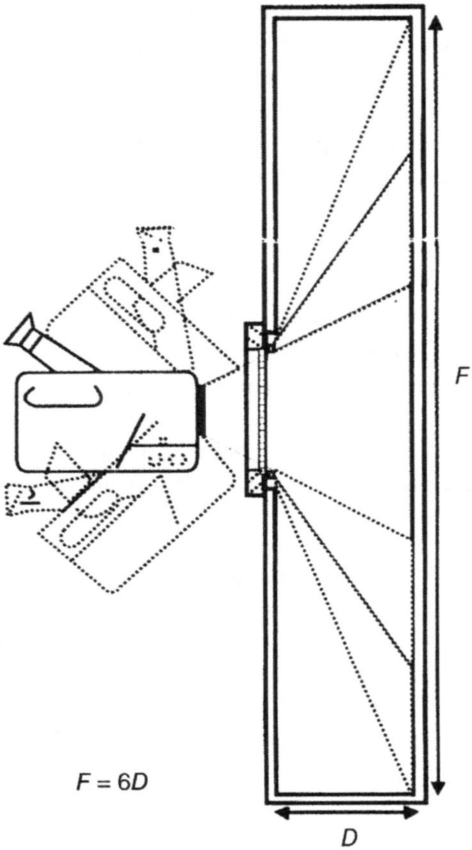


Figure 14.22. Field of view of an infrared window equal to six times the cabinet depth. *Source*: Courtesy of Fluke IR Windows.

- lens: C type, coated crystal, totally insoluble
- thickness: 0.081 in;
- pressure withstand: up to 90 psi
- melting temperature: 1350°C
- IR transmission: up to 96%
- wavelength: 0.13–10.0 μm
- viewing window size: 1.7 in, 2.7 in, or 3.5 in.

Figure 14.23 is another picture of a viewing window for visual inspection and also an IR window combined in one. There is no cover on the IR window. It meets the requirements of IEEE standard [9] testing but is not arc resistant. These are available in sizes ranging from 8.6 in × 6.3 in, 12.0 in × 7.7 in, and 24.0 in × 8.6 in (overall dimensions). The website [10] provides further details.

Figure 14.23. A combination of viewing window and infrared window.

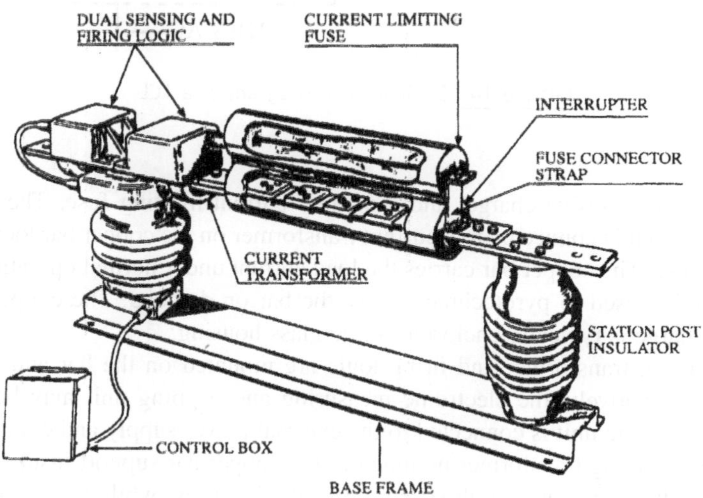


Figure 14.24. Assembly of a TCL (also called FCL).

## 14.7 FAULT CURRENT LIMITERS

A fault current limiter (FCL) or triggered current limiter (TCL) was developed under EPRI project RP1142, though a similar device has existed in Europe for a period of more than 40 years [11]. A TCL limits the fault current in time and magnitude, operating in 1/4 to 1/2 cycles and offering no resistance to the flow of load current.

Figure 14.24 shows the basic assembly of a TCL, and Figure 14.25 is a schematic diagram of its operation. The main components are the firing and sensing logic, a large

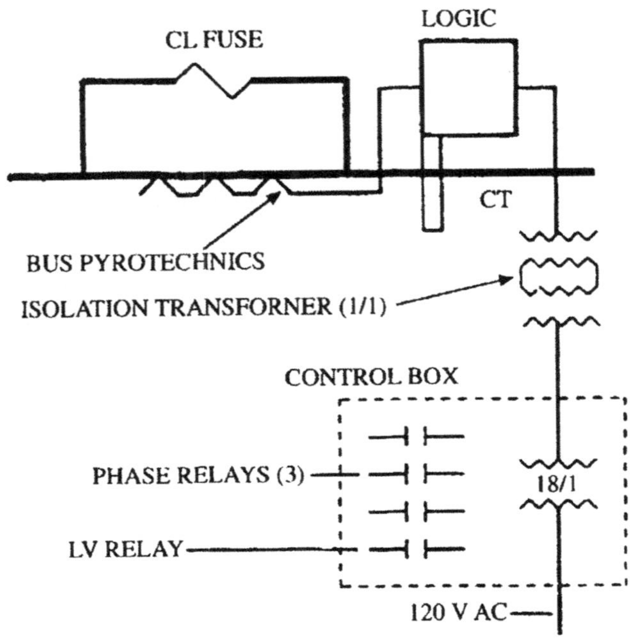


Figure 14.25. Schematic diagram of a TCL.

copper bar, an explosive charge, and a parallel current-limiting fuse. The firing and sensing logic unit is connected to a current transformer on the copper bar located ahead of parallel fuse. The copper bar carries the load current under normal operation. Explosive material is used to pyrotechnically cut the bar on demand. The copper bar with explosive charge is typically enclosed in fiberglass housing.

The current transformer and firing logic are mounted on the bar and are at line potential. Alternatively, the electronic measuring and tripping unit may be remotely mounted. The logic unit is connected to the external power supply and control the unit through the isolating transformer mounted in the copper bar support insulator.

The conductor is cut at high speed, typically 5 mm/μs, while the gas pressure is used to fold the bridging member. The gaps formed in the conductor are not sufficient to interrupt the fault current, and the arcs formed in these gaps generate an arc voltage, which transfers the short-circuit current into a parallel small cross-section fusible element. The current-limiting fuse operates in the conventional manner, providing current-limiting action. The total interrupting time is 1/4 to 1/2 cycles, before the prospective fault reaches its first peak. The proper operation of TCL imposes a number of constraints on fusible element. It must not operate on the fractional current that flows neither through it nor by the inrush and short-circuit currents.

Two types of sensing modes are available, *di*/*dt* sensing and threshold sensing; *di*/*dt* sensing determines the rate at which the current is increasing, while threshold sensing simply responds to the magnitude of the currents; timing diagram is shown in Figure

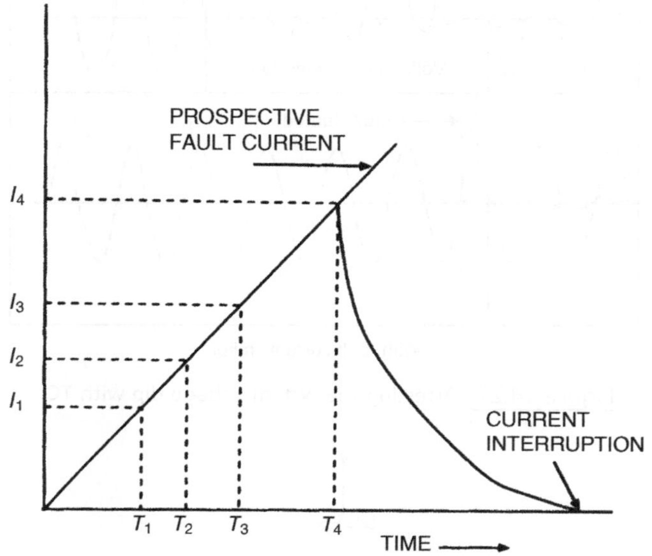


Figure 14.26. The timing diagram of a TCL with magnitude sensing.

TABLE 14.4. Comparisons of $I^2t$ of Various Current Interrupting Devices for 40-kA Fault Current

| Device | $I^2t$ |
|---|---|
| Five-cycle breaker with two-cycle relaying time (seven cycles, 60 Hz) | $(1.87) \times 10^8 A^2s$ |
| Expulsion fuse operating in 1 cycle | $(2.67) \times 10^7 A^2s$ |
| Current limiting fuse | $(10) \times 10^5 A^2s$ |
| TCL | $(1.0) \times 10^5 A^2s$ |

14.26. $I_1$ is the current sensing level equal to the trigger level, $T_1$ is the sensing time delay; $I_2$ is the current at trip initiation after a delay, $T_2 - T_1$ is the trip initiation delay; $I_3$ is the current at commutation to parallel fuse; $T_3 - T_2$ is the commutation time delay to fuse in firing the pyrotechnics; $T_4 - T_3$ is the fuse melting time; $I_4$ is the peak let-through current.

Due to a fast fault clearance, the energy let-though is much reduced. Table 14.4 depicts the energy let-through for a fault current of 40 kA, with various interrupting devices. Figure 14.27 illustrates that the postfault voltage recovers quickly and shut-downs due to fault voltage dips can even be prevented. The arc flash hazard will be much reduced due to this fast operation.

An application in a bypass reactor scheme is shown in Figure 14.28a,b. Consider that a reactor is used to limit the short-circuit current on the 13.8-kV bus so that the existing switchgear can be retained in service. However, the reactor has an adverse impact on the load flow and the voltage regulation. Then, normally, the reactor can be

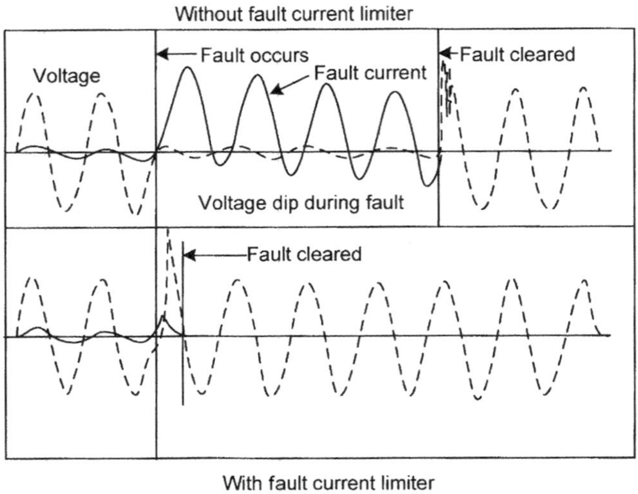


Figure 14.27. Arresting the system voltage dip with TCL.

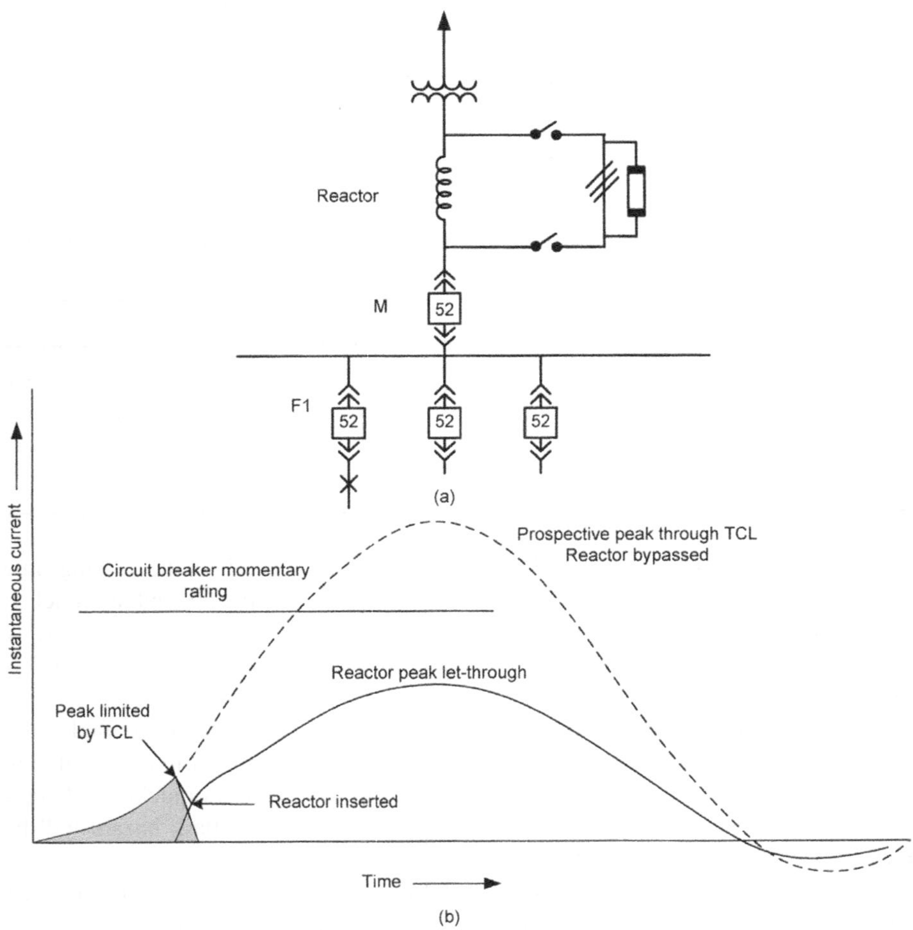


Figure 14.28. (a) TCL used in a reactor bypass scheme; (b) time–current profile, peak limited by TCL, circuit breaker momentary rating, peak through reactor, and also prospective peak current.

bypassed by the TCL, and the load flow situation improved. We accept much higher short-circuit duties on the circuit breakers in this operation. It is opined that the TCL operates so fast that lower rated circuit breakers can be retained in service. Now consider a fault downstream; the TCL can operate to introduce the reactor, which will limit the fault current to acceptable level within the interrupting rating of the switchgear. This situation is shown in Figure 14.28b.

Practically, the following considerations apply:

- It is not always correct that the existing circuit breakers of lower rating will not be subjected to higher first cycle or momentary ratings.
- The fault current through TCL, feeder circuit breaker F1, and main circuit breaker M will be approximately identical, if the fault occurs close to the load side terminals of the feeder circuit breaker. The instantaneous overcurrent protection should not pick up until the reactor is introduced.
- The replacement costs of blown-out TCLs is high.
- The control supply systems for TCL must be duplicated with auto-changeover to ensure reliability.

For further discussions on the limitations of FCLs (TCLs), see Reference [12]. A recent discussion is in CIGRE, "Impact of Fault Current Limiters on Existing Protection Schemes" [13]. From arc flash considerations alone, unit protection systems, like differential relays and AFDs, are better choices. These operate almost equally fast but within their protected zones. As far as other advantages claimed for TCLs of retaining underrated switchgear in service, a caution and complete study is required.

## 14.8 PARTIAL DISCHARGE MEASUREMENTS

The techniques of partial discharge measurements have come a long way and are an effective method of evaluation of electrical insulation and avoidance of a forced shutdown. Some industry statistics indicate that 80% of all failures of plant and equipment are random, and only 20% are age related. These suggest that 80% of prospective failures are not detected by common maintenance practices.

Partial discharge measurement is an accepted indicator to the failure of insulation systems, and can be applied to equipment rated 2.4 kV, though better value is obtained at voltages levels of 4.0 kV and higher.

A partial discharge is an electrical discharge that only partially bridges the insulation between conductors (see Figure 1.1). Sometimes, the term corona is used with this connotation, but this use is imprecise and is being discontinued. The apparent charge $q$ of a partial discharge is defined as that charge which, if injected between the terminals of a test object, will momentarily change the voltage between its terminals by the same amount as the partial discharge. The partial discharge inception voltage is the lowest voltage at which partial discharge levels are observed under specified conditions, when the voltage applied to the test object is gradually increased from a lower value. The partial discharge extinction voltage is defined as the voltage at which the partial discharge exceeding a specified level cease under qualified conditions when the voltage is decreased. See Reference [14] for further terms related to partial discharge testing.

It is recognized that partial discharge activity is unstable. Some factors impacting PD are:

- Voltage, load, temperature, humidity, vibration, and pressure
- Background noise.
- Variations of the PD activity in healthy insulations in machines of the same electrical specifications manufactured by various manufacturers
- Variation in the machines of the same rating of a certain manufacturer over the course of years. When comparing PD test results on different machines, it is necessary that test instrument bandwidth and noise separation techniques are identical, that the type of sensors are identical, and also that the operating voltages of the machines are identical.
- Different test methods can give different results, and absolute limits are difficult to establish.
- PD measurements cannot be expected to detect all the problems in electrical insulation.

It is therefore becomes difficult to establish norms of PD activity, though many such attempts and statistical data exists. A manufacturer's test lab has its own specifications and acceptance levels that will vary with manufacturer to manufacturer. The standards dealing with online measurements of PD avoid stating the specific levels or quantities to determine acceptable limits.

Table 14.5 shows statistical data for air-cooled stators of generators [15].

*Thus, the stress is on trending*. PD measurement carried over a long period will give tangible results provided it is ensured that all dynamics and test conditions remain identical for each test interval. This may not be an easy task.

## 14.8.1 Online versus Offline Measurements

In general, offline measurements are more global in their coverage, yet do not represent the actual operating conditions.

TABLE 14.5. Distribution of $Q_m$ for Air-Cooled Stators of Generators with 80 pF Sensors on the Terminals

| Operating Volts | 2–4 kV | 6–8 kV | 10–12 kV | 13–15 kV | >16 kV |
|---|---|---|---|---|---|
| 25% | 7 mV | 17 mV | 35 mV | 44 mV | 37 mV |
| 50% | 27 | 42 | 88 | 123 | 69 |
| 75% | 100 | 116 | 214 | 246 | 195 |
| 90% | 242 | 247 | 454 | 508 | 615 |

*Source*: Data from Reference [15].

The offline measurements are applied at longer intervals, say, 6–12 months. Many problems can manifest in shorter time span and escape detection.

If a higher discharge activity is detected, then the questions arise: when the problem started, how fast it is degrading and will continue to degrade, and whether immediate attention is required?

On the other hand, continuous monitoring does provide a better trending and eliminates some of the variables discussed above, yet it may not be adequate. For example, partial discharge measurements on motors and generators may apply to a small portion of the windings near the sensors. Some advantages are that no outage is required to conduct the tests, which are conducted under real operating conditions. Dynamics of temperature, humidity, and load current can be correlated to partial discharge activity. Also, an opportunity for remote diagnostic is provided.

PD measurements can be applied to cables, switchgear, rotating machines, transformers, bus bars, and so on. Some IEEE Standards are at References [16–18]. These standards contain extensive bibliography for further reading.

### 14.8.2 Test Methods

As described in Reference [19] for rotating machines, for form-wound stators, the following test methods could be applied:

- online PD measurement or EMI (electromagnetic interference) for PD during normal service
- offline PD test to quantify the PD activity
- transient voltage or corona probe or EMI test to locate PD
- ultrasonic probe test
- blackout or ultraviolet imaging.

Advanced electrical interference technology is required to suppress electrical noise to prevent false indications.

Partial discharge occurs in an air-filled void, and the flow of electrons through the void multiplies. This results in a current pulse of short duration, typically a few nanoseconds. A Fourier transform of the pulse generated shows frequencies up to several hundred MHz.

The test equipment is coupled to equipment through a high voltage capacitor (also Stator Slot Couplers SRC), which passes high frequency PD currents, offering high impedance to power frequency. In metal-clad switchgear, PD sensors can be installed that use the natural capacitance between the CT and main bus, and couples the high frequency signal to the secondary of the CT windings. The PD current that passes through the capacitor creates a voltage that can be displayed on a CRT or spectrum analyzer. Modern detectors are sensitive to many MHz. Every PD will create its own pulse, and the key measurement is the peak magnitude of the pulse $Q_m$. Also, pulse count is equally important, may be sometimes more important than $Q_m$ alone.

The major quantities to be measured are $Q_m$, partial discharge current, PD power loss, pulse discharge repetition rate, and some more. The PD intensity takes into account both the magnitude and the pulse count.

Partial discharge on rotating machines [18] suggest one additional parameter, that is, normalized quantity number (NQN). Mathematically, NQN is defined as:

$$\mathrm{NQN} = \frac{\mathrm{FS}}{G \times N} \times \left[ \frac{\log_{10} P_1}{2} + \sum_{i=2}^{N-1} \log_{10} P_i + \frac{\log_{10} P_N}{2} \right], \tag{14.4}$$

where:

$P_i$ = number of pulses per second in magnitude window $i$
$N$ = number of magnitude windows
$G$ = Gain of partial discharge detector (arithmetic not decibels)
FS = maximum magnitude window in millivolts at unity gain.

Pulses with high repetition rate have less of a contribution to NQN, then the pulses with low repetition rate. Though normalized, the NQN varies with the sensitivity of the PD detector, and, thus, the width and the number of magnitude windows. Figure 14.29 shows a PD measurement, and illustrates that PDI does not depend much upon number of magnitude windows, and overall PD is the sum of positive and negative quantities. This is not true for NQN, which varies up to three times within the chosen range of magnitude windows (see Figure 14.29).

PD in different phases can be measured and compared. This may indicate a defect, when compared with similar results over a period of time.

In a rotating machine, if there is one dominant deterioration mechanism, the PD test can sometimes give the approximate location. The same applies to PD

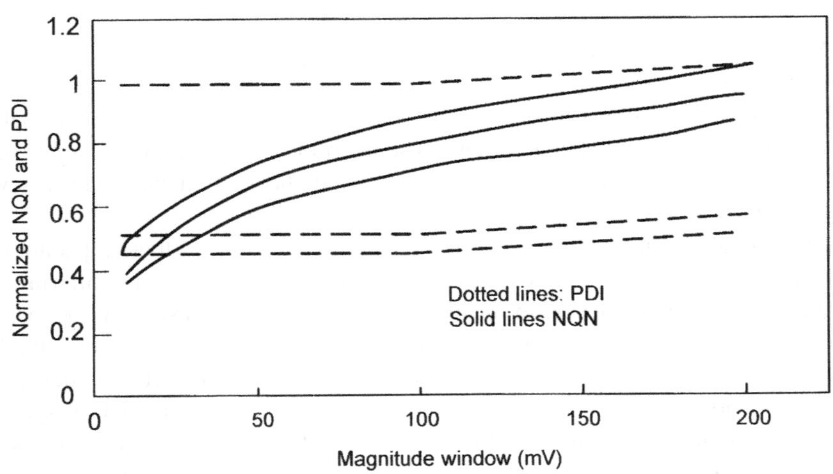


Figure 14.29. Normalized NQN and PDI (see text).

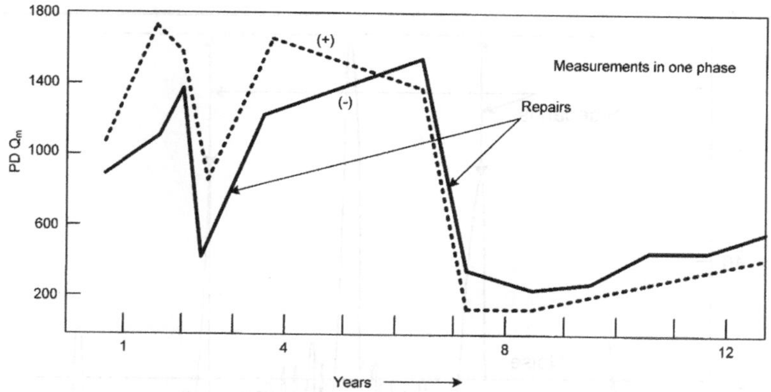


Figure 14.30. PD measurement data of a generator over the course of years.

measurements in cables. The faults have been localized in cable systems with PD measurements to a close resolution, especially with offline testing.

Figure 14.30 shows the PD measurements in a generator over a long period.

### 14.8.3 Current Signature Analysis: Rotating Machines

Current signature analysis (CSA) can detect breaks or cracks in the bars and rotor short-circuit rings of squirrel-cage induction motors. It is done online and close to full load operation. Due to magnetic coupling between the rotor and stator windings, the current fed into an induction motor is not solely dependent upon the power supply voltage and is influenced by the rotor condition. The stator current is measured through a current transformer and analyzed on a spectrum analyzer. Exceptional dynamic range and frequency resolution are needed in the measurements, and current magnitudes are measured in dB. If there are no breaks, there will be no sidebands. A current signature as shown in Figure 14.31 does indicate breaks. Currently, there are no IEEE or IEC guides on performing and interpreting CSA results.

### 14.8.4 Dissipation Factor Tip-Up

The term dielectric is synonymous with insulation. A dielectric or electrical insulation, when placed between two electrodes at different potentials, permits only a small negligible current to flow. The electrical resistivity may be of the order of $10^{20}$–$10^{6}\Omega$cm. Ideally, no current should flow through a perfect dielectric. Practically; an equivalent circuit can be drawn as shown in Figure 14.32 and accompanying phasor diagram. For a perfect dielectric, the phase angle δ, called the *loss angle or dissipation factor*, should be zero. As the insulation deteriorates, this angle increases, and measurements of the loss angle have been used to compare the deterioration over a period of time. We can write:

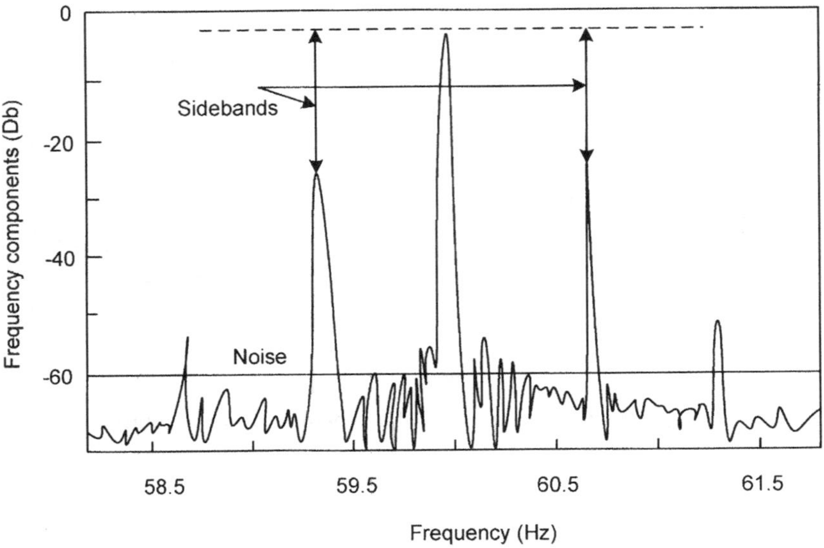


Figure 14.31. CSA (current signal analysis) of a rotating machine showing breaks in rotor bars.

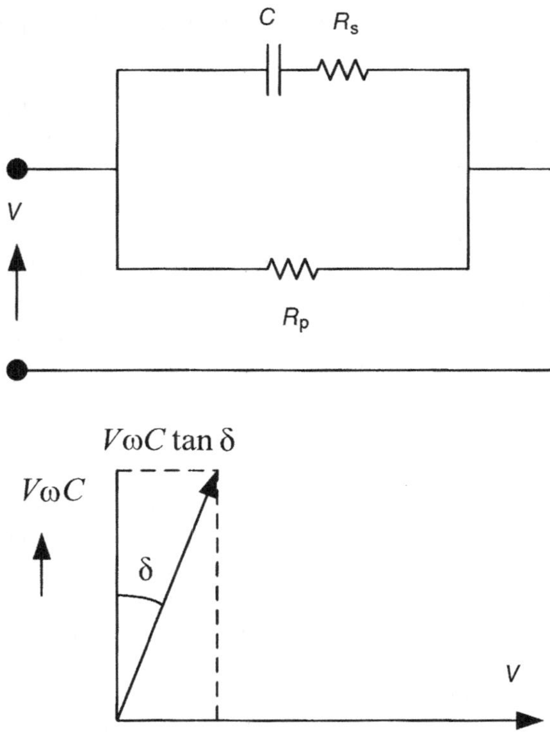


Figure 14.32. Model of a leaky dielectric and loss angle (see text).

$$\begin{aligned}\delta &= \tan^{-1} 2\pi f C_s R_s \quad R_p = 0 \\ &= \tan^{-1}(1/2\pi f C_p R_p) \quad R_s = 0.\end{aligned} \tag{14.5}$$

For insulation materials that have a low dissipation factor, the power factor is identical with it. When the tests are performed at both low and high voltage and increase in value over voltage (tip-up) is evaluated, it is a function of the amount of partial discharge activity in the insulation system.

This section on PD measurements is not exhaustive, and references provide further reading. For arc flash hazard reduction, a prior knowledge of the imminent failure is valuable information. A timely intervention and maintenance can prevent a failure, equipment damage, and unnecessary fault and shutdowns.

Consider two systems, one not so well designed but better maintained and operated and the other much better designed, but essential maintenance and diagnostics neglected. The system that is not so well designed but better maintained and operated can give lesser arc flash hazard (see Chapter 2).

## REVIEW QUESTIONS

1. Describe some limitations of zone interlocking systems. Can such systems be applied to ground fault protection also?
2. Consider a zone interlocking system consisting of two incoming sources connected to sectionalized bus, with bus section circuit breaker normally closed, as discussed in Section 14.2.2. A fault occurs on one of the sources, upstream of a main incoming circuit breaker on a source. Draw a sketch showing the flow of short-circuit currents, and how the selective tripping and zone interlocking is obtained.
3. Compare zone interlocking with bus differential and AFD relaying, especially for multisource distribution systems.
4. Comment on the new breed of low voltage switchgear incorporating differential and zone interlocking through microprocessor concept. At what level does the zone interlocking overtake the differential protection, as discussed in this chapter?
5. Comment upon providing a main fuse on the incoming of an MCC for arc flash reduction. What are the limitations with respect to system grounding and coordination?
6. Write five new innovations in the design of modern low voltage MCCs. Which of these is most important with respect to safety of the operating personnel, and how does it help to reduce arc flash hazard?
7. The statistical data show that 79.5% of arc flash hazard exposure occurs on low voltage MCCs. What could be the main reason for it? Should it be concluded that the MCCs are not properly manufactured?

**8.** The maintenance mode switches are being extensively used to lower the arc flash hazard during maintenance at the cost of risk of some lack of coordination. How it can be ensured that the operating personnel return the switch to "normal" mode after the maintenance work is over? Describe some limitations of these switches. Can an electrical distribution system be designed to lower arc flash hazard without maintenance mode switches?

**9.** Describe five major specifications that an IR arc resistant window should meet.

**10.** State two major limitations in the use of TCLs for arc flash reduction.

**11.** Compare online and offline measurements of PD.

**12.** What is a dissipation factor? What is its significance with respect to the state of an insulation system?

## REFERENCES


1. D.R. Doan, J.K. Slivka, and C.J. Bohrer, "A summary of arc flash hazard assessments and safety improvements," IEEE Trans. Industry Appl., vol. 45, no. 4, pp. 1210–1217, July/August 2009.
2. M. Valdes, T. Papallo, and I. Purkayastha, "The single processor concept for protection and control of circuit breakers in low-voltage switchgear," IEEE Trans. Industry Appl., vol. 40, pp. 932–940, July/August 2004.
3. M. Valdes, R. Narel, P. Hamer, B. Premeriani, and T. Papallo, "Zone based protection for low-voltage systems: zone selective interlocking, bus differential and the single processor concept," in IEEE Pulp and Paper Industry Conference Record, Seattle, Washington, pp. 164–173, June 2008.
4. General Electric, Entellisys™ 4.0 Publication DET-547, November 2007.
5. UL (Underwriters Laboratories) 845, Standard for Motor Control Centers, 2005.
6. D.B. Droucher and D. Mohla, "The final frontier-safety by design in LV motor control assemblies," in Conf. Record, IEEE Pulp and Paper Technical Conference, Seattle, WA, pp. 202–211, 2008.
7. IEEE, C37.20.7. *IEEE Guide for Testing Metal-Enclosed Switchgear Rated up to 38 kV for Internal Arcing Faults*, 2007.
8. R. Kumar, D. Reed, R. Morris, and S. Terry, "Higher withstand MCC for better selective coordination," in IEEE Pulp and Paper Industry Conference Record, San Antonio, Texas, pp. 185–189, June 2010.
9. IEEE Std., C37.20.2. *IEEE Standard for Metal-Clad Switchgear*, 1999.
10. Fluke IR Windows, http://www.fluke.com/fluke/usen/products/IR-windows.
11. Calor-Emag, "Is-Limiter," Calor-Emag AG, Mannheim, Germany, Leaflet No. DEACE 204596E, 1996.
12. J.C. Das, "Limitations of fault current limiters for expansion of electrical distribution systems," IEEE Trans. Industry Appli., vol. 33, no. 4, pp. 1073–1082, 1997.
13. CIGRE, "Impact of fault current limiters on existing protection schemes," CIGRE WG A3.16, Electra, 2008.

14. IEEE Std., C57.124. *IEEE Recommended Practice for Detection and Measurement of Partial Discharge in Dry-Type Transformers*, 1991.

15. V. Warren, "Partial Discharge Testing: A Progress Report," Iris Power, Publication IRMC 2010. See also: http//www.irispower.com.

16. IEEE Std., 1291. *IEEE Guide for Partial Discharge Measurements in Power Switchgear*, 1993.

17. IEEE, C57.113. *IEEE Guide for Partial Discharge Measurements in Liquid Filled Power Transformers and Shunt Reactors*, 1991.

18. IEEE Std., 1434. *IEEE Trail-Use Guide to the Measurement of Partial Discharges in Rotating Machinery*, 2000.

19. G.C. Stone and V. Warren, "Objective methods to interpret partial discharge data in rotating machines," in Conf. Record, IEEE Pulp and Paper Industry Conference, Jacksonville, FL, pp. 115–120, 2005.

# 15

# ARC FLASH HAZARD CALCULATIONS IN DC SYSTEMS

IEEE 1584 Guide does not address arc flash hazard calculations in the DC systems. Some references addressing short-circuits and arcing in DC systems are in References [1–9]. NFPA-70E-2000 includes a brief calculation procedure based on Reference [10]. However, the methodology advocated in NFPA is not adequate and accurate, as will be demonstrated in this chapter. Thus, the current literature is deficient for calculations of arc flash hazard in DC systems.

The steps involved in the calculations are much alike calculations in AC systems, yet there are marked differences, especially with respect to protection, nature of short-circuit currents in the DC systems, and the current interruption devices in DC systems, which operate fast to limit the energy let-through.

- Calculate the short-circuit currents in the DC systems. As we will see, these currents rise at a certain rate, depending upon the DC source.
- Calculate the arcing resistance and the arcing currents.
- Adjust the time–current characteristics of the protective device for rising DC currents.
- Calculate the arcing time.
- Calculate the incident energy.

*Arc Flash Hazard Analysis and Mitigation*, Second Edition. J.C. Das.

Published 2021 by John Wiley & Sons, Inc.

## 15.1 CALCULATIONS OF THE SHORT-CIRCUIT CURRENTS IN DC SYSTEMS

The calculations of short-circuit currents in DC systems is essential for the design and application of distribution and protective apparatuses used in these systems and for arc flash calculations. The DC systems include DC motor drives and controllers, battery power applications, emergency power supply systems for generating stations, data-processing facilities, and computer-based DC power systems and transit systems.

Maximum short-circuit currents should be considered for selecting the rating of the electrical equipment, like cables, buses, and their supports. The high-speed DC protective devices may interrupt the current before the maximum value is reached. It becomes necessary to consider the rate of rise of the current, along with interruption time, in order to determine the maximum current that will be actually obtained. Lower-speed DC protective devices may permit the maximum value to be reached before current interruption. For arc flash calculations, the operation of the protective devices has to be calculated for the arcing currents.

Though the simplified procedures for DC short-circuit current calculation are documented in some publications, these are not well established. There is no ANSI/IEEE standard for calculation of short-circuit currents in DC systems. A General Electric Company publication [1] and ANSI/IEEE standard C37.14 [2] provide some guidelines. IEC standard 61660-1 [11] published in 1997, is the only comprehensive document available on the subject. This standard addresses calculations of short-circuit currents in DC auxiliary installations in power plants and substations, and does not include calculations in other large DC power systems, such as electrical railway traction and transit systems.

The IEC standard describes quasi-steady-state methods for DC systems. The time variation of the characteristics of major sources of DC short-circuit current from initiation to steady-state are discussed, and appropriate estimation curves and procedures are outlined.

A dynamic simulation is an option, however, akin to short-circuit current calculations in AC systems; the simplified methods are easy to use and apply, though rigorously these should be verified by an actual simulation.

Though some literature and calculations examples of short-circuit currents in DC systems according to ANSI/IEEE publications are available, this chapter will use exclusively IEC equations and methodology because of the comprehensive nature of the IEC equations. *This standard uses number of curves and graphical data, all of which is not reproduced. A reader is advised to obtain a copy of this standard for full appreciation of IEC procedures; the standard can be obtained through source cited at Reference [12].*

## 15.2 SOURCES OF DC SHORT-CIRCUIT CURRENTS

Four types of DC sources can be considered:

- lead acid storage batteries
- DC motors

- converters in three-phase bridge configuration
- smoothing capacitors.

Figure 15.1 shows the typical short-circuit current time profiles of these sources, and Figure 15.2 shows the standard approximate function assumed in the IEC standard [11]. The following definitions apply:

$I_k$ = quasi steady-state short-circuit current
$i_p$ = peak short-circuit current
$T_k$ = short-circuit duration
$t_p$ = time to peak
$\tau_1$ = rise time constant
$\tau_2$ = decay time constant.

The function is described by:

$$i_1(t) = i_p \frac{1-e^{-t/\tau_1}}{1-e^{-t_p/\tau_1}} \tag{15.1}$$

$$i_2(t) = i_p[(1-\alpha)e^{-(t-t_p)\tau_2} + \alpha] \qquad (t \geq t_p) \tag{15.2}$$

$$\alpha = I_k / i_p. \tag{15.3}$$

The quasi steady-state current $I_k$ is conventionally assumed as the value at 1 second after the beginning of short-circuits. If no definite maximum is present, as shown in Figure 15.1a for the converter current, then the function is given by Equation (15.1) only.

## 15.3 IEC CALCULATION PROCEDURES

Figure 15.3 shows a hypothetical DC distribution system that has all the four sources of DC short-circuit currents, that is, a storage battery, a charger, a smoothing capacitor, and a DC motor. Two locations of short-circuit are shown: (1) at $F_1$, without a common branch, and (2) at $F_2$, through resistance and inductance, $R_y$ and $L_y$ of the common branch. The short-circuit current at $F_1$ is the summation of short-circuit currents of the four sources, as if these were acting alone through the series resistances and inductances.

For calculation of the short-circuit current at $F_2$, the short-circuit currents are calculated as for $F_1$, but adding $R_y$ and $L_y$ to the series circuit in each of the sources. Correction factors are introduced, and the different time functions are added to the time function of the total current.

Whether it is the maximum or minimum short-circuit current calculation, the loads are ignored (i.e., no shunt branches), and the fault impedance is considered to be zero. For the maximum short-circuit current, the following conditions are applicable:

- The resistance of joints (in bus bars and terminations) is ignored.
- The conductor resistance is referred to 20°C.

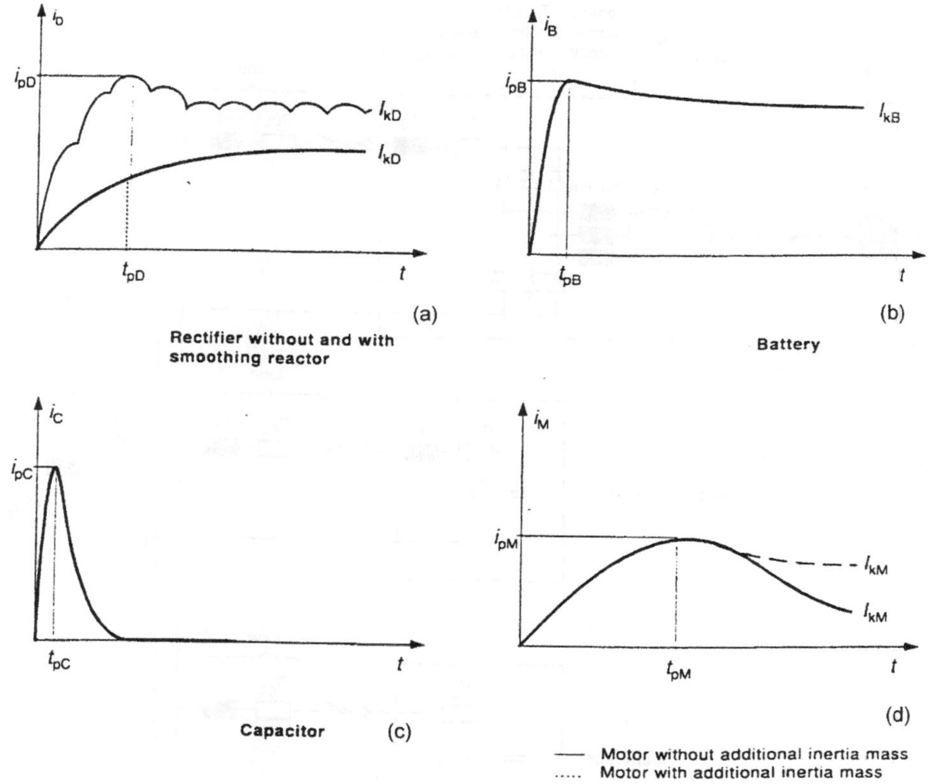


Figure 15.1. Short-circuit profile of various DC sources: (a) rectifier with and without smoothing reactor, (b) battery, (c) capacitor, and (d) DC motor with and without initial inertia mass. *Source*: Reference [11].

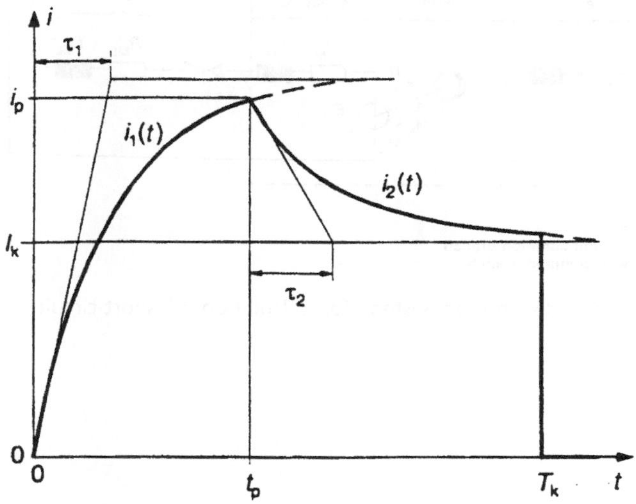


Figure 15.2. Standard approximation of short-circuit function. *Source*: Reference [11].

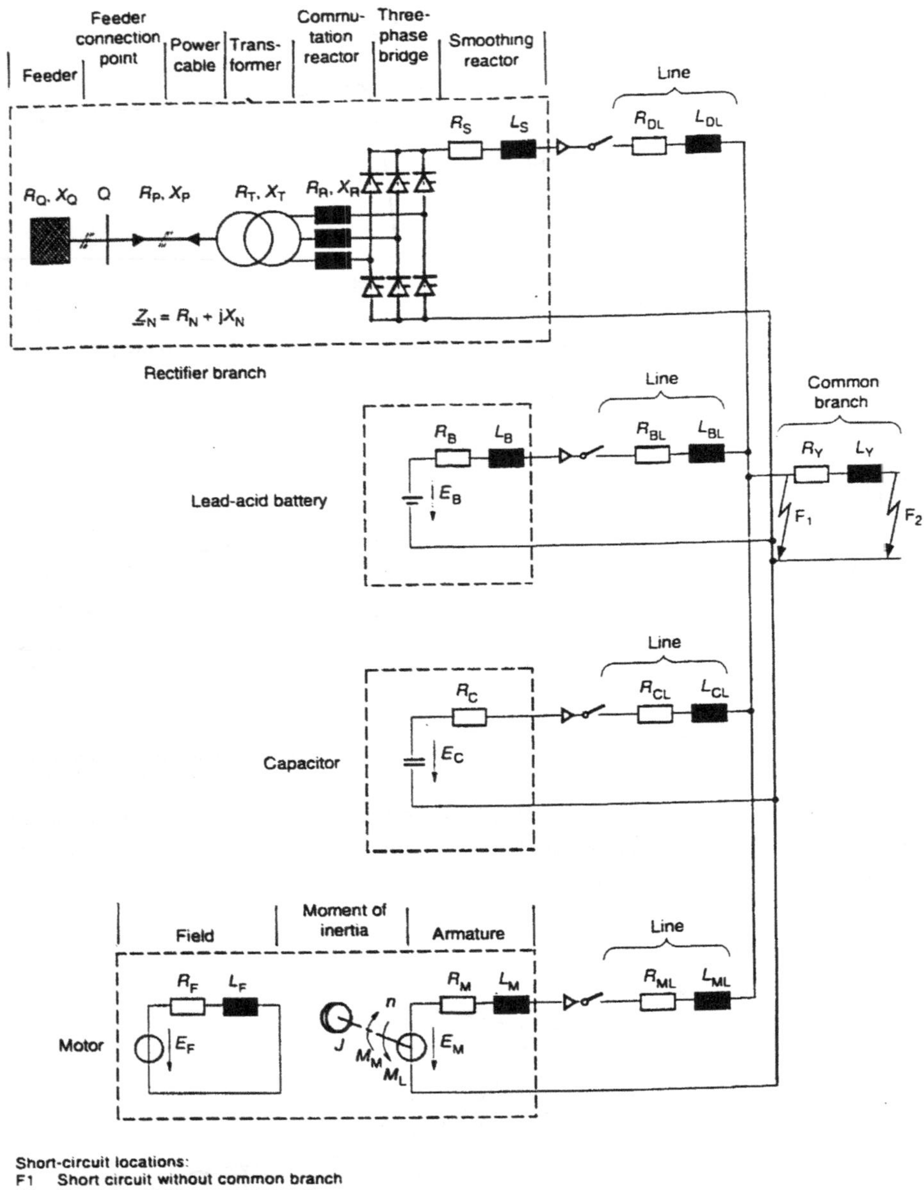


Figure 15.3. A DC distribution system for calculation of short-circuit currents. *Source:* IEC [11].

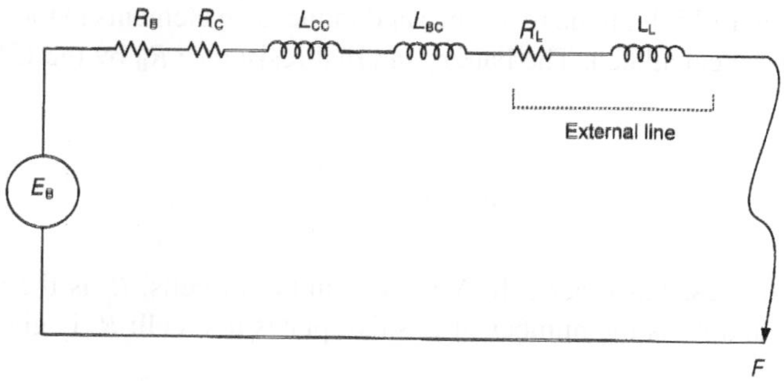


Figure 15.4. Equivalent circuit of short circuit of a battery through external resistance and inductance. *Source:* IEC [11].

- The controls for limiting the rectifier current are not effective.
- The diodes for the decoupling part are neglected.
- The battery is fully charged.
- The current limiting effects of fuses or other protective devices are not taken into account.

## 15.4 SHORT CIRCUIT OF A LEAD ACID BATTERY

The battery short-circuit model is shown in Figure 15.4; $R_B$ is the internal resistance of the battery, $E_B$ is the internal voltage, $R_C$ is the resistance of cell connectors, $L_{CC}$ is the inductance of the cell circuit in H, and $L_{BC}$ is the inductance of the battery cells considered as bus bars. The *internal* inductance of the cell itself is zero. The line resistance and inductance are $R_L$ and $L_L$, respectively. The equivalent circuit is that of a short-circuit of a DC source through equivalent resistance and inductance, that is,

$$iR + L\frac{di}{dt} = E_B. \tag{15.4}$$

The solution of this equation is

$$i = \frac{E_B}{R}(1 - e^{-(R/L)t}). \tag{15.5}$$

The maximum short-circuit current is then

$$I_{Bsc} = \frac{E_B}{R}, \tag{15.6}$$

and the initial maximum rate of rise of the current is given by $di/dt$ at $t = 0$, that is,

$$\frac{di_B}{dt} = \frac{E_B}{L}. \tag{15.7}$$

Referring to Figure 15.4, all the resistance and inductance elements in the battery circuit are required to be calculated. The battery internal resistance $R_B$ by the IEEE method is given by

$$R_B = R_{cell}N = \frac{R_p}{N_p}, \tag{15.8}$$

where $R_{cell}$ is the resistance/per cell, $N$ is the number of cells, $R_p$ is the resistance per positive plate, and $N_p$ is the number of positive plates in a cell; $R_p$ is given by

$$R_p = \frac{V_1 - V_2}{I_2 - I_1}\Omega / \text{positive plate}, \tag{15.9}$$

where $V_1$ is the cell voltage, and $I_1$ is the corresponding rated discharge current per plate. Similarly, $V_2$ is the cell voltage, and $I_2$ is the corresponding rated discharge current per plate at $V_2$.

The following equation for the internal resistance is from Reference [1]:

$$R_B = \frac{E_B}{100 \times I_{8hr}}\Omega, \tag{15.10}$$

where $I_{8hr}$ is the 8-hour ampère rating of the battery to 1.75 V per cell at 25°C. $R_B$ is normally available from manufacturers' data; it is not a constant quantity and depends on the charge state of the battery. A discharged battery will have much higher cell resistance.

To calculate the maximum short-circuit current or the peak current according to IEC, the battery cell resistance $R_B$ is multiplied by a factor of 0.9. Also, if the open-circuit voltage of the battery is unknown, then use $E_B = 1.05U_{nB}$, where $U_{nB} = 2.0$ V/cell for lead acid batteries. The peak current is given by

$$i_{pB} = \frac{E_B}{R_{BBr}}, \tag{15.11}$$

where $i_{pB}$ is the peak short-circuit current from the battery and $R_{BBr}$ is the total equivalent resistance in the battery circuit, with $R_B$ multiplied by a factor of 0.9. The time to peak and the rise time are read from curves in Figure 15.5, based on 1/δ, which is defined as follows:

$$\frac{1}{\delta} = \frac{2}{\dfrac{R_{BBr}}{L_{BBr}} + \dfrac{1}{T_B}}. \tag{15.12}$$

The time constant $T_B$ is specified as equal to 30 ms, and $L_{BBr}$ is the total equivalent inductance to the fault point in the battery circuit. The decay time constant $\tau_{2B}$ is considered to be 100 ms. The quasi steady-state short-circuit current is given by

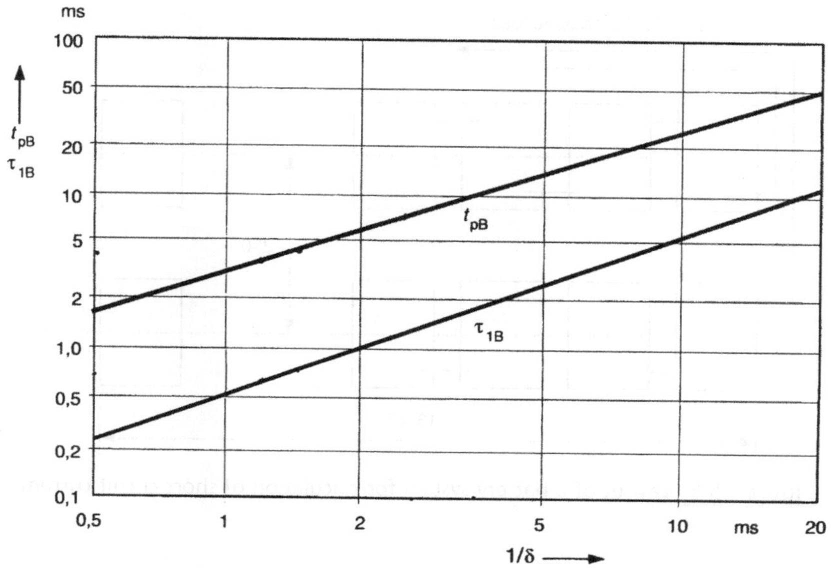


Figure 15.5. Time to peak $t_{pk}$ and rise time constant $\tau_{1B}$ for the short-circuit of a battery. *Source*: IEC [11].

$$I_{kB} = \frac{0.95E_B}{R_{BBr} + 0.1R_B}. \tag{15.13}$$

This expression considers that the battery voltage falls and the internal cell resistance increases after short-circuit. All equations from IEC are in MKS units.

**Example 15.1**

A 60-cell 120-V sealed, valve-regulated, lead acid battery has the following electrical and installation details.

Battery rating = 200 Ah (ampere hour), 8-hour rate of discharge to 1.75 V per cell. Each cell has the following dimensions: height = 7.9 in (= 200 mm), length = 10.7 in (= 272 mm), and width = 6.8 in (= 173 mm). The battery is rack mounted, 30 cells per row, and the configuration is shown in Figure 15.6. Cell interconnectors are 250 KCMIL, diameter = 0.575 in.

Calculate the short-circuit current at the battery terminals. If the battery is connected through a cable of approximately 100 ft length to a circuit breaker, cable resistance 5 mΩ, and inductance 14 μH, calculate the short-circuit current at breaker terminals.

The battery resistance according to Equation (15.10) and considering a cell voltage of 1.75 V per cell is

$$R_B = \frac{E_B}{100 \times I_{8hr}} = \frac{120}{100 \times 200} = 6\ \text{m}\Omega.$$

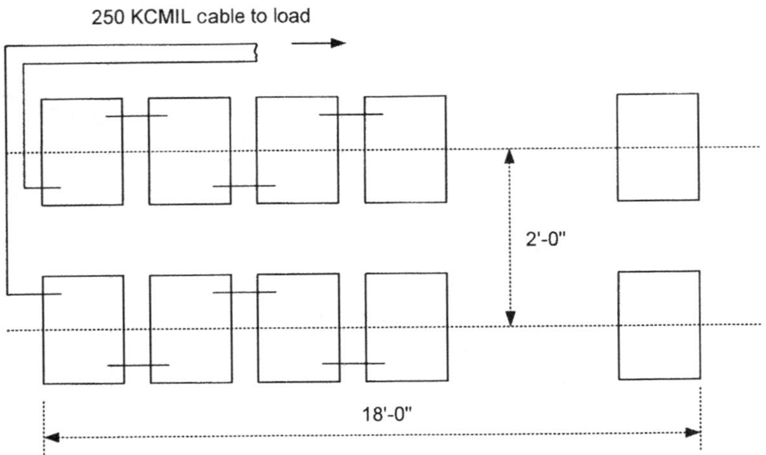


Figure 15.6. Layout of a battery system for calculation of short-circuit current.

The manufacturer supplies the following equation for calculating the battery resistance:

$$R_B = \frac{31 \times E_B}{I_{8hr}} \text{m}\Omega. \tag{15.14}$$

Substituting the values, this gives a battery resistance of 18.6 mΩ. There are three times difference in these values, and the manufacturer's data should be used.

From 15-6, battery connectors have a total length of 28 ft, size 250 KCMIL.

The inductance $L_c$ of the battery circuit is the sum of the inductances of the cell circuit $L_{CC}$ plus the inductance of the battery cells, $L_{CB}$. The inductance of two round conductors of radius $r$, spaced at a distance $d$, is given by the expression:

$$L = \frac{\mu_0}{\pi}\left(0.25 + \text{In}\frac{d}{r}\right), \tag{15.15}$$

where $\mu_0$ is the permeability in vacuum = $4\pi 10^{-7}$ H/m. From Figure 15.6, the distance $d$ = 24 in and $r$, the radius of 250 KCMIL conductor, is 0.2875 in. Substituting the values in Equation (15.12), the inductance is 1.87 μH/m for the *loop* length. Therefore, for an 18-ft loop length in 15-6, the inductance $L_{CC}$ = 10.25 μH.

The inductance of battery cells can be determined by treating each row of cells like a bus bar. Thus, the two rows of cells are equivalent to parallel bus bars at a spacing $d$ = 24 in, the height of the bus bar $h$ = height of the cell = 7.95 in, and the width of the bus bars $w$ = width of the cell = 6.8 in. The expression for inductance of the bus bars in this configuration is:

$$L = \frac{\mu_0}{\pi}\left(\frac{3}{2} + \text{In}\frac{d}{h+w}\right) \tag{15.16}$$

This gives inductance in H per meter loop length. Substituting the values, for an 18-ft loop length, inductance $L_{BC}$ = 4.36 μH. The total inductance is, therefore, 14.61 μH.

The total resistance in the battery circuit, without external cable, is 0.9 × 18.6 + 1.498 = 18.238 mΩ. The battery voltage of 120 V is multiplied by factor 1.05. Therefore, the peak short-circuit current is

$$i_{PB} = \frac{E_B}{R_{BBr}} = \frac{1.05 \times 120}{18.238 \times 10^{-3}} = 6908.6\,\text{A}.$$

1/δ is calculated from Equation (15.15):

$$\frac{1}{\delta} = \frac{2}{\dfrac{18.238 \times 10^{-3}}{14.61 \times 10^{-6}} + \dfrac{1}{30 \times 10^{-3}}} = 1.56\,\text{ms}.$$

From Figure 15.5, the time to peak = 4.3 ms, and the rise time constant is 0.75 ms. The quasi steady-state short-circuit current is

$$I_{kB} = \frac{0.95 \times 120 \times 10^3}{18.238 + 0.1(18.6)} = 5956\,\text{A}.$$

The calculations with external cable added are similarly carried out. The cable resistance is 5 mΩ and inductance is 14 μH. Therefore, $R_{BBr}$ = (0.9)(18.6) + 1.498 + 5 = 23.24 mΩ. This gives a peak current of 5422 A; 1/δ = 2.40 ms and time to peak is 5.4 ms. The rise time constant is 1.3 ms, and the quasi steady-state short-circuit current is 4796 A. The short-circuit current profile is plotted in Figure 15.7a.

## 15.5 SHORT CIRCUIT OF DC MOTORS AND GENERATORS

The resistance and inductance network of the short circuit of a DC machine with a separately excited field is shown in Figure 15.3. The equivalent resistance and inductance are

$$\begin{aligned} R_{MB} &= R_M + R_{ML} + R_y \\ L_{MB} &= L_M + L_{ML} + L_y, \end{aligned} \tag{15.17}$$

where $R_M$ and $L_M$ are the resistance and inductance, respectively, of the armature circuit, including the brushes, $R_{ML}$ and $L_{ML}$ are the resistance and inductance of the conductor in the motor circuit, and $R_y$ and $L_y$ are the resistance and inductance of the common branch, if present. The time constant of the armature circuit up to the short-circuit location, $\tau_M$, is given by

$$\tau_M = \frac{L_{MB}}{R_{MB}}. \tag{15.18}$$

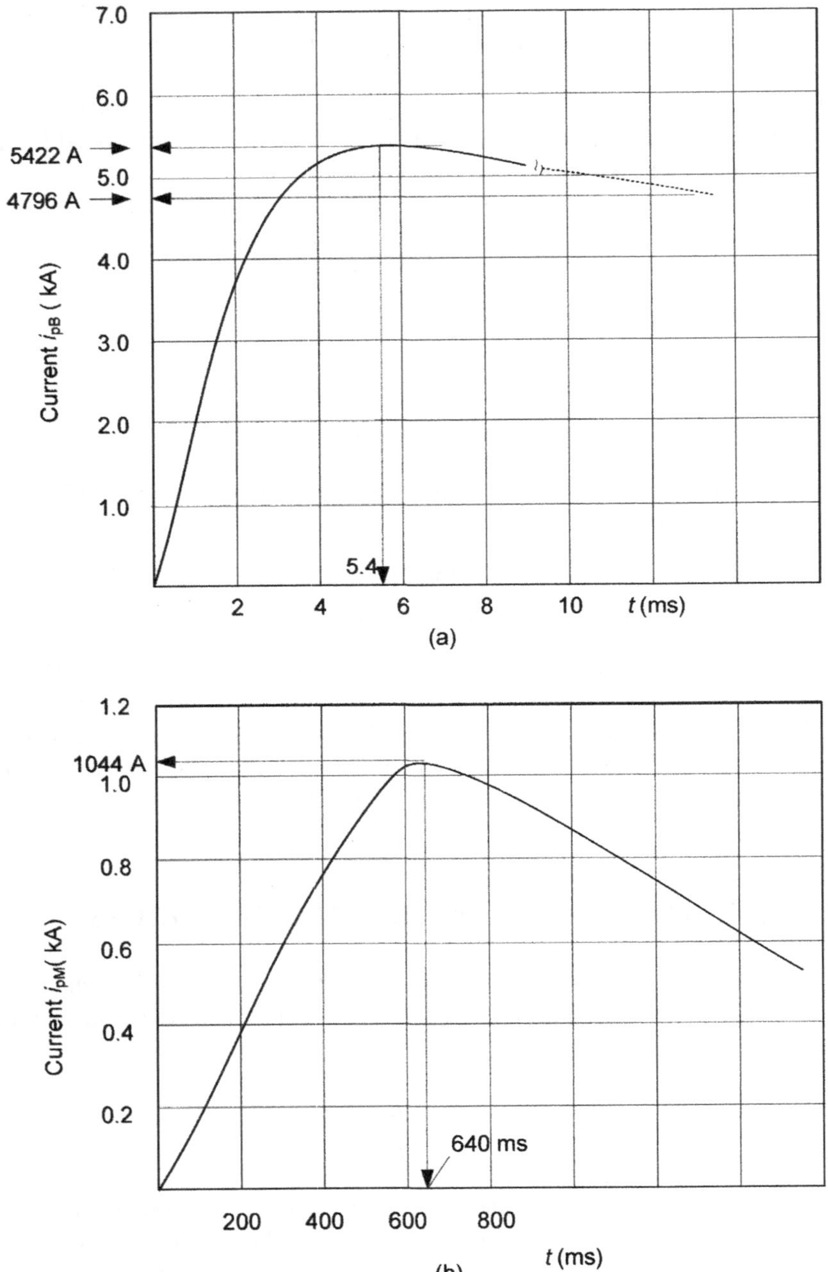


Figure 15.7. Calculated short-circuit time–current profiles: (a) battery, (b) DC motor, (c) rectifier, and (d) capacitor; see Examples 15.1–15.4, respectively.

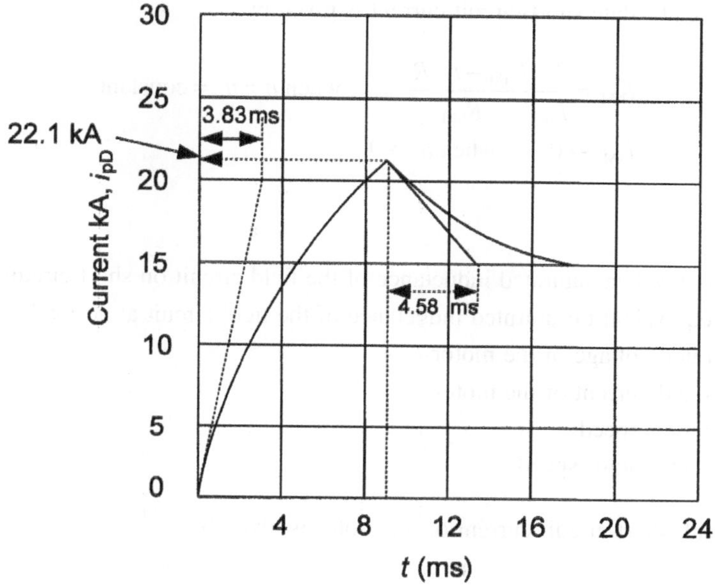


(c)

Rectifier

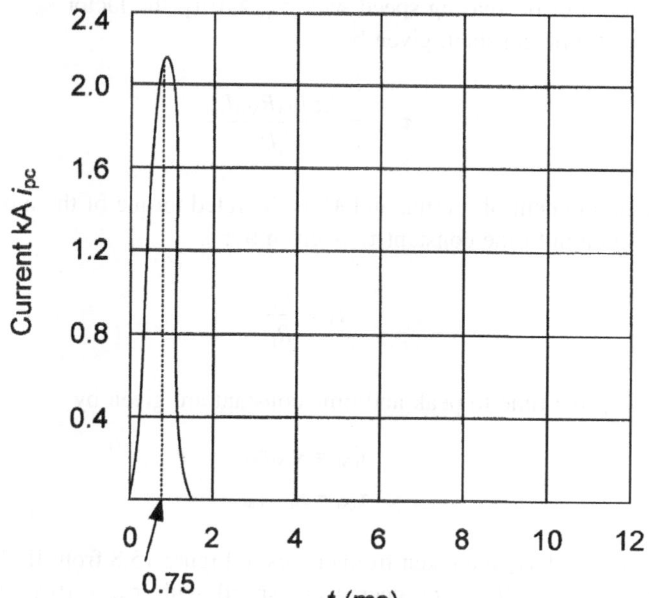


(d)

Figure 15.7. *(Continued)*

The quasi steady-state short-circuit current is given by

$$I_{KM} = \frac{L_F}{L_{0F}} \frac{U_{rM} - I_{rM}R_M}{R_{MB}} \quad \text{when } n = n_n = \text{constant}$$
$$I_{kM} \to 0 \quad \text{when } n \to 0, \tag{15.19}$$

where:

$L_F$ = equivalent saturatcd inductance of the field circuit on short circuit
$L_{0F}$ = equivalent unsaturnted inductance of the field circuit at no load
$U_{rM}$ = rated voltage of the motor
$I_{rM}$ = rated current of the motor
$n$ = motor speed
$n_n$ = rated motor speed.

The peak short-circuit current of the motor is given by

$$i_{pM} = \kappa_M \frac{U_{rM} - I_{rM}R_M}{R_{MB}}. \tag{15.20}$$

At normal speed or decreasing speed with $\tau_{mec} \geq 10\tau_F$, the factor $\kappa_M = 1$, where $\tau_{mec}$ is the mechanical time constant, given by

$$\tau_{mec} = \frac{2\pi J n_0 R_{MB} I_{rM}}{M_r U_{rM}}, \tag{15.21}$$

where $J$ is the moment of inertia, and $M_r$ is the rated torque of the motor.

The field circuit time constant $\tau_F$ is given by:

$$\tau_F = \frac{L_F}{R_F}. \tag{15.22}$$

For $\tau_{mec} \geq 10\tau_F$, the time to peak and time constant are given by

$$t_{pM} = \kappa_{1M}\tau_M$$
$$\tau_{1M} = \kappa_{2M}\tau_M. \tag{15.23}$$

The factors $\kappa_{1M}$ and $\kappa_{2M}$ are taken from curves in Figure 15.8 from IEC, and are dependent on $\tau_F/\tau_M$ and $L_F/L_{0F}$. For decreasing speed with $\tau_{mec} < 10\tau_F$, the factor $\kappa_M$ is dependent on $1/\delta = 2\tau_M$ and $\omega_0$:

$$\omega_0 = \sqrt{\frac{1}{\tau_{mec}\tau_M}\left(1 - \frac{I_{rM}R_M}{U_{rM}}\right)}, \tag{15.24}$$

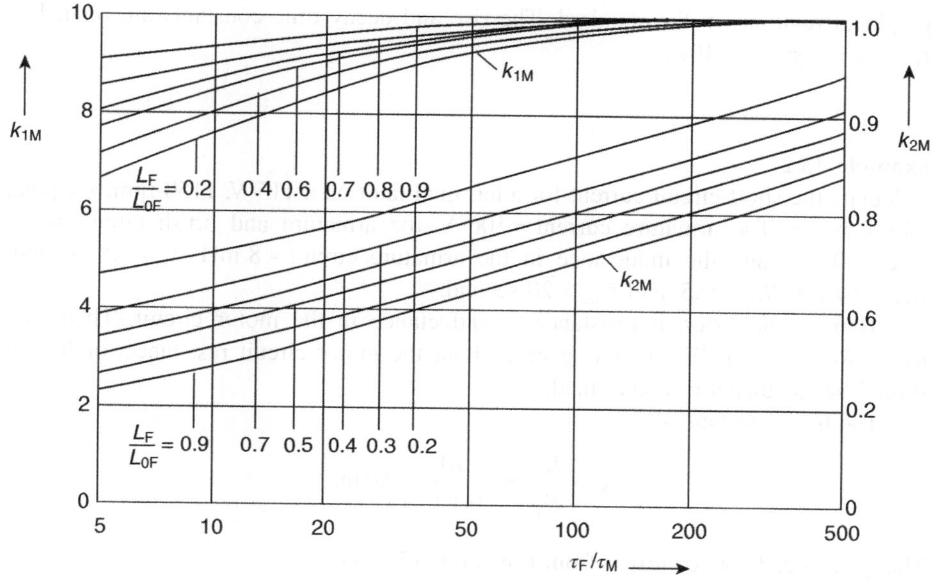


Figure 15.8. Factors $\kappa_{1M}$ and $\kappa_{2M}$ for determining the time to peak $t_{pM}$ and the rise time $\tau_{1M}$ for normal and decreasing speed with $\tau_{mec} \geq 10\tau_F$; short-circuit of a DC motor.

where $\omega_0$ is the undamped natural angular frequency and $\delta$ is the decay coefficient; $\kappa_M$ is derived from the curves in the IEC standard [11].

For decreasing speed with $\tau_{mec} < 10\tau_F$, the time to peak $\tau_M$ is read from a curve in the IEC standard, and the rise time constant is given by

$$\tau_{1M} = \kappa_{3M}\tau_M, \tag{15.25}$$

where the factor $\kappa_{3M}$ is again read from a curve in the IEC standard [11] (not reproduced here).

Regarding the decay time constant $\tau_{2M}$, for nominal speed or decreasing speed with $\tau_{mec} \geq 10\tau_F$:

$$\begin{aligned} \tau_{2M} &= \tau_F \ \text{when } n = n_n = \text{const.} \\ \tau_{2M} &= \frac{L_{0f}}{L_F}\kappa_{4M}\tau_{mec} \ \text{when } n \to 0 \text{ with } \tau_{mec} \geq 10\tau_F. \end{aligned} \tag{15.26}$$

For decreasing speed with $\tau_{mec} < 10\tau_F$:

$$\tau_{2M} = \kappa_{4M}\tau_{mec}, \tag{15.27}$$

where $\kappa_{4M}$ is again read from a curve in the IEC standard, not reproduced here. Thus, the IEC calculation method requires extensive motor data and the use of a number of graphical relations in the standard. The rise and decay time constants are related to $\tau_{mec} < 10\tau_F$ or $\tau_{mec} \geq 10\tau_F$.

**Example 15.2**

Calculate the short-circuit current for a terminal fault on a 115-V, 1150-rpm, six-pole, 15-hp motor. The armature current = 106 A, the armature and brush circuit resistance = 0.1 Ω, and the inductance in the armature circuit = 8 mH; $\tau_F = 0.8$ second, $\tau_{mec} > 10\tau_F$, $L_F/L_{0F} = 0.5$, and $\tau_{mec} = 20$ second.

There is no external resistance or inductance in the motor circuit. Therefore, $R_{MBr} = R_M = 0.10\ \Omega$. IEC is not specific about the motor circuit resistance, or how it should be calculated or ascertained.

The time constant is

$$\tau_M = \frac{L_M}{R_M} = \frac{8\times10^{-3}}{0.10} = 80 \text{ ms}.$$

The quasi-steady-state current from Equation (15.19) is

$$0.5\left(\frac{115-(0.10)(106)}{0.10}\right) = 522 \text{ A}.$$

From Equation (15.20), the peak current is 1044 A, because for $\tau_{mec} > 10\tau_F$, factor $\kappa_M$ in Equation (15.20) = 1. The time to peak and time constant are given by Equation (15.23). From Figure 15.3, and for $\tau_F/\tau_M = 10$ and $L_F/L_{0F} = 0.5$, factor $\kappa_{1M} = 8.3$ and $\kappa_{2M} = 0.370$. Therefore, the time to peak is 640 ms and the time constant $\tau_{1M} = 29.6$ ms.

The short-circuit profile is plotted in Figure 15.7b.

## 15.6 SHORT-CIRCUIT CURRENT OF A RECTIFIER

The typical time–current curve for rectifiers short-circuit is shown in Figure 15.9. The maximum current is reached at one half-cycle after the fault occurs. The peak at half-cycle is caused by the same phenomenon that creates a DC offset in AC short-circuit calculations. The magnitude of this peak is dependent on *X*/*R* ratio, the AC system source reactance, rectifier transformer impedance, and the resistance and reactance through which the current flows in the DC system. The addition of resistance or inductance to the DC system reduces this peak, and, depending on the magnitude of these components, the peak may be entirely eliminated, with a smoothing DC reactor as shown in Figure 15.1a. The region A in Figure 15.9 covers the initial rise of current, the peak current occurs in region B, and region C covers the time after one cycle until the current is interrupted.

The initial rate of rise of the DC short-circuit current for a bolted fault varies with the magnitude of the sustained short-circuit current. The addition of inductance to the DC circuit tends to decrease the rate of rise.

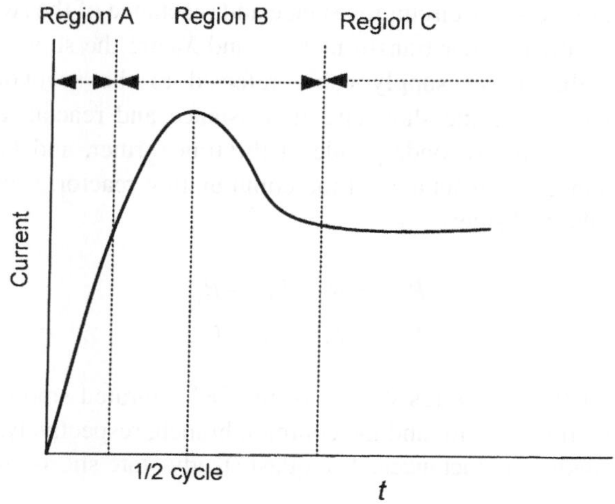


Figure 15.9. Short-circuit current profile of a rectifier.

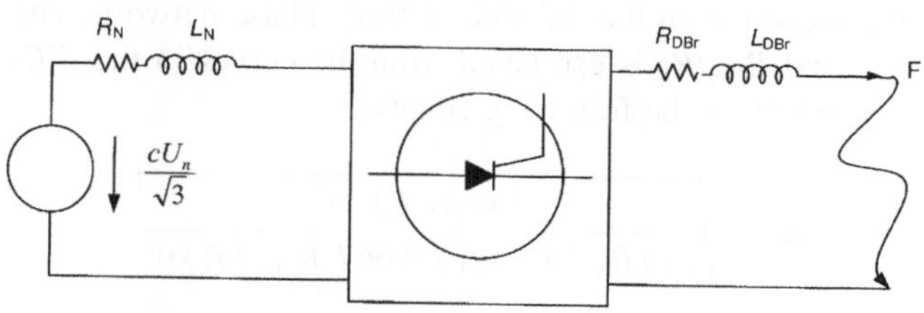


Figure 15.10. Equivalent circuit for the short-circuit current calculation of a rectifier.

The equivalent short-circuit diagram is shown in Figure 15.10. The maximum DC short-circuit current is given by the minimum impedance $Z_{Q\min}$, which is obtained from the maximum short-circuit current $I''_{kQ\max}$ of the AC system:

$$Z_{Q\min} = \frac{cU_n}{\sqrt{3}I''_{KQ\max}}. \tag{15.28}$$

The minimum DC current is given by

$$Z_{Q\max} = \frac{cU_n}{\sqrt{3}I''_{KQ\min}}. \tag{15.29}$$

In Figure 15.10 the resistance and inductances on the AC side are:

$$\begin{aligned} R_N &= R_Q + R_P + R_T + R_R \\ X_N &= X_Q + X_P + X_T + X_R \end{aligned} \tag{15.30}$$

where $R_Q$ and $X_Q$ are the short-circuit resistance and reactance of the AC source referred to the secondary of the rectifier transformer, $R_P$ and $X_P$ are the short-circuit resistance and reactance of the power supply cable referred to the secondary side of the transformer, $R_T$ and $X_T$ are the short-circuit resistance and reactance of the rectifier transformer referred to the secondary side of the transformer, and $R_R$ and $X_R$ are the short-circuit resistance and reactance of the commutating reactor, if present.

Similarly, on the DC side:

$$\begin{aligned} R_{\mathrm{DBr}} &= R_{\mathrm{S}} + R_{\mathrm{DL}} + R_{\mathrm{y}} \\ L_{\mathrm{DBr}} &= L_{\mathrm{S}} + L_{\mathrm{DL}} + L_{\mathrm{y}}, \end{aligned} \tag{15.31}$$

where $R_S$, $R_{DL}$, and $R_y$, are the resistances of the DC saturated smoothing reactor, the conductor in the rectifier circuit, and the common branch, respectively, and $L_S$, $L_{DL}$, and $L_y$ are the corresponding inductances. The quasi steady-state short-circuit current is

$$i_{kD} = \lambda_D \frac{3\sqrt{2}}{\pi} \frac{cU_n}{\sqrt{3}Z_N} \frac{U_{\mathrm{rTLV}}}{U_{\mathrm{rTHV}}}, \tag{15.32}$$

where $Z_N$ is the impedance on the AC side of three-phase network. The factor $\lambda_D$ as a function of $R_N/X_N$ and $R_{DBr}/R_N$ is estimated from the curves in the IEC standard [11]. Alternatively, it is given by the following equation:

$$\lambda_D = \sqrt{\frac{1+(R_N/X_N)^2}{1+(R_N/X_N)^2(1+0.667(R_{\mathrm{DBr}}/R_N))^2}}. \tag{15.33}$$

The peak short circuit current is given by

$$i_{pD} = \kappa_D I_{kD}, \tag{15.34}$$

where the factor $\kappa_D$ is dependent on:

$$\frac{R_N}{X_N}\left[1+\frac{2R_{\mathrm{DBr}}}{3R_N}\right] \text{ and } \frac{L_{\mathrm{DBr}}}{L_N}. \tag{15.35}$$

It is estimated from the curves in the IEC standard [11] (not reproduced) or from the following equation:

$$\kappa_D = \frac{i_{pD}}{I_{kD}} = 1+\frac{2}{\pi} e^{-\left(\frac{\pi}{3}+\phi_D\right)\cot\phi_D} \sin\phi_D\left(\frac{\pi}{2}-\arctan\frac{L_{\mathrm{DBr}}}{L_N}\right), \tag{15.36}$$

where

$$\phi_D = \arctan\frac{1}{\frac{R_N}{X_N}\left(1+\frac{2}{3}\frac{R_{\mathrm{DBr}}}{R_N}\right)}. \tag{15.37}$$

Time to peak $t_{pD}$, when $\kappa_D \geq 1.05$, is given by

$$t_{pD} = (3\kappa_D + 6)\ \text{ms} \quad \text{when} \frac{L_{DBr}}{L_N} \leq 1$$

$$t_{pD} = \left[(3\kappa_D + 6) + 4\left(\frac{L_{DBr}}{L_N} - 1\right)\right] \text{ms} \quad \text{when} \frac{L_{DBr}}{L_N} > 1. \tag{15.38}$$

If $\kappa_D < 1.05$, the maximum current, compared with the quasi steady-state short-circuit current, is neglected, and $t_{pD} = T_k$ (short-circuit duration) is used.

The rise time constant for 50 Hz is given by

$$\tau_{1D} = \left[2 + (\kappa_D - 0.9)\left(2.5 + 9\frac{L_{DBr}}{L_N}\right)\right] \text{ms} \quad \text{when } \kappa_D \geq 1.05$$

$$\tau_{1D} = \left[0.7 + \left[7 - \frac{R_N}{X_N}\left(1 + \frac{2}{3}\frac{L_{DBr}}{L_N}\right)\right]\left(0.1 + 0.2\frac{L_{DBr}}{L_N}\right)\right] \text{ms} \quad \text{when } \kappa_D < 1.05. \tag{15.39}$$

For simplification:

$$\tau_{1D} = \frac{1}{3} t_{pD}. \tag{15.40}$$

The decay time constant $\tau_{2D}$ for 50 Hz is given by

$$\tau_{2D} = \frac{2}{\frac{R_N}{X_N}\left(0.6 + 0.9\frac{R_{DBr}}{R_N}\right)} \text{ms}. \tag{15.41}$$

The time constants for a 60-Hz power system are not given in the IEC standard.

**Example 15.3**

A three-phase rectifier is connected on the AC side to a three-phase, 480–120 V, 100-kVA transformer of percentage $Z_T = 3\%$, $X/R = 4$. The 480-V source short-circuit MVA is 30, and the $X/R$ ratio = 6. The DC side smoothing inductance is 5 μH, and the resistance of the cable connections is 0.002 Ω. Calculate and plot the short-circuit current profile at the end of the cable on the DC side.

Based on the AC side data, the source impedance in series with the transformer impedance referred to the secondary side of the rectifier transformer is

$$R_Q + jX_Q = 0.00008 + j0.00048\ \Omega$$

$$R_T + jX_T = 0.001 + j0.00419\ \Omega.$$

Therefore:

$$R_N + jX_N = 0.0011 + j0.004671\,\Omega.$$

On the DC side:

$$R_{\mathrm{DBr}} = 0.002\,\Omega \quad \text{and} \quad L_{\mathrm{DBr}} = 5\,\mu\mathrm{H}.$$

This gives

$$\frac{R_N}{X_N} = 0.24 \quad \text{and} \quad \frac{R_{\mathrm{DBr}}}{R_N} = 2.0.$$

Calculate $\lambda_{\mathrm{D}}$ from Equation (15.33):

$$\lambda_D = \sqrt{\frac{1+(0.24)^2}{1+(0.24)^2(1+0.667)(2.0)^2}} = 0.897.$$

The quasi-steady-state current is, therefore, from Equation (15.32):

$$I_{\mathrm{kD}} = (0.897)\left(\frac{3\sqrt{2}}{\pi}\right)\left(\frac{1.05\times 480}{\sqrt{3}\times 0.0048}\right)\left(\frac{120}{480}\right) = 18.36\ \mathrm{kA}.$$

To calculate the peak current, calculate the ratios:

$$\frac{R_N}{X_N}\left(1+\frac{2}{3}\frac{R_{\mathrm{DBr}}}{R_N}\right) = (0.24)(1+0.667\times 2) = 0.56$$

$$\frac{L_{\mathrm{DBr}}}{L_N} = \frac{5\times 10^{-6}}{0.0128\times 10^{-3}} = 0.392.$$

Calculate $\kappa_{\mathrm{D}}$ from Equations (15.36) and (15.37). From Equation (15.37):

$$\phi_D = \tan^{-1}\frac{1}{0.24}\left(\frac{1}{1+0.667(2.0)}\right) = 60.75°,$$

and from Equation (15.36), $\kappa_{\mathrm{D}} = 1.204$. Thus, the peak short-circuit current is

$$i_{\mathrm{pD}} = \kappa_{\mathrm{D}} I_{\mathrm{kD}} = 1.204\times 18.36 = 22.10\ \mathrm{kA}.$$

The time to peak is given by Equation (15.38) and is equal to

$$t_{\mathrm{pD}} = (3\kappa_{\mathrm{D}} + 6)\mathrm{ms} = (3\times 1.204 + 6) = 9.62\ \mathrm{ms}.$$

The rise time constant is given by Equation (15.39) and is equal to 3.83 ms, and the decay time constant is given by Equation (15.41) and is equal to 4.58 ms.

The current profile is plotted in Figure 15.7c, which shows the calculated values. The intermediate shape of the curve can be correctly plotted using Equations (15.1) and (15.2). In this example, the IEC equations are for a 50-Hz system. For a 60-Hz system, the peak will occur around 8.3 ms.

## 15.7 SHORT CIRCUIT OF A CHARGED CAPACITOR

The resistance and inductance in the capacitor circuit from Figure 15.3 are

$$R_{\text{CBr}} = R_{\text{C}} + R_{\text{CL}} + R_y$$
$$L_{\text{CBr}} = L_{\text{CL}} + L_y, \tag{15.42}$$

where $R_C$ is the equivalent DC resistance of the capacitor, and $R_{\text{CL}}$ and $L_{CL}$ are the resistance and inductance of a conductor in the capacitor circuit. The steady-state short-circuit current of the capacitor is zero, and the peak current is given by

$$i_{\text{pC}} = \kappa_C \frac{E_C}{R_{\text{CBr}}}, \tag{15.43}$$

where $E_{\text{C}}$ is the capacitor voltage before the short-circuit, and $\kappa_{\text{C}}$ is read from curves in the IEC standard [11] based on:

$$1/\delta = \frac{2L_{\text{CBr}}}{R_{\text{CBr}}}$$
$$\omega_0 = \frac{1}{\sqrt{L_{\text{CBr}}C}}. \tag{15.44}$$

If $L_{\text{CBr}} = 0$, then $\kappa_{\text{C}} = 1$.

The time to peak $t_{\text{pC}}$ is read from curves in the IEC standard [11]. If $L_{\text{CBr}} = 0$, then $t_{\text{pC}} = 0$. The rise time constant is

$$\tau_{1\text{C}} = \kappa_{1\text{C}} t_{\text{pC}}, \tag{15.45}$$

where $\kappa_{1\text{C}}$ is read from curves in IEC. The decay time constant is

$$\tau_{2\text{C}} = \kappa_{2\text{C}} R_{\text{CBr}} C, \tag{15.46}$$

where $\kappa_{2\text{C}}$ is read from curves in IEC standard [11]. The curves for these factors are not reproduced.

**Example 15.4**

A 120-V, 100-μF capacitor has $R_{CBr}$ = 0.05 Ω and $L_{CBr}$ = 10 mH. Calculate the terminal short-circuit profile.

From Equation (15.44):

$$1/\delta = \frac{2\times 10\times 10^{-3}}{0.05} = 0.4.$$

Also,

$$\omega_0 = \frac{1}{\sqrt{10\times 10^{-3}\times 100\times 10^{-6}}} = 1000.$$

From curves in the IEC standard, $\kappa_c$ = 0.92. The peak current from Equation (15.43) is then (0.92) × (120/0.05) = 2208 A. The time to peak from curves in IEC = 0.75 ms, and $\kappa_{1C}$ = 0.58. From Equation (15.45), the rise time constant is (0.58) × (0.75) = 0.435 ms. Also, $\kappa_{2C}$ = 1, and, from Equation (15.46), the decay time constant is 5 μs. The short-circuit current profile is plotted in Figure 15.7d.

## 15.8 TOTAL SHORT-CIRCUIT CURRENT

The total short-circuit current at fault Fl (Figure 15.3) is the sum of the partial short-circuit currents calculated from the various sources. For calculation of the total short-circuit current at F2 (Figure 15.3), the partial currents from each source should be calculated by adding the resistance and inductance of the common branch to the equivalent circuit. A correction factor is then applied. The correction factors for every source are obtained from

$$\begin{aligned} i_{\text{pcor}j} &= \sigma_j i_{\text{p}j} \\ I_{\text{kcor}j} &= \sigma_j I_{\text{k}j}, \end{aligned} \tag{15.47}$$

where the correction factor $\sigma_j$ is described in the IEC standard [11].

**Example 15.5**

The sources (rectifier, battery, motor, and capacitor) in Examples 15.1–15.4 are connected together in a system configuration, as shown in Figure 15.3. Plot the total short-circuit current.

The profiles of partial currents shown in Figure 15.7 are summed. As the time to peak, magnitudes, and decay time constants are different in each case, a graphical approach is taken and the total current profile is shown in Figure 15.11. The peak current is approximately 27.3 kA, and the peak occurs at approximately 9 ms after the fault.

The short-circuit current from the rectifier predominates. The short-circuit current from the capacitor is a high rise pulse, which rapidly decays to zero. The DC motor short-circuits current rises slowly. Smaller DC motors have higher armature inductance,

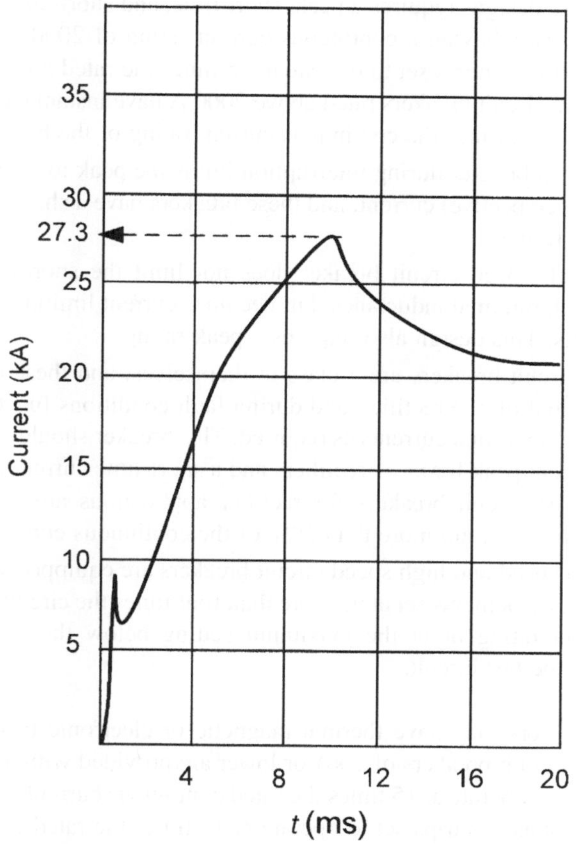


Figure 15.11. Total short-circuit current profile of four partial short-circuit currents in Examples 15.1–15.4.

resulting in a slower rate of current rise. The rectifier current peaks approximately in one half-cycle of the power system frequency. The relative magnitudes of the partial currents can vary substantially, depending on the system configuration. This can give varying profiles of total current and time to peak.

## 15.9 DC CIRCUIT BREAKERS AND FUSES

### 15.9.1 DC Circuit Breakers

The DC breakers may be categorized as follows [2]:

- The general-purpose low voltage DC power circuit breakers do not limit the peak available current and may not prevent the prospective fault current rise to its peak value. These designs requires a peak, short-time, and short-circuit current rating. Circuit breakers having a continuous current rating of 2000 A and below have instantaneous elements set to operate at 15 times the rated continuous current of the breaker. Circuit breakers rated above 2000 A have instantaneous elements set to operate at 12 times the continuous current rating of the breaker.

- A high-speed breaker during interruption limits the peak to a value less than the available (perspective) current, and these breakers have a short-circuit rating and short-time rating.
- A semi-high speed circuit breaker does not limit the short-circuit current on circuits with minimal inductance, but becomes current limiting for highly inductive circuits. This design also requires a peak rating.
- Rectifier circuit breakers are a class in themselves, and these carry the normal current output of one rectifier, and during fault conditions function to withstand or interrupt abnormal currents as required. This breaker should be rated for short-circuit current peak for $n - 1$ rectifiers and a short-time current rating for its own rectifier. The circuit breakers for rectifier applications are fitted with reverse current trips, set at no more than 50% of the continuous current rating.
- Semi-high speed and high speed circuit breakers are equipped with direct acting instantaneous elements set at no more than four times the circuit breaker continuous current rating or at the maximum setting below the available sustained current of the test circuit.

The DC breakers may have thermal magnetic or electronic trip devices, that is, general-purpose circuit breakers of 2 kA or lower are provided with instantaneous tripping elements set to operate at 15 times the rated continuous current, and breakers rated >2 kA have instantaneous trips set to operate at 12 times the rated current.

Two or three poles of a breaker may be connected in series for enhanced interrupting rating. The interrupting capacity of a breaker decreases with increasing DC voltage. The maximum inductance for full interrupting rating in microhenries is specified, and the reduced interrupting rating for higher values of inductance can be calculated. When the breakers are rated for AC as well as DC systems, the interrupting rating on DC systems are much lower.

IEEE Standard [13] provides the preferred ratings, related requirements, and application recommendations for low-voltage AC (635 V and below) and DC (3200 V and below) power circuit breakers. Table 15.1 is based upon this reference.

**Example 15.6**

A general-purpose DC circuit breaker for the fault location Fl in Figure 15.3, the fault current profile shown in Figure 15.11, is selected as follows:

The peak short-circuit current is 27.5 kA, and the quasi-steady state current is approximately 22 kA. The continuous load current of all the sources (200AH DC battery, 15-hp DC motor and 100-kVA rectifier) is approximately 380 A. Therefore, select a 250-V, circuit breaker frame size 600 A, continuous current = 400 A, rated peak

TABLE 15.1. Preferred Ratings of General Purpose DC Power Circuit Breakers with or without Instantaneous Direct Acting Trip Elements

| Circuit Breaker Frame Size (A, DC) | System Nominal Voltage (V, DC) | Rated Maximum Voltage (V, DC) | Rated Peak Current (A, peak) | Rated Maximum Short-Circuit Current or Rated Short-Circuit Current (A) | Maximum Inductance for Full Interrupting Rating (μH) | Load Circuit Stored Energy Factor (kW-s) | Range of Trip Device Current Ratings (A, DC) |
|---|---|---|---|---|---|---|---|
| 600–800 | 250 | 300 | 41,000 | 25,000 | 160 | 50 | 40–800 |
| 1600 | 250 | 300 | 83,000 | 50,000 | 80 | 100 | 200–1600 |
| 2000 | 250 | 300 | 83,000 | 50,000 | 80 | 100 | 200–2000 |
| 3000 | 250 | 300 | 124,000 | 75,000 | 50 | 140 | 2000–3000 |
| 4000 | 250 | 300 | 165,000 | 100,000 | 32 | 160 | 4000 |
| 5000 | 250 | 300 | 165,000 | 100,000 | 32 | 160 | 5000 |
| 6000 | 250 | 300 | 165,000 | 100,000 | 32 | 160 | 6000 |

The peak current rating is only applicable for circuit breakers in solid-state rectifier applications. Rated short-circuit current is applicable only to circuit breakers without instantaneous direct acting trip elements (short-time delay elements or remote relay).

*Source*: Reference [13].

current = 41 kA, rated maximum short-circuit current or short-time current = 25 kA, row 1 of Table 15.1.

The maximum inductance for full interrupting rating is specified as 160 μH. The actual reactance can be calculated from the time constant in 15-11. The peak short-circuit current is 27.5 kA, therefore, the resistance is 4.36 mΩ. The time constant of the current from the rise time is approximately 4 ms, which gives L = 17.4 μH.

If the inductance exceeds the value given in Table 15.1, the reduced interrupting rating is obtained from the expression from IEEE Standard [13]:

$$I = 10^4 \sqrt{20W / L},$$

where $W$ is the value of kW-s specified in Table 15.1, $L$ is the actual inductance in μH, and $I$ is in amperes.

### 15.9.2 DC Rated Fuses

DC rated fuses up to 600 V and interrupting ratings of 200 kA are available. The manufacturers may not publish the short-circuit ratings at all the DC voltages, but if a fuse of required interrupting rating and rated for higher voltage is selected at the lower DC voltage of application, the interrupting capability will be higher.

## 15.10 ARCING IN DC SYSTEMS

Stokes and Oppenlander [7] studied free burning horizontal and vertical arcs between series electrodes in open air. The minimum voltage to maintain an arc depends upon current magnitude, gap width and orientation of electrodes, Figure 15.12a,b showing *V–I* characteristics for horizontal and vertical arcs are based upon this work. In these figures, the transition point is defined as:

$$I_t = 10 + 0.2Z_g, \tag{15.48}$$

where the gap $Z_g$ is in millimeters. The V-I characteristics shows the arc voltage both below and above the transition point. The current voltage relation is given by:

$$V_{arc} = (20 + 0.534Z_g)I_{arc}^{0.12}. \tag{15.49}$$

This gives an arc resistance:

$$R_{arc} = \frac{20 + 0.534Z_g}{I_{arc}^{0.88}}. \tag{15.50}$$

A simplified arc model can be constructed as shown in Figure 15.13. If the arc resistance is zero, this gives the bolted DC current. See also Reference [14].

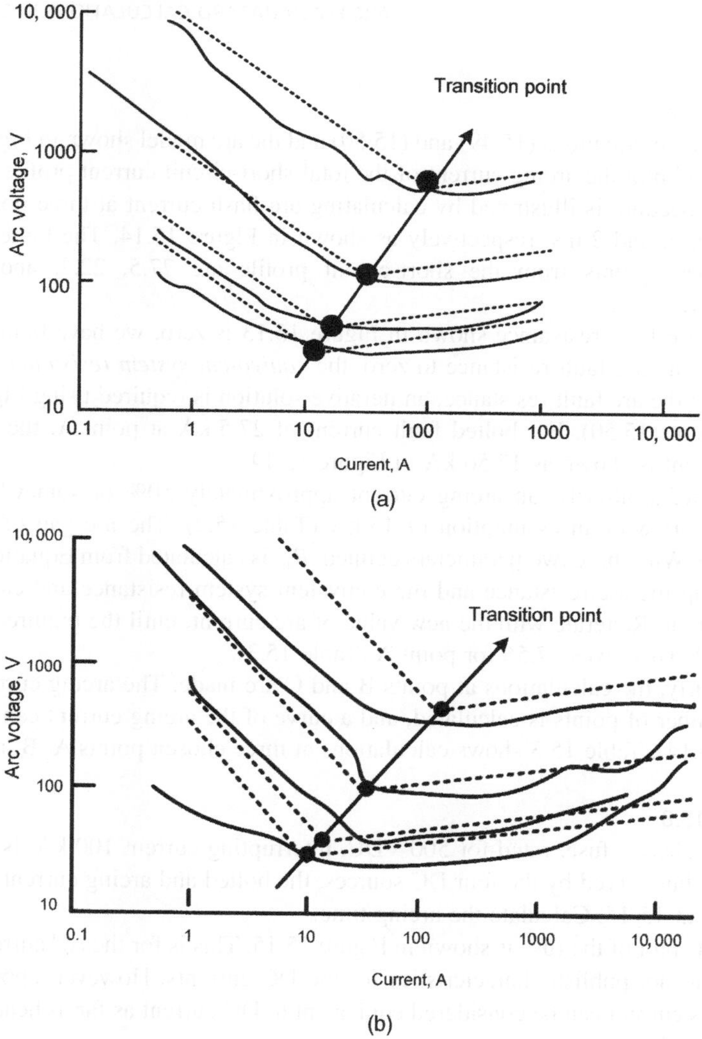


Figure 15.12. (a) Minimum arc voltage for vertical arcs; (b) minimum arc voltage for horizontal arcs. Aluminum electrodes. Continuous lines are measured and broken lines are calculated. Gap widths of curves from top to bottom: 5, 20, 100, and 500 mm, respectively.

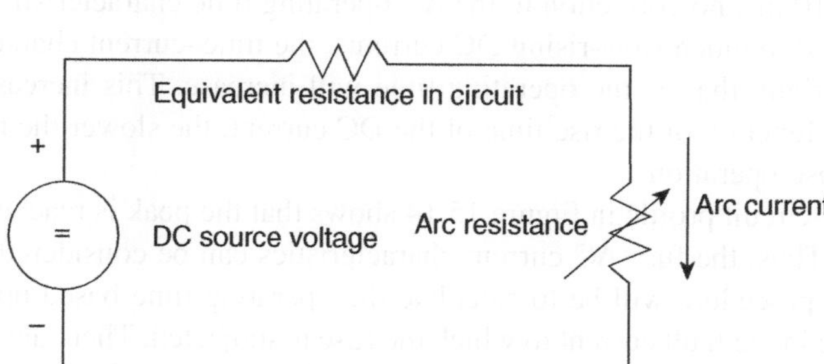


Figure 15.13. A simplified model of arcing in a DC system.

**Example 15.7**

Based upon the Equations (15.49) and (15.50) and the arc model shown in Figure 15.13, calculate and plot the arcing current of the total short-circuit current profile in 15-11.

The procedure is illustrated by calculating arc flash current at three points, A, B, and C at 9, 6, and 2 ms, respectively as shown in Figure 15.14. The bolted currents read at these points from the short-circuit profile are 27.5, 22.2, and 10.0 kA, respectively.

If the arc fault resistance shown in Figure 15.13 is zero, we have bolted current. Thus, equating arc fault resistance to zero, the *equivalent system resistance is known.*

To find the arc fault resistance, an iterative solution is required using Figure 15.13 and Equation (15.50). For bolted fault current of 27.5 kA at point A, the calculated arcing current is shown as 17.56 kA in Figure 15.14.

Assume, arbitrarily, an arcing current approximately 50% or somewhat higher. Here we start with an assumption of 15 kA (Table 15.2). The arc gap of 20 mm is considered. With these two parameters defined, $R_{arc}$ is calculated from Equation (15.50).

Sum up the arc resistance and the equivalent system resistance and calculate the arcing current. Reiterate with the new value of arc current, until the required tolerance is obtained. This gives 17.56 for point A (Table 15.3).

Similarly, the calculations at points B and C are made. The arcing current at sufficient number of points is calculated, and a curve of the arcing current can be plotted (Figure 15.14). Table 15.3 shows calculations at three chosen points A, B, and C.

**Example 15.8**

A 1000-A class L fuse, rated for 500 VDC, interrupting current 100 kA, is applied to a common bus served by the four DC sources, the bolted and arcing current profiles as calculated in 15-14. Calculate the arcing time.

A TCC plot of the fuse is shown in Figure 15.15. This is for the AC current. Manufacturers do not publish characteristics for the DC currents. However, approximately the AC rms current can be considered equivalent to DC current as far as heating effects are concerned.

When subject to DC currents with certain time constants, the device operating time will be higher and the characteristics curve will shift to the right of the current axis. The 60-cycle AC wave rises to peak in 8.33 ms. Therefore, if the DC short-circuit is peaking in 8–10 ms, no correction to the AC operating time characteristics is required. When subjected to much slow-rising DC currents, the time–current characteristics will shift towards right, that is, the operating time will increase. This increased operating time will be a function of the rise time of the DC current, the slower the rise, the more delayed the fuse operation.

The short-circuit profile in Figure 15.14 shows that the peak is reached in approximately 9 ms. Thus, the fuse AC current characteristics can be considered.

A simple procedure will be to calculate the operating time based upon the incremental change in the fault current to which the fuse is subjected. Then take the weighted average to ascertain the average current to which the fuse is subjected.

The peak arcing current is 17.56 kA and occurs at 9 ms. If this current persists unchanged for 39 ms, the fuse will operate. However, at 14.0 ms, it reduces to 14.2 kA.

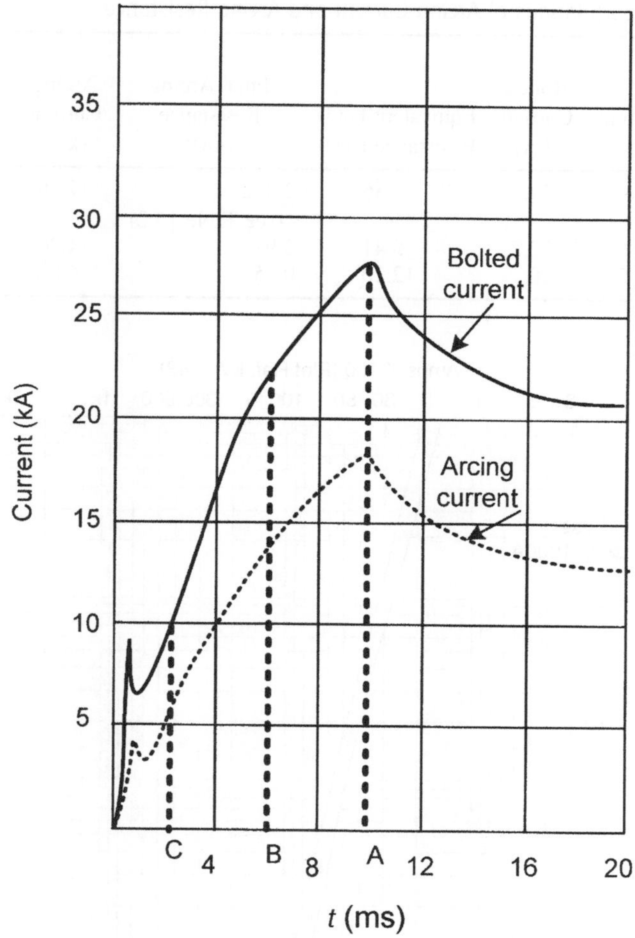


Figure 15.14. Arcing current profile superimposed on the short-circuit current profile of Figure 15.11.

TABLE 15.2. Iterative Calculation of Arcing Current Point A (27.5 kA) in Figure 15.11

| Iteration No. | Arcing Current (kA) | Arcing Resistance (mΩ) | Total Resistance (mΩ) | Calculated Arcing Current |
|---|---|---|---|---|
| 1 | 15 (assume) | 2.83 | 7.19 | 16.68 |
| 2 | 16.68 | 2.57 | 6.93 | 17.30 |
| 3 | 17.30 | 2.497 | 6.856 | 17.50 |
| 4 | 17.50 | 2.472 | 6.832 | 17.56 |

TABLE 15.3. Calculations of Arcing Current and Arcing Resistance

| At Point Marked in Figure 15.14 | Time (ms) | Bolted Current (kA) | Equivalent Fault Resistance (mΩ) | Final Arcing Resistance (mΩ) | Arcing Current (kA) | Arcing Current as % of Bolted Current |
|---|---|---|---|---|---|---|
| A | 9 | 27.5 | 4.36 | 2.472 (see Table 15.2) | 17.56 | 63.8 |
| B | 6 | 22.2 | 5.41 | 2.97 | 14.20 | 63.96 |
| C | 2 | 10 | 12.20 | 0.553 | 6.80 | 68.0 |

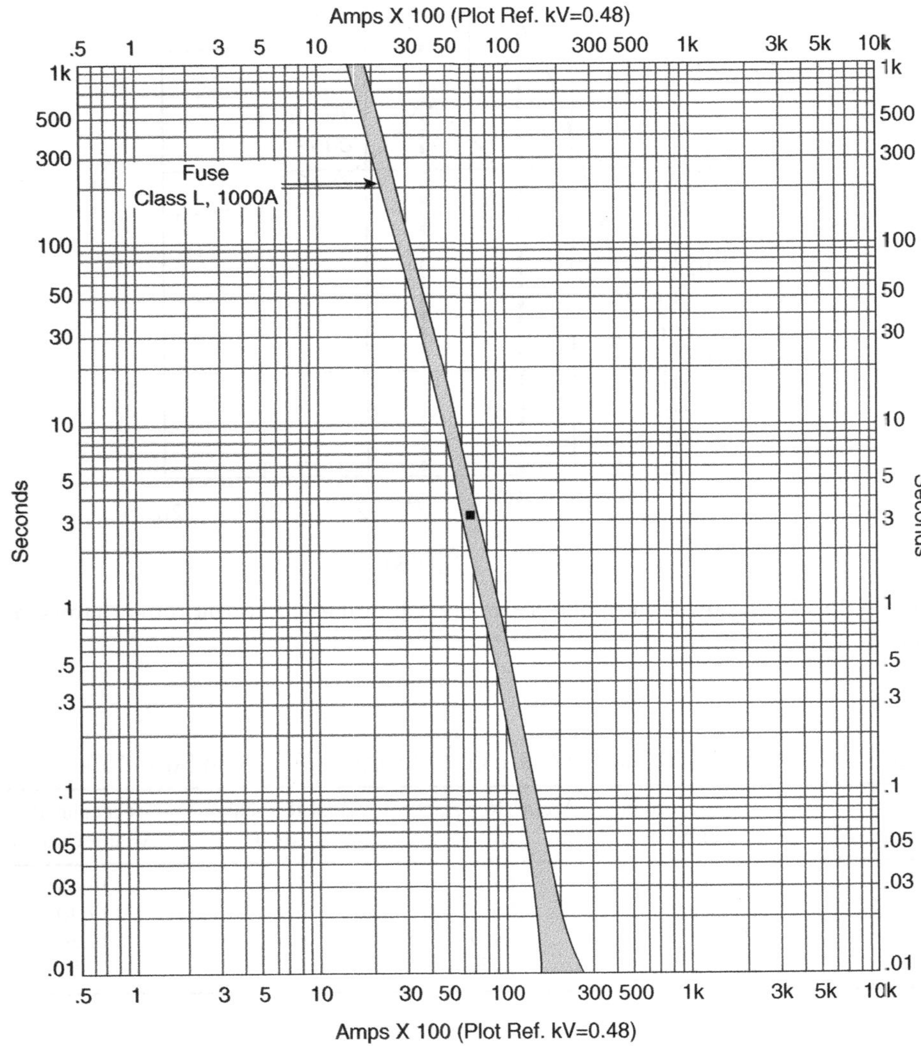


Figure 15.15. Time–current characteristics of a class L, 1000-A fuse.

TABLE 15.4. Calculation of the Average Current through the Fuse (Example 15.8)

| Time (ms) | Arcing Current (kA) | Fuse Operating Time (ms) |
|---|---|---|
| 2 | 6.8 | 300 (ignore) |
| 6 | 14.24 | 70 |
| 9 | 17.56 | 39 |
| 14 | 14.00 | 70 |
| 20 | 13.0 | 90 |
| ≥45 | 12.0 | 100 |

If this current persists for 70 ms the fuse will operate. At 6 ms, the current is 14.20 kA (see Table 15.4).

Thus, in the operating range of the fuse, an average arcing current of 13 kA can be considered between 6 ms to approximately 96 ms, and the fuse operates in approximately 90 ms after the current has risen to 14.2 kA at 6 ms.

Referring to Figure 15.1, the IEC equations give results of peak and quasi-steady-state currents. The capacitor charging current is a pulse, but the currents from the other sources can be sustained. *Assuming that the quasi-steady-state current continues, until it is interrupted, short-circuit duration* $T_k$, *it will be conservative to conclude that the chosen fuse will operate in approximately 90 ms.*

## 15.11 EQUATIONS FOR CALCULATION OF INCIDENT ENERGY IN DC SYSTEMS

The power in DC or single-phase arcs can be expressed as:

$$P_{arc} = V_{arc} I_{arc} = I_{arc}^2 R_{arc}. \tag{15.51}$$

Therefore, the arc energy can be calculated from:

$$E_{arc} = I_{arc}^2 R_{arc} t_{arc} \; J, \tag{15.52}$$

where $t_{arc}$ is the arcing time in seconds. $I_{arc}$ is in amperes. $R_{arc}$ in ohms. This gives the energy in watt-seconds or joules. To convert to calories, multiply by a factor of 0.239

$$E_{arc} = 0.239 I_{arc}^2 R_{arc} t_{arc} \text{ cal}. \tag{15.53}$$

In the open air, the heat transfer depends upon the spherical energy density [14]. Based upon radiant heat transfer:

$$E_s = \frac{E_{arc}}{4\pi d^2} \text{ cal/cm}^2, \tag{15.54}$$

where $d$ is the distance from arc in cm.

TABLE 15.5. Optimum Values of *a* and *k*

| Enclosure | Width (mm) | Height (mm) | Depth (mm) | *a* (mm) | *k* |
|---|---|---|---|---|---|
| Panel board | 305 | 356 | 191 | 100 | 0.127 |
| LV switchgear | 508 | 508 | 508 | 400 | 0.312 |
| MV switchgear | 1143 | 762 | 762 | 950 | 0.416 |

When an arc is initiated in an enclosure, it has the focusing effect on the energy. The spherical density component is replaced by a value $E_1$, which takes into account the focusing effect of the enclosure [14]:

$$E_1 = k\frac{E_{arc}}{a^2 + d^2}. \tag{15.55}$$

The values of *a* and *k* from Ammermann et al. [14] are shown in Table 15.5.

**Example 15.9**

Continuing with Example 15.8, calculate the incident energy release, arcing current profile as shown in 15-14 and protection through a 1000 A, class L fuse, as discussed in Example 15.8.

Referring to Chapter 6, much alike calculations for *decaying* AC short-circuit currents; the procedure can be applied to DC arc flash calculations, with *rising* short-circuit currents also.

We calculated an average arcing current of 13 kA and arcing time of 90 ms in Example 15.8. This gives an arcing resistance of 3.21 mΩ. Then from Equation (15.53):

$$E_{arc} = 0.239(13.0)^2 \times 10^6 \times 3.21 \times 10^{-3} \times -0.09 = 11668.9 \text{ cal}.$$

From Equation (15.54) and considering $d$ = 18 in = 45.7 cm, $E_s$ = 0.444 cal/cm$^2$

In this calculation, we ignored the energy during the first 6 ms, when the current rises to 14.2 kA. Approximately, if an average current of 7.1 kA is considered, with an arcing resistance of 3.13 mΩ, it adds 226 J to the $E_{arc}$ calculated above. This will give $E_s$ = 0.453 cal/cm$^2$.

To calculate the hazard level in an enclosure, use Equation (15.55). This gives:

$$E_1 = 0.127\frac{11668.9}{10^2 + 45.7^2} = 0.68 \text{ cal/cm}^2.$$

The incident energy levels calculated are low.

**Example 15.10**

A three phase rectifier is connected on the AC side to a 2400-V primary system, with a short-circuit current of 30 kA and $X/R$ = 15. The rectifier transformer is rated at 2000 kVA, 2.4-0.48 kV, percentage impedance 5.5, $X/R$ = 7.5. Neglecting the impedance of any smoothing reactor or secondary cable connections, calculate the DC short-circuit current on the 480-V bus, the arcing current, and the incident energy release.

Using the methodology of previous examples and pertinent equations, short-circuit and arcing current profile is shown in Figure 15.16. A gap of 20 mm is considered. Note that the peak short-circuit current is 103.8 kA, and the peak arcing current is 92 kA = 88.6% of the peak short-circuit current. Compare this with NFPA 70E, 2012 calculations, which recommend that the arcing current can be considered = 50% of the short-circuit current. The NFPA recommendations are only approximate based on maximum power transfer theorem, and can lead to serious errors.

Considering an average arcing current of 46 kA, until it reaches its peak value of 92 kA, the arc resistance is 1.06 mΩ. Then the energy release in 11.2 ms, until the arcing current reaches its peak is:

$$(0.239)(46)^2 \times 10^6 \times (1.06) \times 10^{-3} \times (11.2 \times 10^{-3}) = 6004 \text{ cal}$$

At a distance of 18 in, the incident energy release is 0.229 cal/cm$^2$. Practically, this can be achieved if a fast current-limiting semiconductor fuse is provided, which isolates the fault before the peak arcing current of 92 kA is reached. The fuse will operate in less than 1/2 cycle, faster than 11.2 ms arcing time considered above. In fact, the fuse will limit the peak to less than 92 kA and operate even faster, thus the energy release will be much less than calculated above.

Another noteworthy point is that the arcing cannot start when the current is rising from zero value. It has to reach a certain minimum value for the arc to initiate. Thus, the calculation above of averaging the current over the entire period until it reaches its peak is conservative.

If a circuit breaker that is not current limiting and allows the peak to pass through is used, the quasi-steady-state current, say 55 kA in Figure 15.16 can be considered for arc flash calculations. This magnitude of current gives an arcing resistance of 0.902 mΩ. A three-cycle breaker opening time gives: $E_{arc}$ = 32606 cal and $E_s$ = 1.242 cal/cm$^2$. This is higher than the energy calculated with current-limiting fuse practically, the circuit breaker will open in less than three cycles. These calculations are for arcing in the open air. For arcing in an enclosure, low voltage switchgear, the calculated values are:

0.507 cal/cm$^2$ semiconductor fuse

2.76 cal/cm$^2$; three cycle circuit breaker allowing the peak to pass through.

NFPA recommendations are that in an enclosure an energy equal to twice the energy calculated in open air can be used, irrespective of the type of equipment.

## 15.12 PROTECTION OF THE SEMICONDUCTOR DEVICES

The semi-conducting devices like diodes, SCRs, GTOs, cannot be allowed to be damaged on short-circuit currents. Unlike transformers, motors, or cables, these do not have much thermal withstand capability. Invariably, these are protected by high speed $I^2t$ limiting current limiting fuses or circuit breakers. These operate fast within less than

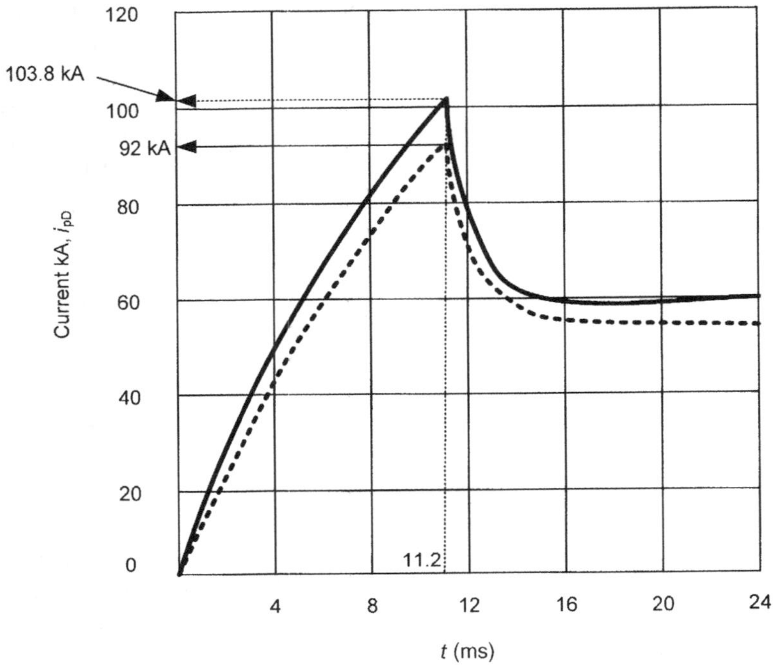


Figure 15.16. Calculated short-circuit and arcing current profiles (see Example 15.10).

a cycle and limit the $I^2t$ let-through. This $I^2t$ of the fuse is coordinated with the device $I^2t$ capability of the semi-conducting device.

A manufacturer's data on fast acting and $I^2t$ limiting fuses for semiconductor devices gives:

- application voltage 450VDC
- current range 35–1000 A
- interrupting rating 79 kA
- for a 1000 A fuse clearing $I^2t$ = 500,000 $A^2$s.

Similar data for a 1200 A fuse is:

- application voltage 500VDC, current range 10–1200 A, interrupting rating 100 kA, for a 1200 A fuse
- clearing $I^2t$ = 100,000 $A^2s$ melting $I^2t$ = 800,000 $A^2$s.

Consider that a semiconductor fuse of 1200A is provided for the protection of semiconductor devices. Then from Equation (15.52), the maximum energy release is:

$$E_{arc} = 900{,}000 \times R_{arc}.$$

Here, we substitute 900,000 $A^2$s from the fuse data.

$R_{arc}$ reduces with the increase of arcing curren; we calculated approximately I mΩ in Example 15.10. Then the maximum energy release is: 215 cal and $E_s$ = 0.008 cal/cm$^2$. In an enclosure approximately double the calculated value = 0.16 cal/cm$^2$.

## 15.12.1 Controlled Converters

The converters with grid control can be called controlled converters, and most of these in practical applications are controlled converters, firing angle adjustable from $0 < \alpha < 180°$ (angles >90° correspond to inverter operation). Yet an adjustable speed drive (ASD) system may have front-end uncontrolled bridge rectifier circuit.

The damage for a short-circuit current in a controlled converter system is limited by grid-control protection schemes. This enables a grid firing circuit to detect abnormal conditions and block grid pulses. The current flow to a converter short-circuit is limited to *one cycle* by the normal action of the grid control protection systems. Figure 15.17

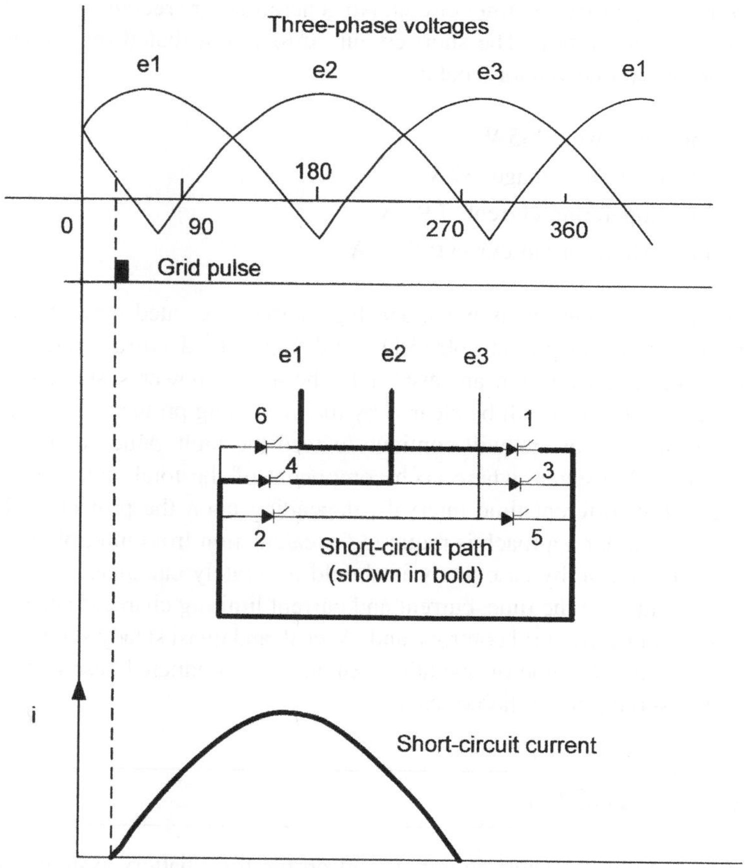


Figure 15.17. Short-circuit current of a converter with grid (gate) control. *Source*: Reference [9].

shows the DC short-circuit in a full-wave SCR converter where the grid protection system is operative [9].

**Example 15.11**

A grid-controlled full-wave converter gives a peak DC short-circuits current of 100 kA. Calculate the incident energy release.

Let the arcing current peak be 85% of the maximum short-circuit current (see example 15.10), the short-circuit profile be a sinusoid, and the average arcing current 54 kA. Then the incident energy release for a period of 16.67 ms is 215.53 cal, and $E_s$ is 0.0082 cal/cm$^2$.

The above analysis and calculations indicate that the arc flash energy in DC systems is rather low because of the requirement of protecting the semiconductor devices through fast acting fuses and circuit breakers or by grid control. Yet a case-by-case analysis should be made, and the results may vary with respect to protective devices used and the system configurations. Figure 15.17b shows that the rise time to peak of the short-circuit current of DC motor is 640 ms; the fault clearance time will be higher compared with a battery or rectifier circuit. No generalities are recommended. One such example is solar generation. The short-circuit current contributed by a solar array is small. According to one vendor's data:

- rated output power: 235 W
- rated open circuit voltage: 37 V
- maximum operating current: 7.97 A
- maximum short-circuit current: 8.54 A.

The short-circuit current is not much higher than the rated current. When such arrays are assembled to provide 600 VDC and 125 A load current ratings, and the assembly is connected through an inverter to the 480-V power system, short-circuit contributed by the inverter will be cleared by the fast acting protective devices, but the current from the solar arrays will continue to feed the fault. Much akin to arc flash calculations in AC systems, where each component of the total short-circuit current may be cleared at different time intervals, depending upon the protective devices in these circuits; a similar approach is required for calculation from multiple DC sources.

To reiterate, a case-by-case analysis should accurately calculate (1) short-circuit and arcing currents, (2) the time–current and current limiting characteristics of protective devices, (3) converter grid controls, and (4) peak and quasi steady state short-circuit currents to calculate the time of operation. An analysis conducted based on guidelines in this chapter should give reliable results.

## REVIEW QUESTIONS

1. Calculate and plot the short-circuit current profile for a battery system with details as follows: lead acid battery, 240 V, 120 cells, 400 A rating at 8-hour rate of

discharge, and 1.75 V per cell at 25°C. Each cell has a length of 15.5 in, width 6.8 in, and height 10 in. The cells are arranged in two-tier configuration in four rows, 30 cells per row. Intercell connectors are of 1 in × 1/2 in cross-section, resistance 0.0321 mΩ/ft. Calculate the battery internal resistance. The battery is connected through a cable of resistance 0.002 Ω and inductance 15 μH. The fault occurs at the end of the battery cable.

2. Calculate and plot the terminal short-circuit current of a DC motor of 50 hp, 230 V, 690 rpm, armature current 178 A, and transient resistance 0.07 Ω. Additional motor data: $\tau_F$ = 1.0s, $J$ = 2 kg/m$^2$, $L_{0F}/L_F$ = 0.3.
3. Calculate and plot the short-circuit current profile for a fault on the DC side of a rectifier system in the following configuration: 480-V, three-phase AC source fault levels 20 kA, $X/R$ = 5, 480–230-V, three-phase 300-kVA rectifier transformer, Z = 3.5%, $X/R$ = 5, and the DC side equivalent resistance and inductance equal to 0.001 Ω and 3 μH, respectively.
4. Sum the partial fault currents calculated in Problems 1, 2, and 3, and calculate the maximum short-circuit current and time to peak. What should be the peak short-circuit rating and interrupting rating of a general-purpose DC circuit breaker?
5. Calculate the arcing current and arcing resistance. Plot the arcing current profile and the bolted current profile together.
6. A 1000-A fuse of the characteristics shown in Figure 15.15 is applied as an incoming to the common bus being fed by the three sources of Problems 1, 2, and 3. Calculate the arcing time.
7. Continuing with Problem 6, calculate the incident energy for a fault in low voltage DC panel.

## REFERENCES


1. General Electric Company, *GE Industrial Power System Data Book*, General Electric, Schenectady, NY, 1978.
2. IEEE, Standard C37.14. *IEEE Standard for Low-Voltage DC Power Circuit Breakers Used in Enclosures*, 2002.
3. IEEE, Standard 946. *DC Auxiliary Power Systems for Generating Stations*, 1992.
4. AIEE Committee Report, "Maximum short-circuit current of DC motors and generators. Transient characteristics of DC motors and generators," AIEE Trans., vol. 69, pp. 146–149, 1950.
5. A.T. McClinton, E.L. Brancato, and R. Panoff, "Maximum short-circuit current of DC motors and generators. Transient characteristics of DC motors and generators," AIEE Trans., vol. 68, pp. 1100–1106, 1949.
6. A.G. Darling and T.M. Linville, "Rate of rise of short-circuit current of DC motors and generators," AIEE Trans., vol. 71, pp. 314–325, 1952.
7. A.D. Stokes and W.T. Oppenlander, "Electrical arcs in open air," J. Phys. D Appl. Phys., vol. 24, no. 1, pp. 26–35, Jan. 1991.

8. R. Wilkins, "Simple improved equations for arc flash hazard analysis," in Proc. IEEE Electrical Safety Forum, pp. 1–12, Aug. 30, 2004.
9. IEEE Standard, 551. *IEEE Recommended Practice for Calculating Short-Circuit Currents in Industrial and Commercial Power Systems*, 2006.
10. D.R. Doan, "Arc flash calculations for exposures to DC systems," IEEE Trans. Industry Appl., vol. 46, no. 6, pp. 2299–2302, Nov./Dec. 2010.
11. IEC, Standard 61660-1. *Short-Circuit Currents in DC Auxiliary Installations in Power Plants and Substations*, 1997, corrigendum, 2000.
12. International Electrotechnical Commission, http://www.iec.ch.
13. IEEE Standard, C37.16. *IEEE Standard for Preferred Ratings, Related Requirements, And Application Recommendations for Low-Voltage AC (635 V and Below) and DC (3200 V and Below) Power Circuit Breakers*, 2009.
14. R.F. Ammermann, T. Gammon, P.K. Sen, and J. Nelson, "DC-arc models and incident energy calculations," IEEE Trans. Industry Appl., vol. 46, no. 5, pp. 1810–1819, Sept./Oct. 2010.

# 16

# APPLICATION OF ETHERNET AND IEC 61850 COMMUNICATIONS

Modern industrial and utility power systems are complex. From the protection, communication, and control view points, a variety of intelligent electronic devices (IEDs) must be connected together and operate in synchronism with reliability. The automation requires high-speed data transfer and reliability for effective control and managements. Consider, for example, Modbus RTU Protocol. It is a master–slave protocol, and can address up to 254 slaves, primarily defined on RS485, but can be operated on the Ethernet. The data is addressed via 2-byte registers and Modbus packet can transmit up to 120 registers per message.

The Distributed Network Protocol (DNP) was developed by Harris Distributed Automation Products in Calgary, Canada, in 1993. The DNP user group is a forum of 300 users. The present version is DNP3, which is defined in three distinct levels. Level 1, DNP V3.00 implementation, DNP-L1, has the least functionality and is for simple IEDs, and DNP V3.00 level 2 describes a subset of protocols slightly larger than level 1. DNP-L3 describes a set of protocols larger than level 2; it has the most functionality for SCADA master station communication front-end processors. Some benefits are interoperability between multilayer devices, fewer protocols to support in the field, reduced software costs, and shorter delivery schedules. It is implemented by various

*Arc Flash Hazard Analysis and Mitigation*, Second Edition. J.C. Das.

Published 2021 by John Wiley & Sons, Inc.

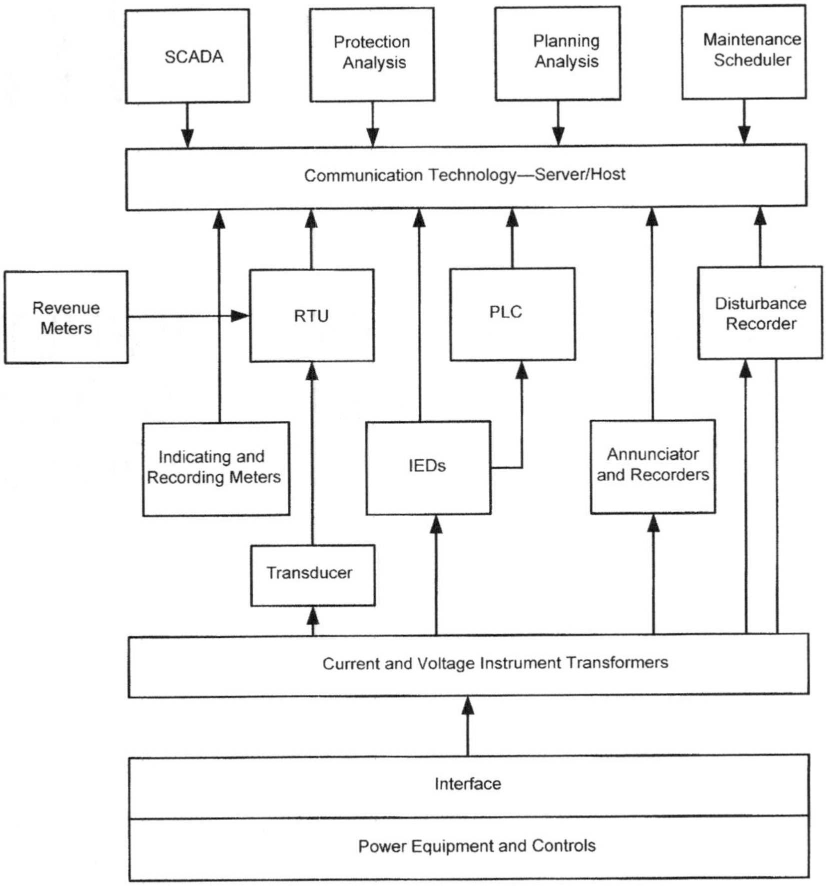


Figure 16.1. Power station substation automation (SA) functional diagram.

vendors and is based upon IEC 870-5 standards, and is in process of becoming an IEEE standard. It allows multiple masters, and layered protocols allow mix-and-match features. It has address capability of over 65,000 devices and 4 × 109 data points of each data type. Figure 16.1 is a basic block circuit diagram of substation automation.

## 16.1 IEC 61850 PROTOCOL

IEC 61850 [1] is the international standard for substation/plant automation systems; it defines the communication between devices in a substation or facility and the related system requirements and supports all automation functions and their engineering. The technical approach makes IEC 61850 flexible and adaptable to future needs. It has become synonymous with:

TABLE 16.1. Description of IEC 16850 Part Numbers

| Part Number | Description |
|---|---|
| 1 | Introduction and overview |
| 2 | Glossary |
| 3 | General requirements |
| 4 | System and project management |
| 5 | Communication requirements for functions and device models |
| 6 | Configuration language for electrical substation IEDs |
| | Abstract communication services (ACSI) |
| 7.1 | Principles and models |
| 7.2 | ACSI |
| | Data Models |
| 7.3 | Common data classes |
| 7.4 | Compatible logical node classes and data classes |
| | Mapping to real communication networks (SCSM) |
| 8.1 | Mapping to MMS and to ISO/IEC 8802-3 |
| 9.1 | Sampled values over serial unidirectional multidrop point-to-point link |
| 9.2 | Sampled values over ISO 8802-3 |
| | Testing |
| 10 | Conformance testing |

- designator for the substation secondary systems with high degree of integration
- reduced costs
- greater flexibility
- communication networks replacing hardwired connections
- plug-and play functionality
- reduced construction and commissioning times.

IEC 16850 is a large document consisting of 1850 pages; experts from 20 countries participated in its development. A description of the various parts is shown in Table 16.1.

## 16.2 MODERN IEDs

Developments in IED hardware design and developments of high-speed peer-to-peer communication protocols have resulted in a new generation of IEDs. These protective and control relays (described as MMPRs in previous chapters) have the capability of accepting multilevels of current and voltage inputs and to analyze these values at increased speeds. The advantage of using these microprocessor-based IEDs are simplifications of device-to-device wiring, cost reduction, and reliability (Chapter 7).

IEDs can be classified by their functions. Common types of IEDs include relays, circuit breaker controllers, voltage regulators, and so on. An IED can perform more than one function, taking advantage of microprocessor technology. An IED may have an operating system, like Linux, running in it.

An efficient way of reduction in device-to-device wiring is to use high-speed peer-to-peer IEC 61850 Generic Object Oriented Substation Event (GOOSE) messaging between protective relays. It is a fast, reliable, interoperable device-to-device communication. GOOSE messaging has the following characteristics:

- It has time critical data, like trips, blocks, and interlocks.
- The data transfer is initiated only on occurrence of an event, say a trip, closing or opening of a breaker, change in the status of an arc flash maintenance mode switch.
- Data can be sent periodically for self-test and reliability.
- Primarily, it is local; but application to wide area network (WAN) is possible. A WAN may be defined as a computer network that covers a broad area. It can be a network whose communication links regional or cross-metropolitan boundaries, that is, utility networks. On the other hand, a local area network (LAN) is a computer network covering a small physical area like a substation or a facility.
- It can also support virtual local area network that are created and configured to handle bandwidth more efficiently and provide additional network security.
- It is user-defined, a set of data that is "published" on detection of a change in any of the contained data items.

## 16.3 SUBSTATION ARCHITECTURE

A substation network is connected to the outside wide area network (WAN) via a secure gateway. Outside remote operators and control centers can use abstract communication service interface (ACSI) defined in part 7-2 of IEC 61850 to query and control devices in the substation. A substation bus is utilized as a medium bandwidth Ethernet, which carries all the ACSI requests and responses and substation event messages. There is another kind of bus called process bus for communications in each bay. A process bus connects the IEDs to the traditional devices like merge units and is realized as a high-bandwidth Ethernet network. A substation usually has one global substation bus but multiple process buses, one for each bay. Interactions in a substation fall under three categories:

- data gathering and setting
- data monitoring and reporting
- event logging.

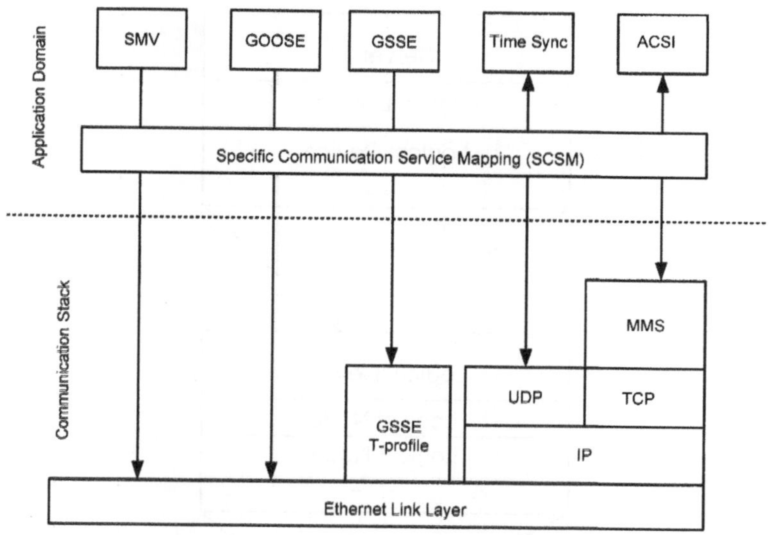


Figure 16.2. IEC 61850 communication profiles.

In IEC standard, all inquiries and control activities toward physical devices are modeled as getting or setting the values of the corresponding data attributes, and data monitoring/reporting provides an efficient way to track system status.

## 16.4 IEC 61850 COMMUNICATION STRUCTURE

IEC defines a rather complicated communication structure (Figure 16.2). Five kinds of protocols are defined in the standard:

1. ACSI
2. GOOSE
3. the generic substation status even (GSSE)
4. the sampled measured value multicast profile (SMV)
5. the time-synchronized profile.

ACSI is the primary interface in IEC. It is the interface by which applications talk with servers, and is an important part of logical connections between two nodes. An object-oriented approach is adopted in the design of ACSI, which includes a hierarchal and comprehensive data model and a set of available services in each class in this data model. Though the data model is defined outside the scope of ACSI, actually, it is a part of it. Figure 16.3 shows the hierarchical data models of IEC. The server is the topmost component in this hierarchy. It serves as the common point of physical devices and logical objects. Each server hosts several files or logical devices. A logical device

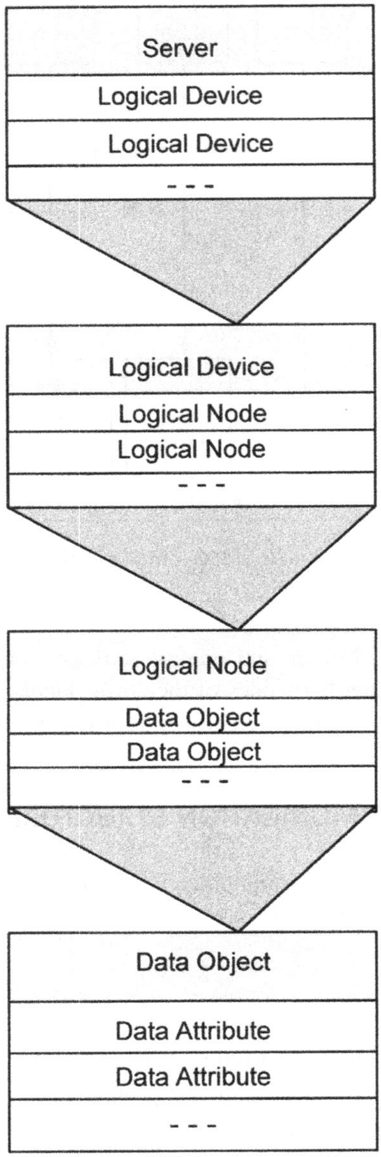


Figure 16.3. Hierarchy of IEC 61850 data model.

TABLE 16.2. Logical Node Groups

| Logical Node | Description |
|---|---|
| L | System LN (2) |
| P | Protection (28) |
| R | Protection related (10) |
| C | Control (5) |
| G | Generic (3) |
| I | Interfacing and archiving (4) |
| A | Automatic control (4) |
| M | Metering and measurements (8) |
| S | Sensor and monitoring (4) |
| X | Switchgear (2) |
| T | Instrument transformers (2) |
| Y | Power transformers (4) |
| Z | Further power system equipment (15) |

Examples:
PDIF: Differential protection.
RBRF: Breaker failure.
MMXU: Measurement unit.
YPTR: Power transformer.

is the logical correspondence of a physical device—it is basically a group of logical nodes.

## 16.5 LOGICAL NODES

Information is exchanged between all IEDs, which comprise the system, and, more precisely, the data are exchanged between the functions and subfunctions resting in the devices. The smallest part of the function that exchanges data is called logical node (LN) in IEC 61850. The logical node groups, first letter listed, are shown in Table 16.2.

Data exchanged between logical nodes is modeled as data objects. A logical node usually contains several data objects. Each data object is an instance of DATA class and has a common data type. Data attributes are typed and restricted by some functional constraints. Functional constraints provide a way to organize all data attributes in a logical node by functions. Figure 16.4 shows the circuit breaker XCBR information tree. Figure 16.5 shows the anatomy of IEC 61850 object names.

## 16.6 ETHERNET CONNECTION

IEC 61850 uses an Ethernet connection as the physical medium of communication between IEDs. Logical I/O via Ethernet communication is used in place of traditional hardwired systems to exchange information between protective IEDs. This may include

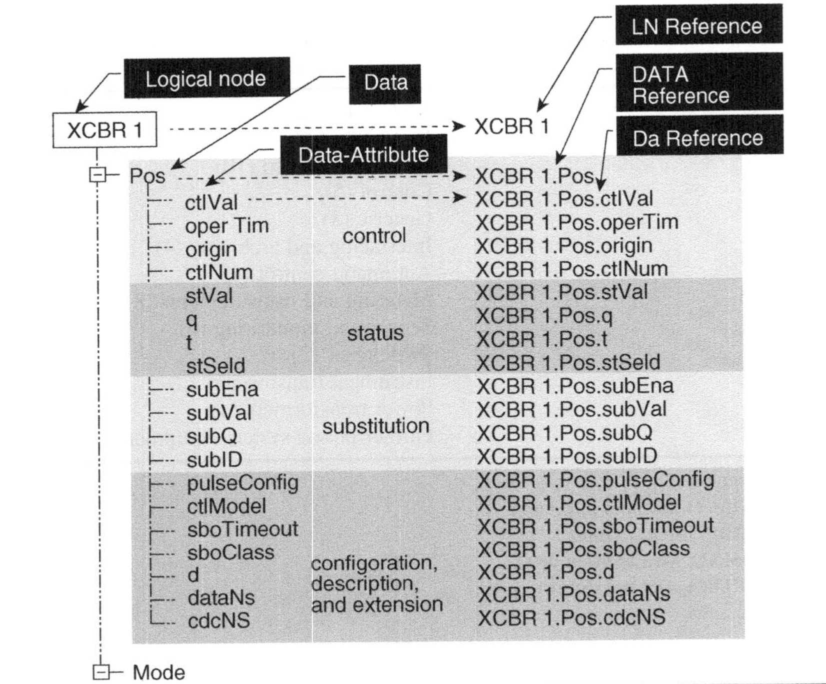


Figure 16.4. Circuit breaker Model-XCBR information tree.

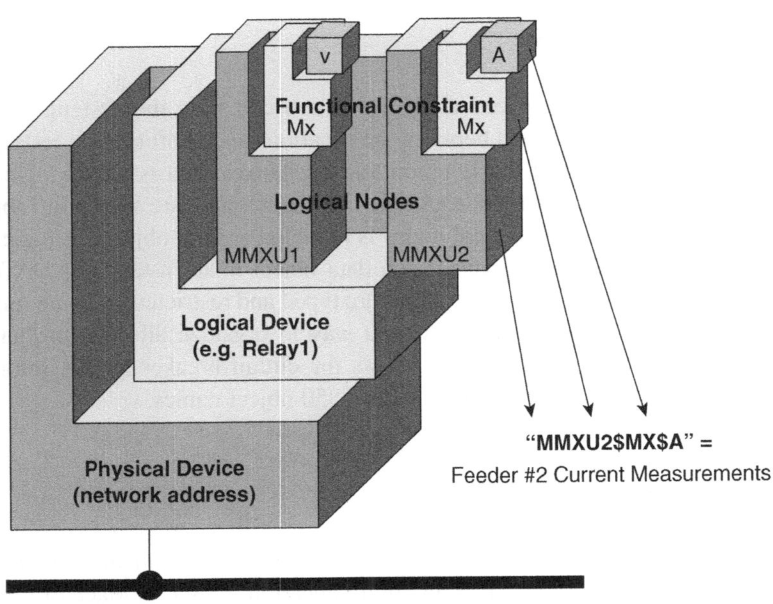


Figure 16.5. Anatomy of IEC 61850 object nomenclature.

connected device I/O, protective elements statuses, and programmable logic states. The implementations are able to send messages between protective relays at speed of 1–4 ms. Analogue data can be exchanged through GOOSE messaging. Some applications are:

- high-speed bus transfer schemes
- switching set groups
- load shedding
- bus protection
- breaker failure initiate
- transfer tripping
- remote start–stop commands
- blocking and tripping schemes.

As an example consider the zone interlocking scheme described in Figure 14.3. If IEDs and Ethernet connections are used, the blocking signal from the feeder IEDs to the main breaker IED can be transmitted fast:

| | |
|---|---|
| IEC61850 100 Mbps LAN response time | 2–4 ms |
| Feeder relay (IED) response time | 2 ms |
| Main circuit breaker (IED) response time | 2 ms |
| Input recognition time | 0–2 ms |
| Output recognition time | 2–4 ms |
| Total | 8–14 ms |

This will further reduce the arc flash incident energy release.

Also, the use of Ethernet allows simultaneous use of multiple protocols on the same hardware, for example, Modbus and DNP.

IEC introduced a common language that can be used to exchange information, manufacturer independent. Export of IED's description into a common XML-based language is allowed, and the so-called ICD file (IED Capability Description) contains all information about the IED, which allows the user to configure a GOOSE message. This configuration can be performed by IEC61850 System Configurator. All ICD files get imported into the IEC61850 System Configurator and the GOOSE message can be programmed by specifying the sender (publisher) and receivers (subscribers) of a message. The whole descriptions of the system, including description of GOOSE messages, get stored in a Substation Configuration Description (SCD) file. Each proprietary tool is able to import this SCD file and extract the information needed for the IED.

Figure 16.6 shows that one device publishes information, and the subscriber devices receive it. The reaction of each subscriber depends on its configuration and

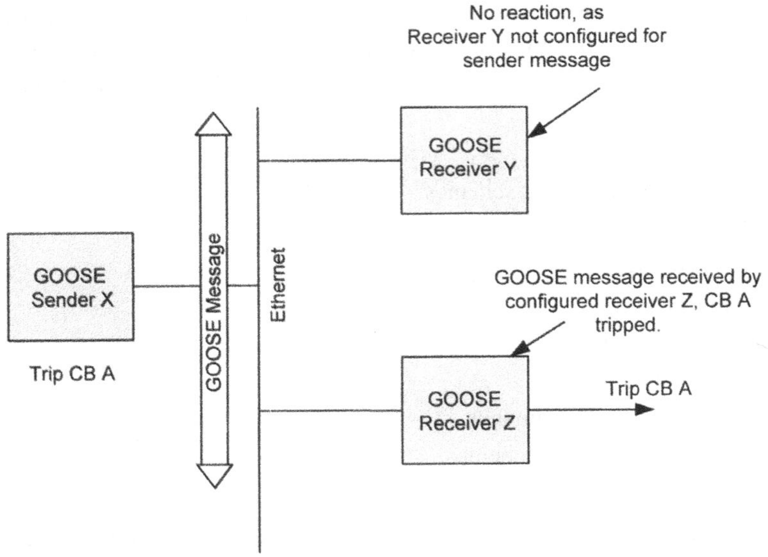


Figure 16.6. GOOSE messaging.

functionality. This figure shows that the sensor *X* trip command, "trip CB A," has no reaction on GOOSE receiver *Y*, but GOOSE receiver *Z* is configured to receive it and trip the breaker. Note that the figure shows only one receiver responding, but a GOOSE message can be received and used by many receivers.

Replacing hardwired connections with digital communications require new understanding and practices—a transition from physical terminations to logical interconnections between IEDs that open up a vast opportunity of data communications and controls. It is fast, deterministic, and automatically connected to predefined logical terminations between the publisher and receivers. Experience indicates that over 50% of copper terminations, associated cost and labor, and potentialities for mistakes are eliminated. The key concepts are summarized:

- *Interoperability.* The ability of system components to function effectively with other components, including communication among multiple vendor components using GOOSE protocols.
- *Interchangeability.* The ability to interchange a system component with another manufacturer's component that fulfills the same functions.
- *Vendor Independent.* Selection of system components not dependent upon one specific supplier.
- *Substation Architecture.* See Section 16.8 below.
- *Networking Features.* Compliance with IEEE 1613 [2, 3] and IEC 61850.

## 16.7 NETWORKING MEDIA

Networking media can be divided into two categories:

1. twisted pair copper
2. fiber optic

The major considerations of selection between these two options are:

- The "Optical Power Budget," which can be described as the maximum permissible attenuation of the light signal as it travels from the transmitter to receiver while still permitting reliable communication
- The distance involved, which is a major consideration
- The speed of communication, which is defined in MB(megabits) per second
- The cost and reliability.

Ethernet over fiber is becoming the medium of choice where long distances are involved and immunity from electromagnetic interference (EMI) is desired. The cost difference between copper and fiber cables is not much; however, the fiber Ethernet transceivers are more expensive. Table 16.3 provides some basic idea of the optical power budget and the distances. The terms used in this table are explained further in the following sections.

### 16.7.1 Copper Twisted Shielded and Unshielded

The interface is defined by speed, the modulation type (Base), and the physical interface. For example, T or TX is twisted pair and FL or FX is fiber. Ten Base T and 100 Base TX are the most common twisted pair copper media standards. The cable can be shielded or unshielded twisted pair. Unshielded twisted pair (UTP) has several categories 1 through 7, with their specific fields of applications. The cable consists of four pairs of wires, color coded and terminated in RJ45 connectors; the maximum permitted length is 100 m for the unshielded and 150 m for the shielded (STP) cable. In practical

TABLE 16.3. Port Description, Typical Distance, and Power Budget

| Port Type | Description | Typical Distance | Power Budget |
|---|---|---|---|
| 10/100 Base T | 10/100-MB RJ45 copper unshielded | 100 m | |
| 10/100 Base T | 10/100-MB RJ45 copper shielded | 150 m | |
| 100 Base FX | 100 MB multimode SC fiber optic (full duplex) | 2 km | 14 dB |
| 100 Base FX | 100-MB single-mode SC fiber optic | 20 km | 12.5 dB |
| 100 Base FX | 100-MB single-mode SC fiber optic | 70 km | 32.5 dB |
| 1000 Base FX | 1-GB, single-mode 1550 LC fiber optic | 70 km | 20 dB |

installations, the distance is a serious limitation, and copper twisted pair is not in much use.

## 16.7.2 Fiber Optic Cable

The fiber optic cable is classified into multimode and single mode. Obviously, the single-mode fiber can be used for much longer distances and gives less attenuation. Both the multimode and the single-mode fiber cables can support a wide range of wavelengths, notably 820, 1300, and 1550 nm. Figure 16.7 shows single-mode and multimode fiber cables; the outer clad is 120 μm in both cases. However, the cores are of vastly different diameters. This figure shows transmission through these two cable types. The differences are that for a range of light injection angles, there is a total reflection of the light being transmitted down the core—analogous to a perfect mirrored surface. Multimode has a larger core diameter and supports multiple injection angles, resulting in a substantial amount of input pulse spreading. A single-mode fiber continuously focuses the light into the center. As a result, the single-mode fiber has less attenuation for a given wavelength of light.

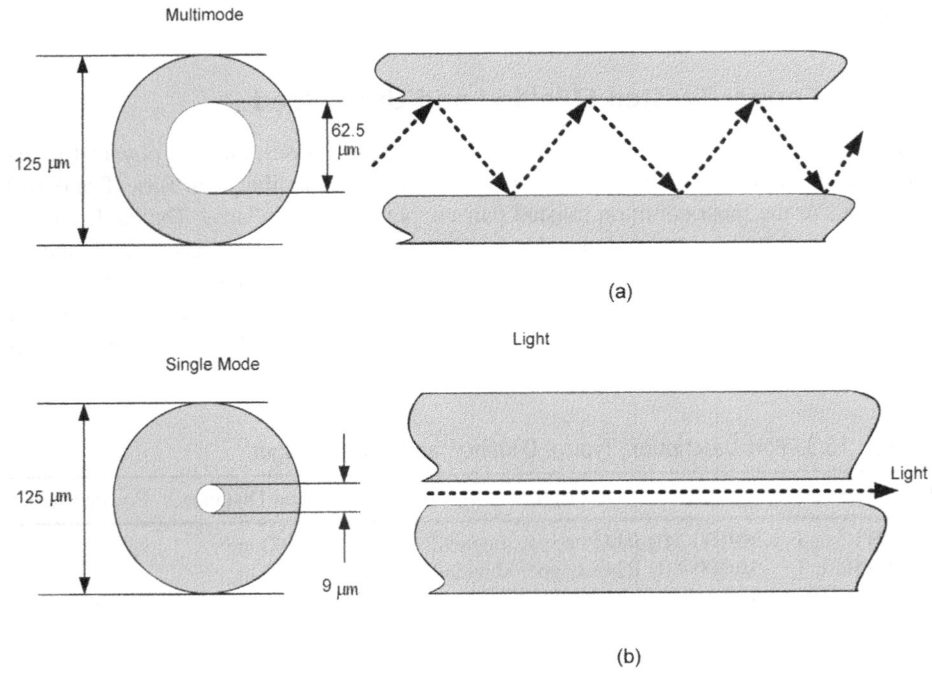


Figure 16.7. (a) Fiber optic multimode cable cross-section and light transmission; (b) fiber optic single-mode cable cross-section and light transmission.

The maximum distance can be calculated by worst-case optical power budget (OPB). The OPB can be considered as the maximum permitted attenuation of light signal, while permitting reliable communication.

$$OPB_{worst} = OPB - (1 + n)dB,$$

where $n$ = number of pairs, $n$dB gives insertion loss of $n$ pairs, and 1 dB is the loss for LED aging. Then the worst-case OPB can be divided by cable loss, say for a 62.5/125 μm, it will be of the order of 2.8 dB/km, while for 9/125 μm (single mode, 1550 nm), it is 0.2 to 0.25 dB/kM.

A half-duplex fiber optic system provides communication in both directions but only one direction at a time. The transmitted signal must stop before a response can be sent back. A full duplex or double duplex allows simultaneous communications in both directions. IEDs can send and receive data simultaneously over the link [4].

## 16.8 NETWORK TOPOLOGIES

Some network topologies are:

- star
- redundant star
- mesh
- ring.

The port formed by connecting one Ethernet switch to another is often called an uplink port. Figure 16.8a,b shows a redundant star and ring configurations, respectively, using redundant ports on the IEDs. In a single star, a single point failure causes a loss of communication; additional Ethernet switches are required, and the network recovery time is 5–6 ms per Ethernet switch. A redundant star provides higher availability at the cost of additional hardware.

A ring architecture, Figure 16.8b provides network redundancy, and using proprietary techniques, has a failure recovery time of 5 ms per Ethernet switch. Integrated Ethernet switches can be provided in the IEDs. The IED is internally connected to the Ethernet switch through internal hardware. The last IED can be connected to the first IED to form a ring network. When one of the component fails, it should be detected and the communication path reconfigured. IEEE standard performs this function through Rapid Spanning Tree Protocol (RSTP). The protocol sends messages to the various nodes in the network to detect the broken path and reconfigure. The "link pulses" are sent when idle, and used to detect connectivity and communication capability. When a receiver loses a link, it indicates a problem, and the communication path is routed in the other direction around the ring. The link loss function also allows IEDs to recover, where one of the two fiber optic cables connected to the IEDs is damaged. Upon detection of the broken transmit fiber; IEDs will switch to the secondary port.

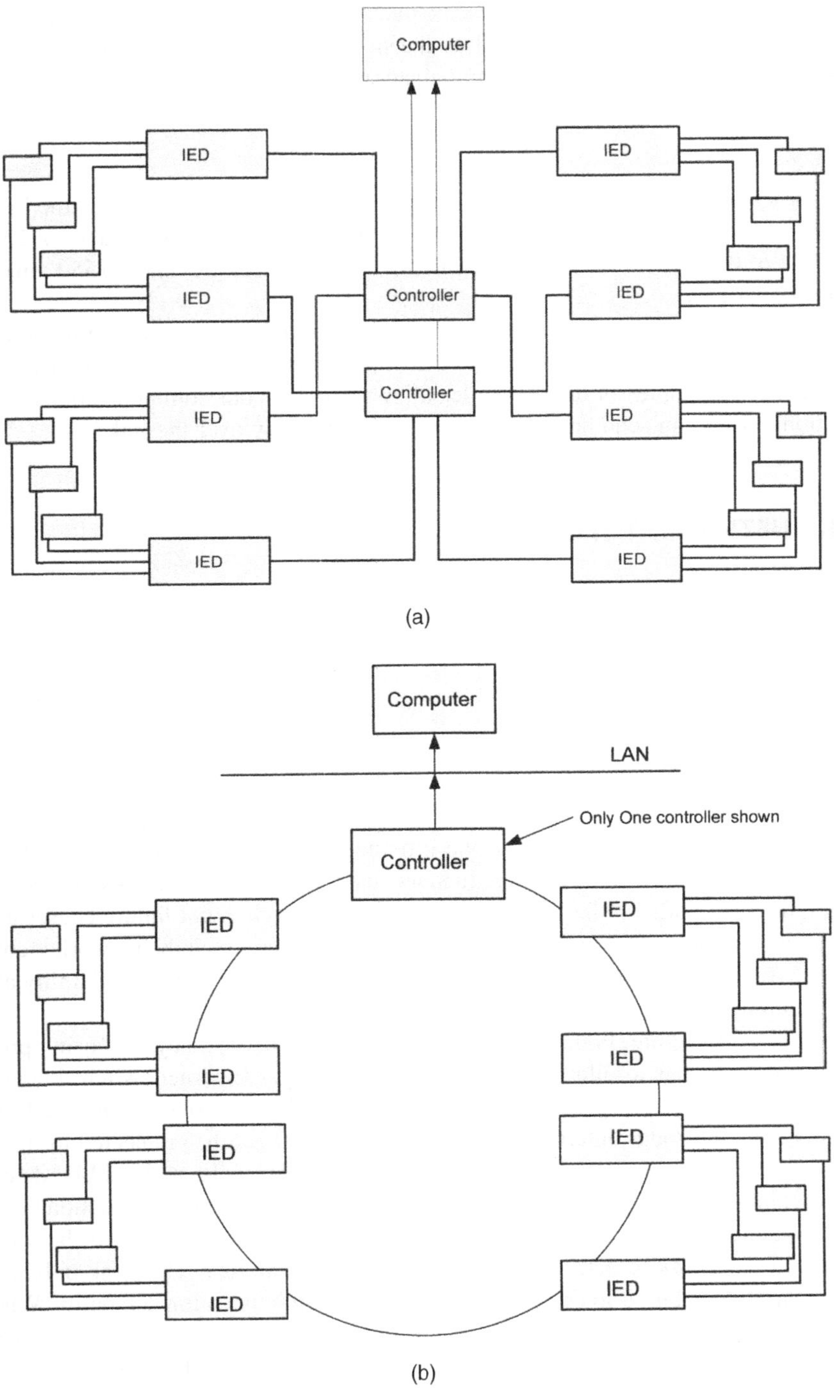


Figure 16.8. (a) Redundant star architecture using redundant ports on IEDs; (b) ring architecture using redundant ports on IEDs.

Architectures using Ethernet switches require one Ethernet switch for approximately 12 IEDs when using redundant fiber optic communications.

### 16.8.1 Prioritizing GOOSE Messages

Quality of service signifies prioritizing the traffic, so that the critical data is processed first. Considering a trip command and metering data communications—the trip command must be prioritized. IEC61850 GOOSE messaging provides a priority setting with eight levels of property. Processed in an Ethernet switch, the message with the highest priority is moved to the front of the queue. Reliability of critical messages is improved by using redundant Ethernet networks, redundant Ethernet ports, and power supplies.

IEEE Power System Relaying Committee Working Group WG119 [2, 3] lays down the following recommendations when using IEC 61850 for critical applications:

- Multiple switches are connected in a ring, so that there are two paths for any switch port used by a relay to any other such switch port. Ethernet switches include RTSP, by which the switches use a default path without calculating messages in the loop. One link in the loop is blocked. If the ring suffers a break or one switch fails, the switches can detect the path lose, unblock the spare path, and set up new routing.
- Many GOOSE-capable relays have primary and failover communication ports. Provide two switches in two groups; connect the relay primary port to one group, and the failover port to another switch group.

### 16.8.2 Technoeconomical Justifications

- Cost reduction by replacing many control cables with fiber optic links for alarm and control signals
- Reduced construction for trenches and cable raceways for new substations
- Maintenance cost reduction by taking advantage of multifunction relays to reduce the number of IEDs
- Panel-to-panel cable reduction using extensive communication infrastructure and GOOSE messaging
- Reduced control building and associated costs
- Reduction and elimination of auxiliary relays reduce failure points, panel size and wiring
- Pretested systems, as a part of quality assurance requirements
- Reduction of devices like push buttons, mechanical relays, and selector switches.

References [5–9] provide further reading.

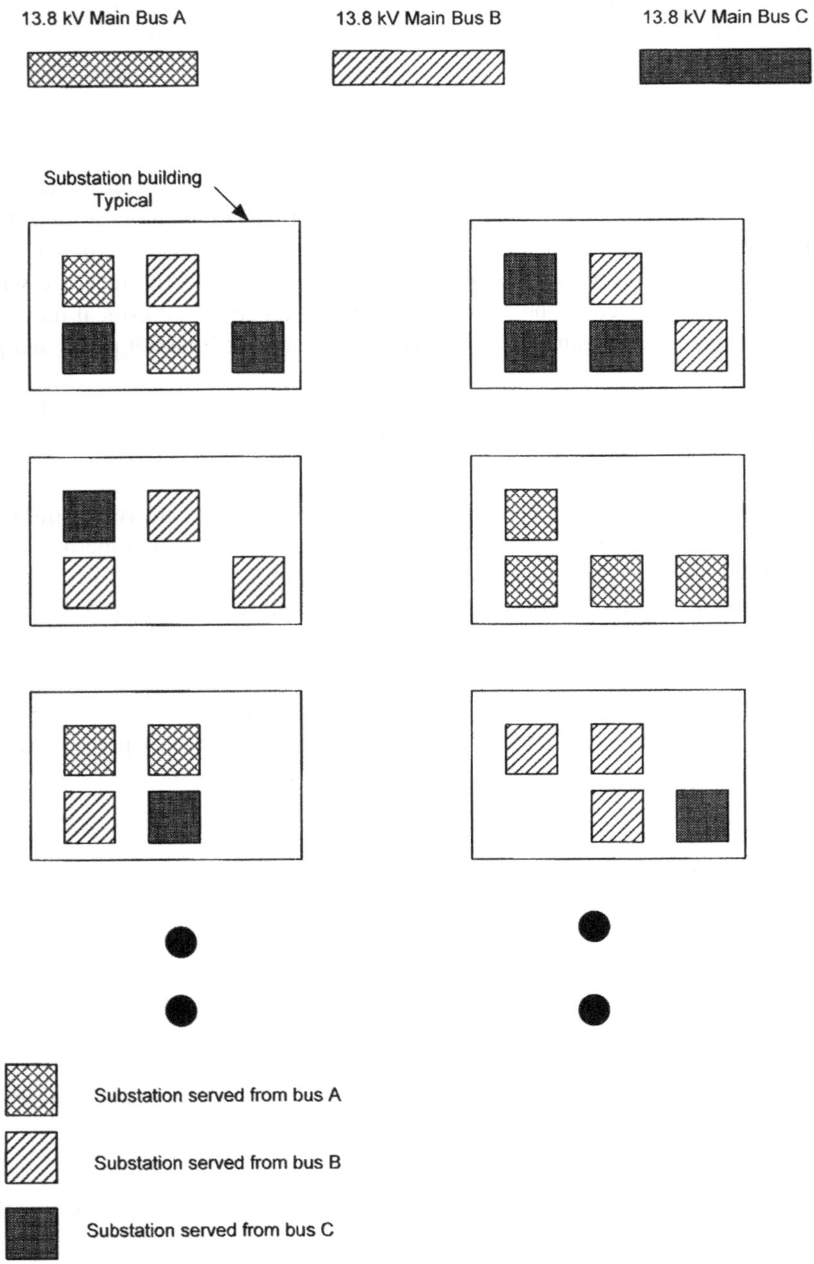


Figure 16.9. A schematic picture of substation layouts and connections in a large industrial distribution system.

## 16.9 APPLICATION TO ARC FLASH RELAYING AND COMMUNICATIONS

Consider a large industrial distribution system, load demand of 90 MVA, served through three main 13.8-kV buses, A, B, C, located some distance apart. These buses have plant generation and utility interconnections and are interconnected through reactors connected to a synchronizing bus. There are a total of 150 substations in the distribution system, and there are 35 substation buildings. Assume that all substations are indoor type and a substation building may house substations connected to three 13.8-kV buses (Figure 16.9). The substation buildings are spread over an area of 2 mi$^2$.

Consider that there are 150 smart IEDs, one in each substation. The desired operation is that as follows below.

One hundred fifty IEDs should trip 13.8 kV breakers if the maintenance mode switches hard-wired into the IED is actuated for live maintenance of the equipment (Chapter 14) and a fault occurs in the maintenance mode. Also, the metering data on the secondary of each substation is required to be communicated to a central control location along with status indications of the maintenance mode switches and some other breaker status. The system can be implemented by hard wired trips, transducers, and controllers; which will require considerable wiring, labor, and terminations. Also, remote tripping of breakers over long distances may need auxiliary relays extending the power supplies or providing redundant batteries and chargers.

Alternatively, the system can be configured with IEC GOOSE network in ring formation, using about 35 managed Ethernet switches and a couple of controllers. The controllers will be programmed to poll the IEDs, and data can be presented in a web-based HMI. A switch can be installed and programmed at the door of each substation so that it acts simultaneously on all the IEDs in a substation to bring alternate settings into operation, and a worker can safely enter the substation. Chapter 1 shows that when arc flash hazard is high, this is associated with much larger arc flash boundary. Bringing alternate settings on all the substations in a building by mere flip of a switch ensures that arc flash hazard, as well the arc flash boundary, is reduced to safe levels. The status of any IED, its settings, and metering data and breaker status, can be accessed on any computer connected to LAN. The system is expandable, and new devices can be added, with their unique addresses and programming.

## REVIEW QUESTIONS

1. Why is Ethernet ring architecture preferred? What is RSTP?
2. Describe four characteristics of a GOOSE message.

## REFERENCES

1. IEC61850 Ed. 1.0, *Communication Networks and Systems in Substations—All Parts*, 2011.

2. IEEE Standard, IEEE 1613. *Environmental and Testing Requirements for Communications Networking Devices Installed in Electrical Power Substations-2009.*
3. IEEE Standard, IEEE 1613a. *Environmental and Testing Requirements for Communications Networking Devices Installed in Electrical Power Substations, Amendment 1*, 2011.
4. C. Wester and M. Adamiak, "Practical applications of Ethernet in substations and industrial facilities," in Conf. Record, IEEE Pulp and Paper Industry Technical Conference, Nashville, TN, June 2011.
5. IEEE PSRC Working Group, "Redundancy considerations for protective relaying systems," in Conf. Record, Texas A&M Protective Relaying Conference, April 2010.
6. K.P. Brand, "The standard IEC61850 as prerequisite for intelligent applications in substations," in IEEE Power Engineering Society General Meeting, vol. 1, pp. 714–718, June 2004.
7. J. McDonald, "Substation automation and integration and availability of information," IEEE Power and Energy Magazine, pp. 22–31, March/April 2003.
8. D. Dolezilek, *IEC 61850: What You Need to Know about Functionality and Practical Implementation*, Schweitzer Engineering Laboratories, Pullman, WA, 2004, 2005, http://www.selinc.com.
9. E. Atienza, *Testing and Troubleshooting IEC 61850 GOOSE-Based Control and Protection Schemes*, Schweitzer Engineering Laboratories, Pullman, WA, 2004, 2005, http://www. selinc.com.

# APPENDIX A: STATISTICS AND PROBABILITY APPLIED TO ELECTRICAL ENGINEERING

In Chapter 3, we introduced statistical and probabilistic terms such as *mean deviation*, *standard deviation*, *skewness*, and *Kurtosis*, least square line, and *normal distribution*. This appendix provides a basic understanding.

## A.1 MEAN MODE AND MEDIAN

The frequency distribution data can be represented by a histogram, frequency polygon, frequency curve, bar chart, and pie diagrams. The arithmetic mean is given for $n$ numbers, $x_1, x_2 \ldots x_n$ is given by:

$$x_{\mathrm{m}} = \frac{\sum x}{n}. \tag{A.1}$$

If $x_1$ occurs $f_1$ times, $x_2$, occurs $f_2$ times,…., and $f_n$ occurs $n$ times then the arithmetic mean is:

$$x_{\mathrm{m}} = \frac{\sum fx}{\sum f}. \tag{A.2}$$

If $a$ is the assumed arithmetic mean and $d$ is deviation of the variate $x$ from $a$, we can write:

*Arc Flash Hazard Analysis and Mitigation*, Second Edition. J.C. Das.

Published 2021 by John Wiley & Sons, Inc.

$$\frac{\sum fd}{\sum f} = \frac{\sum f(x-a)}{\sum f} = x_m + \frac{a\sum f}{\sum f}$$
$$x_m = a + \frac{\sum f(x-a)}{\sum f} = a + \frac{\sum fd}{\sum f}. \tag{A.3}$$

Median is the measure of the central item of the distribution when it is arranged in ascending or descending order. For $n$ = odd, it is given by;

$$M_d = \frac{n+1}{2} th \quad \text{item.} \tag{A.4}$$

For even frequency, there are two middle terms, and the mean of $n/2$ and $(n/2 + 1)$ terms gives the median.

Mode is defined as the size of the variable in a population that occurs most frequently

$$\text{Mean} - \text{Mode} = 3(\text{Mean} - \text{Median}). \tag{A.5}$$

## A.2 MEAN AND STANDARD DEVIATION

Average deviation or mean deviation is defined as the mean of the absolute values of the deviations of a given set of numbers from their arithmetic mean.

$$\text{Mean Deviation} = \frac{\sum_{n=1}^{n} f_n |x_n - x_m|}{\sum f}. \tag{A.6}$$

Where $x_1, x_2 \ldots x_n$ is a set of numbers with frequencies $f_1, f_2, \ldots f_n$. Standard deviation is defined as square root of the mean of the square of the deviation from the arithmetic mean:

$$SD = \sigma = \sqrt{\frac{\sum_{n=1}^{n} f_n (x_n - x_m)^2}{\sum f}}. \tag{A.7}$$

The square of the SD, $\sigma^2$, is called variance. It is also called the second moment about the mean and denoted by $\mu_2$. The coefficient of variance is given by;

$$\frac{\sigma}{x_m} \times 100. \tag{A.8}$$

The standard deviation can be found form the following expression:

$$\sigma = \sqrt{\frac{\sum fd^2}{\sum f} - \left(\frac{\sum fd}{\sum f}\right)^2}. \tag{A.9}$$

## A.3 SKEWNESS AND KURTOSIS

Skewness is opposite of symmetry. In a symmetrical series, the mode, the median, and $x_{am}$ are the same. Figure A.1a,b show positive and negative skewness. Define coefficient of skewness by:

$$\frac{\text{Mean} - \text{Mode}}{\sigma}. \tag{A.10}$$

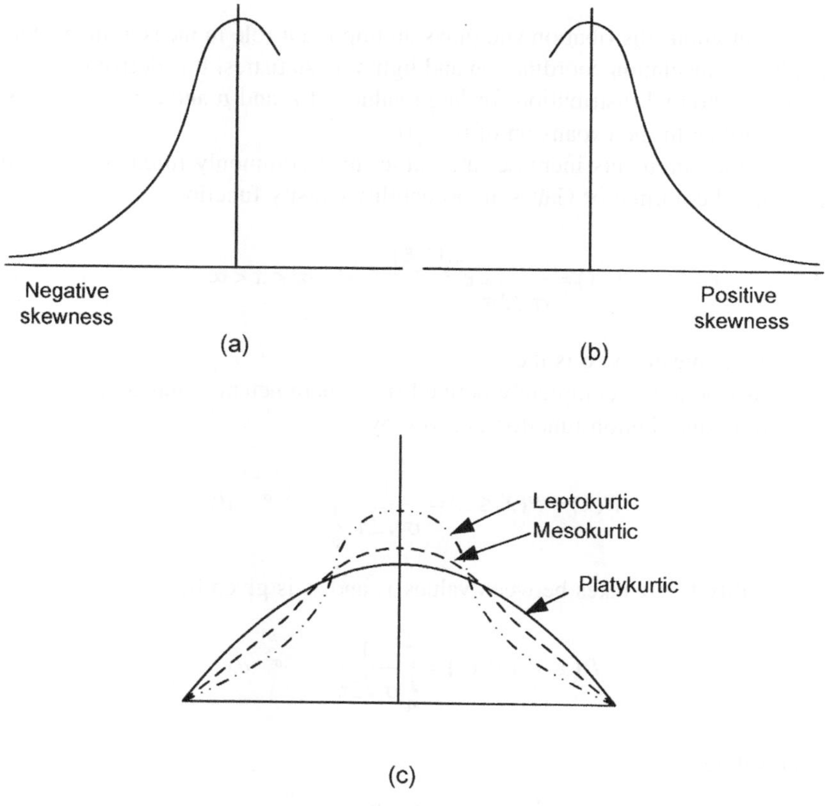


Figure A.1. Illustration of (a) positive and (b) negative skewness. (c) Illustration of kurtosis.

Measure of Kurtosis is given by:

$$\beta_2 = \frac{\mu_4}{\mu_2^2}$$

where

$$\mu_2 = \frac{\sum (x - x_{am})^2}{\sum f} \tag{A.11}$$

$$\mu_4 = \frac{\sum (x - x_{am})^4}{\sum f}.$$

If $\beta_2$ is = 3, the curve is normal, or mesokurtic. If $\beta_2$ is >3, the curve is peaked or leptokurtic, if $\beta_2$ is <3, the curve is flat topped or platykuric (Figure A.1c).

## A.4 NORMAL OR GAUSSIAN DISTRIBUTION

This is a continuous distribution and plays an important role in measurement statistics. It is applied in insulation coordination and lightning statistics. It is derived as the limiting form of binomial distribution for large values of $n$ and $p$ and $q$ is not very small; $p$, $q$, and $n$ relate to the expansion of $(q + p)^n$.

As the measurements increase, the curve most commonly fitted is a bell-shaped curve, called the normal or Gaussian probability density function.

$$p(x) = \frac{1}{\sigma\sqrt{2\pi}} e^{-\frac{1}{2}\left(\frac{x-\mu}{\sigma}\right)^2} \qquad -\infty < x < \infty. \tag{A.12}$$

Where μ is the mean and σ is the SD

The distribution is completely defined if the parameters σ and μ are known. The corresponding distribution function is given by:

$$P(x) = P(X \le x) = \frac{1}{\sigma\sqrt{2\pi}} \int_{-\infty}^{x} e^{-\frac{1}{2}\left(\frac{x-\mu}{\sigma}\right)^2} dx. \tag{A.13}$$

The probability that $x$ takes between values $x_1$ and $x_2$ is given by:

$$P(x_1 < x < x_2) = \int_{x_1}^{x_2} \frac{1}{\sigma\sqrt{2\pi}} e^{-\frac{(x-\mu)^2}{2\sigma^2}} dx. \tag{A.14}$$

If we substitute:

$$z = \frac{x - \mu}{\sigma}. \tag{A.15}$$

Where $z$ = standardized variable corresponding to $x$, then the mean of $z = 0$ and the variance is 1. In such a case, the density function becomes:

$$f(z) = \frac{1}{\sqrt{2\pi}} e^{-\frac{z^2}{2}}. \tag{A.16}$$

This is referred as *standard normal density function.* The corresponding distribution function is:

$$P(z) = P(Z \leq z) = \frac{1}{\sqrt{2\pi}} \int_{-\infty}^{z} e^{-\frac{z^2}{2}} dz. \tag{A.17}$$

A graph of the density function is shown in Figure A.2, and the properties of Gaussian distribution are shown in many texts on statistics [1, 2]:

- The curve is symmetrical about the $y$-axis. The mean, median and mode coincide at the origin.
- The area of the curve is equal to the total number of observations

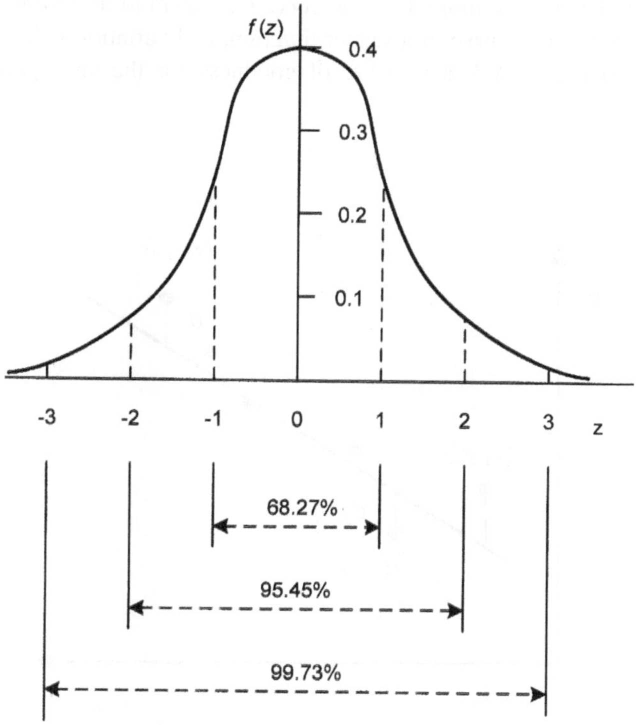


Figure A.2. Gaussian or normal distribution graph of the density function.

- $f(z)$ decreases rapidly as $z$ increases numerically. The curve extends on either side of the origin.

In Figure A.2, the areas within 1, 2, and 3 standard deviations of the mean are indicated.

$$P(-1 \leq Z \leq 1) = 0.6827.$$

About 68.27% of the values lie between ($\mu - \sigma$) and ($\mu + \sigma$)

$$P(-2 \leq Z \leq 2) = 0.9545.$$

About 95.45% of the values lie between ($\mu - 2\sigma$) and ($\mu + 2\sigma$),

$$P(-3 \leq Z \leq 3) = 0.9973\%.$$

About 99.73% of the values lie between ($\mu - 3\sigma$) and ($\mu + 3\sigma$)

The total area under the curve is 1, divided into two equal parts.

## A.5 CURVE FITTING: LEAST SQUARE LINE

For some given data points, more than one curve may seem to fit. Intuitively, it will be hard to fit an appropriate curve in a scatter diagram, and variation will exist.

Referring to Figure A.3, a measure of goodness for the appropriate fit can be described as:

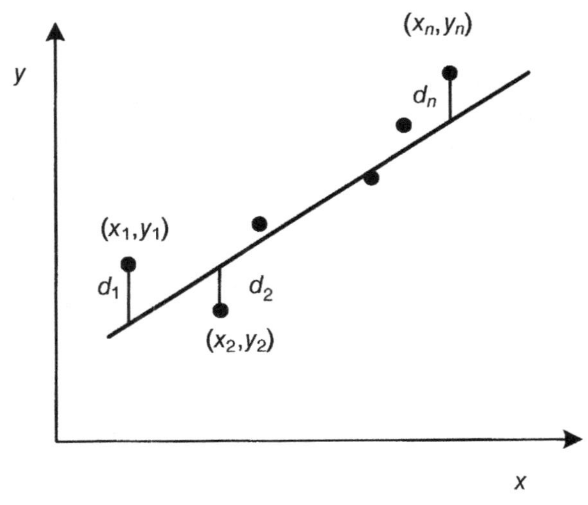


Figure A.3. Criteria of fitting a least square line in a scatter plot.

$$d_1^2 + d_2^2 + \ldots + d_n^2 = a \quad \text{minimum.} \tag{A.18}$$

A curve meeting these criteria is said to fit the data in the *least square sense* and is called a least square regression curve, or simply a least square curve—straight line or parabola. The least square line imitating the points $(x_1.y_1)$, ..$(x_n, y_n)$ has the equation:

$$y = a + bx. \tag{A.19}$$

The constants $a$ and $b$ are determined from solving simultaneous equations, which are called the normal equations for the least square line:

$$\begin{aligned} \sum y &= an + b\sum x \\ \sum xy &= a\sum x + b\sum x^2. \end{aligned} \tag{A.20}$$

This gives:

$$\begin{aligned} a &= \frac{\left(\sum y\right)\left(\sum x^2\right) - \left(\sum x\right)\left(\sum xy\right)}{n\sum x^2 - \left(\sum x\right)^2} \\ b &= \frac{n\sum xy - \left(\sum x\right)\left(\sum y\right)}{n\sum x^2 - \left(\sum x\right)^2}. \end{aligned} \tag{A.21}$$

The constant b can be written as:

$$b = \frac{\sum (x - x_{\text{mean}})(y - y_{\text{mean}})}{\sum (x - x_{\text{mean}}{}^2)}. \tag{A.22}$$

This yields,

$$y_{\text{mean}} = a + bx_{\text{mean}}. \tag{A.23}$$

From above, we can also write the least square line as:

$$y - y_{\text{mean}} = b(x - x_{\text{mean}}) = \frac{\sum (x - x_{\text{mean}})(y - y_{\text{mean}})}{\sum (x - x_{\text{mean}})^2}(x - x_{\text{mean}}). \tag{A.24}$$

Similarly, for $x$ on $y$

$$x - x_{\text{mean}} = \frac{\sum (x - x_{\text{mean}})(y - y_{\text{mean}})}{\sum (y - y_{\text{mean}})^2}(y - y_{\text{mean}}). \tag{A.25}$$

The least square line can be written in terms of variance and covariance. The sample variance and covariance are given by:

$$S_x^2 = \frac{\sum (x - x_{\text{mean}})^2}{n}$$
$$S_y^2 = \frac{\sum (y - y_{\text{mean}})^2}{n} \qquad \text{(A.26)}$$
$$S_{xy} = \frac{\sum (x - x_{\text{mean}})(y - y_{\text{mean}})}{n}.$$

In terms of these, the least square lines of $y$ on $x$ and $x$ on y

$$y - y_{\text{mean}} = \frac{S_{xy}}{S_x^2}(x - x_{\text{mean}})$$
$$x - x_{\text{mean}} = \frac{S_{xy}}{S_y^2}(y - y_{\text{mean}}). \qquad \text{(A.27)}$$

A sample correlation coefficient can be defined as:

$$r = \frac{S_{xy}}{S_x S_y} \qquad \text{(A.28)}$$

**Example A.1**

Given the data points $(x, y)$ as:

(1, 1), (3, 2), (4, 5), (6, 7), (7, 6), (9, 8), (12, 10), (15, 16), fit a least square line with $x$ as independent variable, $y$ as dependent variable.

Table A.1 shows the various steps of calculations. Substituting the values arrived in this table in Equation (A.20), we have:

$$8a + 57b = 55$$
$$57a + 561b = 543,$$

$n$ = number of samples = 8. Solving these equations, $a$ = -0.0775, $b$ = 0.975. Therefore, the least square line is:

$$y = -0.0775 + 0.975x$$

If $x$ is considered as the dependent variable, and $y$ as the independent variable, then:

$$8c + 55d = 57$$
$$55c + 535d = 543.$$

Solution of which gives $c$ = 0.507 and $d$ = 0.963.

TABLE A.1. Fitting the Least Square Line, Example A.1

| $x$ | $y$ | $x^2$ | $xy$ | $y^2$ |
|---|---|---|---|---|
| 1 | 1 | 1 | 1 | 1 |
| 3 | 2 | 9 | 6 | 4 |
| 4 | 5 | 16 | 20 | 25 |
| 6 | 7 | 36 | 42 | 49 |
| 7 | 6 | 49 | 42 | 36 |
| 9 | 8 | 81 | 72 | 64 |
| 12 | 10 | 144 | 120 | 100 |
| 15 | 16 | 225 | 240 | 256 |
| $\sum x = 57$ | $\sum y = 55$ | $\sum x^2 = 561$ | $\sum xy = 543$ | $\sum y^2 = 535$ |

## REFERENCES

1. F.M. Dekking, C. Kraaikamp, H.P. Loupaä, L.E. Meester, and A. Modern, *Introduction to Probability and Statistics (Springer Texts in Statistics)*, Springer, London, 2007.
2. J.L. Devore, *Probability and Statistics for Engineering and Sciences*, Brooks/Cole Publishing Co. (Duxbury Press), Pacific Grove, CA, 2007.

# APPENDIX B: TABLES FOR QUICK ESTIMATION OF INCIDENT ENERGY AND PPE IN ELECTRICAL SYSTEMS

This appendix provides tables for quick estimation of incident energy release and personal protective equipment (PPE) required for low voltage motor control centers (MCCs), and switchgears (480-V) and medium voltage (2.4, 4.16, and 13.8 kV) switchgears commonly used in industrial distribution systems. The three-phase short-circuit levels covered are from 10 to 50 kA for low voltage equipment and 10 to 40 kA for the medium voltage switchgears. The tables provide calculations with arcing times from 0.05 seconds to 2 seconds.

Note that for incident energy release above 40 cal/cm$^2$, the PPE category is shown as "Danger" in the tables. Though this term is used throughout the book, it is not defined as such in NFPA 70E. The industry and computer software manufacturers for arc flash hazard calculations have coined this term to indicate incident energy levels exceeding 40 cal/cm$^2$. This term may be somewhat misleading in the sense that unprotected exposure above 1.2 cal/cm$^2$ can also be dangerous. Although arc flash outfits rated above 100 cal/cm$^2$ are available, the vendors do not recommend their use when calculated incident energy levels exceed 40 cal/cm$^2$; see Chapter 2.

The electrode configuration (EC) is HCB in all cases. The calculations are independent of system grounding.

*Arc Flash Hazard Analysis and Mitigation*, Second Edition. J.C. Das.

Published 2021 by John Wiley & Sons, Inc.

## Summary of Tables for Quick Estimate of Incident Energy, Hazard Risk Category, and PPE in Three-Phase, 60 Hz Electrical Systems

| Table No | Description of Electrical Equipment | Bolted Three-Phase Short-Circuit Current | Breaker Interrupting Time |
|---|---|---|---|
| B-1 | 480-V MCCs | 10 | Included in |
| B-2 | 480-V MCCs | 20 | arcing time |
| B-3 | 480-V MCCs | 30 | for 480-V |
| B-4 | 480-V MCCs | 40 | systems |
| B-5 | 480-V MCCs | 50 | See tables |
| B-6 | 480-V, low voltage switchgear | 10 | |
| B-7 | 480-V, low voltage switchgear | 20 | |
| B-8 | 480-V, low voltage switchgear | 30 | |
| B-9 | 480-V, low voltage switchgear | 40 | |
| B-10 | 480-V, low voltage switchgear | 50 | |
| B-11 | 2.4/4.16-kV MV switchgear | 10 | 3 cycle |
| B-12 | 2.4/4.16-kV MV switchgear | 20 | 3 cycle |
| B-13 | 2.4/4.16-kV MV switchgear | 30 | 3 cycle |
| B-14 | 2.4/4.16-kV MV switchgear | 40 | 3 cycle |
| B-15 | 13.8-kV MV switchgear | 10 | 3 cycle |
| B-16 | 13.8-kV MV switchgear | 20 | 3 cycle |
| B-17 | 13.8-kV MV switchgear | 30 | 3 cycle |
| B-18 | 13.8-kV MV switchgear | 40 | 3 cycle |
| B-19 | 2.4/4.16-kV MV switchgear | 10 | 5-cycle |
| B-20 | 2.4/4.16-kV MV switchgear | 20 | 5-cycle |
| B-21 | 2.4/4.16-kV MV switchgear | 30 | 5-cycle |
| B-22 | 2.4/4.16-kV MV switchgear | 40 | 5-cycle |
| B-23 | 13.8-kV MV switchgear | 10 | 5-cycle |
| B-24 | 13.8-kV MV switchgear | 20 | 5-cycle |
| B-25 | 13.8-kV MV switchgear | 30 | 5-cycle |
| B-26 | 13.8-kV MV switchgear | 40 | 5-cycle |

TABLE B.1. Arc Flash Calculations in 480-V, Three-Phase, 60-Hz Ungrounded Systems: Low Voltage Motor Control Centers, Dimensions: 30″W × 25″D × 90″H

| kV | Bolted Three-Phase Current (kA) | Arcing Current (kA) | Arcing Time (s) | Gap (mm) | Working Distance (in) | Arc Flash Boundary (in) | Incident Energy (cal/cm$^2$) | PPE Class |
|---|---|---|---|---|---|---|---|---|
| 0.48 | 10 | 6.31 | 0.050 | 25 | 18 | 21 | 1.56 | 1 |
| 0.48 | 10 | 6.31 | 0.100 | 25 | 18 | 29 | 3.12 | 1 |
| 0.48 | 10 | 6.31 | 0.200 | 25 | 18 | 41 | 6.24 | 2 |
| 0.48 | 10 | 6.31 | 0.300 | 25 | 18 | 50 | 9.36 | 3 |
| 0.48 | 10 | 6.31 | 0.400 | 25 | 18 | 57 | 12.50 | 3 |
| 0.48 | 10 | 6.31 | 0.500 | 25 | 18 | 64 | 15.6 | 3 |
| 0.48 | 10 | 6.31 | 1.00 | 25 | 18 | 90 | 31.20 | 4 |
| 0.48 | 10 | 6.31 | 1.25 | 25 | 18 | 100 | 39 | 4 |
| 0.48 | 10 | 6.31 | 1.5 | 25 | 18 | 110 | 46.8 | Danger |
| 0.48 | 10 | 6.31 | 1.75 | 25 | 18 | 118 | 54.6 | Danger |
| 0.48 | 10 | 6.31 | 2.0 | 25 | 18 | 126 | 62.4 | Danger |

Bolted three-phase short-circuit current = 10 kA.

TABLE B.2. Arc Flash Calculations in 480-V, Three-Phase, 60-Hz Ungrounded Systems: Low Voltage Motor Control Centers, Dimensions: 30″W × 25″D × 90″H

| kV | Bolted Three-Phase Current (kA) | Arcing Current (kA) | Arcing Time (s) | Gap (mm) | Working Distance (in) | Arc Flash Boundary (in) | Incident Energy (cal/cm$^2$) | PPE Class |
|---|---|---|---|---|---|---|---|---|
| 0.48 | 20 | 14.78 | 0.050 | 25 | 18 | 29 | 3.15 | 1 |
| 0.48 | 20 | 14.78 | 0.100 | 25 | 18 | 41 | 6.5 | 2 |
| 0.48 | 20 | 14.78 | 0.200 | 25 | 18 | 58 | 13 | 3 |
| 0.48 | 20 | 14.78 | 0.300 | 25 | 18 | 71 | 19.5 | 3 |
| 0.48 | 20 | 14.78 | 0.400 | 25 | 18 | 82 | 26 | 4 |
| 0.48 | 20 | 14.78 | 0.500 | 25 | 18 | 92 | 32.5 | 4 |
| 0.48 | 20 | 14.78 | 1.00 | 25 | 18 | 129 | 65 | Danger |
| 0.48 | 20 | 14.78 | 1.25 | 25 | 18 | 144 | 81.2 | Danger |
| 0.48 | 20 | 14.78 | 1.5 | 25 | 18 | 157 | 97.4 | Danger |
| 0.48 | 20 | 14.78 | 1.75 | 25 | 18 | 176 | 113.7 | Danger |
| 0.48 | 20 | 14.78 | 2.0 | 25 | 18 | 181 | 129.9 | Danger |

Bolted three-phase short-circuit current = 20 kA.

TABLE B.3. Arc Flash Calculations in 480-V, Three-Phase, 60-Hz Ungrounded Systems: Low Voltage Motor Control Centers, Dimensions: 30″W × 25″D × 90″H

| kV | Bolted Three-Phase Current (kA) | Arcing Current (kA) | Arcing Time (s) | Gap (mm) | Working Distance (in) | Arc Flash Boundary (in) | Incident Energy (cal/cm$^2$) | PPE Class |
|---|---|---|---|---|---|---|---|---|
| 0.48 | 30 | 21.5 | 0.050 | 25 | 18 | 36 | 4.8 | 2 |
| 0.48 | 30 | 21.5 | 0.100 | 25 | 18 | 50 | 9.6 | 3 |
| 0.48 | 30 | 21.5 | 0.200 | 25 | 18 | 71 | 19.2 | 3 |
| 0.48 | 30 | 21.5 | 0.300 | 25 | 18 | 86 | 28.8 | 4 |
| 0.48 | 30 | 21.5 | 0.400 | 25 | 18 | 99 | 38.4 | 4 |
| 0.48 | 30 | 21.5 | 0.500 | 25 | 18 | 111 | 48.0 | Danger |
| 0.48 | 30 | 21.5 | 1.00 | 25 | 18 | 156 | 96.1 | Danger |
| 0.48 | 30 | 21.5 | 1.25 | 25 | 18 | 174 | 120.1 | Danger |
| 0.48 | 30 | 21.5 | 1.5 | 25 | 18 | 191 | 144.1 | Danger |
| 0.48 | 30 | 21.5 | 1.75 | 25 | 18 | 206 | 168.1 | Danger |
| 0.48 | 30 | 21.5 | 2.0 | 25 | 18 | 220 | 192.2 | Danger |

Bolted three-phase short-circuit Current = 30 kA.

TABLE B.4. Arc Flash Calculations in 480-V, Three-Phase, 60-Hz Ungrounded Systems: Low Voltage Motor Control Centers, Dimensions: 30″W × 25″D × 90″H

| kV | Bolted Three-Phase Current (kA) | Arcing Current (kA) | Arcing Time (s) | Gap (mm) | Working Distance (in) | Arc Flash Boundary (in) | Incident Energy (cal/cm$^2$) | PPE Class |
|---|---|---|---|---|---|---|---|---|
| 0.48 | 40 | 26.49 | 0.050 | 25 | 18 | 40 | 6.1 | 2 |
| 0.48 | 40 | 26.49 | 0.100 | 25 | 18 | 57 | 12.2 | 3 |
| 0.48 | 40 | 26.49 | 0.200 | 25 | 18 | 80 | 24.4 | 3 |
| 0.48 | 40 | 26.49 | 0.300 | 25 | 18 | 97 | 36.6 | 4 |
| 0.48 | 40 | 26.49 | 0.400 | 25 | 18 | 112 | 48.5 | 4 |
| 0.48 | 40 | 26.49 | 0.500 | 25 | 18 | 125 | 61 | Danger |
| 0.48 | 40 | 26.49 | 1.00 | 25 | 18 | 176 | 121.9 | Danger |
| 0.48 | 40 | 26.49 | 1.25 | 25 | 18 | 196 | 152.4 | Danger |
| 0.48 | 40 | 26.49 | 1.5 | 25 | 18 | 215 | 182.9 | Danger |
| 0.48 | 40 | 26.49 | 1.75 | 25 | 18 | 231 | 213.4 | Danger |
| 0.48 | 40 | 26.49 | 2.0 | 25 | 18 | 247 | 243.8 | Danger |

Bolted three-phase short-circuit current = 40 kA.

TABLE B.5. Arc Flash Calculations in 480-V, Three-Phase, 60-Hz Ungrounded Systems: Low Voltage Motor Control Centers, Dimensions: 30″W × 25″D × 90″H

| kV | Bolted Three-Phase Current (kA) | Arcing Current (kA) | Arcing Time (s) | Gap (mm) | Working Distance (in) | Arc Flash Boundary (in) | Incident Energy (cal/cm$^2$) | PPE Class |
|---|---|---|---|---|---|---|---|---|
| 0.48 | 50 | 31.5 | 0.050 | 25 | 18 | 43 | 7.08 | 2 |
| 0.48 | 50 | 31.5 | 0.100 | 25 | 18 | 61 | 14.2 | 3 |
| 0.48 | 50 | 31.5 | 0.200 | 25 | 18 | 86 | 28.3 | 4 |
| 0.48 | 50 | 31.5 | 0.300 | 25 | 18 | 105 | 42.5 | Danger |
| 0.48 | 50 | 31.5 | 0.400 | 25 | 18 | 120 | 56.7 | Danger |
| 0.48 | 50 | 31.5 | 0.500 | 25 | 18 | 134 | 70.8 | Danger |
| 0.48 | 50 | 31.5 | 1.00 | 25 | 18 | 189 | 141.7 | Danger |
| 0.48 | 50 | 31.5 | 1.25 | 25 | 18 | 211 | 177.1 | Danger |
| 0.48 | 50 | 31.5 | 1.5 | 25 | 18 | 231 | 212.5 | Danger |
| 0.48 | 50 | 31.5 | 1.75 | 25 | 18 | 249 | 247.9 | Danger |
| 0.48 | 50 | 31.5 | 2.0 | 25 | 18 | 266 | 283.3 | Danger |

Bolted three-phase short-circuit current = 50 kA.

**TABLE B.6. Arc Flash Calculations in 480-V, Three-Phase, 60-Hz Low Voltage Switchgear, 36"W x 90"H x 48"D, EC = HCB**

| kV | Bolted Three-Phase Current (kA) | Arcing Current (kA) | Arcing Time (s) | Gap (mm) | Working Distance (in) | Arc Flash Boundary (in) | Incident Energy (cal/cm$^2$) | PPE Class |
|---|---|---|---|---|---|---|---|---|
| 0.48 | 10 | 7.07 | 0.050 | 32 | 24 | 22 | 1.04 | 1 |
| 0.48 | 10 | 7.07 | 0.100 | 32 | 24 | 31 | 2.07 | 1 |
| 0.48 | 10 | 7.07 | 0.200 | 32 | 24 | 44 | 4.15 | 2 |
| 0.48 | 10 | 7.07 | 0.300 | 32 | 24 | 54 | 6.22 | 2 |
| 0.48 | 10 | 7.07 | 0.400 | 32 | 24 | 62 | 8.3 | 2 |
| 0.48 | 10 | 7.07 | 0.500 | 32 | 24 | 70 | 10.4 | 3 |
| 0.48 | 10 | 7.07 | 1.00 | 32 | 24 | 98 | 20.7 | 3 |
| 0.48 | 10 | 7.07 | 1.25 | 32 | 24 | 109 | 25.9 | 4 |
| 0.48 | 10 | 7.07 | 1.5 | 32 | 24 | 120 | 31.1 | 4 |
| 0.48 | 10 | 7.07 | 1.75 | 32 | 24 | 129 | 36.3 | 4 |
| 0.48 | 10 | 7.07 | 2.0 | 32 | 24 | 138 | 41.5 | 4 |

Bolted three-phase short-circuit current = 10 kA.

TABLE B.7. Arc Flash Calculations in 480-V, Three-Phase, 60-Hz Low Voltage Switchgear, 36"W x 90"H x 48"D, EC = HCB

| kV | Bolted Three-Phase Current (kA) | Arcing Current (kA) | Arcing Time (s) | Gap (mm) | Working Distance (in) | Arc Flash Boundary (in) | Incident Energy (cal/cm$^2$) | PPE Class |
|---|---|---|---|---|---|---|---|---|
| 0.48 | 20 | 14.26 | 0.050 | 32 | 24 | 30 | 1.92 | 1 |
| 0.48 | 20 | 14.26 | 0.100 | 32 | 24 | 43 | 3.84 | 1 |
| 0.48 | 20 | 14.26 | 0.200 | 32 | 24 | 60 | 7.67 | 2 |
| 0.48 | 20 | 14.26 | 0.300 | 32 | 24 | 75 | 11.5 | 3 |
| 0.48 | 20 | 14.26 | 0.400 | 32 | 24 | 84 | 15.3 | 3 |
| 0.48 | 20 | 14.26 | 0.500 | 32 | 24 | 94 | 19.2 | 3 |
| 0.48 | 20 | 14.26 | 1.00 | 32 | 24 | 133 | 38.4 | 4 |
| 0.48 | 20 | 14.26 | 1.25 | 32 | 24 | 148 | 48 | Danger |
| 0.48 | 20 | 14.26 | 1.5 | 32 | 24 | 162 | 57.5 | Danger |
| 0.48 | 20 | 14.26 | 1.75 | 32 | 24 | 177 | 66.9 | Danger |
| 0.48 | 20 | 14.26 | 2.0 | 32 | 24 | 186 | 76.2 | Danger |

Bolted three-phase short-circuit current = 20 kA.

TABLE B.8. Arc Flash Calculations in 480-V, Three-Phase, 60-Hz Low Voltage Switchgear, 36″W x 90″H x 48″D, EC = HCB

| kV | Bolted Three-Phase Current (kA) | Arcing Current (kA) | Arcing Time (s) | Gap (mm) | Working Distance (in) | Arc Flash Boundary (in) | Incident Energy ($cal/cm^2$) | PPE Class |
|---|---|---|---|---|---|---|---|---|
| 0.48 | 30 | 20.16 | 0.050 | 32 | 24 | 37 | 2.84 | 1 |
| 0.48 | 30 | 20.16 | 0.100 | 32 | 24 | 52 | 5.68 | 2 |
| 0.48 | 30 | 20.16 | 0.200 | 32 | 24 | 73 | 11.4 | 3 |
| 0.48 | 30 | 20.16 | 0.300 | 32 | 24 | 89 | 17 | 3 |
| 0.48 | 30 | 20.16 | 0.400 | 32 | 24 | 102 | 22.7 | 3 |
| 0.48 | 30 | 20.16 | 0.500 | 32 | 24 | 114 | 28.4 | 4 |
| 0.48 | 30 | 20.16 | 1.00 | 32 | 24 | 161 | 56.8 | Danger |
| 0.48 | 30 | 20.16 | 1.25 | 32 | 24 | 179 | 71 | Danger |
| 0.48 | 30 | 20.16 | 1.5 | 32 | 24 | 196 | 85.2 | Danger |
| 0.48 | 30 | 20.16 | 1.75 | 32 | 24 | 212 | 99.4 | Danger |
| 0.48 | 30 | 20.16 | 2.0 | 32 | 24 | 226 | 113.6 | Danger |

Bolted three-phase short-circuit current = 30 kA.

TABLE B.9. Arc Flash Calculations in 480-V, Three-Phase, 60-Hz Low Voltage Switchgear, 36″W x 90″H x 48″D, EC = HCB

| kV | Bolted Three-Phase Current (kA) | Arcing Current (kA) | Arcing Time (s) | Gap (mm) | Working Distance (in) | Arc Flash Boundary (in) | Incident Energy (cal/cm$^2$) | PPE Class |
|---|---|---|---|---|---|---|---|---|
| 0.48 | 40 | 26.09 | 0.050 | 32 | 24 | 41 | 3.61 | 1 |
| 0.48 | 40 | 26.09 | 0.100 | 32 | 24 | 48 | 7.21 | 2 |
| 0.48 | 40 | 26.09 | 0.200 | 32 | 24 | 82 | 14.4 | 3 |
| 0.48 | 40 | 26.09 | 0.300 | 32 | 24 | 100 | 21.8 | 3 |
| 0.48 | 40 | 26.09 | 0.400 | 32 | 24 | 115 | 28.8 | 4 |
| 0.48 | 40 | 26.09 | 0.500 | 32 | 24 | 129 | 36.1 | 4 |
| 0.48 | 40 | 26.09 | 1.00 | 32 | 24 | 181 | 72.1 | Danger |
| 0.48 | 40 | 26.09 | 1.25 | 32 | 24 | 202 | 90.1 | Danger |
| 0.48 | 40 | 26.09 | 1.5 | 32 | 24 | 221 | 108.2 | Danger |
| 0.48 | 40 | 26.09 | 1.75 | 32 | 24 | 238 | 126.2 | Danger |
| 0.48 | 40 | 26.09 | 2.0 | 32 | 24 | 254 | 144.2 | Danger |

Bolted three-phase short-circuit current = 40 kA.

TABLE B.10. Arc Flash Calculations in 480-V, Three-Phase, 60-Hz Low Voltage Switchgear, 36″W x 90″H x 48″D, EC = HCB

| kV | Bolted Three-Phase Current (kA) | Arcing Current (kA) | Arcing Time (s) | Gap (mm) | Working Distance (in) | Arc Flash Boundary (in) | Incident Energy (cal/cm$^2$) | PPE Class |
|---|---|---|---|---|---|---|---|---|
| 0.48 | 50 | 31.04 | 0.050 | 32 | 24 | 45 | 4.19 | 2 |
| 0.48 | 50 | 31.04 | 0.100 | 32 | 24 | 63 | 8.39 | 3 |
| 0.48 | 50 | 31.04 | 0.200 | 32 | 24 | 88 | 16.8 | 3 |
| 0.48 | 50 | 31.04 | 0.300 | 32 | 24 | 108 | 25.2 | 4 |
| 0.48 | 50 | 31.04 | 0.400 | 32 | 24 | 124 | 33.5 | 4 |
| 0.48 | 50 | 31.04 | 0.500 | 32 | 24 | 158 | 41.9 | Danger |
| 0.48 | 50 | 31.04 | 1.00 | 32 | 24 | 195 | 83.9 | Danger |
| 0.48 | 50 | 31.04 | 1.25 | 32 | 24 | 217 | 104.8 | Danger |
| 0.48 | 50 | 31.04 | 1.5 | 32 | 24 | 238 | 125.8 | Danger |
| 0.48 | 50 | 31.04 | 1.75 | 32 | 24 | 257 | 146.8 | Danger |
| 0.48 | 50 | 31.04 | 2.0 | 32 | 24 | 274 | 167.7 | Danger |

Bolted three-phase short-circuit current = 50 kA.

TABLE B.11. Arc Flash Calculations in Medium Voltage, Three-Phase, 60-Hz, 2.4- or 4.16-kV Switchgear, EC = HCB, Dimensions 26″W × 96″D × 95″H

| kV | Bolted Three-Phase Current (kA) | Arcing Current (kA) | Relay Time (s) | Breaker Time (cycle) | Arcing Time (s) | Gap (mm) | Working Distance (in) | Arc Flash Boundary (in) | Incident Energy (cal/cm$^2$) | PPE Class |
|---|---|---|---|---|---|---|---|---|---|---|
| 2.4 or 4.16 | 10 | 8.76 | 0.0042 | 3 cy | 0.054 | 104 | 36 | 35 | 1.13 | 1 |
| 2.4 or 4.16 | 10 | 8.76 | 0.0125 | 3 cy | 0.063 | 104 | 36 | 38 | 1.3 | 1 |
| 2.4 or 4.16 | 10 | 8.76 | 0.025 | 3 cy | 0.075 | 104 | 36 | 42 | 1.56 | 1 |
| 2.4 or 4.16 | 10 | 8.76 | 0.05 | 3 cy | 0.100 | 104 | 36 | 50 | 2.8 | 1 |
| 2.4 or 4.16 | 10 | 8.76 | 0.10 | 3 cy | 0.150 | 104 | 36 | 63 | 3.11 | 1 |
| 2.4 or 4.16 | 10 | 8.76 | 0.20 | 3 cy | 0.250 | 104 | 36 | 84 | 5.19 | 2 |
| 2.4 or 4.16 | 10 | 8.76 | 0.30 | 3 cy | 0.350 | 104 | 36 | 102 | 7.27 | 2 |
| 2.4 or 4.16 | 10 | 8.76 | 0.40 | 3 cy | 0.450 | 104 | 36 | 118 | 9.34 | 3 |
| 2.4 or 4.16 | 10 | 8.76 | 0.50 | 3 cy | 0.550 | 104 | 36 | 133 | 11.4 | 3 |
| 2.4 or 4.16 | 10 | 8.76 | 0.60 | 3 cy | 0.65 | 104 | 36 | 146 | 13.5 | 3 |
| 2.4 or 4.16 | 10 | 8.76 | 0.70 | 3 cy | 0.75 | 104 | 36 | 159 | 15.6 | 3 |
| 2.4 or 4.16 | 10 | 8.76 | 0.80 | 3 cy | 0.85 | 104 | 36 | 171 | 17.6 | 3 |
| 2.4 or 4.16 | 10 | 8.76 | 1.00 | 3 cy | 1.05 | 104 | 36 | 193 | 21 | 3 |
| 2.4 or 4.16 | 10 | 8.76 | 1.25 | 3 cy | 1.30 | 104 | 36 | 218 | 27 | 4 |
| 2.4 or 4.16 | 10 | 8.76 | 1.50 | 3 cy | 1.55 | 104 | 36 | 242 | 32.2 | 4 |
| 2.4 or 4.16 | 10 | 8.76 | 1.75 | 3 cy | 1.80 | 104 | 36 | 264 | 37.4 | 4 |
| 2.4 or 4.16 | 10 | 8.76 | 2.00 | 3 cy | 2.05 | 104 | 36 | 284 | 42.6 | 4 |

Bolted three-phase short-circuit current = 10 kA, 3-cycle breakers.

TABLE B.12. Arc Flash Calculations in Medium Voltage Three-Phase, 60-Hz, 2.4- or 4.16-kV Switchgear, EC = HCB, Dimensions 26″W × 96″D × 95″H

| kV | Bolted Three-Phase Current (kA) | Arcing Current (kA) | Relay Time (s) | Breaker Time (cycle) | Arcing Time (s) | Gap (mm) | Working Distance (in) | Arc Flash Boundary (in) | Incident Energy (cal/cm$^2$) | PPE Class |
|---|---|---|---|---|---|---|---|---|---|---|
| 2.4 or 4.16 | 20 | 17.37 | 0.0042 | 3 cy | 0.054 | 104 | 36 | 53 | 2.36 | 1 |
| 2.4 or 4.16 | 20 | 17.37 | 0.0125 | 3 cy | 0.063 | 104 | 36 | 58 | 2.72 | 1 |
| 2.4 or 4.16 | 20 | 17.37 | 0.025 | 3 cy | 0.075 | 104 | 36 | 64 | 3.26 | 1 |
| 2.4 or 4.16 | 20 | 17.37 | 0.05 | 3 cy | 0.100 | 104 | 36 | 76 | 4.35 | 2 |
| 2.4 or 4.16 | 20 | 17.37 | 0.10 | 3 cy | 0.150 | 104 | 36 | 96 | 6.52 | 2 |
| 2.4 or 4.16 | 20 | 17.37 | 0.20 | 3 cy | 0.250 | 104 | 36 | 129 | 10.9 | 3 |
| 2.4 or 4.16 | 20 | 17.37 | 0.30 | 3 cy | 0.350 | 104 | 36 | 157 | 15.2 | 3 |
| 2.4 or 4.16 | 20 | 17.37 | 0.40 | 3 cy | 0.450 | 104 | 36 | 181 | 19.6 | 3 |
| 2.4 or 4.16 | 20 | 17.37 | 0.50 | 3 cy | 0.550 | 104 | 36 | 204 | 23.9 | 3 |
| 2.4 or 4.16 | 20 | 17.37 | 0.60 | 3 cy | 0.65 | 104 | 36 | 224 | 28.3 | 4 |
| 2.4 or 4.16 | 20 | 17.37 | 0.70 | 3 cy | 0.75 | 104 | 36 | 244 | 32 | 4 |
| 2.4 or 4.16 | 20 | 17.37 | 0.80 | 3 cy | 0.85 | 104 | 36 | 262 | 36.9 | 4 |
| 2.4 or 4.16 | 20 | 17.37 | 1.00 | 3 cy | 1.05 | 104 | 36 | 296 | 45.6 | Danger |
| 2.4 or 4.16 | 20 | 17.37 | 1.25 | 3 cy | 1.30 | 104 | 36 | 335 | 56.4 | Danger |
| 2.4 or 4.16 | 20 | 17.37 | 1.50 | 3 cy | 1.55 | 104 | 36 | 371 | 67.4 | Danger |
| 2.4 or 4.16 | 20 | 17.37 | 1.75 | 3 cy | 1.80 | 104 | 36 | 404 | 78.2 | Danger |
| 2.4 or 4.16 | 20 | 17.37 | 2.00 | 3 cy | 2.05 | 104 | 36 | 436 | 89.1 | Danger |

TABLE B.13. Arc Flash Calculations in Medium Voltage, Three-Phase, 60-Hz, 2.4- or 4.16-kV Switchgear, EC = HCB, Dimensions 26″W × 96″D × 95″H

| kV | Bolted Three-Phase Current (kA) | Arcing Current (kA) | Relay Time (s) | Breaker Time (cycle) | Arcing Time (s) | Gap (mm) | Working Distance (in) | Arc Flash Boundary (in) | Incident Energy (cal/cm$^2$) | PPE Class |
|---|---|---|---|---|---|---|---|---|---|---|
| 2.4 or 4.16 | 30 | 25.68 | 0.0042 | 3 cy | 0.054 | 104 | 36 | 68 | 3.59 | 1 |
| 2.4 or 4.16 | 30 | 25.68 | 0.0125 | 3 cy | 0.063 | 104 | 36 | 74 | 4.15 | 2 |
| 2.4 or 4.16 | 30 | 25.68 | 0.025 | 3 cy | 0.075 | 104 | 36 | 82 | 4.97 | 2 |
| 2.4 or 4.16 | 30 | 25.68 | 0.05 | 3 cy | 0.100 | 104 | 36 | 97 | 6.63 | 2 |
| 2.4 or 4.16 | 30 | 25.68 | 0.10 | 3 cy | 0.150 | 104 | 36 | 123 | 9.95 | 3 |
| 2.4 or 4.16 | 30 | 25.68 | 0.20 | 3 cy | 0.250 | 104 | 36 | 165 | 16.6 | 3 |
| 2.4 or 4.16 | 30 | 25.68 | 0.30 | 3 cy | 0.350 | 104 | 36 | 200 | 23.2 | 3 |
| 2.4 or 4.16 | 30 | 25.68 | 0.40 | 3 cy | 0.450 | 104 | 36 | 231 | 29.8 | 3 |
| 2.4 or 4.16 | 30 | 25.68 | 0.50 | 3 cy | 0.550 | 104 | 36 | 260 | 36.5 | 3 |
| 2.4 or 4.16 | 30 | 25.68 | 0.60 | 3 cy | 0.65 | 104 | 36 | 286 | 43.1 | Danger |
| 2.4 or 4.16 | 30 | 25.68 | 0.70 | 3 cy | 0.75 | 104 | 36 | 311 | 49.7 | Danger |
| 2.4 or 4.16 | 30 | 25.68 | 0.80 | 3 cy | 0.85 | 104 | 36 | 334 | 56.4 | Danger |
| 2.4 or 4.16 | 30 | 25.68 | 1.00 | 3 cy | 1.05 | 104 | 36 | 378 | 69.6 | Danger |
| 2.4 or 4.16 | 30 | 25.68 | 1.25 | 3 cy | 1.30 | 104 | 36 | 428 | 86.2 | Danger |
| 2.4 or 4.16 | 30 | 25.68 | 1.50 | 3 cy | 1.55 | 104 | 36 | 473 | 102.8 | Danger |
| 2.4 or 4.16 | 30 | 25.68 | 1.75 | 3 cy | 1.80 | 104 | 36 | 516 | 119.4 | Danger |
| 2.4 or 4.16 | 30 | 25.68 | 2.00 | 3 cy | 2.05 | 104 | 36 | 556 | 136 | Danger |

Bolted three-phase short-circuit current = 30 kA, three-cycle breakers.

TABLE B.14. Arc Flash Calculations in Medium Voltage Three-Phase, 60-Hz, 2.4- or 4.16-kV Switchgear, EC = HCB, Dimensions 26″W × 96″D × 95″H

| kV | Bolted Three-Phase Current (kA) | Arcing Current (kA) | Relay Time (s) | Breaker Time (cycle) | Arcing Time (s) | Gap (mm) | Working Distance (in) | Arc Flash Boundary (in) | Incident Energy (cal/cm$^2$) | PPE Class |
|---|---|---|---|---|---|---|---|---|---|---|
| 2.4 or 4.16 | 40 | 33.55 | 0.0042 | 3 cy | 0.054 | 104 | 36 | 80 | 4.8 | 2 |
| 2.4 or 4.16 | 40 | 33.55 | 0.0125 | 3 cy | 0.063 | 104 | 36 | 87 | 5.54 | 2 |
| 2.4 or 4.16 | 40 | 33.55 | 0.025 | 3 cy | 0.075 | 104 | 36 | 97 | 6.64 | 2 |
| 2.4 or 4.16 | 40 | 33.55 | 0.05 | 3 cy | 0.100 | 104 | 36 | 115 | 8.86 | 3 |
| 2.4 or 4.16 | 40 | 33.55 | 0.10 | 3 cy | 0.150 | 104 | 36 | 145 | 13.3 | 3 |
| 2.4 or 4.16 | 40 | 33.55 | 0.20 | 3 cy | 0.250 | 104 | 36 | 195 | 22.1 | 3 |
| 2.4 or 4.16 | 40 | 33.55 | 0.30 | 3 cy | 0.350 | 104 | 36 | 237 | 31 | 4 |
| 2.4 or 4.16 | 40 | 33.55 | 0.40 | 3 cy | 0.450 | 104 | 36 | 274 | 39.9 | 4 |
| 2.4 or 4.16 | 40 | 33.55 | 0.50 | 3 cy | 0.550 | 104 | 36 | 307 | 48.7 | Danger |
| 2.4 or 4.16 | 40 | 33.55 | 0.60 | 3 cy | 0.65 | 104 | 36 | 336 | 57.6 | Danger |
| 2.4 or 4.16 | 40 | 33.55 | 0.70 | 3 cy | 0.75 | 104 | 36 | 368 | 66.4 | Danger |
| 2.4 or 4.16 | 40 | 33.55 | 0.80 | 3 cy | 0.85 | 104 | 36 | 395 | 75.6 | Danger |
| 2.4 or 4.16 | 40 | 33.55 | 1.00 | 3 cy | 1.05 | 104 | 36 | 447 | 93 | Danger |
| 2.4 or 4.16 | 40 | 33.55 | 1.25 | 3 cy | 1.30 | 104 | 36 | 505 | 115.2 | Danger |
| 2.4 or 4.16 | 40 | 33.55 | 1.50 | 3 cy | 1.55 | 104 | 36 | 560 | 137.3 | Danger |
| 2.4 or 4.16 | 40 | 33.55 | 1.75 | 3 cy | 1.80 | 104 | 36 | 610 | 159.5 | Danger |
| 2.4 or 4.16 | 40 | 33.55 | 2.00 | 3 cy | 2.05 | 104 | 36 | 658 | 181.6 | Danger |

Bolted three-phase short-circuit current = 30 kA, three-cycle breakers.

TABLE B.15. Arc Flash Calculations in Medium Voltage Three-Phase, 60-Hz, 13.8-kV Switchgear EC = HCB, Dimensions 26″W × 96″D × 95″H

| kV | Bolted Three-Phase Current (kA) | Arcing Current (kA) | Relay Time (s) | Breaker Time (cycle) | Arcing Time (s) | Gap (mm) | Working Distance (in) | Arc Flash Boundary (in) | Incident Energy (cal/cm$^2$) | PPE Class |
|---|---|---|---|---|---|---|---|---|---|---|
| 13.8 | 10.0 | 9.04 | 0.0042 | 3 cy | 0.054 | 152 | 36 | 39 | 1.34 | 1 |
| 13.8 | 10.0 | 9.04 | 0.0125 | 3 cy | 0.063 | 152 | 36 | 42 | 1.55 | 1 |
| 13.8 | 10.0 | 9.04 | 0.025 | 3 cy | 0.075 | 152 | 36 | 47 | 1.86 | 1 |
| 13.8 | 10.0 | 9.04 | 0.05 | 3 cy | 0.100 | 152 | 36 | 56 | 2.47 | 1 |
| 13.8 | 10.0 | 9.04 | 0.10 | 3 cy | 0.150 | 152 | 36 | 71 | 3.71 | 1 |
| 13.8 | 10.0 | 9.04 | 0.20 | 3 cy | 0.250 | 152 | 36 | 97 | 6.90 | 2 |
| 13.8 | 10.0 | 9.04 | 0.30 | 3 cy | 0.350 | 152 | 36 | 119 | 8.6 | 3 |
| 13.8 | 10.0 | 9.04 | 0.40 | 3 cy | 0.450 | 152 | 36 | 138 | 11.1 | 3 |
| 13.8 | 10.0 | 9.04 | 0.50 | 3 cy | 0.550 | 152 | 36 | 156 | 13.6 | 3 |
| 13.8 | 10.0 | 9.04 | 0.60 | 3 cy | 0.65 | 152 | 36 | 173 | 16.10 | 3 |
| 13.8 | 10.0 | 9.04 | 0.70 | 3 cy | 0.75 | 152 | 36 | 188 | 18.6 | 3 |
| 13.8 | 10.0 | 9.04 | 0.80 | 3 cy | 0.85 | 152 | 36 | 203 | 21 | 3 |
| 13.8 | 10.0 | 9.04 | 1.00 | 3 cy | 1.05 | 152 | 36 | 231 | 26 | 4 |
| 13.8 | 10.0 | 9.04 | 1.25 | 3 cy | 1.30 | 152 | 36 | 262 | 32.2 | 4 |
| 13.8 | 10.0 | 9.04 | 1.50 | 3 cy | 1.55 | 152 | 36 | 292 | 38.4 | 4 |
| 13.8 | 10.0 | 9.04 | 1.75 | 3 cy | 1.80 | 152 | 36 | 319 | 44.5 | Danger |
| 13.8 | 10.0 | 9.04 | 2.00 | 3 cy | 2.05 | 152 | 36 | 345 | 50.7 | Danger |

Bolted three-phase short-circuit current = 10 kA, 3-cycle breakers.

TABLE B.16. Arc Flash Calculations in Medium Voltage Three-Phase, 60-Hz, 13.8-kV Switchgear EC = HCB, Dimensions 26″W × 96″D × 95″H

| kV | Bolted Three-Phase Current (kA) | Arcing Current (kA) | Relay Time (s) | Breaker Time (cycle) | Arcing Time (s) | Gap (mm) | Working Distance (in) | Arc Flash Boundary (in) | Incident Energy (cal/cm$^2$) | PPE Class |
|---|---|---|---|---|---|---|---|---|---|---|
| 13.8 | 20 | 18.11 | 0.0042 | 3 cy | 0.054 | 152 | 36 | 61 | 2.85 | 1 |
| 13.8 | 20 | 18.11 | 0.0125 | 3 cy | 0.063 | 152 | 36 | 68 | 3.29 | 1 |
| 13.8 | 20 | 18.11 | 0.025 | 3 cy | 0.075 | 152 | 36 | 74 | 3.94 | 1 |
| 13.8 | 20 | 18.11 | 0.05 | 3 cy | 0.100 | 152 | 36 | 88 | 5.27 | 2 |
| 13.8 | 20 | 18.11 | 0.10 | 3 cy | 0.150 | 152 | 36 | 112 | 7.9 | 2 |
| 13.8 | 20 | 18.11 | 0.20 | 3 cy | 0.250 | 152 | 36 | 152 | 13.2 | 3 |
| 13.8 | 20 | 18.11 | 0.30 | 3 cy | 0.350 | 152 | 36 | 188 | 18.4 | 3 |
| 13.8 | 20 | 18.11 | 0.40 | 3 cy | 0.450 | 152 | 36 | 218 | 23.7 | 3 |
| 13.8 | 20 | 18.11 | 0.50 | 3 cy | 0.550 | 152 | 36 | 246 | 29 | 4 |
| 13.8 | 20 | 18.11 | 0.60 | 3 cy | 0.65 | 152 | 36 | 272 | 34.2 | 4 |
| 13.8 | 20 | 18.11 | 0.70 | 3 cy | 0.75 | 152 | 36 | 297 | 39.5 | 4 |
| 13.8 | 20 | 18.11 | 0.80 | 3 cy | 0.85 | 152 | 36 | 320 | 44.8 | Danger |
| 13.8 | 20 | 18.11 | 1.00 | 3 cy | 1.05 | 152 | 36 | 361 | 55.3 | Danger |
| 13.8 | 20 | 18.11 | 1.25 | 3 cy | 1.30 | 152 | 36 | 414 | 68.5 | Danger |
| 13.8 | 20 | 18.11 | 1.50 | 3 cy | 1.55 | 152 | 36 | 460 | 87.6 | Danger |
| 13.8 | 20 | 18.11 | 1.75 | 3 cy | 1.80 | 152 | 36 | 504 | 94.8 | Danger |
| 13.8 | 20 | 18.11 | 2.00 | 3 cy | 2.05 | 152 | 36 | 545 | 108.6 | Danger |

Bolted three-phase short-circuit current = 20 kA, 3-cycle breakers.

TABLE B.17. Arc Flash Calculations in Medium Voltage Three-Phase, 60-Hz, 13.8-kV Switchgear EC = HCB, Dimensions 26″W × 96″D × 95″H

| kV | Bolted Three-Phase Current (kA) | Arcing Current (kA) | Relay Time (s) | Breaker Time (cycle) | Arcing Time (s) | Gap (mm) | Working Distance (in) | Arc Flash Boundary (in) | Incident Energy (cal/cm$^2$) | PPE Class |
|---|---|---|---|---|---|---|---|---|---|---|
| 13.8 | 30 | 27.2 | 0.0042 | 3 cy | 0.054 | 152 | 36 | 79 | 4.44 | 2 |
| 13.8 | 30 | 27.2 | 0.0125 | 3 cy | 0.063 | 152 | 36 | 87 | 5.11 | 2 |
| 13.8 | 30 | 27.2 | 0.025 | 3 cy | 0.075 | 152 | 36 | 97 | 6.14 | 2 |
| 13.8 | 30 | 27.2 | 0.05 | 3 cy | 0.100 | 152 | 36 | 115 | 8.18 | 3 |
| 13.8 | 30 | 27.2 | 0.10 | 3 cy | 0.150 | 152 | 36 | 147 | 12.7 | 3 |
| 13.8 | 30 | 27.2 | 0.20 | 3 cy | 0.250 | 152 | 36 | 200 | 20.5 | 3 |
| 13.8 | 30 | 27.2 | 0.30 | 3 cy | 0.350 | 152 | 36 | 245 | 28.6 | 4 |
| 13.8 | 30 | 27.2 | 0.40 | 3 cy | 0.450 | 152 | 36 | 285 | 36.8 | 4 |
| 13.8 | 30 | 27.2 | 0.50 | 3 cy | 0.550 | 152 | 36 | 321 | 45 | Danger |
| 13.8 | 30 | 27.2 | 0.60 | 3 cy | 0.65 | 152 | 36 | 355 | 53.2 | Danger |
| 13.8 | 30 | 27.2 | 0.70 | 3 cy | 0.75 | 152 | 36 | 387 | 61.4 | Danger |
| 13.8 | 30 | 27.2 | 0.80 | 3 cy | 0.85 | 152 | 36 | 418 | 69.6 | Danger |
| 13.8 | 30 | 27.2 | 1.00 | 3 cy | 1.05 | 152 | 36 | 475 | 85.9 | Danger |
| 13.8 | 30 | 27.2 | 1.25 | 3 cy | 1.30 | 152 | 36 | 540 | 106.4 | Danger |
| 13.8 | 30 | 27.2 | 1.50 | 3 cy | 1.55 | 152 | 36 | 600 | 126.8 | Danger |
| 13.8 | 30 | 27.2 | 1.75 | 3 cy | 1.80 | 152 | 36 | 657 | 147.3 | Danger |
| 13.8 | 30 | 27.2 | 2.00 | 3 cy | 2.05 | 152 | 36 | 771 | 167.7 | Danger |

Bolted three-phase short-circuit current = 30 kA, 3-cycle breakers.

TABLE B.18. Arc Flash Calculations in Medium Voltage Three-Phase, 60-Hz, 13.8-kV Switchgear, EC = HCB, Dimensions 26"W × 96"D × 95"H

| kV | Bolted Three-Phase Current (kA) | Arcing Current (kA) | Relay Time (s) | Breaker Time (cycle) | Arcing Time (s) | Gap (mm) | Working Distance (in) | Arc Flash Boundary (in) | Incident Energy (cal/cm$^2$) | PPE Class |
|---|---|---|---|---|---|---|---|---|---|---|
| 13.8 | 40 | 35.95 | 0.0042 | 3 cy | 0.054 | 152 | 36 | 95 | 6 | 2 |
| 13.8 | 40 | 35.95 | 0.0125 | 3 cy | 0.063 | 152 | 36 | 104 | 6.92 | 2 |
| 13.8 | 40 | 35.95 | 0.025 | 3 cy | 0.075 | 152 | 36 | 116 | 8.3 | 3 |
| 13.8 | 40 | 35.95 | 0.05 | 3 cy | 0.100 | 152 | 36 | 138 | 11.2 | 3 |
| 13.8 | 40 | 35.95 | 0.10 | 3 cy | 0.150 | 152 | 36 | 176 | 16.6 | 3 |
| 13.8 | 40 | 35.95 | 0.20 | 3 cy | 0.250 | 152 | 36 | 240 | 27.7 | 4 |
| 13.8 | 40 | 35.95 | 0.30 | 3 cy | 0.350 | 152 | 36 | 294 | 38.7 | 4 |
| 13.8 | 40 | 35.95 | 0.40 | 3 cy | 0.450 | 152 | 36 | 342 | 49.8 | Danger |
| 13.8 | 40 | 35.95 | 0.50 | 3 cy | 0.550 | 152 | 36 | 386 | 60.9 | Danger |
| 13.8 | 40 | 35.95 | 0.60 | 3 cy | 0.65 | 152 | 36 | 427 | 72.0 | Danger |
| 13.8 | 40 | 35.95 | 0.70 | 3 cy | 0.75 | 152 | 36 | 465 | 83 | Danger |
| 13.8 | 40 | 35.95 | 0.80 | 3 cy | 0.85 | 152 | 36 | 501 | 94.1 | Danger |
| 13.8 | 40 | 35.95 | 1.00 | 3 cy | 1.05 | 152 | 36 | 570 | 116 | Danger |
| 13.8 | 40 | 35.95 | 1.25 | 3 cy | 1.30 | 152 | 36 | 648 | 144 | Danger |
| 13.8 | 40 | 35.95 | 1.50 | 3 cy | 1.55 | 152 | 36 | 720 | 171.6 | Danger |
| 13.8 | 40 | 35.95 | 1.75 | 3 cy | 1.80 | 152 | 36 | 789 | 199.3 | Danger |
| 13.8 | 40 | 35.95 | 2.00 | 3 cy | 2.05 | 152 | 36 | 853 | 227 | Danger |

Bolted three-phase short-circuit current = 40 kA, 3-cycle breakers.

TABLE B.19. Arc Flash Calculations in Medium Voltage, Three-Phase, 60-Hz, 2.4- or 4.16-kV Switchgear, EC = HCB, Dimensions 26″W × 96″D × 95″H

| kV | Bolted Three-Phase Current (kA) | Arcing Current (kA) | Relay Time (s) | Breaker Time (cycle) | Arcing Time (s) | Gap (mm) | Working Distance (in) | Arc Flash Boundary (in) | Incident Energy (cal/cm$^2$) | PPE Class |
|---|---|---|---|---|---|---|---|---|---|---|
| 2.4 or 4.16 | 10 | 8.29 | 0.0042 | 5 cy | 0.088 | 104 | 36 | 46 | 1.81 | 1 |
| 2.4 or 4.16 | 10 | 8.29 | 0.0125 | 5 cy | 0.096 | 104 | 36 | 48 | 1.98 | 1 |
| 2.4 or 4.16 | 10 | 8.29 | 0.025 | 5 cy | 1.080 | 104 | 36 | 52 | 2.24 | 1 |
| 2.4 or 4.16 | 10 | 8.29 | 0.05 | 5 cy | 0.133 | 104 | 36 | 58 | 2.76 | 1 |
| 2.4 or 4.16 | 10 | 8.29 | 0.10 | 5 cy | 0.183 | 104 | 36 | 70 | 3.8 | 1 |
| 2.4 or 4.16 | 10 | 8.29 | 0.20 | 5 cy | 0.283 | 104 | 36 | 90 | 5.88 | 2 |
| 2.4 or 4.16 | 10 | 8.29 | 0.30 | 5 cy | 0.383 | 104 | 36 | 108 | 7.95 | 2 |
| 2.4 or 4.16 | 10 | 8.29 | 0.40 | 5 cy | 0.483 | 104 | 36 | 123 | 10 | 3 |
| 2.4 or 4.16 | 10 | 8.29 | 0.50 | 5 cy | 0.583 | 104 | 36 | 137 | 12.1 | 3 |
| 2.4 or 4.16 | 10 | 8.29 | 0.60 | 5 cy | 0.683 | 104 | 36 | 150 | 14.2 | 3 |
| 2.4 or 4.16 | 10 | 8.29 | 0.70 | 5 cy | 0.783 | 104 | 36 | 163 | 16.3 | 3 |
| 2.4 or 4.16 | 10 | 8.29 | 0.80 | 5 cy | 0.883 | 104 | 36 | 175 | 18.3 | 3 |
| 2.4 or 4.16 | 10 | 8.29 | 1.00 | 5 cy | 1.080 | 104 | 36 | 196 | 22.5 | 3 |
| 2.4 or 4.16 | 10 | 8.29 | 1.25 | 5 cy | 1.333 | 104 | 36 | 222 | 27.7 | 4 |
| 2.4 or 4.16 | 10 | 8.29 | 1.50 | 5 cy | 1.583 | 104 | 36 | 245 | 32.9 | 4 |
| 2.4 or 4.16 | 10 | 8.29 | 1.75 | 5 cy | 1.833 | 104 | 36 | 266 | 38.1 | 4 |
| 2.4 or 4.16 | 10 | 8.29 | 2.00 | 5 cy | 2.083 | 104 | 36 | 287 | 43.4 | Danger |

Bolted three-phase short-circuit current = 10 kA, 5-cycle breakers.

TABLE B.20. Arc Flash Calculations in Medium Voltage, Three-Phase, 60-Hz, 2.4- or 4.16-kV Switchgear, EC = HCB, Dimensions 26″W × 96″D × 95″H

| kV | Bolted Three-Phase Current (kA) | Arcing Current (kA) | Relay Time (s) | Breaker Time (cycle) | Arcing Time (s) | Gap (mm) | Working Distance (in) | Arc Flash Boundary (in) | Incident Energy (cal/cm$^2$) | PPE Class |
|---|---|---|---|---|---|---|---|---|---|---|
| 2.4 or 4.16 | 20 | 17.37 | 0.0042 | 5 cy | 0.088 | 104 | 36 | 70 | 3.74 | 1 |
| 2.4 or 4.16 | 20 | 17.37 | 0.0125 | 5 cy | 0.096 | 104 | 36 | 74 | 4.15 | 2 |
| 2.4 or 4.16 | 20 | 17.37 | 0.025 | 5 cy | 1.080 | 104 | 36 | 79 | 4.69 | 2 |
| 2.4 or 4.16 | 20 | 17.37 | 0.05 | 5 cy | 0.133 | 104 | 36 | 90 | 5.78 | 2 |
| 2.4 or 4.16 | 20 | 17.37 | 0.10 | 5 cy | 0.183 | 104 | 36 | 108 | 7.95 | 2 |
| 2.4 or 4.16 | 20 | 17.37 | 0.20 | 5 cy | 0.283 | 104 | 36 | 139 | 12.3 | 3 |
| 2.4 or 4.16 | 20 | 17.37 | 0.30 | 5 cy | 0.383 | 104 | 36 | 165 | 16.6 | 3 |
| 2.4 or 4.16 | 20 | 17.37 | 0.40 | 5 cy | 0.483 | 104 | 36 | 189 | 21 | 3 |
| 2.4 or 4.16 | 20 | 17.37 | 0.50 | 5 cy | 0.583 | 104 | 36 | 211 | 25.3 | 4 |
| 2.4 or 4.16 | 20 | 17.37 | 0.60 | 5 cy | 0.683 | 104 | 36 | 231 | 29.7 | 4 |
| 2.4 or 4.16 | 20 | 17.37 | 0.70 | 5 cy | 0.783 | 104 | 36 | 250 | 34 | 4 |
| 2.4 or 4.16 | 20 | 17.37 | 0.80 | 5 cy | 0.883 | 104 | 36 | 268 | 38.4 | 4 |
| 2.4 or 4.16 | 20 | 17.37 | 1.00 | 5 cy | 1.080 | 104 | 36 | 301 | 47.1 | Danger |
| 2.4 or 4.16 | 20 | 17.37 | 1.25 | 5 cy | 1.333 | 104 | 36 | 340 | 57.9 | Danger |
| 2.4 or 4.16 | 20 | 17.37 | 1.50 | 5 cy | 1.583 | 104 | 36 | 375 | 68.8 | Danger |
| 2.4 or 4.16 | 20 | 17.37 | 1.75 | 5 cy | 1.833 | 104 | 36 | 438 | 79.7 | Danger |
| 2.4 or 4.16 | 20 | 17.37 | 2.00 | 5 cy | 2.083 | 104 | 36 | 440 | 90.5 | Danger |

Bolted three-phase short-circuit current = 20 kA, 5-cycle breakers.

TABLE B.21. Arc Flash Calculations in Medium Voltage, Three-Phase, 60-Hz, 2.4- or 4.16-kV Switchgear, EC = HCB, Dimensions 26″W × 96″D × 95″H

| kV | Bolted Three-Phase Current (kA) | Arcing Current (kA) | Relay Time (s) | Breaker Time (cycle) | Arcing Time (s) | Gap (mm) | Working Distance (in) | Arc Flash Boundary (in) | Incident Energy (cal/cm$^2$) | PPE Class |
|---|---|---|---|---|---|---|---|---|---|---|
| 2.4 or 4.16 | 30 | 25.68 | 0.0042 | 5 cy | 0.088 | 104 | 36 | 90 | 5.78 | 2 |
| 2.4 or 4.16 | 30 | 25.68 | 0.0125 | 5 cy | 0.096 | 104 | 36 | 94 | 6.38 | 2 |
| 2.4 or 4.16 | 30 | 25.68 | 0.025 | 5 cy | 1.080 | 104 | 36 | 101 | 7.16 | 2 |
| 2.4 or 4.16 | 30 | 25.68 | 0.05 | 5 cy | 0.133 | 104 | 36 | 114 | 8.82 | 3 |
| 2.4 or 4.16 | 30 | 25.68 | 0.10 | 5 cy | 0.183 | 104 | 36 | 138 | 12.1 | 3 |
| 2.4 or 4.16 | 30 | 25.68 | 0.20 | 5 cy | 0.283 | 104 | 36 | 177 | 16.8 | 3 |
| 2.4 or 4.16 | 30 | 25.68 | 0.30 | 5 cy | 0.383 | 104 | 36 | 211 | 25.4 | 4 |
| 2.4 or 4.16 | 30 | 25.68 | 0.40 | 5 cy | 0.483 | 104 | 36 | 241 | 32 | 3 |
| 2.4 or 4.16 | 30 | 25.68 | 0.50 | 5 cy | 0.583 | 104 | 36 | 269 | 38.7 | 4 |
| 2.4 or 4.16 | 30 | 25.68 | 0.60 | 5 cy | 0.683 | 104 | 36 | 295 | 45.3 | Danger |
| 2.4 or 4.16 | 30 | 25.68 | 0.70 | 5 cy | 0.783 | 104 | 36 | 319 | 51.9 | Danger |
| 2.4 or 4.16 | 30 | 25.68 | 0.80 | 5 cy | 0.883 | 104 | 36 | 342 | 58.6 | Danger |
| 2.4 or 4.16 | 30 | 25.68 | 1.00 | 5 cy | 1.080 | 104 | 36 | 365 | 71.8 | Danger |
| 2.4 or 4.16 | 30 | 25.68 | 1.25 | 5 cy | 1.333 | 104 | 36 | 434 | 88.4 | Danger |
| 2.4 or 4.16 | 30 | 25.68 | 1.50 | 5 cy | 1.583 | 104 | 36 | 479 | 105 | Danger |
| 2.4 or 4.16 | 30 | 25.68 | 1.75 | 5 cy | 1.833 | 104 | 36 | 522 | 121.6 | Danger |
| 2.4 or 4.16 | 30 | 25.68 | 2.00 | 5 cy | 2.083 | 104 | 36 | 569 | 138.2 | Danger |

Bolted three-phase short-circuit current = 30 kA, 5-cycle breakers.

TABLE B.22. Arc Flash Calculations in Medium Voltage Three-Phase, 60-Hz, 2.4- or 4.16-kV Switchgear, EC = HCB, Dimensions 26″W × 96″D × 95″H

| kV | Bolted Three-Phase Current (kA) | Arcing Current (kA) | Relay Time (s) | Breaker Time (Cycle) | Arcing Time (s) | Gap (mm) | Working Distance (in) | Arc Flash Boundary (in) | Incident Energy (cal/cm$^2$) | PPE Class |
|---|---|---|---|---|---|---|---|---|---|---|
| 2.4 or 4.16 | 40 | 33.55 | 0.0042 | 5 cy | 0.088 | 104 | 36 | 106 | 7.73 | 2 |
| 2.4 or 4.16 | 40 | 33.55 | 0.0125 | 5 cy | 0.096 | 104 | 36 | 112 | 8.46 | 3 |
| 2.4 or 4.16 | 40 | 33.55 | 0.025 | 5 cy | 1.080 | 104 | 36 | 120 | 9.59 | 3 |
| 2.4 or 4.16 | 40 | 33.55 | 0.05 | 5 cy | 0.133 | 104 | 36 | 135 | 14.8 | 3 |
| 2.4 or 4.16 | 40 | 33.55 | 0.10 | 5 cy | 0.183 | 104 | 36 | 168 | 16.2 | 3 |
| 2.4 or 4.16 | 40 | 33.55 | 0.20 | 5 cy | 0.283 | 104 | 36 | 209 | 25.1 | 3 |
| 2.4 or 4.16 | 40 | 33.55 | 0.30 | 5 cy | 0.383 | 104 | 36 | 249 | 33.9 | 4 |
| 2.4 or 4.16 | 40 | 33.55 | 0.40 | 5 cy | 0.483 | 104 | 36 | 285 | 42.2 | Danger |
| 2.4 or 4.16 | 40 | 33.55 | 0.50 | 5 cy | 0.583 | 104 | 36 | 318 | 51.7 | Danger |
| 2.4 or 4.16 | 40 | 33.55 | 0.60 | 5 cy | 0.683 | 104 | 36 | 348 | 60.5 | Danger |
| 2.4 or 4.16 | 40 | 33.55 | 0.70 | 5 cy | 0.783 | 104 | 36 | 377 | 69.4 | Danger |
| 2.4 or 4.16 | 40 | 33.55 | 0.80 | 5 cy | 0.883 | 104 | 36 | 404 | 78.2 | Danger |
| 2.4 or 4.16 | 40 | 33.55 | 1.00 | 5 cy | 1.080 | 104 | 36 | 455 | 95.9 | Danger |
| 2.4 or 4.16 | 40 | 33.55 | 1.25 | 5 cy | 1.333 | 104 | 36 | 513 | 118 | Danger |
| 2.4 or 4.16 | 40 | 33.55 | 1.50 | 5 cy | 1.583 | 104 | 36 | 567 | 140 | Danger |
| 2.4 or 4.16 | 40 | 33.55 | 1.75 | 5 cy | 1.833 | 104 | 36 | 617 | 162 | Danger |
| 2.4 or 4.16 | 40 | 33.55 | 2.00 | 5 cy | 2.083 | 104 | 36 | 664 | 184 | Danger |

Bolted three-phase short-circuit current = 40 kA, 5-cycle breakers.

TABLE B.23. Arc Flash Calculations in Medium Voltage, Three-Phase, 60-Hz, 13.8-kV Switchgear, EC = HCB, Dimensions 26″W × 96″D × 95″H

| kV | Bolted Three-Phase Current (kA) | Arcing Current (kA) | Relay Time (s) | Breaker Time (cycle) | Arcing Time (s) | Gap (mm) | Working Distance (in) | Arc Flash Boundary (in) | Incident Energy (cal/cm$^2$) | PPE Class |
|---|---|---|---|---|---|---|---|---|---|---|
| 13.8 | 10 | 9.02 | 0.0042 | 5 cy | 0.088 | 152 | 36 | 51 | 2.16 | 1 |
| 13.8 | 10 | 9.02 | 0.0125 | 5 cy | 0.096 | 152 | 36 | 54 | 2.36 | 1 |
| 13.8 | 10 | 9.02 | 0.025 | 5 cy | 1.080 | 152 | 36 | 59 | 2.67 | 1 |
| 13.8 | 10 | 9.02 | 0.05 | 5 cy | 0.133 | 152 | 36 | 66 | 3.29 | 1 |
| 13.8 | 10 | 9.02 | 0.10 | 5 cy | 0.183 | 152 | 36 | 80 | 4.53 | 2 |
| 13.8 | 10 | 9.02 | 0.20 | 5 cy | 0.283 | 152 | 36 | 105 | 7.0 | 2 |
| 13.8 | 10 | 9.02 | 0.30 | 5 cy | 0.383 | 152 | 36 | 126 | 9.48 | 3 |
| 13.8 | 10 | 9.02 | 0.40 | 5 cy | 0.483 | 152 | 36 | 144 | 12.0 | 3 |
| 13.8 | 10 | 9.02 | 0.50 | 5 cy | 0.583 | 152 | 36 | 162 | 14.4 | 3 |
| 13.8 | 10 | 9.02 | 0.60 | 5 cy | 0.683 | 152 | 36 | 178 | 16.90 | 3 |
| 13.8 | 10 | 9.02 | 0.70 | 5 cy | 0.783 | 152 | 36 | 193 | 19.40 | 3 |
| 13.8 | 10 | 9.02 | 0.80 | 5 cy | 0.883 | 152 | 36 | 208 | 21.9 | 3 |
| 13.8 | 10 | 9.02 | 1.00 | 5 cy | 1.080 | 152 | 36 | 235 | 26.8 | 4 |
| 13.8 | 10 | 9.02 | 1.25 | 5 cy | 1.333 | 152 | 36 | 266 | 33 | 4 |
| 13.8 | 10 | 9.02 | 1.50 | 5 cy | 1.583 | 152 | 36 | 296 | 39.2 | 4 |
| 13.8 | 10 | 9.02 | 1.75 | 5 cy | 1.833 | 152 | 36 | 323 | 45.4 | Danger |
| 13.8 | 10 | 9.02 | 2.00 | 5 cy | 2.083 | 152 | 36 | 349 | 51.5 | Danger |

Bolted three-phase short-circuit current = 10 kA, 5-cycle breakers.

TABLE B.24. Arc Flash Calculations in Medium Voltage, Three-Phase, 60-Hz, 13.8-kV Switchgear, EC = HCB, Dimensions 26″W × 96″D × 95″H

| kV | Bolted Three-Phase Current (kA) | Arcing Current (kA) | Relay Time (s) | Breaker Time (cycle) | Arcing Time (s) | Gap (mm) | Working Distance (in) | Arc Flash Boundary (in) | Incident Energy (cal/cm$^2$) | PPE Class |
|---|---|---|---|---|---|---|---|---|---|---|
| 13.8 | 20 | 18.11 | 0.0042 | 5 cy | 0.088 | 152 | 36 | 81 | 4.59 | 2 |
| 13.8 | 20 | 18.11 | 0.0125 | 5 cy | 0.096 | 152 | 36 | 86 | 5.03 | 2 |
| 13.8 | 20 | 18.11 | 0.025 | 5 cy | 1.080 | 152 | 36 | 92 | 5.69 | 2 |
| 13.8 | 20 | 18.11 | 0.05 | 5 cy | 0.133 | 152 | 36 | 105 | 7.01 | 2 |
| 13.8 | 20 | 18.11 | 0.10 | 5 cy | 0.183 | 152 | 36 | 127 | 9.64 | 3 |
| 13.8 | 20 | 18.11 | 0.20 | 5 cy | 0.283 | 152 | 36 | 165 | 14.9 | 3 |
| 13.8 | 20 | 18.11 | 0.30 | 5 cy | 0.383 | 152 | 36 | 198 | 20.20 | 3 |
| 13.8 | 20 | 18.11 | 0.40 | 5 cy | 0.483 | 152 | 36 | 228 | 25.3 | 4 |
| 13.8 | 20 | 18.11 | 0.50 | 5 cy | 0.583 | 152 | 36 | 255 | 30.70 | 4 |
| 13.8 | 20 | 18.11 | 0.60 | 5 cy | 0.683 | 152 | 36 | 281 | 36.00 | 4 |
| 13.8 | 20 | 18.11 | 0.70 | 5 cy | 0.783 | 152 | 36 | 305 | 41.20 | Danger |
| 13.8 | 20 | 18.11 | 0.80 | 5 cy | 0.883 | 152 | 36 | 328 | 46.50 | Danger |
| 13.8 | 20 | 18.11 | 1.00 | 5 cy | 1.080 | 152 | 36 | 371 | 57.0 | Danger |
| 13.8 | 20 | 18.11 | 1.25 | 5 cy | 1.333 | 152 | 36 | 420 | 70.2 | Danger |
| 13.8 | 20 | 18.11 | 1.50 | 5 cy | 1.583 | 152 | 36 | 466 | 83.4 | Danger |
| 13.8 | 20 | 18.11 | 1.75 | 5 cy | 1.833 | 152 | 36 | 509 | 96.5 | Danger |
| 13.8 | 20 | 18.11 | 2.00 | 5 cy | 2.083 | 152 | 36 | 550 | 109.7 | Danger |

Bolted three-phase short-circuit current = 20 kA, 5-cycle breakers.

TABLE B.25. Arc Flash Calculations in Medium Voltage, Three-Phase, 60-Hz, 13.8-kV Switchgear, EC = HCB, Dimensions 26″W × 96″D × 95″H

| kV | Bolted Three-Phase Current (kA) | Arcing Current (kA) | Relay Time (s) | Breaker Time (cycle) | Arcing Time (s) | Gap (mm) | Working Distance (in) | Arc Flash Boundary (in) | Incident Energy (cal/cm$^2$) | PPE Class |
|---|---|---|---|---|---|---|---|---|---|---|
| 13.8 | 30 | 27.20 | 0.0042 | 5 cy | 0.088 | 152 | 36 | 113 | 4.6 | 2 |
| 13.8 | 30 | 27.20 | 0.0125 | 5 cy | 0.096 | 152 | 36 | 124 | 5.0 | 2 |
| 13.8 | 30 | 27.20 | 0.025 | 5 cy | 1.080 | 152 | 36 | 176 | 5.6 | 2 |
| 13.8 | 30 | 27.20 | 0.05 | 5 cy | 0.133 | 152 | 36 | 218 | 6.9 | 2 |
| 13.8 | 30 | 27.20 | 0.10 | 5 cy | 0.183 | 152 | 36 | 303 | 9.5 | 3 |
| 13.8 | 30 | 27.20 | 0.20 | 5 cy | 0.283 | 152 | 36 | 474 | 15 | 3 |
| 13.8 | 30 | 27.20 | 0.30 | 5 cy | 0.383 | 152 | 36 | 647 | 20 | 3 |
| 13.8 | 30 | 27.20 | 0.40 | 5 cy | 0.483 | 152 | 36 | 822 | 25 | 4 |
| 13.8 | 30 | 27.20 | 0.50 | 5 cy | 0.583 | 152 | 36 | 997 | 30 | 4 |
| 13.8 | 30 | 27.20 | 0.60 | 5 cy | 0.683 | 152 | 36 | 1173 | 35 | 4 |
| 13.8 | 30 | 27.20 | 0.70 | 5 cy | 0.783 | 152 | 36 | 1350 | 41 | Danger |
| 13.8 | 30 | 27.20 | 0.80 | 5 cy | 0.883 | 152 | 36 | 1528 | 46 | Danger |
| 13.8 | 30 | 27.20 | 1.00 | 5 cy | 1.080 | 152 | 36 | 1879 | 56 | Danger |
| 13.8 | 30 | 27.20 | 1.25 | 5 cy | 1.333 | 152 | 36 | 2333 | 69 | Danger |
| 13.8 | 30 | 27.20 | 1.50 | 5 cy | 1.583 | 152 | 36 | 2783 | 82 | Danger |
| 13.8 | 30 | 27.20 | 1.75 | 5 cy | 1.833 | 152 | 36 | 3236 | 95 | Danger |
| 13.8 | 30 | 27.20 | 2.00 | 5 cy | 2.083 | 152 | 36 | 3690 | 108 | Danger |

Bolted three-phase short-circuit current = 30 kA, 5-cycle breakers.

TABLE B.26. Arc Flash Calculations in Medium Voltage, Three-Phase, 60-Hz, 13.8-kV Switchgear, EC = HCB, Dimensions 26″W × 96″D × 95″H

| kV | Bolted Three-Phase Current (kA) | Arcing Current (kA) | Relay Time (s) | Breaker Time (cycle) | Arcing Time (s) | Gap (mm) | Working Distance (in) | Arc Flash Boundary (in) | Incident Energy (cal/cm$^2$) | PPE Class |
|---|---|---|---|---|---|---|---|---|---|---|
| 13.8 | 40 | 35.95 | 0.0042 | 5 cy | 0.088 | 152 | 36 | 127 | 9.65 | 3 |
| 13.8 | 40 | 35.95 | 0.0125 | 5 cy | 0.096 | 152 | 36 | 134 | 10.6 | 3 |
| 13.8 | 40 | 35.95 | 0.025 | 5 cy | 1.080 | 152 | 36 | 144 | 12 | 3 |
| 13.8 | 40 | 35.95 | 0.05 | 5 cy | 0.133 | 152 | 36 | 199 | 14.7 | 3 |
| 13.8 | 40 | 35.95 | 0.10 | 5 cy | 0.183 | 152 | 36 | 258 | 20.3 | 3 |
| 13.8 | 40 | 35.95 | 0.20 | 5 cy | 0.283 | 152 | 36 | 310 | 31.3 | 4 |
| 13.8 | 40 | 35.95 | 0.30 | 5 cy | 0.383 | 152 | 36 | 357 | 42.4 | Danger |
| 13.8 | 40 | 35.95 | 0.40 | 5 cy | 0.483 | 152 | 36 | 399 | 53.5 | Danger |
| 13.8 | 40 | 35.95 | 0.50 | 5 cy | 0.583 | 152 | 36 | 439 | 64.5 | Danger |
| 13.8 | 40 | 35.95 | 0.60 | 5 cy | 0.683 | 152 | 36 | 497 | 75.6 | Danger |
| 13.8 | 40 | 35.95 | 0.70 | 5 cy | 0.783 | 152 | 36 | 513 | 86.7 | Danger |
| 13.8 | 40 | 35.95 | 0.80 | 5 cy | 0.883 | 152 | 36 | 513 | 97.8 | Danger |
| 13.8 | 40 | 35.95 | 1.00 | 5 cy | 1.080 | 152 | 36 | 580 | 119.9 | Danger |
| 13.8 | 40 | 35.95 | 1.25 | 5 cy | 1.333 | 152 | 36 | 658 | 147.6 | Danger |
| 13.8 | 40 | 35.95 | 1.50 | 5 cy | 1.583 | 152 | 36 | 730 | 175.3 | Danger |
| 13.8 | 40 | 35.95 | 1.75 | 5 cy | 1.833 | 152 | 36 | 797 | 202.9 | Danger |
| 13.8 | 40 | 35.95 | 2.00 | 5 cy | 2.083 | 152 | 36 | 861 | 230.6 | Danger |

Bolted three-phase short-circuit current = 40 kA, 5-cycle breakers.

# INDEX



*Arc Flash Hazard Analysis and Mitigation*, Second Edition. J.C. Das.

Published 2021 by John Wiley & Sons, Inc.

# IEEE Press Series on Power Engineering

**Series Editor: M. E. El-Hawary,** Dalhousie University, Halifax, Nova Scotia, Canada

The mission of IEEE Press Series on Power Engineering is to publish leading- edge books that cover the broad spectrum of current and forward-looking technologies in this fast-moving area. The series attracts highly acclaimed authors from industry/academia to provide accessible coverage of current and emerging topics in power engineering and allied fields. Our target audience includes the power engineering professional who is interested in enhancing their knowledge and perspective in their areas of interest.

1. *Electric Power Systems: Design and Analysis, Revised Printing*
Mohamed E. El-Hawary

2. *Power System Stability*
Edward W. Kimbark

3. *Analysis of Faulted Power Systems*
Paul M. Anderson

4. *Inspection of Large Synchronous Machines: Checklists, Failure Identification, and Troubleshooting*
Isidor Kerszenbaum

5. *Electric Power Applications of Fuzzy Systems*
Mohamed E. El-Hawary

6. *Power System Protection*
Paul M. Anderson

7. *Subsynchronous Resonance in Power Systems*
Paul M. Anderson, B.L. Agrawal, and J.E. Van Ness

8. *Understanding Power Quality Problems: Voltage Sags and Interruptions*
Math H. Bollen

9. *Analysis of Electric Machinery*
Paul C. Krause, Oleg Wasynczuk, and S.D. Sudhoff

10. *Power System Control and Stability, Revised Printing*
Paul M. Anderson and A.A. Fouad

*Arc Flash Hazard Analysis and Mitigation*, Second Edition. J.C. Das.

Published 2021 by John Wiley & Sons, Inc.

11. *Principles of Electric Machines with Power Electronic Applications, Second Edition*
Mohamed E. El-Hawary

12. *Pulse Width Modulation for Power Converters: Principles and Practice*
D. Grahame Holmes and Thomas A. Lipo

13. *Analysis of Electric Machinery and Drive Systems, Second Edition*
Paul C. Krause, Oleg Wasynczuk, and Scott D. Sudhoff

14. *Risk Assessment for Power Systems: Models, Methods, and Applications*
Wenyuan Li

15. *Optimization Principles: Practical Applications to the Operation of Markets of the Electric Power Industry*
Narayan S. Rau

16. *Electric Economics: Regulation and Deregulation*
Geoffrey Rothwell and Tomas Gomez

17. *Electric Power Systems: Analysis and Control*
Fabio Saccomanno

18. *Electrical Insulation for Rotating Machines: Design, Evaluation, Aging, Testing, and Repair, Second Edition*
Greg C. Stone, Ian Culbert, Edward A. Boulter, and Hussein Dhirani

19. *Signal Processing of Power Quality Disturbances*
Math H. J. Bollen and Irene Y. H. Gu

20. *Instantaneous Power Theory and Applications to Power Conditioning*
Hirofumi Akagi, Edson H. Watanabe, and Mauricio Aredes

21. *Maintaining Mission Critical Systems in a 24/7 Environment*
Peter M. Curtis

22. *Elements of Tidal-Electric Engineering*
Robert H. Clark

23. *Handbook of Large Turbo-Generator Operation and Maintenance, Second Edition*
Geoff Klempner and Isidor Kerszenbaum

24. *Introduction to Electrical Power Systems*
Mohamed E. El-Hawary

25. *Modeling and Control of Fuel Cells: Distributed Generation Applications*
M. Hashem Nehrir and Caisheng Wang

26. *Power Distribution System Reliability: Practical Methods and Applications*
Ali A. Chowdhury and Don O. Koval

27. *Introduction to FACTS Controllers: Theory, Modeling, and Applications*
Kalyan K. Sen and Mey Ling Sen

28. *Economic Market Design and Planning for Electric Power Systems*
James Momoh and Lamine Mili

29. *Operation and Control of Electric Energy Processing Systems*
James Momoh and Lamine Mili

30. *Restructured Electric Power Systems: Analysis of Electricity Markets with Equilibrium Models*
Xiao-Ping Zhang

31. *An Introduction to Wavelet Modulated Inverters*
S. A. Saleh and M. Azizur Rahman

32. *Control of Electric Machine Drive Systems*
Seung-Ki Sul

33. *Probabilistic Transmission System Planning*
Wenyuan Li

34. *Electricity Power Generation: The Changing Dimensions*
Digambar M. Tagare

35. *Electric Distribution Systems*
Abdelhay A. Sallam and Om P. Malik

36. *Practical Lighting Design with LEDs*
Ron Lenk and Carol Lenk

37. *High Voltage and Electrical Insulation Engineering*
Ravindra Arora and Wolfgang Mosch

38. *Maintaining Mission Critical Systems in a 24/7 Environment, Second Edition*
Peter M. Curtis

39. *Power Conversion and Control of Wind Energy Systems*
Bin Wu, Yongqiang Lang, Navid Zargari, and Samir Kouro

40. *Integration of Distributed Generation in the Power System*
Math H. J. Bollen and Fainan Hassan

41. *Doubly Fed Induction Machine: Modeling and Control for Wind Energy Generation*
Gonzalo Abad, Jesús López, Miguel Rodrigues, Luis Marroyo, and Grzegorz Iwanski

42. *High Voltage Protection for Telecommunications*
Steven W. Blume

43. *Smart Grid: Fundamentals of Design and Analysis*
James Momoh

44. *Electromechanical Motion Devices, Second Edition*
Paul Krause, Oleg Wasynczuk, and Steven Pekarek

45. *Electrical Energy Conversion and Transport: An Interactive Computer-Based Approach, Second Edition*
George G. Karady and Keith E. Holbert

46. *ARC Flash Hazard and Analysis and Mitigation*
J. C. Das

47. *Handbook of Electrical Power System Dynamics: Modeling, Stability, and Control*
Mircea Eremia and Mohammad Shahidehpour

48. *Analysis of Electric Machinery and Drive Systems, Third Edition*
Paul Krause, Oleg Wasynczuk, Scott Sudhoff, and Steven Pekarek

49. *Extruded Cables for High-Voltage Direct-Current Transmission: Advances in Research and Development*
Giovanni Mazzanti and Massimo Marzinotto

50. *Power Magnetic Devices: A Multi-Objective Design Approach*
S. D. Sudhoff

51. *Risk Assessment of Power Systems: Models, Methods, and Applications, Second Edition*
Wenyuan Li

52. *Practical Power System Operation*
Ebrahim Vaahedi

53. *The Selection Process of Biomass Materials for the Production of Bio-Fuels and Co-Firing*
Najib Altawell

54. *Electrical Insulation for Rotating Machines: Design, Evaluation, Aging, Testing, and Repair, Second Edition*
Greg C. Stone, Ian Culbert, Edward A. Boulter, and Hussein Dhirani

55. *Principles of Electrical Safety*
Peter E. Sutherland

56. *Advanced Power Electronics Converters: PWM Converters Processing AC Voltages*
Euzeli Cipriano dos Santos Jr. and Edison Roberto Cabral da Silva

57. *Optimization of Power System Operation, Second Edition*
Jizhong Zhu

58. *Power System Harmonics and Passive Filter Designs*
J. C. Das

59. *Digital Control of High-Frequency Switched-Mode Power Converters*
Luca Corradini, Dragan Maksimovic, Paolo Mattavelli, and Regan Zane

60. *Industrial Power Distribution, Second Edition*
Ralph E. Fehr, III

61. *HVDC Grids: For Offshore and Supergrid of the Future*
Dirk Van Hertem, Oriol Gomis-Bellmunt, and Jun Liang

62. *Advanced Solutions in Power Systems: HVDC, FACTS, and Artificial Intelligence*
Mircea Eremia, Chen-Ching Liu, and Abdel-Aty Edris

63. *Operation and Maintenance of Large Turbo-Generators*
Geoff Klempner and Isidor Kerszenbaum

64. *Electrical Energy Conversion and Transport: An Interactive Computer-Based Approach*
George G. Karady and Keith E. Holbert

65. *Modeling and High-Performance Control of Electric Machines*
John Chiasson

66. *Rating of Electric Power Cables in Unfavorable Thermal Environment*
George J. Anders

67. *Electric Power System Basics for the Nonelectrical Professional*
Steven W. Blume

68. *Modern Heuristic Optimization Techniques: Theory and Applications to Power Systems*
Kwang Y. Lee and Mohamed A. El-Sharkawi

69. *Real-Time Stability Assessment in Modern Power System Control Centers*
Savu C. Savulescu

70. *Optimization of Power System Operation*
Jizhong Zhu

71. *Insulators for Icing and Polluted Environments*
Masoud Farzaneh and William A. Chisholm

72. *PID and Predictive Control of Electric Devices and Power Converters Using MATLAB®/Simulink®*
Liuping Wang, Shan Chai, Dae Yoo, Lu Gan, and Ki Ng

73. *Power Grid Operation in a Market Environment: Economic Efficiency and Risk Mitigation*
Hong Chen

74. *Electric Power System Basics for the Nonelectrical Professional, Second Edition*
Steven W. Blume

75. *Energy Production Systems Engineering*
Thomas Howard Blair

76. *Model Predictive Control of Wind Energy Conversion Systems*
Venkata Yaramasu and Bin Wu

77. *Understanding Symmetrical Components for Power System Modeling*
J. C. Das

78. *High-Power Converters and AC Drives, Second Edition*
Bin Wu and Mehdi Narimani

79. *Current Signature Analysis for Condition Monitoring of Cage Induction Motors: Industrial Application and Case Histories*
William T. Thomson and Ian Culburt

80. *Introduction to Electric Power and Drive Systems*
Paul Krause, Oleg Wasynczuk, Timothy O'Connell, and Maher Hasan

81. *Instantaneous Power Theory and Applications to Power Conditioning, Second Edition*
Hirofumi Akagi, Edson Hirokazu Watanabe, and Mauricio Aredes

82. *Practical Lighting Design with LEDs, Second Edition*
Ron Lenk and Carol Lenk

83. *Introduction to AC Machine Design*
Thomas A. Lipo

84. *Advances in Electric Power and Energy Systems: Load and Price Forecasting*
Mohamed E. El-Hawary

85. *Electricity Markets: Theories and Applications*
Jeremy Lin and Fernando H. Magnago

86. *Multiphysics Simulation by Design for Electrical Machines, Power Electronics, and Drives*
Marius Rosu, Ping Zhou, Dingsheng Lin, Dan Ionel, Mircea Popescu, Frede Blaabjerg, Vandana Rallabandi, and David Staton

87. *Modular Multilevel Converters: Analysis, Control, and Applications*
Sixing Du, Apparao Dekka, Bin Wu, and Navid Zargari

88. *Electrical Railway Transportation Systems*
Morris Brenna, Federica Foiadelli, and Dario Zaninelli

89. *Energy Processing and Smart Grid*
James A. Momoh

90. *Handbook of Large Turbo-Generator Operation and Maintenance, Third Edition*
Geoff Klempner and Isidor Kerszenbaum

91. *Advanced Control of Doubly Fed Induction Generator for Wind Power Systems*
Dehong Xu, Frede Blaabjerg, Wenjie Chen, and Nan Zhu

92. *Electric Distribution Systems, Second Edition*
Abdelhay A. Sallam and Om P. Malik

93. *Power Electronics in Renewable Energy Systems and Smart Grid: Technology and Applications*
Bimal K. Bose

94. *Distributed Fiber Optic Sensing and Dynamic Rating of Power Cables*
Sudhakar Cherukupalli and George J Anders

95. *Power System and Control and Stability, Third Edition*
Vijay Vittal, James D. McCalley, Paul M. Anderson, and A. A. Fouad

96. *Electromechanical Motion Devices: Rotating Magnetic Field-Based Analysis and Online Animations, Third Edition*
Paul Krause, Oleg Wasynczuk, Steven D. Pekarek, and Timothy O'Connell

97. *Applications of Modern Heuristic Optimization Methods in Power and Energy Systems*
Kwang Y. Lee and Zita A. Vale

98. *Handbook of Large Hydro Generators: Operation and Maintenance*
Glenn Mottershead, Stefano Bomben, Isidor Kerszenbaum, and Geoff Klempner

99. *Advances in Electric Power and Energy: Static State Estimation*
Mohamed E. El-Hawary

100. *Arc Flash Hazard Analysis and Mitigation, Second Edition*
J.C. Das

Printed and bound by CPI Group (UK) Ltd, Croydon, CR0 4YY
17/09/2026
14964318-0007